Luigia Sabbatini, Elvira De Giglio (Eds.)
Polymer Surface Characterization

Also of interest

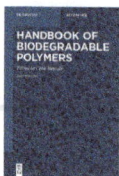

Polymer Surface Characterization

Edited by
Luigia Sabbatini, Elvira De Giglio

2nd Edition

DE GRUYTER

Editors
Prof. Dr. Luigia Sabbatini
Chemistry Department
University of Bari Aldo Moro
Via Orabona 4
70126 Bari
Italy
luigia.sabbatini@uniba.it

Prof. Dr. Elvira De Giglio
Chemistry Department
University of Bari Aldo Moro
Via Orabona 4
70126 Bari
Italy
elvira.degiglio@uniba.it

ISBN 978-3-11-070104-3
e-ISBN (PDF) 978-3-11-070109-8
e-ISBN (EPUB) 978-3-11-070114-2

Library of Congress Control Number: 2021952136

Bibliographic information published by the Deutsche Nationalbibliothek
The Deutsche Nationalbibliothek lists this publication in the Deutsche Nationalbibliografie;
detailed bibliographic data are available on the Internet at http://dnb.dnb.de.

© 2022 Walter de Gruyter GmbH, Berlin/Boston
Cover image: Micro Discovery/Corbis Documentary/Getty Images
Typesetting: Integra Software Services Pvt. Ltd.
Printing and binding: CPI books GmbH, Leck

www.degruyter.com

Preface

Polymers permeate our everyday life more than any other materials. It is difficult to think of another material as versatile and irreplaceable as polymers in terms of applications, both in consolidated and in emerging technological areas. Their use spans from the most trivial consumer goods to the most advanced microelectronics, packaging, miniaturized pieces for artificial organs, high-performance coating in aerospace industry, etc.

Technological demand, moreover, requires materials not only with peculiar bulk properties but also with specific surface chemistry and structure, both generally different from the bulk one. The combination of these two requisites makes a polymeric material suitable and unique for the job it is called to do.

In this respect, surfaces in polymers, more than in any other class of materials, can be quite easily handled and engineered to a high level of sophistication by fully exploiting their chemical reactivity. Furthermore, apart from the changes intentionally promoted on polymer surfaces to tailor new application-oriented properties, the importance of polymer dynamics occurring on surfaces whenever they are exposed to the environment (real world) or put in contact with another material (interface) is self-evident: dramatic changes of structural and thermodynamic properties in the confinement area can occur, and these dramatically reflect on properties such as bonding, friction, adsorption, wettability, inertness, etc. Moreover, aging modifies the surface chemistry of polymers (as well as of any other material), thus provoking a remarkable deterioration of their properties.

In all the above cases, knowledge of both the surface chemistry and the structure is mandatory to understand the behavior of polymeric materials, monitor the surface processing, and tailor and develop new surface modification strategies. A wealth of analytical techniques is available for the purpose; they are surface-specific, and the analyst/researcher needs a basic knowledge of their principles and procedures in order to choose the most appropriate way to solve the specific problem.

In the *first edition* of this book, techniques were selected that are well suited for characterization of surfaces/interfaces of thin polymer-based films but also of more general applicability in materials science. Basic principles, operative conditions, applications, performance, and limiting features were supplied, together with general information on instrumental apparatus.

This *second edition*, other than to revive the same techniques, fully updated as to application fields and/or instrumental details and/or operative conditions, and relevant references, is enriched with a couple of new chapters dealing with Near-Edge X-Ray Absorption Fine Structure Spectroscopy (NEXAFS), and Nanoindentation, respectively.

Each chapter of the book is devoted to one technique and is self-consistent; the end-of-chapter references would allow the reader a quick access to more detailed information.

https://doi.org/10.1515/9783110701098-202

In **Chapter 1**, an introductory chapter, a section devoted to nanostructured polymers has been introduced: the exceptionally high surface to volume ratio of these materials is the key factor of their properties, therefore is quite evident the importance of their surface characterization. Chapters 2, 3, and 4 deal with techniques with great potential in the definition of surface chemical composition of polymers as highlighted and documented by selected application examples. In **Chapter 2**, more details on Ambient Pressure X-Ray Photoelectron Spectroscopy (APXPS) have been supplied; **Chapter 3**, completely new, is devoted to NEXAFS: basic principles, detection modes and significant applications such as polymer fingerprinting, functional groups identification and study of pendant groups orientation at surfaces of complex polymers are reported. Moreover, the value of C1s NEXAFS spetromicroscopy in many studies of both the bulk and the surface of polymers is described.

Chapter 4 provides insight on Time-of-flight Secondary Ions Mass Spectrometry (TOF-SIMS); in this second edition, how sputter process can be modeled by molecular dynamics computer simulations to better understand its complexity has been highlighted; moreover, information on new gas cluster ion sources and their performances are reported. **Chapter 5** is dedicated to Advances in Attenuated Total Reflection Fourier Transform Infrared Spectroscopy (ATR-FTIR), which provides information on polymer chemical structure with varying surface thicknesses, of the order of tenths of a micrometer. New applications in the study of hydrogels surface properties are described as well as in-situ studies of tribological phenomena. **Chapter 6** deals with atomic force microscopy (AFM), a technique endowed with high-resolution capability to map a variety of material properties on the nanoscale. Evidence is supplied of how probing mechanical properties leads to the identification of important polymer surface processes such as phase segregation, continuity of phases, and dispersion, which play a major role in determining macroscopic properties. New sections have been introduced: one devoted to mapping of mechanical properties by AFM in order to obtain quantitative information on polymers viscoelastic properties; a second one describing the various modes available in AFM to probe polymers electrical properties. Characterization of surface morphology is addressed in **Chapter 7**, which is focused on electron microscopy. Scanning electron microscopy (SEM) is described in detail, and some general information on transmission electron microscopy (TEM) is given. The point of strength of electron microscopy characterization, well highlighted in this chapter, is that the surface morphology of a polymer system is always closely connected with its chemical composition and processing conditions. A more profound discussion on the multidimensional information, which can be obtained by SEM measurements, is provided.

Chapter 8 deals with spectroscopic ellipsometry: after the description of the basics of the method, an investigation of optical, structural, and thermodynamic properties of polymers in the whole spectral range (UV-VIS-IR) is discussed. Some interesting industrial applications such as those in the quality control of roll-to-roll fabrication process or heterogeneity control of micropatterned films are also reported.

Chapter 9 addresses the issue of wettability and its measurements; contact angle measurements and its relationships with surface energy and wetting behavior are clearly described as well as advanced technological processes to develop polymer surfaces with tunable hydrophobic/hydrophilic character. Recent developments and applications are reported. **Chapter 10**, completely new, ends the book with an insight on Nanoindentation, a technique which provides quantitaive, absolute measurements of polymer mechanical properties with nanoscale spatial resolution; the way this information can be obtained and its influence on polymer performances are described.

The editors gratefully acknowledge the invaluable contributions of the authors and, as it is behind every edited book, the kind invitation and precious support of the publisher, in particular, the enthusiasm of Kristine Berber-Nerlinger and Vivien Schubert, and the qualified cooperation of Mervin Ebenezer at De Gruyter.

Contents

Polymer Surface Characterization – 2nd Edition

Stefania Cometa
Senior Researcher, Jaber Innovation srl,
Via Calcutta 8, 00144 Roma, Italy
stefania.cometa@jaber.it

Luigia Sabbatini
Full Professor, Department of Chemistry,
University of Bari Aldo Moro, Via Orabona 4,
70126 Bari, Italy
luigia.sabbatini@uniba.it

Elvira De Giglio
Associate Professor, Department of
Chemistry, University of Bari Aldo Moro,
Via Orabona 4, 70126 Bari, Italy
elvira.degiglio@uniba.it

Nicoletta Ditaranto
Assistant Professor, Department of
Chemistry, University of Bari Aldo Moro,
Via Orabona 4, 70126 Bari, Italy
nicoletta.ditaranto@uniba.it

Adam Hitchcock
Professor, Department of Chemistry and
Chemical Biology, McMaster University, AN
Bourns Science Building, 1280 Main Street
West, Hamilton, Ontario, L8S 4M1 Canada
aph@mcmaster.ca

Beat Keller
ANABE GmbH, Rebhaldenstrasse 12, 8340
Hinwil, Switzerland
beat.keller@web.de

Marco Consumi
Researcher, Department of Biotechnology,
Chemistry and Pharmacy, University of Siena,
Via A. Moro, 2, 53100 Siena, Italy
marco.consumi@unisi.it

Gemma Leone
Researcher, Department of Biotechnology,
Chemistry and Pharmacy, University of Siena,
Via A. Moro, 2, 53100 Siena, Italy
gemma.leone@unisi.it

Agnese Magnani
Associate Professor, Department of
Biotechnology, Chemistry and Pharmacy,
University of Siena, Via A. Moro, 2, 53100
Siena, Italy
agnese.magnani@unisi.it

Filippo Mangolini
Assistant Professor, Materials Science and
Engineering Program and Texas Materials
Institute, J. Mike Walker Department of
Mechanical Engineering, The University of
Texas at Austin, 204 E. Dean Keeton, Stop
C2200Austin, TX 78712-1591, USA
Filippo.Mangolini@austin.utexas.edu

Antonella Rossi
Full Professor, Department of Chemical and
Geological Sciences, University of Cagliari,
Cittadella Universitaria, 09042 Monserrato
(CA), Italy
rossi@unica.it

Dalia Yablon, PhD
Founder, SurfaceChar LLC, Greater Boston,
MA, USA
dalia.yablon@surfacechar.com

Miroslav Slouf
Research Fellow, Head of the Department of
Polymer Morphology, Institute of
Macromolecular Chemistry, Czech Academy
of Sciences, Heyrovskeho nam. 2, 16206
Prague 6, Czech Republic
slouf@imc.cas.cz

Frantisek Lednicky
Professor Emeritus,Institute of
Macromolecular Chemistry, Czech Academy
of Sciences, Heyrovskeho nam. 2, 16206
Prague 6, Czech Republic
frantisek.lednicky@iex.cz

https://doi.org/10.1515/9783110701098-204

Petr Wandrol
Applications Manager for SEM and Small
Dual-Beam, Thermo Fisher Scientific,
Vlastimila Pecha 12, 627 00 Brno,
Czech Republic
petr.wandrol@thermofisher.com

Taťana Vacková
Assistant professor, Department of Materials
Engineering, Faculty of Mechanical
Engineering, Czech Technical University in
Prague, Karlovo náměstí 13, 121 35,
Prague, Czech Republic
Tatana.Vackova@fs.cvut.cz

Eva Bittrich
Scientist, "Center for Macromolecular
Structure Analysis (CMSA)," Leibniz-Institut
für Polymerforschung Dresden e.V.
Hohe Straße 6, 01069 Dresden, Germany
bittrich-eva@ipfdd.de

Klaus-Jochen Eichhorn
Leibniz-Institut für Polymerforschung
Dresden e.V., Hohe Straße 6, 01069 Dresden,
Germany

Fabio Palumbo
Researcher, National Research Council,
Institute of Nanotechnology, c/o Department
of Chemistry, University of Bari Aldo Moro,
Via Orabona 4, 70126 Bari, Italy
fabio.palumbo@cnr.it

Rosa Di Mundo
Researcher, Dipartimento di Ing. Civile,
Ambientale, del Territorio, Edile e di Chimica-
DICATECh, Politecnico Di Bari, Via Orabona 4,
70125 Bari, Italy
rosa.dimundo@poliba.it

Elisabetta Tranquillo
Post-Doctoral Researcher, Department of
Mechanical, Aerospace and Civil Engineering
and Henry Royce Institute, The University of
Manchester, Manchester M13 9PL, UK
elisabettatra90@gmail.com

Antonio Gloria
Researcher, Institute of Polymers,
Composites and Biomaterials – National
Research Council of Italy,V.le J.F. Kennedy
54 – Mostra d'Oltremare Pad. 20, 80125
Naples, Italy
antonio.gloria@cnr.it

Marco Domingos
Associate Professor in Bioprinting and
Regenerative Medicine, Department of
Mechanical, Aerospace and Civil Engineering
and Henry Royce Institute, School of
Engineering, Faculty of Science and
Engineering, The University of Manchester,
Manchester M13 9PL, UK
marco.domingos@manchester.ac.uk

Stefania Cometa, Luigia Sabbatini

1 Introductory remarks on polymers and polymer surfaces

1.1 Why polymers?

1.1.1 Generality

Polymers are everywhere in our everyday life. Both synthetic and natural polymers have unique properties (chemical, mechanical, electrical, and thermal), which make them not replaceable in the peculiar job they are called to do. The most impressive aspect in the world of polymers is the enormous variety of applications, which span from everyday consumer goods (housewares, toys, bottles, packaging, textiles, and furniture), large use electrical and electronic items (computer, cable insulation, and household appliances), construction (housing, pipes, adhesives, coatings, and insulation), transportation (tires, appliance parts, hardware, and carpeting), medical and hospital furniture and goods (gloves, syringes, catheters, and bandage) to high-performance, sophisticated materials for aerospace, bullet-proof vests, nonflammable fabric, artificial organs, degradable device for controlled release of drugs or chemicals, high power-density batteries, high-strength cables for oceanic platforms, etc. A recent comprehensive encyclopedia reports the main applications and the continually evolving technologies involving use of polymeric materials [1].

From a chemical point of view, polymers are high-molecular-weight compounds made up of simple repeating units called monomers. Molecular weights range from thousands to millions atomic mass units which make impossible writing a definitive molecular structure for these materials. Polymer structures are therefore represented by enclosing in brackets the repeating unit (monomer) and placing "n" as subscript (Fig. 1.1A).

Polymers having more than one kind of repeating units are called copolymers; the units can be distributed either randomly in the chains or in an ordinate alternate sequence or in blocks. Moreover, as to the arrangement of chains, the polymer structure may be linear, branched, network, stars, ladder, and dendrimer fashion (Fig. 1.1B). Complex structures, such as networks, are formed when cross-linking occurs, i.e., the formation of covalent bonds between polymer chains; this process leads to a remarkable increase in molecular weight and this in turn strongly influences chemical, physical and mechanical properties of the material as well as its processability.

https://doi.org/10.1515/9783110701098-001

-[CH₂-CH₂]-   poly(ethylene)   **A**

-[CH₂-CH₂-O]-   poly(ethylene oxide)

-[CH₂-CH]-   poly(styrene)

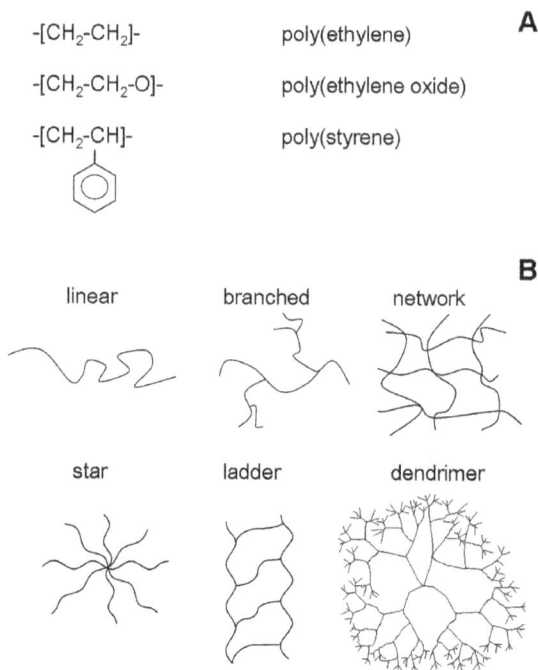

Fig. 1.1: (A) Conventional representation of polymers structures and (B) arrangement of chains in the polymer structure.

1.1.2 Synthesis

The first synthetic polymer which was produced on a commercial scale was bakelite, a phenol–formaldehyde resin (by its inventor, the Belgian chemist Leo Baekeland [2]), at the beginning of 1900; however, it is the German chemist Hermann Staudinger to be rightly recognized as the father of the chemistry of macromolecules for his studies on the relationship between structure of polymers and their properties he received the Nobel Prize in Chemistry in 1953. In those years, the polymer chemistry experienced the most significant advances and the beginning of a true revolution in the polymer industry thanks to the work of Giulio Natta in Italy who synthesized, for the first time, polymers having controlled stereochemistry [3] by using the coordination catalysts just developed by the German chemist Karl Ziegler [4]. The two scientists were awarded with the Nobel Prize in Chemistry in 1963. It was clearly demonstrated that polymers with similar chemical composition but exhibiting different molecular orientation have different morphology and their mechanical properties too may differ markedly. It is therefore vital, in view of commercial exploitation of polymer materials, the full knowledge and control of the polymerization process. In brief, two reaction mechanisms are possible: step reaction (generally, a condensation reaction) and chain reaction (generally, an addition reaction). In the former, polyfunctional monomers react randomly to give dimers, trimers, oligomers that can react with other monomers o among themselves: monomers are consumed rapidly but polymer chains do not grow so much (slow increase of the molecular weight). Polyamides (e.g., nylon 66) and polyesters (e.g., polyurethane) are produced by this reaction mechanism. In the chain reaction, on the other hand, an *initiator* (generally a free radical) starts the reaction by opening the double bond of a vinyl monomer and adding to it, with one electron remaining unpaired. The reaction *propagates* through the successive linking of a monomer to the end of the growing chain; finally, two free radicals react to end each other's growth activity (*termination*). Monomers are consumed more slowly with respect to step reactions but there is a considerable quick increase of the chain molecular weight. Polytetrafluoroethylene (Teflon), polyethylene, and polystyrene are important examples of polymers obtained by chain-reaction mechanism.

In polymer synthesis, it can be assumed that each chain reacts independently. Therefore, the bulk polymer is characterized by a wide distribution of molecular weights and chain lengths and these factors primarily determine the properties of polymers. The degree of polymerization (DP) is the number of repeat units (included chain ends) in the chain, and gives a measure of molecular weight; due to the presence, in any polymer, of chains of varying lengths, the average degree of polymerization, $\overline{DP}$, is better used. Moreover, in order to characterize the distribution of polymer lengths in a sample, two parameters are defined: *number average* (M_n) and *weight average* (M_w) molecular weight. The number average is just the sum of individual molecular weights divided by the number of polymers. The weight average is proportional to the square of the molecular weight. Therefore, the weight average is always larger than the number average (unless all molecules possess the same weight; then $M_n = M_w$).

1.1.3 Classification and nomenclature

Polymers are classified in many ways, the most common classifications being: i) based on source (natural, semisynthetic, synthetic); ii) based on structure (linear, branched chain, cross linked or network); iii) based on molecular forces, which determine mechanical properties (elastomers, fibers, thermoplastic polymers, thermosetting polymers); iv) based on polymerization mode (addition and condensation polymers, nowadays also referred to as chain growth polymers and step growth polymers, respectively).

The reported classifications, while useful from a practical point of view, in some cases are a rough simplification. For example, it must be underlined that not all the step reactions are condensation processes and, on the other hand, not all the chain reactions are addition processes; this explains why IUPAC classified polymerization processes [5] in a more comprehensive way with respect to the above simplified scheme.

At the same time, an IUPAC Commission elaborated a series of rules for polymer nomenclature [6]. This is quite a complicate matter: some details and significant examples have been simply and clearly reviewed in many textbooks on polymers; see, for example [7–10], but this is outside the scope of this book. However, readers should be familiar with the most widely accepted terminology and therefore some general information is provided. Polymers (and copolymers) are grouped in families according to the functional group(s) present in the repeating units when their structure is easily identifiable: polyesters, polyethers, polyimides, polyamides, polyamideimides, etc. In presence of structures that cannot be easily defined, (co) polymer nomenclature is based on the name of the starting monomer(s). This approach is referred to as source-based system and is more widely used than the structure-based IUPAC nomenclature. Moreover, when the starting monomers contain a carbon–carbon double bond, the resulting polymers are classified as *vinyl polymers*, in all the other cases they are referred to as *nonvinyl polymers*: the nomenclature of latter may be quite complex and the source-based approach once again is more generally used than IUPAC rules. Depending on the approach used, terminology may be quite different; just to give an example: $(-CH_2-CH_2-)_n$ is named polyethylene or poly(methylene) in the source-based and IUPAC approaches respectively and, similarly, $(-CH_2CHCOOH-)_n$ is named poly(acrylic acid) or poly(1-carboxylatoethylene) respectively. Given the complexity of polymer nomenclature, its abbreviations and/ or trade terms have been and still are widely employed both in the scientific literature and in the industrial world. Many of these terms are so popular and widely recognized that they are commonly used in the everyday languages too (PVC, PE, PET, HDPE, LDPE, PMMA, Teflon, etc.).

1.1.4 Morphology and properties

Chemical constitution and the way molecules self-assemble in a solid on the micro- and macroscopic scale, together with molecular weight, are important factors in determining the ultimate mechanical and thermal properties of polymers, which means their applications and processability.

As to assembly and orientation of molecules, polymers can be considered crystalline or amorphous materials. Crystalline materials have highly ordered structures with molecules arranged in repeating patterns. Amorphous materials, by contrast, show no long-range order, with molecules arranged randomly and the chains tangled and curved around one another. As a matter of fact, polymers never achieve complete crystallinity and they are more conveniently categorized as amorphous and semicrystalline, the latter morphology being characterized by tangled and disordered regions surrounding crystalline zones (Fig. 1.2).

Semicrystalline polymers have true melting temperatures (T_m) at which the ordered regions break up and become disordered. In contrast, the amorphous regions

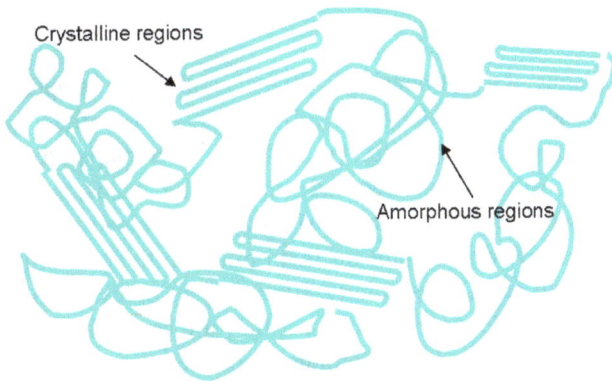

Fig. 1.2: Schematic representation of a semicrystalline polymer.

soften over a relatively wide temperature range (always lower than T_m) known as the glass transition temperature (T_g). Fully amorphous polymers do not exhibit T_m, of course, but all polymers exhibit T_g. Above these temperatures, polymers are liquid-like.

Polymers are usually in the solid state at room temperature. Above a certain temperature, all polymers soften and, in the case of thermoplastics, are able to flow, having liquid-like order, i.e., they are disordered and in the melt state (for common polymers this usually occurs in the range of 100–250 °C).

When the temperature drops, their density increases. Depending on the polymer, there are two possibilities: (1) polymers with irregular molecular structure (atactic PS, atactic PP, PMMA) solidify keeping their disordered microstructure and forming a stiff but brittle amorphous solid called polymer glass; (2) polymers which have a regular structure at the molecular scale (PE, PEO, isotactic PP, isotactic PS, PA, PTFE, PETP) partially crystallize.

Reporting in a graph the specific volume versus temperature dependence, behavior (1) presents a change of slope in a continuous fashion at the specific temperature of the transition and for this reason it is called glass transition temperature, T_g (Fig. 1.3, left). On the other hand, behavior (2) leads to a semicrystalline material

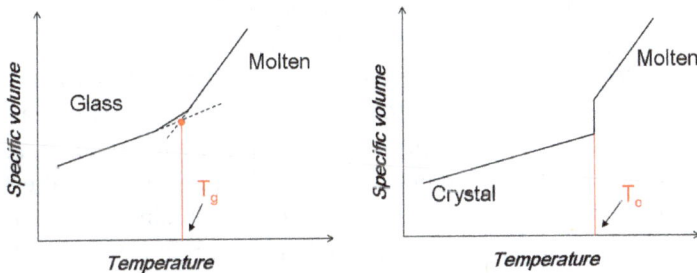

Fig. 1.3: Cooling down from the melt state of a polymer to glass or to crystal.

in an abrupt manner at a characteristic temperature called crystallization temperature, T_c (Fig. 1.3, right).

For polymers (unlike crystals of small molecules), the crystallization temperature T_c might differ from the melting temperature, T_m. Usually T_c is lower than T_m of several degrees, due to the undercooling effect. Furthermore, both temperatures depend on the rate of cooling/heating. This is because long polymer chains, unlike small molecules, have decreased flexibility and are easily trapped in kinetically arrested states. For this reason, it is very difficult for them to attain their absolute thermodynamic equilibrium state of full crystallinity. High degrees of crystallinity can be attained in the case of very slow cooling rates. If they are cooled abruptly enough, they freeze in the fully disordered state forming a polymer glass. Thermoset polymers and gels lack organization at all scales and are inherently amorphous.

It is worth noting that melt using processes for the fabrication of polymers usually lead to a surface morphology different from that of the bulk of the material. The so-called skin-core effects, also observable in solution-prepared polymers as well as in natural polymers, are treated in details in Chapter 7.

Chemical structure and morphology are tightly related in polymers and both relate to main bulk properties: mechanical characteristics, thermal stability, chemical resistance, flame resistance, degradability, electrical conductivity, and nonlinear optical properties.

A variety of polymer applications depend on their unique mechanical properties like tensile, compressive, and flexural strength, elasticity, plasticity, and toughness, which are governed by intermolecular forces binding the polymer chains. Thermally stable is defined a polymer which does not decompose below 400 °C; thermal stability is related to bond energy and this explains why, for example, high-performance materials for aerospace industry which have to withstand extreme temperatures comprise aromatic compounds based polymers. Polymers with high halogen content exhibit both a good flame and chemical (solvent) resistance but, once again, morphology plays a role. Durability of polymers is one of the blessings of these materials, but it is also a curse from the point of view of environmental awareness; this justifies the many efforts spent in the last years by materials scientists in capturing the fundamentals of polymer degradation mechanisms, in order to design and synthesize materials with tuned degradability, also capable of controlled release. An interesting class of polymers, characterized by an extended conjugated system and by a high degree of crystallinity, conveniently doped, exhibit electrical conductivity comparable to that of metals; this property, together with the much lower density and the higher processability with respect to metals, make such conducting polymers extremely attractive for many applications including low-weight batteries, light-emitting diodes, thin film transistors, etc. Polymers containing chromophores are able to promote a change in the refractive index when interacting with a light-induced electric field, exhibit the so-called nonlinear optic (NLO) properties. These organic materials comprise one or more aromatic systems in conjugated positions, leading to charge transfer systems; functionalization of both ends of the π conjugated system with appropriate electron donor and acceptor groups can increase the asymmetric electronic distribution in either or both the ground and excited states, thus leading to an increased optical nonlinearity. Organic nonlinear optical materials have potentially high nonlinearities and rapid response in electrooptic effect compared to inorganic NLO materials, foreseeing a great impact on information technology and industrial applications.

To complete the panorama of the enormous impact that advanced polymeric materials can have in the everyday life, new generation nanostructured polymers should be mentioned (as deeply evidenced in Section 1.2.3). Various techniques have been developed to produce nanosized polymer materials (polymer nanoparticles, nanocapsules, nanorods, and nanobrushes) such as solvent evaporation, supercritical fluid technology, salting-out, dialysis, micro-emulsion, mini-emulsion, surfactant-free emulsion, and interfacial polymerization. A review [11] covers the general description of the preparation of polymer nanostructures and the detailed description of the crucial parameters involved in techniques designed to obtain the desired properties. More recently, a toolbox of available methods for the preparation and characterization of polymeric nanoparticles was supplied by Crucho et al. [12].

Polymers properties can be substantially modified and strategically tuned by the addition of the so-called additives. These substances are categorized on the basis of their function as, for example, plasticizers, nucleating agents, flame retardant, cross-linking agents, slip agents, lubricants, and reinforcing fillers, odorants. They can be completely miscible or completely immiscible to polymer bulk and are often added in a combination. On considering the extent of additives classes, it appears immediately clear as the wise use of different formulations of additives can remarkably extend the variety of polymers of technological interest.

Just to make an example, the classical polymer processing techniques normally involve extrusion, film casting, film blowing, blow molding and injection molding. The processing rate in most of these techniques is limited by the appearance of flow instabilities, collectively known as melt fracture phenomena [13, 14]. Linear viscoelastic rheology and non-linear rheology are dominant factors for good polymer processing. Additives aimed to better processing of polymers are proficiently exploited; they can be classified as acid neutralizers, antiblocks, pigments, antioxidants, compatibilizers, etc.

The improvement of many physical and mechanical properties of polymers (such as process viscosity, modulus, fracture toughness, and coefficient of thermal expansion) is efficiently carried out by the introduction of new types of additives, mainly intended for reinforcement of polymer hosts. This reinforcement can be comprised of:

1. Low-molecular-weight crystalline compounds, that simultaneously improve the impact or fracture toughness, increase or do not otherwise detrimentally decrease the modulus, and enhance the processability (through reduced viscosity);

2. thermotropic liquid crystalline polymers (TLCPs), which are able to reduce the melt viscosity of the thermoplastics and can also improve their mechanical properties [15–17];
3. carbon, glass, basalt, aramid, or natural fiber reinforcement [18]. Fiber-reinforced polymers (FRP) are today used in almost every type of advanced engineering structures: aircraft, helicopters, automobiles, sports goods, civil infrastructures such as bridges and buildings, and so on;
4. nanoparticle, nanofiber or nanotube enriched polymers: nanotechnology represents the emerging research area in innovative plastic-based devices. Nanofibers, in particular, have been used for a wide range of applications such as tissue engineering [19], filter media [20], reinforcement in composites [21] and micro/nanoelectromechanical systems (MEMS/NEMS) [22]. Such fibers can be made from polymers, carbon, and semiconductors into the form of continuous nanofibers, nanofibrous networks, or short nanowires and nanotubes. Moreover, the reinforcement can also occur using particles of spherical shape as well as plate-like and fibrous nanofillers. As a result, nanofibers because of their excellent mechanical properties coupled with very high aspect ratio theoretically are the ideal candidates for reinforcing a polymer matrix.

In all these cases, a strong interface between the reinforcing phase and the host polymer matrix is always desirable in order to achieve desired properties.

1.2 Why to investigate a polymer surface?

The efforts of the last 20 years to quantify the chemical and physical properties of synthetic or naturally derived soft matter systems is mainly linked to the need to engineer polymers to a high level of sophistication. Today, the synthetic chemistry can allow for excellent control over the bulk polymer properties. On the other hand, the importance of polymer dynamics occurring on surface is self-evident: the nature of polymer surfaces allows to control or to modulate properties such as wetting, adsorption, bonding, friction, surface energy, compatibility with other substrates, chemical inertness or reactivity, presence of contaminants, surface morphology, crystallinity, or roughness.

Polymers, due to their very instable nature, can undergo dramatic changes of structural and thermodynamic properties in the confinement area, that may be produced by relatively weak variations of external conditions, such as application of external fields, changes in temperature, moisture, pH, and mechanical stresses. Therefore, surface rearrangements of the polymer chains as a consequence of environmental or interfacial interactions are often unpredictable. In this respect, the dynamics of polymer chains near surfaces has attracted growing attention and will be analyzed in the following paragraphs. Many analytical techniques have been devoted to the recognition and systematic investigation of the chemistry, morphology and evolution of polymers surfaces.

In this book, the peculiarities, differences, and the recent progresses of the analytical tools adopted by polymer scientists to investigate polymer surfaces will be described and critically discussed.

1.2.1 Nature and dynamics of polymer surfaces

Soft materials such as polymers, especially amorphous or semicrystalline ones, cannot be simply considered as solids. They must be seen as semiliquids or very viscous liquids. Moreover, they are characterized by weak aggregation linkages between macromolecules, if compared to conventional engineering materials such as inorganic systems, metals, ceramics, or semiconductors.

In this respect, the molecular motions occurring inside a polymer result in a wide range of cases:

1. the translation motion of entire macromolecules, which is traduced into a mutual flowing of polymer chains or, on the contrary, in a closed and blocked chains packing;
2. the cooperative wriggling and jumping of segments of molecules, which is in turn traduced into flexibility and elasticity;
3. the motion of a few atoms along the main chain or side groups, which allows the three-dimensional reorganization of the chains;
4. finally, the vibrational motions of the bonds between atoms of the polymer, common to all the molecules, from the simplest (i.e., low-molecular-weight compounds) to the most complicated one (i.e., synthetic polymers or copolymers, natural macromolecules, etc.).

These motions have a direct effect on many of properties in bulk, melt and solution of polymers.

As far as the surfaces are concerned, it appears clear that the reorganization of polymer surfaces is strongly related to the characteristics of the interfacing medium (air, solvents, vacuum, other polymers or totally different materials, etc.). In this respect, dynamics on surfaces or interfaces must be treated as a different object of study, when relevant to soft matter.

In detail, we will focus our attention on vibrational dynamics, changes of thermodynamic properties, rotation of functional groups in response to external stimuli and inter- or outdiffusion of segments or chains from the bulk toward the surface, especially in polymer blends and in copolymers.

1.2.1.1 Vibrational dynamics of macromolecules

When we discuss about the different vibrational processes involving macromolecules, we must consider that, as far as polymers are concerned, these processes could cover an extremely broad range of characteristic time scales, spanning from about 100 femtoseconds up to years.

Theoretically speaking, the basic concepts discussed for molecular vibrations of finite molecules can be applied to polymer chains. The model of finite molecules consists of particles endowed with mass and held together by electrical forces such as covalent, dipole–dipole, and Van der Waals interactions. Atoms in molecules are treated as particles in which all the mass is concentrated at a point. In particular, Wilson's GF matrix method has been developed in the past [23, 24]. Briefly, it uses internal coordinates that are bond stretches, in-plane angle bending, torsions, and waggings. For a molecule having n atoms there are (3n−6) vibrational degrees of freedom or (3n−5) if the molecule is linear. Using any set of generalized coordinates, the problem can be reduced to writing the secular equations to be solved to get eigenfrequencies and eigenvectors. However, as the polymer molecules are of infinite length, the order of the matrices becomes infinite. Polymers are subject to complex intra- and intermolecular interactions combined with many intramolecular degrees of motional freedom. In order to reduce the problem to one of workable dimensions, the screw symmetry of the polymer chain is exploited. In view of these facts, the number of mathematical and theoretical models aimed to describe polymers vibrational dynamics is very high and cannot be fully covered herein [25–36].

Today, computer simulations constitute an excellent tool for the investigation of the dynamic properties of complex systems [37–40]. These methods are usually ab initio molecular dynamics schemes [41, 42]. Molecular dynamics (MD) simulations can provide, within statistical averages, results that can be used to test the quality of the predictions of an approximate theory applied to the same model. If theory and simulations disagree, theory must be improved.

In order to obtain experimental results, several techniques have been used to characterize the conformation of polymeric systems. These include infrared (IR) absorption, Raman spectra, inelastic neutron scattering (INS), X-ray diffraction (XRD), nuclear magnetic resonance (NMR), Circular dichroism (CD), and optical rotatory dispersion (ORD). In particular, IR spectroscopy is a very powerful analytical technique used to obtain information about molecular structure of polymers by measuring the frequency of IR radiation needed to excite vibrations in molecular bonds. Infrared spectroscopy in attenuated total reflection (ATR-FTIR) couples the analytical method of infrared spectroscopy with the physical phenomena of total internal reflection (i.e., reflection and refraction of electromagnetic radiation at an interface of two media having different indices of refraction) to restrict the analyzed volume on the surface region of the sample. A deeper investigation about this spectroscopic technique will be supplied in Chapter 5.

1.2.1.2 Changes in thermodynamic properties on the surface

Starting from the classical physicochemical theories typically developed for metals or ceramics, the thermodynamic properties of solids have been then extended also to rigid polymers. However, the mobility of the constituents of the soft matter represented a considerable complication in this extension, due to the greater degrees of freedom of such liquid-like systems. Indeed, at room temperature, polymers are much more mobile than inorganic materials.

The assessment of the classical thermodynamic properties of the most common polymer systems is reported by Li et al. [43]. The polymer macromolecules can be thought as segmental objects joined together by covalent bonds. These segments continuously move to reach the configurational entropy, which determines the equilibrium state having the minimum free energy. A flexible polymer can adopt a large number of configurations of equal energy. Because of the opposite signs of the energy and entropy, maximizing the entropy is equivalent to minimizing the free energy. The effect of thermal fluctuations are then accounted for by minimizing the free energy with respect to all possible configurations.

From a statistical point of view, a long polymer chain can adopt a large number of microscopic conformation states: therefore, the long-chain conformational entropy, S, is described by the following formula:

$$S = k_b \cdot \ln \Omega$$

with k_b being the Boltzmann constant and Ω the number of long-chain conformation states. If a polymer long chain is deformed, whether compressed or elongated, its conformational entropy will be changed and accordingly its free energy will be changed as well.

On the basis of these considerations, we can argue that the thermodynamic of soft matter, such as polymers, is a very complicated field of investigation. In addition, when we focalize our attention to the confinement area of polymers, as well as for all the other atomic or molecular solids, other complications arise: in general, the surface represents a break of the symmetry experienced in the bulk, indeed dramatically changing the thermodynamic properties of the material (e.g., the melting point or the surface energy). The statistical thermodynamics of surfaces are often treated by first determining the surface configuration that minimizes the internal energy [44].

While much of the surface science of hard matter focuses on the atomic-scale structure, the study of soft surfaces involves the large-scale nature of the interface and especially its thermal fluctuations.

Polymer surfaces and, in general, solid phase surfaces, result to be intrinsically less energetically favorable than the bulk, due to the fact that the surface molecules have more energy compared with the molecules in the bulk of the material, since the last ones experience balanced forces of interaction with all the neighbor molecules, while surface ones interact attractively with the underlying layer molecules otherwise there would be a driving force for surfaces to be created, removing the bulk of the material (see sublimation).

The surface energy may therefore be defined as the excess energy at the surface of a material compared to the bulk. In the bulk – in the absence of strong enthalpic interactions – polymer chains typically take on a so-called random coil conformation, where the physical orientation of each repeat unit along the chain's backbone is essentially uncorrelated with the orientations of the chain's other repeat units. The presence of a surface, however, places tight constraints upon the conformations of chains near the surface such that the random coil conformation is perturbed. To alleviate this entropically unfavorable situation and lower the overall free energy of the system, it has been predicted [45] that polymer chains in the vicinity of the surface will segregate their chain ends to the boundary, thereby avoiding the required "reflection" at the material boundary and so minimizing the loss in conformational entropy.

An outstanding question regarding the surface tension is whether this property depends solely on properties of species localized within a few angstroms of the surface, or whether the contributions are more delocalized because of mobility in the fluid. Much of the complexity arises from surface orientation effects, which are especially pronounced in some copolymers and other inherently ordered systems such as liquid crystalline oligomers. Studies of surface free energies of solid monolayers are relevant for the understanding of fluids, and these and other data can be applied to the interpretation of polymer and copolymer melts.

The surface energy of polymers is a very important property that determines their wettability, adhesion, adsorption especially for applications such as coating, printing, gluing, biocompatible interfacing, self-cleaning, patterning, and microfluidics. The breaking of symmetry experimented in the confinement area strongly influences the melting point of polymers near the surface [46–48] and, in the case of amorphous polymers, the mobility of the polymer chains also changes. As a consequence, a fundamental question is if the glass transition temperature, T_g, or polymer mobility at a free surface is significantly different from that in the bulk.

Liu et al. [49] largely discussed about this question, reporting previous studies of different authors, with different conclusions. In particular, Keddie et al. studied the thermal expansion of polystyrene (PS), showing that the T_g of a polymer surface is much less than that of the bulk [50], while Wallace and coworkers [51] found no indication of an increased mobility at the surface, and, in fact, indicated that the mobility at the free surface may be suppressed. Xie et al. [52] initially concluded that there was no evidence of increased mobility, but subsequent experiments suggested the presence of a surface layer with enhanced mobility [53]. Kajiyama and coworkers [54] concluded that the surface mobility of PS was enhanced only for low-molecular-weight polymers where the preferential segregation of chain ends to the surface was sufficiently large to enhance the mobility. For higher molecular weight polymers no enhanced mobility at the free surface was found. In response to these not clear conclusions, Liu and coworkers investigated the relaxation of PS segments at the free surface as a function of temperature and they showed that the relaxation of the polymer within the first nanometer from the free surface is parallel to the surface or planar, whereas in the first 10 nm the dominant relaxation occurs normal to the free surface. Full relaxation of the surface was found for temperatures above the bulk T_g. The studies presented here have quantitatively shown that the segmental mobility of high-molecular-weight, amorphous polymer chains at a free surface is not significantly different from that in the bulk. Chain ends can enhance the surface mobility; however, in order to have high enough chain-end concentrations, the molecular weight of the polymer must be sufficiently low to have an effect.

As far as glassy polymers are concerned, Priestley et al. demonstrated that the rate of structural relaxation of poly(methyl methacrylate) near surfaces and interfaces with respect to that of bulk was reduced by a factor of 2 at a free surface and by a factor of 15 at a silica substrate interface [55]. Many studies have attempted to provide a physical understanding of changes in T_g induced by spatial confinement. Several factors have been invoked as being responsible for differences in structural and dynamical behavior of confined polymers compared to the bulk. These include geometry (e.g., pore size and shape) as well as specific interfacial interactions (e.g., hydrogen bonding and electrostatic forces) that can affect in a significant manner both chain conformation and segmental mobility. For weakly adsorbing surfaces, both positive and negative shifts in T_g are reported, yet these results still remain quite controversial [56]. Several methods can be used to obtain surface parameters related to the interfacial free energy and other thermodynamic interaction measures such as the enthalpy of adsorption or displacement [57–60].

A deep knowledge of the thermodynamic properties of polymer surfaces is the basis for the development of smart surfaces. It has been recognized that characterizing the surface thermodynamic properties (e.g., hydrophobicity–hydrophilicity balance, electrokinetic properties, adhesion, and thermal behavior) are vitally important for defining suitable applications and developing new markets. Awaja et al. reviewed the main research efforts relevant to the study of polymer adhesion with a special focus on chemical, mechanical and thermodynamic mechanisms linked to this property [61]. Köstler and coworkers focused their attention on the study of surface thermodynamic properties and wetting behavior of polyelectrolyte multilayers that can be applied in biomedical or electronic devices [62], while Luzinov and coworkers demonstrated the importance of understanding the relationships between the bulk properties of pristine polymeric materials and their surface characteristics, in order to design materials with smart surface behavior [63]. The dependence of surface free energy on molecular orientation in polycarbonate films, as determined by contact angle measurements, was demonstrated by Stachewicz, and coworkers [64]; Geoghegan et al. summarized the main theoretical and experimental aspects of wetting and dewetting phenomena relevant to polymer films [65]. Wetting or measurements of the contact angle of a liquid test droplet on a solid surface represent a straightforward technique revealing surface energetic information inaccessible by the surface spectroscopies [66] These information resulted crucial especially in the ink, paint or adhesive research fields [67, 68]. Contact angle measurements at polymer surfaces can be carried out by several different methods [69]. Polymer surface energetics and contact angle measurements will be described in detail in Chapter 9.

1.2.1.3 Rotation of functional groups on polymer backbones in response to different environmental conditions

In the past, the surface energetic properties of solids were assumed to be independent of the environment (polar and nonpolar) in which they are immersed. Successively, Holly and Refojo [70], taking contact angle hysteresis experiments into consideration, concluded that hydrogel surfaces changed their polarity according to the nature of the environment. Since then, such surface modification phenomena have been investigated by employing the X-ray photoelectron spectroscopy (XPS) analysis [71, 72] and dynamic contact angle measurements [73, 74].

Today, it is well accepted that polymer chains are not "frozen" or immobile entities, and environmentally induced changes in the surface structure have been observed on a variety of polymers, not only on those exposing relatively mobile surface layers or containing polar moieties. The polymer chains mobility is an inherent property of the soft matter: the polymer configuration instantaneously changes. This enhanced mobility is also achieved through the rotation of functional groups along the polymer chain backbone rather than the long-chain segmental motion of macromolecules. From a mathematical point of view, adjacent bonds can be described by two angles in spherical coordinates: the zenith angle θ and the azimuth φ. The zenith angle represents the angle between adjacent bonds, and the azimuth represents the rotations about the zenith.

Here we provide only a brief presentation of the major models of polymer rotation modes:
1. Rotational isomeric model: In this model, the bond angle with respect to the adjacent one is allowed to take distinct rotational positions (usually about three).
2. Freely rotating chain: The rotational restriction is removed, and the azimuth can take any value, but the zenith angle is fixed.
3. Worm-like chain: A continuous version of the freely rotating chain, with vanishing bond lengths.
4. Freely jointed chain: A sequence of freely rotating chains looses its bond angle correlation after one Kuhn length (i.e., in polymer melts or solutions, the ratio of mean square end-to-end distance, R_0^2, and the fully extended size, R_{max}). When the bond sequence is replaced by a sequence of Kuhn lengths, two adjacent Kuhn lengths have uncorrelated bond angles. They are like rods joined

freely without any angular restrictions (both in φ and θ). In these models the rod represents a Kuhn length only in the distance. The rods are assumed to be of fixed length and massless. The mass of the local collection of monomers (equivalent to one Kuhn length) is replaced by a spherical bead at the joints.

5. Bead spring chain: In this model, as Kuhn length represents a collection of bonds, a collection of Kuhn lengths is represented by a spring. However, the spring is not a replacement of a fixed distance (i.e., a Kuhn length), but represents a variable distance.

Considering the polymer surfaces, the surface chains can respond to different surrounding conditions [75]. Given sufficient mobility, polymer surfaces will reorient or restructure in response to their local microenvironment so as to minimize their interfacial free energy with the surrounding phase. In particular, side chains of polymers resulted in the most inclined to reorganization. The specific spatial arrangement of functional groups interfaced with a contacting medium is called the surface configuration. A typical rearrangement takes place when the polymer surfaces experience a change in polarity of solvents or interfacing media (Fig. 1.4). In this case, for example, when a polymer having polar functional groups experiences a polar medium, a rearrangement of these groups on surface is thermodynamically advantageous. On the contrary, when in contact with an apolar medium, these groups tend to minimize the interaction with this opposing environment.

An example of the influence of reorientation of functional groups on the features of the polymer surfaces is the configuration of atoms in a macromolecule of gelatin gel. Gelatin is known to contain hydrophilic groups; however, gelatin gels are surprisingly hydrophobic. This is because all the hydrophilic groups at the gelatin/air interface are oriented toward the bulk. Thus, it is not the configuration of hydrophilic groups with respect to the polymer backbone but rather the surface configuration of those groups that dictates wettability. The reorientation of surface-

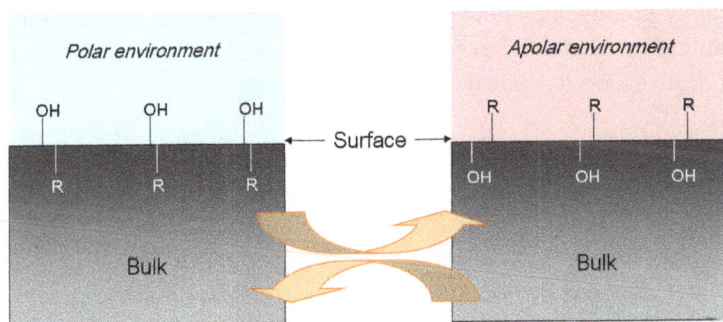

Fig. 1.4: Surface arrangement of functional groups of a polymer interfaced with different contact media.

state functional groups in response to a change in the interfacing environment was called surface configuration change by H. Yasuda et al. [76–78]. Wei and coworkers have proposed a recent review, comprehensive of the different applications and the recent advances of stimuli-responsive polymers [79].

One of the most useful tools used to investigate the orientation of molecular chain was the reflection–absorption Fourier-transform infrared spectroscopy (RA-FTIR) [80]. In principle, if there is a change accompanied by a reorientation of the dipole moment in the conformation of polymer chain, this difference can be detected through the RA-FTIR spectroscopy. For example, when a p-polarized IR beam is incident at a grazing angle to the substrate, the absorption for those dipole moments oriented normally to the substrate will be significantly enhanced. An increase or decrease in the intensity of a particular IR absorption peak is therefore indicative of a change in the molecular chain orientation. This feature is of great benefit when examining the orientation of a specific group within the bulk polymer film. For example, Lu and Mi [81] studied the chain orientation states of the polyacrylamide (PAL) by analyzing the vibrational signals of molecular groups using the polarized RA-FTIR with the grazing-angle incidence. They found a gradiently varied bi-layered structure of the PAL thin films on the Au substrate: indeed, PAL chains in the layer adjacent to the Au substrate showed different orientational orderings with respect to the chains present on the top layer. In Fig. 1.5, the RA-FTIR spectra (in the range of amide I) of the PAL films are reported, with eight different thicknesses (9.8–559 nm), collected using the s- (Panel A) and p-polarized (Panel B) lights, respectively. As the film thickness increased, the splitting of the amide I band gradually was evident with one band at ~1,690 cm^{-1} and the other band at ~1,658 cm^{-1}. This band splitting was significantly more evident using the s-polarized than the p-polarized light. These spectral features clearly pointed out that the state of chain packing in proximity of the interface was quite different from that far from the interface.

More recently, Koziol et al. [82] employed polarized FTIR to investigate a human tissue sample, mainly fibrous and collagen-rich, calculating Herman's orientation function and obtaining the azimuthal angle of selected vibrational modes with very intriguing and innovative tools to study biological tissues.

1.2.1.4 Surface interdiffusion or segregation of copolymers or polymer blends

In several applications, copolymers or polymer blends (i.e., physical mixtures of two or multicomponent polymers) are required, in order to obtain enhanced properties and functions. Moreover, as described in Section 1.1.4, additives are also commonly added into polymeric materials used in our daily life in order to tune their properties: plasticizers, fillers, compatibilizers, antistatics, reinforcement or aesthetic enhancers, pigments, antioxidizing agents, etc.

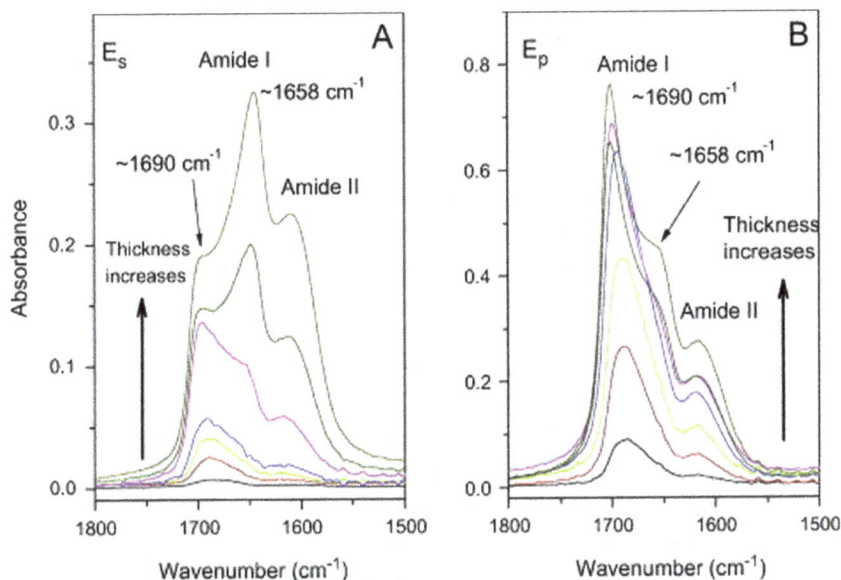

Fig. 1.5: (Panel A) s- and (Panel B) p-polarized RA-FTIR spectra in the amide I range (1,800–1,500 cm^{-1}) of PAL thin films with different thicknesses (i.e., 9.8, 39, 73, 104, 170, 269, 375, and 559 nm) coated on Au substrates with grazing angle (85°) incidence. Reproduced by Lu X, Mi M. Gradiently varied chain packing/orientation states of polymer thin films revealed by polarization-dependent infrared absorption. Eur Polym J 2015, 63, 247–254, Copyright 2015, Elsevier [81].

Since many polymers are nonmiscible, the constitutive components of block co-polymers or blends tend to segregate into different domains giving raise to the so-called phase separation. In copolymers, the covalent bonds of these components prevent the total phase separation from extending over a large length scale. Instead, only small domains with typical sizes in the range of tens of nanometers are formed. Phase separation or segregation phenomena lead to a different and often unpredictable surface polymer composition. Moreover, when polymer additives are added into the polymer host, they sometimes migrate toward the surface and other times stay in bulk, according to different requirements. All these phenomena are particularly important for commercial polymer films, since these systems normally comprise multiple polymer species, solvents and additives, leading to numerous complex phase separation morphologies, such as bi-continuous structure, islands or holes. Possible sources of phase segregation can be involved in polymer synthesis or formulation – composition, molecular weight, film thickness, solvents, additives – or can be linked to processing successive to their synthesis – changes in the external environment, pressure, temperature, interaction with other interfaces, etc. Xue et al. [83] summarized the fundamental theory and factors influencing phase separation in polymer thin films, in order to develop new strategies for the utilization of phase separation to generate ordered pattern in polymer thin film systems. Recently, Stein et al. [84] evidenced the importance of the consequences of surface enrichment of one constituent (or segment type) in polymer blends (or block copolymers). The important conclusion of this review was that, even if simulations can guide in understanding the thermodynamic principles that control the attraction of polymers toward surfaces, the ability to anticipate the surface disposition of complex materials remains difficult. Anyway, many investigations have been performed to gain much more ability to predict surface phenomena relevant to polymer blends. For example, Huang et al. [85] demonstrated that phase separation could be mitigated by increasing the number of polymer species (n) in the blend. They ascribed this suppression to the high mixing entropy and a kinetic steric effect blocking like-polymer aggregation during film formation.

As far as the surface migration of polymer additives is concerned, this feature can be considered both a problem and a desired phenomenon: in the latter case, surface-active additives are intentionally added to the polymer host, so the enhancement of additive migration on surface must be reached. However, for both the cases, the control and the establishment of principles for the selection of additives and for the optimized processing conditions to enhance or to reduce the surface localization of these molecules play a primary role in the complex polymer systems manufacture.

In polymer systems, also the surface concentration of additives is normally different from their bulk concentration. There are a lot of factors that influence the surface migration of additives in polymers: Wu and Fredrickson [86] and Minnikanti and Archer [87, 88] proposed a linear response theory that explains this effect in terms of a so-called entropic attraction of chain ends to surfaces and thermodynamic features of the blend near a surface and in bulk.

In the recent years, the migration of additives to the surface is perceived as dangerous to human health and the environment, since additives can undesirably lead to human exposure via, for example, food contact materials, such as packaging or, in general, they could contaminate soil, air, water, and food, as recently summarized by Hahladakis [89].

Surface segregation phenomena in segmented or random copolymers, polymer blends and polymer hybrid composites have been intensively studied by surface-sensitive spectroscopies, by contact angle measurements and more recently by scanning probe techniques and mass spectrometry. Some significant examples of case studies will be illustrated in the following chapters.

1.2.2 Surface modification of polymers

Polymeric materials are designed for a given application primarily considering their favorable bulk properties (including thermal stability, mechanical strength, solvent resistance, cost, etc.). In addition to these fundamental characteristics, the selected polymers must also offer surface characteristics, which could be modulated to the intended application. Today, to achieve this, the strategy of surface modification represents a common approach to obtain the ultimately desired properties for the myriads of polymer applications. Polymer surfaces can thereby be changed to reach a variety of goals, i.e., increasing adhesion, improving wettability, enhancing processability, reducing friction, reducing susceptibility to harsh chemicals or environmental agents, increasing dye absorption, etc. A brief excursus of the most important surface modifications of polymers is reported below.

1.2.2.1 Improvement of wettability

The wettability, i.e., the ability of a liquid to maintain contact with a solid surface, is a fundamental property of solid surfaces both in the natural and technological world. The investigation of the polymer wetting phenomena has been the subject of intensive studies both in the past (i.e., the spraying of a paint, the penetration of ink in paper, the liquid absorbency or repellence of cloths, etc.) and in the recent research (including microfluidic technologies, regulation of protein adsorption, and biomaterials design). In particular, as far as the biomaterials research is concerned, the surface wettability represents a property of primary importance [90, 91]. Polymers can be easily processed due to their good plasticity or fluidity after being heated, emulsified, mixed, or dissolved in an appropriate solvent. Indeed, various methods for modification of wetting behavior of polymer surfaces have been proposed. Super-hydrophilic surfaces with extremely wetting water contact angle values ($0° < CA < 10°$) were obtained by a random nanoporous structure and uniform micropattern made of poly(methyl methacrylate) (PMMA) thermoplastic coatings on microextruded substrates [92]. On the other hand, fluorinated polymers and copolymers are usually synthesized for obtainment of superhydrophobic surfaces [93–95]. However, hazardous fluorinated materials limit their applications at the present stage. Therefore, different polymer coatings are recently proposed to obtain superhydrophobic surfaces [96].

The *surface chemical composition* of a polymeric system, determining its surface free energy, has great influence on wettability [97, 98]. In the case of segmented or random copolymers as well as polymer blends, as reported in a previous paragraph, these materials spontaneously rearrange to have on the surface the low-energy nonwottable segments. The wettability of polymer surfaces could be tuned over a wide range by selecting polymer brushes with appropriate functional groups. A deeper discussion of surface grafting of macromolecules by post-polymerization modification techniques was exposed by Galvin and Genzel [99]. The wettability modification of a surface may be also achieved by different chemical and/or physical treatments [100–103], including plasma treatment with the appropriate gases to render a surface hydrophilic or hydrophobic [104, 105]. Plasma is a highly reactive mixture of gases, such as argon, oxygen, hydrogen, or a mixture of these gases, consisting of a great concentration of ions, electrons, radicals, and other neutral species [106]. Plasma treatments provide an efficient, economic, environmentally friendly, and versatile technique for modifying the surface properties of materials, in particular polymers.

The *surface topographic structure* is also an important factor that influences the wettability. On the basis of these principles, intensive studies have been made to obtain super-hydrophobicity by improving the surface roughness [107–111]. It has been recognized that the cooperation between the surface chemical compositions and the topographic structures is crucial to construct special wettability. In this respect, a promising strategy to regulate wetting behavior is the combination of the surface patterning and the chemical surface modification. Therefore, polymer substrates can be patterned using physical methods to selectively degrade hydrophobic regions in order to result in surface portions of enriched hydrophobicity. Many elegant methods have been developed to construct micro- and nanostructures, such as laser ablation and photolithography-based microfabrication [112–114], chemical vapor deposition (CVD) [115, 116], domain-selective oxygen plasma treatment [117], and sol−gel method [118].

Only to cite some representative examples of the most significant surface wettability modifications, IBM researchers developed a method to produce patterns with modified wettability via the deprotection of poly(tert-butyl acrylate) (PtBA) films to poly(acrylic acid) (PAA) using acids generated during lithographic patterning of the resist [119]. Wettability patterning could be achieved by masking only defined regions during a first modification, introducing thus hydrophobic pendant groups, and then backfilling in a second modification step with hydrophilic groups [120]. Moreover, Huck and coworkers modified the wetting behavior of polyelectrolyte brushes simply by exchanging different ions into the brush [121].

Worth mentioning in the control of the wetting phenomena is the ability to transform a superhydrophilic coating (contact angle close to 0°) into a superhydrophobic one (contact angle 165°), by functionalizing with 1 H,1 H,2 H,2 H-perfluorodecyltriethoxysilane the films based on poly(allyl amine hydrochloride) (PAH) and poly (sodium phosphate) (PSP), as reported by Lopez-Torres et al. [122]. The films were combined by means of the layer-by-layer (LbL) technique and then functionalized by chemical vapor disposition (CVD) technique (see Fig. 1.6).

Finally, a recent and very inspiring work of Das and coworkers [123] browsed the foremost topographical and chemical strategies proposed to carry out bio-inspired antiwetting interfaces, evidencing that most of all are difficult to be applied in the real world. Beyond these techniques, the design of chemically reactive porous polymeric coatings was found to be a promising approach for developing smart and durable bioinspired interfaces.

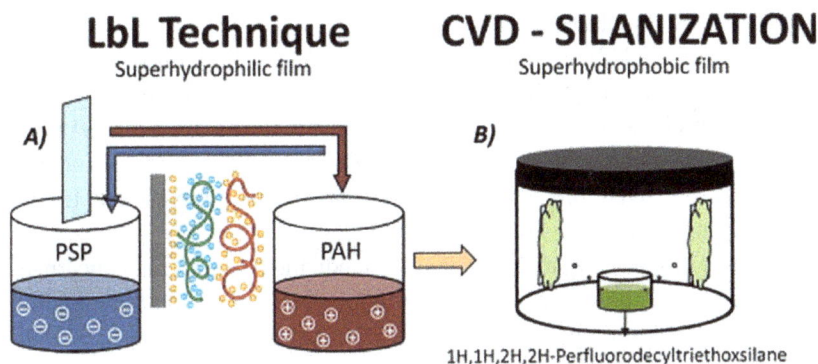

Fig. 1.6: Schematic picture of the techniques employed to make a superhydrophilic film (A) and to transform it in a superhydrophobic one (B). Reproduced by Lopez-Torres D, Elosua C, Hernaez M, Goicoechea J, Arregui FJ, From superhydrophilic to superhydrophobic surfaces by means of polymeric layer-by-layer films. Appl Surf Sci 2015, 351, 1081–1086, Copyright 2015, Elsevier [122].

1.2.2.2 Improvement of porosity or roughening

Porous polymer surfaces are required in applications such as separation processes, filters, foams, membranes, chromatography, biomaterials, paper coatings, and superconductors. Capability of mass transfer, liquid retention, improved opacity, lighter weight, enhanced gloss, controlled spread and imbibition, and control of heat conduction are properties strictly related to porous substrates. In all these contests, the pore size, distribution, geometry, and stability could be critical features. Porous resins are widely used in chromatography [124]. Moreover, the presence of a variety of functional groups on polymer surface, which enable a specific sorption of various chemical species, allows the ability to operate as chromatographic separation systems [125]. Porous membranes are used in dialysis, desalination, filtration, gas separation, reverse osmosis, microfiltration, etc. Porous materials are also employed as drug delivery systems (DDS) [126]. Polymers such as poly(urethanes), poly(siloxanes) or poly(vinylic) systems have been employed to control the release of pharmacologically active agents. Various typologies of pores (i.e., open, closed, transport, and blind pores) allow the polymer to adsorb and release drugs in a more reproducible and predictable manner. Pharmaceutically exploited porous adsorbents include mesoporous, macroporous, microporous and nanoporous scales. Nanoporous systems represent today the new challenge in pharmaceutical research [127]. When a porous polymeric DDS is placed in contact with an appropriate medium, the release of drug to medium must be preceded by the drug dissolution from surface and successively by diffusion through the liquid filled channels or pores.

A controlled porosity is also imperative in superabsorbent polymers, i.e., polymers able to absorb very large amount of liquid, mostly employed in the market of personal disposable hygiene absorbent products (baby diapers, adult protective underwear, and sanitary napkins) [128]. In all these fields, hydrogels, i.e., polymers or copolymers opportunely cross-linked and able to absorb great amounts of water without dissolving in this liquid, became the main class of polymers employed. The pore structure in these materials is fundamental: a connected microporosity in the dry hydrogel is required, since it increases the hydrogel sorption capacity for capillary retention effects. The hydrogel interconnected, molecular-sized pores are of great interest also for absorption, separation, and catalytic applications.

As far as tissue engineering and regenerative medicine research are concerned, the design, production, and characterization of porous networks result demanding tasks: large pores and highly porous structures are required to promote in vivo direct osteogenesis and to allow vascularization and proper oxygen supply to the cells, while smaller pores result in osteochondral ossification. The interconnectivity of such pores is also significant since phenomena as cell migration, ingrowth, vascularization and nutrient diffusion for cell survival are inhibited if pores are not interconnected, even if the structures possess an elevated porosity.

Different procedures have been developed in order to obtain porous polymeric systems:

1. Solvent casting/particle leaching: This technique is based on the dispersion of a porogen, which is usually an inorganic salt, into a polymer solution. The solvent evaporation produces a polymer–porogen network. The solute particles are subsequently leached or dissolved away by immersing the material in a selective solvent, resulting in the formation of a porous network [129]. The surface of these porous systems is particularly important, since the solvent evaporation starts at the interface polymer/air as well as the first step of pore formation in the solvent for salt removal is governed by surface rearrangement of polymer surface in the selected liquid.

2. Freeze-drying: It consists in freezing the material and then reducing the surrounding pressure to allow the frozen water in the polymer to pass directly from the solid to the gas phase, so the porosity is generated from the direct sublimation of water [130].

3. Gas foaming: The use of blowing agent, in gaseous state, i.e., a gaseous porogen, generates a high porous system. The porosity depends upon the amount of gas dissolved in the polymer. This fabrication method requires no leaching step and uses no harsh chemical solvents, but usually requires high pressures and/or temperatures [131].

4. Electrospinning: It is a simple, versatile and widely accepted technique to produce fibers ranging from nm to μm, mostly in solvent solution and some in melt form. In this process, a high voltage is used to create an electrically charged jet of polymer solution or melt, which forms polymer fibers after drying or solidification [132]. Today, more than 200 polymers are used for electrospinning [133–136]. Scanning Probe and Electron microscopies (SPM and SEM) represent the most employed techniques to ascertain the surface size, orientation and uniformity of the electrospun nanofibers [137, 138], as well as X-ray microcomputed tomography analysis [139].

5. Phase inversion: In this technique, the principal pore-forming process is the employment of a ternary system, i.e., polymer, solvent, and non-solvent. Pore size and distribution can be controlled for each specific application depending on the choice of ternary system and preparation parameters [140].
6. Microfabrication processes or computer-aided additive biofabrication: These represent the newest and non-conventional scaffold manufacturing processes. The computer-aided fabrication processes include: *direct writing, fused deposition modeling, selective laser sintering, stereolithography, dimensional printing,* etc. The pore size and geometry of the scaffold is entirely computer-controlled. A complete presentation of these techniques was summarized by Bartolo and coworkers [141]. More recently, enormous strides have been made, coming to conceive the biofabrication of heterogeneous artificial tissues [142]. All these innovative processes should require a deep investigation on the surface properties of the processed polymers.

On the other hand, the surface roughness represents another important property of polymer surfaces.

For instance, in the biomaterial field, adhesion, proliferation, differentiation and overall cell viability on polymer surfaces can be significantly affected by roughness [143, 144]. It could be increased by mechanical treatments, wet-chemical treatments with strong acids or bases, or by exposure to flames or corona discharge. At the nanometer scale, roughness could be obtained by grafting spacers having different molecular weights or different lengths to the polymer surface [145].

For other practical purposes, surface finish is crucial both for aesthetics and functional features of commercial polymers. In the case of reinforced thermoplastic materials, i.e., thermoplastic polymers reinforced with special additives during the molding step, surface finish variables must be considered in choosing the optimum material, in terms of additives and processing techniques. Aesthetics are important in some appearance products (glossy, shiny, lustrous): specular or metallic-like surfaces may be often desired; therefore, surface roughness must be controlled regards to the standard and variable molding parameters. Not only aesthetic but also functional features are linked to the surface finish, when parameters as wear resistance, coefficient of friction or adhesion to other materials (plastics-on-plastics, plastics-on-metals, etc.) are affected by the surface smoothness or roughness.

1.2.2.3 Improvement of adhesion

Adhesion can be defined as the chemical (i.e., bond) or physical (i.e., adsorption) anchoring between two adjacent materials. Adhesion strength is measured as the cohesive force involved when separation occurs within two materials.

A typical embodiment where adhesion is a critical property is the coating of a substrate. The factors contributing to such property are essentially chemical or mechanical.

The durability of a coating onto the substrate is of primary importance in many applications: the coating must sustain mechanical and thermal stresses, as well as elastoplastic distortions. If adhesion is poor, the localized break of the coating/substrate interface takes place, leading to blistering or even complete spalling-off of the coating. This is particularly true if the coating is subject to corrosion or moisture: indeed, in an oxidative or humid environment, the tendency for the film to crack or peel from the substrate may be well aggravated.

Often, much more than two layers of different materials are required in order to develop a complex technological device. An example is supplied by solar cells, where usually a multilayer structure, such as that reported in Fig. 1.7, is proposed [146].

Usually, improvement of adhesion could be carried out by surface preparation of the substrate, for example through contamination removal by different cleaning procedures (aimed to remove protective layers, organic contaminants, adsorbed layers, dust and fingerprints) or by surface activation [147].

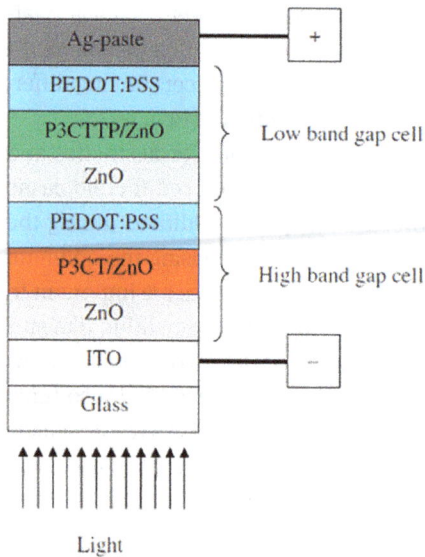

Fig. 1.7: An example of tandem polymer solar cell comprising eight layers. Acronyms used in the figure: PEDOT:PSS = poly(ethylenedioxythiophene):polystyrenesulphonic acid; P3CTTP/ZnO poly(carboxyterthiophene-co-diphenylthienopyrazine) with ZnO nanoparticles; P3CT/ZnO = poly(3-carboxydithiophene) with ZnO nanoparticles; ITO = indium tin oxide. Reproduced from Krebs FC. Fabrication and processing of polymer solar cells: A review of printing and coating techniques. Solar Energy Materials & Solar Cells 2009, 93, 394–412, Copyright 2009, Elsevier [146].

Different approaches can be adopted to alter the substrate surface and thus to enhance adhesion:

1. *Mechanical*: Roughening surfaces by grinding, brushing, blasting or similar methods creates enough sites where glue or paint can physically grab on. The cutting action on the surface during such a treatment also may cut polymer chains, creating open or pendant bonds exposed at the surface. This could allow for improved chemical bonding on the surface.
2. *Chemical*: Treating surfaces with chemicals can alter the surface chemistry to improve the surface energy by providing active chemical bonds or groups on the polymer surface.
3. *Plasma treatments*: Surface preparation of substrates can be carried out by plasma processes. Surface activation and contamination removal can be attained, without creating hazardous by-products and without changing bulk properties. Surface activation by plasma treatment is performed by using the right mixture of gases generating excited species which react with the polymer surface to create functional groups such as hydroxyl (–OH), carbonyl (–C=O), carboxyl (–COOH), or amino (NH_x), which exhibit high polarity and change the base/acid interaction at the polymer surface. Different researchers highlighted the usefulness of plasma treatments for adhesion improvement [148, 149]. The plasma processes are employed not only for adhesion promotion, but also in a very wide variety of fields (micro- and optoelectronics, biomedical field, etc.) and with different aims (contamination removal, surface activation, polymerization, etching, cross-linking, etc.). In this respect, this chapter is not comprehensive of this matter and in-depth dissertations on plasma processes are reported elsewhere [106].

Both mechanical and chemical improvements of adhesion can be monitored and evaluated by a number of analytical techniques, such as atomic force microscopy (AFM) and SEM for a morphological evaluation of the surface modification, or FTIR, SIMS (secondary ions mass spectrometry), and XPS for detection of new functional groups originated on surface.

1.2.2.4 Interaction of polymer with biological environment: biocompatibility

The biomaterial field probably represents the research field where the boundaries between the traditional sciences, such as chemistry, physic, mathematic, and biology are increasingly superimposing and connecting with those of engineering and medicine, in order to find new and more sophisticated solutions to medical problems and finally to develop products to facilitate the recovery of the human body.

Biomaterials can be considered all the materials that interface with a biological system: starting from catheters, prostheses, fracture fixation devices, intraocular lenses or soft contact lens, pacemakers, artificial hearts, cardiac assist devices, and, more recently, drug delivery systems, microfluidics, biochips, robotics, etc. It is difficult to find a device which is not fabricated, totally or partially, by using polymeric materials. Traditionally a biomaterial, as defined by Park et al., was considered a material used to replace part of a living system or to be located in intimate contact with living tissue [150]. More recently, biomaterials are thought as materials that could deliberately elicit a particular response from the body to achieve a therapeutic goal. The effectiveness of a biomaterial placed in contact with human fluids or cells is strongly dependent on its interactions with this biological environment, which determine how it is accepted or rejected.

In general, a controlled biological response is primarily governed by the chemistry and topography of the outermost layers of biomaterials. The properties of interest for biomaterial surfaces include the chemical composition, the hydrophilicity/hydrophobicity balance, the morphology (i.e., the domain structure) and the topography (i.e., the surface roughness, planarity, and feature dimensions).

No common rules exist in selecting surface biomaterial properties: in the case of stents, for instance, no cell adhesion is desirable, while in tissue engineering applications the main effort is to promote cell adhesion and proliferation on polymer surfaces [151]. For polymeric biomaterials, surface engineering is often required: the polymer surface, effectively, can be physically, chemically, or biologically modified by different techniques and for different purposes, without affecting the key bulk physical properties of polymers. Surface engineering approaches to develop improved biomaterial performances can be generally divided into physical–chemical and biological methods.

Examples of physical–chemical methods are:

1) selective etching, to increase the surface roughness, change the surface chemical constitution, degrade low molecular weights which migrate to the surface and mitigate residual surface stresses [152]. It can be carried out by acids or bases as well as by irradiation treatments exposed below;

2) ionizing irradiation treatments, such as plasma [153], ion [154], electron beam [155], laser treatments [156], UV irradiation [157], used to develop micro- and nano-patterning of biomedical surfaces or to create new surface functional groups;

3) film deposition, i.e., self-assembled monolayer (SAM) deposition [158] or Langmuir–Blodgett films [159];

4) surface grafting of functional groups by wet chemical reactions, such as fluorination, organosilanization [160], or post-polymerization modification (PPM) [99]. PPM incorporates functionalities into polymer chains by decorating already-synthesized polymers with an appropriate chemical and/or biological species.

5) Electrochemical deposition of polymers on metal substrates, by electropolymerization or electrophoretic deposition, to obtain bioactive, anticorrosion or antimicrobial properties [161].

On the other hand, examples of biological treatments, which are probably the most forefront surface modification techniques of polymeric biomaterials, can be divided into these main classes:
1) Covalent attachment of bioactive compounds to biologically functionalize polymer surfaces. It is possible to produce complex brush systems with position-dependent distributions of the chemical modifiers. A number of different types of biologically active molecules have been covalently incorporated into polymer systems and/or grafted on their surface: aminoacids [162, 163], short sequences of peptides [164] or specific proteins and enzymes [165–167], antibodies [168], lipids [169], nucleotides or polynucleotides [170], etc.
2) Surface absorption, embedding or non-covalent immobilization of biologically active molecules, by using hydrogels [171–174], foams, or scaffolds [175], micro- or nanoparticles or micelles [176, 177], biodegradable/bioerodible polymer matrices [178–180], etc.
3) Cell pre-seeding on polymer biomaterials [181].

In any case, it is clear that an in-depth characterization of biomaterial surfaces is crucial to relate their modifications with changes in biological response: all the information gained from surface analysis may allow for tailoring future improvements in surface modification.

1.2.2.5 Improvement of conductivity

Plastics are usually conceived as insulating materials. Nevertheless, the improvement of polymer conductivity is one of the most representative examples of the possibility to act on specific physical properties, tailoring for a potential integration of polymers in electronic devices.

The use of polymers in electronics technology has several advantages over conventional inorganic (i.e., silicon-based) materials: lower costs and ease of manufacturing, large flexibility, and variety of substrates and modifications. Indeed, the polymer structural order can be varied from purely amorphous to highly crystalline as well as its shape can range from fibers to films, roads, or to nano-sized objects.

The majority of polymers do not possess an inherent conductivity; however, we have seen in a previous paragraph that polymers characterized by a conjugated backbone can exhibit inherent conductivity. In double bonds, the electrons are localized in one σ and one π bond; in the last one, the electrons are more mobile and can flow over the whole system, giving rise to an electron delocalization [182]. While metals are able to conduct electricity via mobile electrons in the conduction band that travel across atoms in the lattice, an equivalent situation in organic materials is provided by the delocalized π-electrons. Only a few numbers of polymers are inherently conductive (see Fig. 1.8).

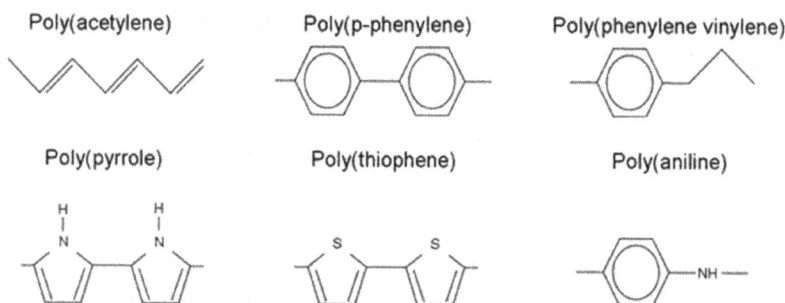

Poly(acetylene)   Poly(p-phenylene)   Poly(phenylene vinylene)

Poly(pyrrole)   Poly(thiophene)   Poly(aniline)

Fig. 1.8: Main examples of electrically conductive polymers.

However, electron conjugation is not enough to make a polymer conductive. Therefore, polymer material needs to be doped for electron flow to occur. The main strategy to improve the conductivity of a polymer is the oxidation or reduction by electron acceptors or donors, resulting in p- or n-type doped materials, respectively. The formation of a complex between the polymer and the dopant produces a considerable increase in electron mobility. Electrical conductivities can be varied by as much as 15 orders of magnitude by changing dopant concentrations so that electronic property control is feasible over the entire range from insulator to semiconductor. Although typically doping conductive polymers involves oxidation or reduction, chemical, electrochemical, photochemical and ion implantation methods are also available [183, 184]. Moreover, different additives can be employed in order to improve the conductivity of polymers, developing the so-called nanocomposite conductive materials: carbon fibers, carbon black, carbon nanotubes or nanorods, graphene oxide, layered titanates, rutiles, phosphates or phosphonates, metallic-based nanoparticles, etc. [185].

Modern applications of conductive polymers are in biosensors [186], electronic noses [187, 188], lab-on-a-chip [189], biomaterials [190], diagnostic devices [191], solar cells [192], etc.

A variety of techniques have been widely used to delineate the physical properties of the conjugated polymers. For example, electrochemical characterization provides information regarding influence of the electrolyte and pH on growth and on electrochemical properties of conducting polymers [193].

Since often the interfacing of different polymers is carried out in these integrated devices, as well as surface doping or surface modification of polymers is necessary, the outmost layer properties and features become an important issue. A typical surface technique currently employed to analyze the nanoscale properties of thin films of conducting polymers is AFM, considering information gained by its extension called phase imaging (PI-AFM). For example, in the case of poly(bithiophene) (PBT), deposited under potentiostatic and potentiodynamic conditions, the morphologies of the films prepared using the two techniques resulted quite similar by conventional AFM, while the phase contrast measurements revealed profound differences in the mechanisms of potentiostatic and potentiodynamic electropolymerization [194]. Also spectroscopic techniques have been successfully employed to characterize conductive polymers: Meana-Esteban et al. pointed out an experimental approach to correlate the Raman and the ATR spectra from azulene-based polymers according to the effective conjugation coordinate theory, observing that the IR spectra of the conducting state showed new doping induced infrared active vibrations [195].

Considering that the abovementioned are only two of the copious examples of surface investigations performed on conductive polymers, it must be emphasized that surface characterization plays a significant role for the correct application of polymers to electronic devices.

1.2.3 Nanostructured polymers

Even if scientists have not already unanimously settled on a precise definition of nanomaterials, the definition of the International Organization for Standards (ISO) (ISO/TS 80004-2:2008, 2008) was relevant to materials having one or more dimensions in a scale below 100 nm.

The exceptionally high surface-area-to-volume ratio (S/V), the key feature of nanostructured materials, is the reason of their unique or empathized characteristics, such as optical, magnetic, electrical, and reactivity properties. In this scenario, the phenomena occurring on the surface of nanomaterials have an impact on their final properties significantly higher compared to those of the corresponding materials at normal scale. Therefore, it is evident the importance to study the surface properties of nanomaterials.

Nanomaterials strongly attract the scientist's attentions since they opened new worlds of possibility in terms of functionalities and technologies, respect to the traditional structures. Anyway, even if nanomaterials are recognized to provide great benefits, the potential side effects on human health and the environment, directly linked to their nanosizes, represent the main brake on the widespread use of such systems [196].

Among the different nanomaterials, polymeric nanomaterials have gained considerable attention, mainly due to their tunable chemical and physical properties, flexibility and softness, as well as their ability to be modified on surface, which makes them useful for the many different applications.

Polymeric nanomaterials can be classified in two main categories:
1. Polymer nanoparticles (PNPs)
2. Polymer nanofibers (PNFs)

even if other kinds of geometries, such as nanocubes, nanosheets, nanorods, nanoflowers, and many more can be obtained.

Depending on the method of preparation or synthetic rout, PNPs or PNFs can be obtained.

PNPs can be prepared using "bottom-up" approaches, where the polymer nanostructure production involve the self-assembly at a molecular scale from tailored monomer building blocks, or "top-down" processes, i.e., the size reduction materials to nanosize, usually by physical methods (emulsion evaporation or diffusion methods, nanoprecipitation, etc.) or making nanostructures by templating [197].

In Fig. 1.9, a scheme of the bottom-up and top-down processes for PNPs production is reported.

PNFs usually are developed by "top-down" processes, in which macroscopic polymer resins or solutions are converted into nanofibers through various techniques, such as electrospinning, wet spinning, solution blow spinning, melt blowing, and melt electrospinning [198].

As far as PNPs is concerned, they usually range in the 10–1,000 nm interval [199]. They can be distinguished into nanospheres, where the polymer chains are uniformly located both in the surface and in the core of the particles, or nanocapsules or nanovehicles, i.e., hollow particles typically used as delivery systems.

On the other hand, PNFs are defined as objects with two external dimensions on the scale of 1–100 nm, with the third dimension being significantly larger, as reported in the ISO 80004-2 [200].

Even if nanosized polymers have found applications in the very different and new industrial sectors, here we summarize the most recent and impressive applications, knowing that this list is by no means exhaustive:

1) *Analysis and sensing applications*: in this field, nanomaterials represent the perfect candidates due to their sizes comparable with those of biological molecules, allowing for extreme miniaturization. Currently, polymer integrated inorganic nanomaterials (also called "polymer/inorganic nanohybrids") are considered particularly attractive for analytical biosensors, as recently reviewed [201].

2) *Nanostructured conducting polymers*: these peculiar class of nanomaterials are found applications in supercapacitors, sensors, corrosion protection, electrochromic devices, photocatalysis, etc. [202]. On the other hand, the latest advances on the use of nanostructured conducting polymers are related to biomedical applications, such as controlled drug release, artificial muscles, and tissue engineering, thanks to their stimuli responsiveness. Moreover, these nanomaterials have proven to be effective in several environmental applications (extraction and preconcentration of trace amount pollutants from complex matrices, degradation of organic contaminants through photoelectrocatalysis, etc.) [203]. For pollutant detection, an intriguing example is linked to polymer-functionalized nanostructures, which contain specific chemical functional groups that can lead to increased surface activity and reactivity, association, and phase separation [204].

Preformed polymers

Nanocapsules

Nanospheres

Synthetic polymers

Synthetic polymers/
natural polymers

Nanoprecipitation

Emulsion

Coacervation

Interfacial
deposition

Emulsion
diffusion

Emulsion
evaporation

Top-down approach

Polymeric nanoparticles

Interfacial
polymerization

Molecular
inclusion

Interfacial
polycondensation

Emulsion
polymerization

Nanocapsules

Emulsion

Nanospheres

Monomer

Bottom-up approach

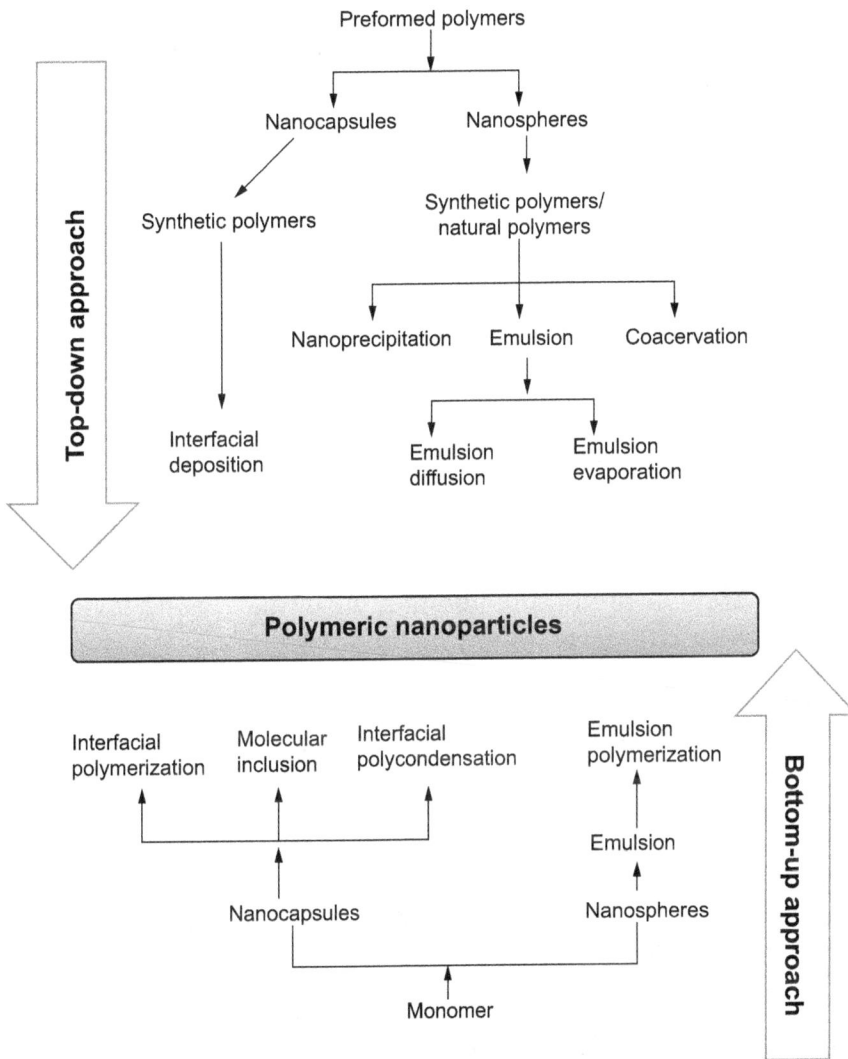

Fig. 1.9: A scheme of the top-down and bottom-up approaches for PNPs production. Reproduced from Krishnaswamy K, Orsat V. Sustainable Delivery Systems through Green Nanotechnology. In Nano- and Microscale Drug Delivery Systems, Grumezescu AM Ed. Elsevier, Amsterdam Netherlands, 2017, 17–32, Copyright 2017, Elsevier [197].

3) *Food nanotechnology*: as recently reviewed by Ramachandraiah and Hong [205], polymeric nanomaterials represent a promising tool in the challenge of building sustainable food systems, in all the plant-derived food product value chain (i.e., crop farming, processing, and storage).

4) *Building applications*: polymers commonly utilized in construction applications (pipes and fittings, foundations, roofing, flooring, paneling, roads and

insulation, etc.) have been opportunely shaped or synthesized in nanostructured features, finding applications such as air filtration, thermal energy storage, sound absorption, etc. [206].

5) *Drug delivery systems*: one of the most relevant and topical examples of application of nanostructured polymers in DDS are polymeric micelles, combining various block polymeric moieties, employed in pre-clinical studies for delivering poorly soluble chemotherapeutic agents, to achieve enhanced therapeutic activity, as recently reviewed by Ghosh and Biswas [207]. Finally, as demonstrated by the ongoing pandemic coronavirus disease (COVID-19), vaccination has proven to play a central role in overcoming of this infection. In this respect, in the last decade, there has been a considerable progress in the rational design and synthesis of polymer nanoparticles for vaccine development. A recent review [208] provides latest progresses on rational designs of polymeric nanostructures loaded with antigens and immunostimulatory molecules to develop innovative PNP-based vaccines to supply prompt responses to awful pandemic threats.

1.2.4 Possibility of predicting polymer performances by means of surface characterization techniques

Today there is a huge demand for new functional properties of polymer interfaces, in many academic and industrial branches, such as biotechnology, nanotechnology, biomaterials, electronics, and coating applications. Even if an almost complete description of methods for the characterization of bulk and surface properties of homopolymers is supplied by several books or databases, increasingly more complex structures (such as hybrid organic–inorganic materials, multilayer films, nanomaterials, copolymers, or polymer blends) require more specialized characterization for end-use performance evaluation. Obviously, it is difficult to accurately predict end-use performance characteristics of the final product using tabulated data of individual components. As a result, an accurate characterization of these new systems is necessary, rather than using model polymers or components. In particular, knowing the surface properties and evolution of polymer materials when in contact with different environments can allow to understand the response of the material to external stimuli and ultimately to choose between the myriads of polymers and polymer combinations. If a rational design approach can be used to plan the synthetic conditions of a polymer-based material, feedback data obtainable from surface characterization in conjunction with molecular modelling programs yield a prediction of the desired properties.

Predict the performances of a polymer-based device is clearly linked to the knowledge of molecular processes occurring on surface, since those can drive the molecular, microscopic and macroscopic structure and the resulting properties of the material.

Different studies have been devoted to develop a theoretical understanding of behavior of surfaces and interfaces in multiconstituent polymers, starting to examine pure component surfaces. These data can be used to either validate or construct predictive thermodynamic theories of polymeric interphase properties. An example is supplied by Fuentes et al., who proposed a model study to predict polymer/substrate compatibility employing contact angle measurements at high temperature directly performed between different molten thermoplastics on smooth glass fibers and plates [209]. Moreover, in nanoclay-reinforced polymers for high-tech applications, the surface characterization of the hybrid material, performed by X-ray diffraction and SEM, was employed to evaluate the impact of nanoclay reinforcement on hardness and dynamic-mechanical properties. Both mathematical regression models and measured values have been employed for theoretical prediction [210]. Surface properties are also important in order to predict the polymer service lifetime: in order to monitor the polymer changes with environmental exposures from the early stages of degradation, surface characterization techniques are required. Many analytical tools provide actual solutions in corrosion, hydro-, photo-, and thermal degradation studies, giving information about the changes of functional groups by means of degradation processes. An interesting study of polymer oxo-biodegradation is reported in literature, evidencing the need of analytical tools to monitor the degradation process under simulated test conditions [210].

In the industrial polymer research, the ability to predict the polymer performances is pressing: the main goal is to develop novel products able to meet specific end-use performance requirements, differentiating them from competitors but maintaining them cost-effective to produce. In this respect, a surface characterization allows to preliminary select the right polymer for the right application, optimize the technological parameters of production, solve processing failures (such as degradation, not correct mixture of additives, phase separation, aging, contamination, and presence of reaction by-products), meet regulatory requirements, envisage suitable treatments for polymer modification, validate cleanliness, assess leachability/toxicity, predict shelf-life and finally collect all the obtained information to aid in the rational design of surface properties to accumulate a very large scale data sets and reference data, for interlaboratory comparisons. On the other hand, in academic field the main goal is to find new materials to be successively transferred to the industrial field, supplying more technologically advanced properties: examples are self-cleaning ability, switchability, superior tribological/mechanical properties, biodegradability, long-term stability, corrosion protection, and super-hydrophobicity or super-hydrophilicity.

Recently, a pioneering tendency is to study reactions occurring on surface under conditions similar to the real ones, since a surface analysis may not be often able to simulate the real conditions of polymeric materials daily applications, due to the operative restrictions required by many techniques. In this respect, new operating conditions are provided, for example, using ambient pressure X-ray photoelectron spectroscopy (AP-XPS or APPES), ultra-low-energy secondary ion mass spectrometry, environmental scanning electron microscopy (E-SEM), etc. A deep dissertation on the operative conditions, applications, possibilities and limiting features of the main surface techniques employed in polymer science is supplied in the course of this book.

References

[1]   Mishra M. Encyclopedia of Polymer Applications, 1st, Boca Raton, Taylor& Francis, CRC Press, 2019.

[2]   Baekeland LH. The chemical constitution of resinous phenolic condensation products. Ind Eng Chem 1913, 5, 506–511.

[3]   Natta G, Pino P, Corradini P, Danusso F, Mantica E, Mazzanti G, Moraglio G. Crystalline high polymers of α-olefins. J Am Chem Soc 1955, 77, 1708–1710.

[4]   Ziegler K, Holzkamp E, Breil H, Martin H. Das Mulheimer normaldruck polyathylen-verfahren. Angew Chem 1955, 67, 541–547.

[5]   IUPAC. Basic classification and definitions of polymerization reactions. Pure Appl Chem 1994, 66, 2483–2486.

[6]   Metanomski WV. IUPAC. Compendium of Macromolecular Nomenclature, Oxford, UK, Blackwell Science, 1991.

[7]   Young RJ, Lovell PA. Introduction to Polymers, 3rd, Boca Raton, Florida, USA, CRC Press, 2011.

[8]   Hiemenz PC, Lodge TP. Olymer Chemistry, 2nd, Boca Raton, Florida, USA, CRC Press, 2007.

[9]   Billmeyer FW Jr. Textbook of Polymer Science, 3rd, Manhattan, USA, John Wiley & Sons, 1984.

[10]   Stevens MP. Polymer Chemistry: An Introduction, 3rd, Oxford, UK, Oxford University Press, 1999.

[11]   Prasad Raoa J, Geckelera KE. Polymer nanoparticles: Preparation techniques and size-control parameters. Prog Polym Sci 2011, 36(7), 887–913.

[12]   Crucho CIC, Barros MT. Polymeric nanoparticles: A study on the preparation variables and characterization methods. Mater Sci Eng C 2017, 80(2017), 771–784.

[13]   Ramamurthy AV. Wall slip in viscous fluids and influence of materials of construction. J Rheol 1986, 30, 337–357.

[14]   Larson RG. Instabilities in viscoelastic flows. Rheol Acta 1992, 31, 213–263.

[15]   Mucha M. Polymer as an important component of blends and composites with liquid crystals. Progr Polym Sci 2003, 28(5), 837–873.

[16]   Pawlikowski GT, Dutta D, Weiss RA. Molecular composites and self-reinforced liquid crystalline polymer blends. Ann Rev Mater Sci 1991, 21, 159–184.

[17]   Yonezawa J, Martin SM, Macosko CW, Ward MD. Rheology and morphology of smectic liquid crystal/polymer blends. Macromolecules 2004, 37(17), 6424–6643.

[18]   Rosato DV, Rosato DV. Reinforced Plastics Handbook, Oxford (UK), Elsevier Advanced Technology, 2005.

[19]   Oprea AE, Ficai A, Andronescu E. Electrospun nanofibers for tissue engineering applications. In: Holban AM, Grumezescu AM, ed, Materials for Biomedical Engineering, Amsterdam, Elsevier, 2019, 77–95.

[20]   Lv D, Zhu M, Jiang Z, Jiang S, Zhang Q, Xiong R, Huang C. Green electrospun nanofibers and their application in air filtration. Macromol Mater Eng 2018, 303(12), 1800336.

[21]   Zakaria MR, Akil HM, Kudus MHA, Ullah F, Javed F, Nosbi N. Hybrid carbon fiber-carbon nanotubes reinforced polymer composites: A review. Composites B: Eng 2019, 176, 107313.

[22]   Ejeian F, Azadi S, Razmjou A, Orooji Y, Kottapalli A, Warkiani ME, Asadnia M. Design and
        applications of MEMS flow sensors: A review. Sens Actuat A 2019, 295, 483–502.
[23]   Wilson EB, Wells AJ. Infra-Red and Raman Spectra of Polyatomic Molecules XIV. Propylene.
        J Chem Phys 1941, 9, 319–322.
[24]   Wilson EB, Decius JC, Cross PC. Molecular Vibrations, New York, NY, USA, McGraw-Hill, 1955.
[25]   Volkenstein MV. Configurational Statistics of Polymer Chains. New York, NY, USA,
        Interscience, 1963.
[26]   Flory PG. Statistical Mechanics of Chain Molecules. New York, NY, USA, Interscience, 1969.
[27]   Gotlib Y, Darinskii AA, Svetlov Y. Physical Kinetics of Macromolecules, Saint Petersburg,
        Russia, Khimia, 1986.
[28]   Grosberg A, Khokhlov AR. Giant Molecules: Here, There, and Everywhere . . . San Diego, CA,
        USA, Academic, 1997.
[29]   Doi M, Edwards SF. The Theory of Polymer Dynamics, New York, NY, USA, Clarendon Press,
        Oxford university Press, 1986.
[30]   Fischer E, Kimmich R, Fatkullin N, Yatsenko G. Segment diffusion and flip-flop spin diffusion
        in entangled polyethyleneoxide melts: A field-gradient NMR diffusometry study. Phys Rev E
        2000, 62, 775–782.
[31]   Dutcher JR, Marangoni AG. Soft Materials: Structure And Dynamics, New York, NY, USA, CRC
        Press, 2004.
[32]   Tzoumanekas C, Theodorou DN. From atomistic simulations to slip-link models of entangled
        polymer melts: Hierarchical strategies for the prediction of rheological properties. Curr Opin
        Solid State Mater Sci 2006, 10(2), 61–72.
[33]   Allegra G, Raos G, Vacatello M. Theories and simulations of polymer-based nanocomposites:
        From chain statistics to reinforcement. Progr Polym Sci 2008, 33(7), 683–731.
[34]   Khalatur PG. Molecular dynamics simulations in polymer science: Methods and main results.
        In: Matyjaszewski K, Möller M, ed, Polymer Science: A Comprehensive Reference,
        Amsterdam, Elsevier, 2012, 417–460.
[35]   Xiang Y, Zhong D, Wang P, Yin T, Zhou H, Yu H, Baliga C, Qu S, Yang W. A physically based
        visco-hyperelastic constitutive model for soft materials. J Mech Phys Sol 2019, 128, 208–218.
[36]   Muthukumar M. Trends in polymer physics and theory. Progr Polym Sci 2020, 100, 101184.
[37]   Chen L, Pilania G, Batra R, Huan TD, Kim C, Kuenneth C, Ramprasad R. Polymer informatics:
        Current status and critical next steps. Mater Sci Eng R Rep 2021, 144, 100595.
[38]   Zhao J, Wu L, Zhan C, Shao Q, Guo Z, Zhang L. Overview of polymer nanocomposites:
        Computer simulation understanding of physical properties. Polymer 2017, 133, 272–287.
[39]   Yan LT, Xie XM. Computational modeling and simulation of nanoparticle self-assembly in
        polymeric systems: Structures, properties and external field effects. Progr Polym Sci 2013,
        38(2), 269–405.
[40]   Bochicchio D, Pavan GM, Molecular modelling of supramolecular polymers. Adv Phy 2018,
        3(1), DOI: 10.1080/23746149.2018.1436408.
[41]   Mavrantzas V, Harmandaris V. Molecular dynamics simulations of polymers. In: Kotelyanskii M,
        Theodorou DN, ed, Simulation Methods for Polymers, Boca Raton, CRC Press, 2004, Chapter 6,
        61 pages.
[42]   Gartner TE, Jayaraman A. Modeling and simulations of polymers: A roadmap.
        Macromolecules 2019, 52(3), 755–786.
[43]   Li YC, Wang CP, Liu XJ. Assessment of thermodynamic properties in pure polymers. Comput
        Coupl Phase Diagr Thermochem 2008, 32, 217–226.
[44]   Safran SA. Statistical Thermodynamics of Surfaces, Interfaces, and Membranes, Boca Raton,
        CRC Press, 2003.

[45] Schneider K. Mechanical properties of polymers at surfaces and interfaces. In: Stamm M, ed, Polymer Surfaces and Interfaces, Berlin, Heidelberg, Springer, 2008.

[46] Fuoss PH, Norton LJ, Brennan S. X-ray scattering studies of the melting of lead surfaces. Phys Rev Lett 1988, 60(20), 2046–2049.

[47] Mailaender L, Dosch H, Peisl J, Johnson RL. Near-surface critical x-ray scattering from Fe$_3$Al. Phys Rev Lett 1990, 64, 2527–2530.

[48] Kumar SK, Vacatello M, Yoon DY. Off-lattice monte carlo simulations of polymer melts confined between two plates. J Chem Phys 1988, 89, 5206–5215.

[49] Liu Y, Russell TP, Samant MG, Stohr J, Brown HR, Cossy-Favre A, Diaz J. Surface relaxations in polymers. Macromolecules 1997, 30, 7768–7771.

[50] Keddie J, Jones RAL, Cory RA. Interface and surface effects on glass transition temperature in thin polymer films. Faraday Discuss 1994, 98, 219–230.

[51] Wallace WE, Van Zanten JH, Wu WL. Influence of an impenetrable interface on a polymer glass-transition temperature. Phys Rev E 1995, 52, R3329–32.

[52] Xie L, DeMaggio GB, Frieze WE, DeVries J, Gildey DW, Hristov HA, Yee AF. Positronium formation as a probe of polymer surfaces and thin films. Phys Rev Lett 1996, 74, 4947–4950.

[53] DeMaggio GB, Frieze WE, Gildey DW, Zhu M, Hristov HA, Yee AF. Interface and surface effects on the glass transition in thin polystyrene films. Phys Rev Lett 1997, 78, 1524–1527.

[54] Kajiyama T, Tanaka K, Takahara A. Surface molecular motion of monodisperse polystyrene films. Macromolecules 1997, 30, 280–285.

[55] Priestley RD, Ellison CJ, Broadbelt LJ, Torkelson JM. Structural relaxation of polymer glasses at surfaces, interfaces, and in between. Science 2005, 15, 456–459.

[56] Carelli C, Young RN, Jones RAL, Sferrazza M. Chain length effects on confined polymer/polymer interfaces. Europhys Lett 2006, 75, 274–280.

[57] Qi D, Ilton M, Forrest JA. Measuring surface and bulk relaxation in glassy polymers. Eur Phys J E 2011, 34, 56.

[58] Glynos E, Johnson KJ, Frieberg B, Chremos A, Narayanan S, Sakellariou G, Green PF. Free surface relaxations of star-shaped polymer films. Phys Rev Lett 2017, 119, 227801.

[59] Torres MF, La Rocca CE, Braunstein LA. Fluctuations of a surface relaxation model in interacting scale free networks. Physica A 2016, 463, 182–187.

[60] Bailey EJ, Winey KI. Dynamics of polymer segments, polymer chains, and nanoparticles in polymer nanocomposite melts: A review. Progr Polym Sci 2020, 105, 101242.

[61] Awaja F, Gilbert M, Kelly G, Fox B, Pigram PJ. Adhesion of polymers. Progr Polym Sci 2009, 34(9), 948–968.

[62] Köstler S, Delgado AV, Ribitsch V. Surface thermodynamic properties of polyelectrolyte multilayers. J Coll Interf Sci 2005, 286, 339–348.

[63] Luzinov I, Minko S, Tsukruk VV. Adaptive and responsive surfaces through controlled reorganization of interfacial polymer layers. Prog Polym Sci 2004, 29(7), 635–698.

[64] Stachewicz U, Li S, Bilotti E, Barber AH. Dependence of surface free energy on molecular orientation in polymer films. Appl Phys Lett 2012, 100, 094104–094104-4.

[65] Geoghegan M, Krausch G. Wetting at polymer surfaces and interfaces. Progr Polym Sci 2003, 28(2), 261–302.

[66] Kung CH, Sow PK, Zahiri B, Mérida W. Assessment and Interpretation of surface wettability based on sessile droplet contact angle measurement: Challenges and opportunities. Adv Mater Interfaces 2019, 6, 1900839.

[67] Etzler FM. Determination of the surface free energy of solids. Rev Adhes Adhes 2013, 1(43), 3–45.

[68] Aydemir C, Altay BN, Akyol M. Surface analysis of polymer films for wettability and ink adhesion. Color Res Appl 2021, 46, 489–499.

[69] Klee D, Hocker H. Advances in polymer science: Biomedical application/polymer blends. Biomed Appl Polym Blends 1999, 149, 1–55.
[70] Holly FJ, Refojo MF. Wettability of hydrogels. I. Poly (2-hydroxyethyl methacrylate). J Biomed Mater Res 1975, 9(3), 315–326.
[71] Ratner BD. Biomaterials: Interfacial phenomena and applications, ACS advances in chemistry series, cooper and peppas. Eds, Amer Chem Soc Washington, DC, USA, 1982, 199, 9.
[72] Ratner BD. Surface structure of polymers for biomedical applications. Makromolekulare Chemie Macromolecular Symposia 1988, 19, 163–178.
[73] Ruckenstein E, Gourisankar SV. Surface restructuring of polymeric solids and its effect on the stability of the polymer-water interface. J Colloid Interface Sci 1986, 109, 557–566.
[74] Lee SH, Ruckenstein E. Stability of polymeric surfaces subjected to ultraviolet irradiation. J Colloid Interface Sci 1987, 117, 172–178.
[75] Andrade JD, Gregonis DE, Smith LM. Polymer Surface Dynamics. In: Surface and Interfacial Aspects of Biomedical Polymers. Surface Chemistry & Physics, Springher, USA, Plenum Press, 1985, Vol. 1, 15–41.
[76] Yasuda H, Charlson EJ, Charlson EM, Yasuda T, Miyama M, Okuno T. Dynamics of surface property change in response to changes in environmental conditions. Langmuir 1991, 7, 2394–2400.
[77] Yasuda T, Okuno T, Yasuda H. Contact angle of water on polymer surfaces. Langmuir 1994, 10, 2435–2439.
[78] Yasuda H, Sharma AK, Yasuda T. Effect of orientation and mobility of polymer molecules at surfaces on contact angle and its hysteresis. J Polym Sci A 1981, 19, 1285–1291.
[79] Wei M, Gao Y, Li X, Serpe MJ. Stimuli-responsive polymers and their applications. Polym Chem 2017, 8, 127–143.
[80] Suëtaka W. Surface Infrared and Raman Spectroscopy: Methods and Applications, New York (US), Springer Science & Business Media, 2013, 270.
[81] Lu X, Mi M. Gradiently varied chain packing/orientation states of polymer thin films revealed by polarization-dependent infrared absorption. Eur Polym J 2015, 63, 247–254.
[82] Koziol P, Liberda D, Kwiatek WM, Wrobel TP. Macromolecular orientation in biological tissues using a four-polarization method in FT-IR imaging. Anal Chem 2020, 92(19), 13313–13318.
[83] Xue L, Zhang J, Han Y. Phase separation induced ordered patterns in thin polymer blend films. Prog Polym Sci 2012, 37, 564–594.
[84] Stein GE, Laws TS, Verduzco R. Tailoring the attraction of polymers toward surfaces. Macromolecules 2019, 52(13), 4787–4802.
[85] Huang Y, Yeh JW, Yang ACM. "High-entropy polymers": A new route of polymer mixing with suppressed phase separation. Materialia 2021, 15, 100978.
[86] Wu DT, Fredrickson GH. Effect of architecture in the surface segregation of polymer blends. Macromolecules 1996, 29, 7919–7930.
[87] Agrawal G, Negi YS, Pradhan S, Dash M, Samal SK. Wettability and contact angle of polymeric biomaterials. In: Tanzi MC, Farè S, ed, Characterization of Polymeric Biomaterials, Sawston, Cambridge (UK), Woodhead Publishing, 2017, 57–81.
[88] Minnikanti VS, Archer LA. Surface migration of branched molecules: Analysis of energetic and entropic factors. J Chem Phys 2005, 123, 144902.
[89] Hahladakis JN, Velis CA, Weber R, Iacovidou E, Purnell P. An overview of chemical additives present in plastics: Migration, release, fate and environmental impact during their use, disposal and recycling. J Hazard Mater 2018, 344, 179–199.
[90] Yul LY, Liu X, Vogler EA, Donahue HJ. Systematic variation in osteoblast adhesion and phenotype with substratum surface characteristics. J Biomed Mater Res A 2003, 68(3), 504–511.

[91] Xu LC, Siedlecki CA. Effects of surface wettability and contact time on protein adhesion to biomaterial surfaces. Biomaterials 2007, 28(22), 3273–3283.

[92] Dumond JJ, Low HL. Long-Lasting superhydrophilic polymers via multiscale topographies. ACS Appl Polym Mater 2021, 3(1), 233–242.

[93] Imae T. Fluorinated polymers. Curr Opin Colloid Interface Sci 2003, 8(3), 307–314.

[94] Li Q, Yan Y, Yu M, Song B, Shi S, Gong Y. Synthesis of polymeric fluorinated sol–gel precursor for fabrication of superhydrophobic coating. Appl Surf Sci 2016, 367, 101–108.

[95] Peng Z, Song J, Gao Y, Liu J, Lee C, Chen G, Wang Z, Chen J, Leung MKH. A fluorinated polymer sponge with superhydrophobicity for high-performance biomechanical energy harvesting. Nano Energy 2021, 85, 106021.

[96] Li X, Li B, Li Y, Sun J. Nonfluorinated, transparent, and spontaneous self-healing superhydrophobic coatings enabled by supramolecular polymers. Chem Eng J 2021, 404, 126504.

[97] Woodward JT, Gwin H, Schwartz DK. Contact angles on surfaces with mesoscopic chemical heterogeneity. Langmuir 2000, 16, 2957–2961.

[98] Sun T, Song W, Jiang L. Control over the responsive wettability of poly(N-isopropylacrylamide) film in a large extent by introducing an irresponsive molecule. Chem Comm 2005, 1723–1725.

[99] Garvin CJ, Genzel J. Applications of surface-grafted macromolecules derived from post-polymerization modification reactions. Progr Polym Sci 2012, 37, 871–906.

[100] Manoudis PN, Karapanagiotis I. Modification of the wettability of polymer surfaces using nanoparticles. Progr Org Coat 2014, 77(2), 331–338.

[101] Maji D, Lahiri SK, Das S. Study of hydrophilicity and stability of chemically modified PDMS surface using piranha and KOH solution. Surf Interf Anal 2012, 44(1), 62–69.

[102] Wanamaker CL, Neff BS, Nejati-Namin A, Spatenka ER, Yang ML. Effect of chemical and physical modifications on the wettability of polydimethylsiloxane surfaces. J Chem Educ 2019, 96(6), 1212–1217.

[103] Gang Z, Wang D, Gu ZZ, Möhwald H. Fabrication of superhydrophobic surfaces from binary colloidal assembly. Langmuir 2005, 21, 9143–9148.

[104] Correia DM, Nunes-Pereira J, Alikin D, Kholkin AL, Carabineiro SAC, Rebouta L, Rodrigues MS, Vaz F, Costa CM, Lanceros-Méndez S. Surface wettability modification of poly(vinylidene fluoride) and copolymer films and membranes by plasma treatment. Polymer 2019, 169, 138–147.

[105] Iqbal M, Dinh DK, Abbas Q, Imran M, Sattar H, Ul Ahmad A. Controlled surface wettability by plasma polymer surface modification. Surfaces 2019, 2, 349–371.

[106] Rossnagel SM, Cuomo JJ, Westwood WD. Handbook of Plasma Processing Technology, Fundamentals, Etching, Deposition, and Surface Interactions, Noyes Publication, Westwood, NJ, USA, 1990.

[107] López AB, De La Cal JC, Asua JM. From fractal polymer dispersions to mechanically resistant waterborne superhydrophobic coatings. Polymer 2017, 124, 12–19.

[108] Maghsoudi K, Vazirinasab E, Momen G, Jafari R. Advances in the fabrication of superhydrophobic polymeric surfaces by polymer molding processes. Ind Eng Chem Res 2020, 59(20), 9343–9363.

[109] Formentín P, Marsal LF. Hydrophobic/Oleophilic structures based on macroPorous silicon: Effect of topography and fluoroalkyl silane functionalization on wettability. Nanomaterials 2021, 11, 670.

[110] Guriyanova S, Semin B, Rodrigues TS, Butt HJ, Bonaccurso E. Hydrodynamic drainage force in a highly confined geometry: Role of surface roughness on different length scales. Microfluid Nanofluid 2010, 8(5), 653–663.

[111] Tang J, Liu B, Gao L, Wang W, Liu T, Su G. Impacts of surface wettability and roughness of styrene-acrylic resin films on adhesion behavior of microalgae Chlorella sp. Colloids Surf B: Biointerfaces 2021, 199, 111522.

[112] Fu Y, Soldera M, Wang W, Milles S, Deng K, Voisiat B, Nielsch K, Lasagni AF. Wettability control of polymeric microstructures replicated from laser-patterned stamps. Sci Rep 2020, 10, 22428.

[113] Toosi SF, Moradi S, Hatzikiriakos SG. Fabrication of Micro/Nano patterns on polymeric substrates using laser ablation methods to control wettability behaviour: A critical review. Rev Adhes Adhes 2017, 5(1), 55–78.

[114] Atthi N, Dielen M, Sripumkhai W, Pattamang P, Meananeatra R, Saengdee P, Thongsook O, Ranron N, Pankong K, Uahchinkul W, Supadech J, Klunngien N, Jeamsaksiri W, Veldhuizen P, Ter Meulen JM. Fabrication of high aspect ratio micro-structures with superhydrophobic and oleophobic properties by using large-area roll-to-plate nanoimprint lithography. Nanomaterials 2021, 11, 339.

[115] Wang M, Wang X, Moni P, Liu A, Kim DH, Jo WJ, Sojoudi H, Gleason KK. CVD polymers for devices and device fabrication. Adv Mater 2017, 29, 1604606.

[116] Silverio V, Canane PAG, Cardoso S. Surface wettability and stability of chemically modified silicon, glass and polymeric surfaces via room temperature chemical vapor deposition. Colloids Surf A Physicochem Eng Asp 2019, 570, 210–217.

[117] Teshima K, Sugimura H, Inoue Y, Takai O, Takano A. Transparent ultra water-repellent poly (ethylene terephthalate) substrates fabricated by oxygen plasma treatment and subsequent hydrophobic coating. Appl Surf Sci 2005, 244(1–4), 619–622.

[118] Kaya AST, Cengiz U. Fabrication and application of superhydrophilic antifog surface by sol-gel method. Prog Org Coat 2019, 126, 75–82.

[119] Husemann M, Morrison M, Benoit D, Frommer J, Mate CM, Hinsberg WD, Hedrick JL, Hawker CJ. Manipulation of surface properties by patterning of covalently bound polymer brushes. J Am Chem Soc 2000, 122, 1844–1845.

[120] Hensarling RM, Doughty VA, Chan JW, Patton DL. "Clicking" polymer brushes with thiol-yne chemistry: Indoors and out. J Am Chem Soc 2009, 131, 14673–14675.

[121] Azzaroni O, Brown AA, Huck WTS. Tunable wettability by clicking counterions into polyelectrolyte brushes. Adv Mater 2007, 19, 151–154.

[122] Lopez-torres D, Elosua C, Hernaez M, Goicoechea J, Arregui FJ. From superhydrophilic to superhydrophobic surfaces by means of polymeric Layer-by-Layer films. Appl Surf Sci 2015, 351, 1081–1086.

[123] Das A, Shome A, Manna U. Porous and reactive polymeric interfaces: An Emerging avenue for achieving durable and functional bio-inspired wettability. Mater Chem A 2021, 9, 824–856.

[124] Zhang J, Chen J, Peng S, Peng S, Zhang Z, Tong Y, Miller PW, Yan X-P. Emerging porous materials in confined spaces: From chromatographic applications to flow chemistry. Chem Soc Rev 2019, 48, 2566–2595.

[125] Eder K, Huber CG, Buchmeiser MR. Surface-Functionalized, Ring-Opening metathesis polymerization-derived monoliths for anion-exchange chromatography. Macromol Rapid Comm 2007, 28(20), 2029–2032.

[126] Obayemi JD, Jusu SM, Salifu AA, Ghahremani S, Tadesse M, Uzonwanne VO, Soboyejo WO. Degradable porous drug-loaded polymer scaffolds for localized cancer drug delivery and breast cell/tissue growth. Mater Sci Eng: C 2020, 112, 110794.

[127] Xie Y, Hillmyer MA. Nanostructured polymer monoliths for biomedical delivery applications. ACS Appl Bio Mater 2020, 3(5), 3236–3247.

[128] Buchholz FL, Graham AT. Modern Sperabsorbent Polymer Technology, New York, NY, USA, Wiley-VCH, 1998.

[129] Prasad A, Sankar MR, Katiyar V. State of art on solvent casting particulate leaching method for orthopedic scaffolds fabrication. Mater Today Proc 2017, 4(2), 898–907.

[130] Grenier J, Duval H, Barou F, Lv P, David B, Letourneur D. Mechanisms of pore formation in hydrogel scaffolds textured by freeze-drying. Acta Biomaterialia 2019, 94, 195–203.

[131] Costantini M, Barbetta A, Gas foaming technologies for 3D scaffold engineering. In: Deng Y, Kuiper J, ed, Functional 3D Tissue Engineering Scaffolds, Woodhead Publishing, 2018, 127–149.

[132] Xie X, Chen Y, Wang X, Xu X, Shen Y, Khan AR, Aldalbahi A, Fetz AE, Bowlin GL, El-Newehy M, Mo X. Electrospinning nanofiber scaffolds for soft and hard tissue regeneration. J Mater Sci Tech 2020, 59, 243–261.

[133] Law JX, Liau LL, Saim A, Yang Y, Idrus R. Electrospun collagen nanofibers and their applications in skin tissue engineering. Tissue Eng Regen Med 2017, 14, 699–718.

[134] Enderami SE, Ahmadi SF, Mansour RN, Abediankenari S, Ranjbaran H, Mossahebi-Mohammadi M, Salarinia R, Mahboudi H. Electrospun silk nanofibers improve differentiation potential of human induced pluripotent stem cells to insulin producing cells. Mater Sci Eng C 2020, 108, 110398.

[135] Puppi P, Piras AM, Detta N, Dinucci D, Chiellini F. Poly (lactic-co-glycolic acid) electrospun fibrous meshes for the controlled release of retinoic acid. Acta Biomaterialia 2010, 6(4), 1258–1268.

[136] Kai D, Jin G, Prabhakaran MP, Ramakrishna S. Electrospun synthetic and natural nanofibers for regenerative medicine and stem cells. Biotechnol J 2013, 8(1), 59–72.

[137] Efimov AE, Agapova OI, Safonova LA, Bobrova MM, Parfenov VA, Koudan EV, Pereira FDAS, Bulanova EA, Mironov VA, Agapov II. 3D scanning probe nanotomography of tissue spheroid fibroblasts interacting with electrospun polyurethane scaffold. eXPRESS Polym Lett 2019, 13(7), 632–641.

[138] Zhigalina VG, Zhigalina OM, Ponomarev II, Skupov KM, Razorenov DY, Ponomarev I, Kiseleva NA, Leitinger G. Electron microscopy study of new composite materials based on electrospun carbon nanofibers. Cryst Eng Comm 2017, 19, 3792–3800.

[139] Rawal A, Shukla S, Sharma S, Singh D, Lin Y-M, Hao J, Rutledge GC, Vásárhelyi L, Kozma G, Kukovecz A, Janovák L. Metastable wetting model of electrospun mats with wrinkled fibers. Appl Surf Sci 2021, 551, 149147.

[140] Sadman K, Delgado DE, Won Y, Wang O, Gray KA. Shull KR versatile and high-throughput polyelectrolyte complex membranes via phase inversion. ACS Appl Mater Interf 2019, 11(17), 16018–16026.

[141] Bàrtolo PJ, Domingos M, Patrìcio T, Cometa S, Mironov V. Biofabrication Strategies for Tissue Engineering. In: Fernandes P, Bártolo PJ, eds, Advances on Modeling in Tissue Engineering Series: Computational Methods in Applied Sciences, 1st, Springer, 2011, Vol. 20, 137–176.

[142] Ambhorkar P, Rakin RH, Wang Z, Kumar H, Kim K. Biofabrication strategies for engineering heterogeneous artificial tissues. Addit Manuf 2020, 36, 101459.

[143] Xu C, Yang F, Wang S, Ramakrishna S. *In vitro* study of human vascular endothelial cell function on materials with various surface roughness. J Biomed Mater Res 2004, 71A, 154–161.

[144] Metwally S, Ferraris S, Spriano S, Krysiak ZJ, Kaniuk L, Marzec MM, Kim SK, Szewczyk PK, Gruszczyński A, Sarna MW, Karbowniczek JF, Bernasik A, Kar-Narayan S, Stachewicz U. Surface potential and roughness controlled cell adhesion and collagen formation in electrospun PCL fibers for bone regeneration. Mater Des 2020, 194, 108915.

[145] Chung TW, Liu DZ, Wang SY, Wang SS. Enhancement of the growth of human endothelial cells by surface roughness at nanometer scale. Biomaterials 2003, 24, 4655–4661.

[146] Krebs FC. Fabrication and processing of polymer solar cells: A review of printing and coating techniques. Solar Energy Mater Sol Cells 2009, 93, 394–412.

[147] Johansson KS. Surface Modification of Plastics in Kutz M Ed. Applied Plastics Engineering Handbook (Second Edition), Processing, Materials, and Applications. Plastics Design Library, Amsterdam, Netherlands, William Andrew Publishing, Elsevier, 2017, 443–487.

[148] Liston EM, Martinu L, Wertheimer MR. Plasma surface modification of polymers for improved adhesion: A critical review. J Adhes Sci Tech 1993, 7(10), 1091–1127.

[149] Hegemann D, Brunner H, Oehr C. Plasma treatment of polymers for surface and adhesion improvement. Nucl Instr Meth Phys Res B 2003, 208, 281–286.

[150] Park JB, Lake RS. Biomaterials, an Introduction, 2nd, NewYork, NY, USA, Plenum Press, Springer, 1992.

[151] Guney A, Kara F, Ozgen O, Aksoy EA, Hasirci V, Hasirci N. Surface modification of polymeric biomaterials. In: Taubert A, Mano JF, Rodríguez-Cabello JC, eds, Biomaterials Surface Science, Weinheim, Germany, Wiley-VCH Verlag GmbH & Co. KGaA, 2013, 89–158.

[152] Mijovic JS, Koutsky JA. Etching of polymeric surfaces: A review. Polym Plast Technol Eng 1977, 9(2), 139–179.

[153] Plasma Deposition, Treatment, and Etching of Polymers: The Treatment and Etching of Polymers, D'Agostino R, Flamm DL, Auciello O, ed, Netherlands, Elsevier, Amsterdam, 2012.

[154] Satriano C, Carnazza S, Guglielmino S, Marletta G. Differential cultured fibroblast behavior on plasma and ion-beam-modified polysiloxane surfaces. Langmuir 2002, 18(24), 9469–9475.

[155] Kim SM, Fan H, Cho YJ, Eo MY, Park JH, Kim BN, Lee BC, Lee SK. Electron beam effect on biomaterials I: Focusing on bone graft materials. Biomater Res 2015, 19(10).

[156] Laser Surface Modification of Biomaterials: Techniques and Applications, Vilar R, ed, Sawston Cambridge (UK), Woodhead Publishing, 2016.

[157] Ahad AU, Bartnik A, Fiedorowicz H, Kostecki J, Brabazon D, Korczyc D, Ciach T. Surface modification of polymers for biocompatibility via exposure to extreme ultraviolet (EUV) radiation. J Biomed Mater Res A 2014, 102(9), 3298–3310.

[158] Self-assembling Biomaterials (1st Edition) Molecular Design. Characterization and Application in Biology and Medicine, Azevedo HS, Da Silva RMP, ed, Sawston Cambridge (UK), Woodhead Publishing, 2018.

[159] Ariga K. Don't forget Langmuir–Blodgett films 2020: Interfacial nanoarchitectonics with molecules, materials, and living objects. Langmuir 2020, 36(26), 7158–7180.

[160] Goddard JM, Hotchkiss JH. Polymer surface modification for the attachment of bioactive compounds. Progr Polym Sci 2007, 32(7), 698–725.

[161] Cometa S, Bonifacio MA, Mattioli-Belmonte M, Sabbatini L, De Giglio E. Electrochemical strategies for titanium implant polymeric coatings: The why and how. Coatings 2019, 9(4), 268.

[162] De Giglio E, Cometa S, Calvano CD, Sabbatini L, Zambonin PG, Colucci S, Di Benedetto A, Colaianni G. A new titanium biofunctionalized interface based on poly(pyrrole-3-acetic acid) coating: Proliferation of osteoblast-like cells and future perspectives. J Mater Sci Mater Med 2007, 18(9), 1781–1789.

[163] De Giglio E, Cometa S, Cioffi N, Torsi L, Sabbatini L. Analytical investigations on Poly(Acrilic Acid) coatings electrodeposited on titanium based implants: A versatile approach to biocompatibility enhancement. Anal Bioanal Chem 2007, 389, 2055–2063.

[164] De Giglio E, Sabbatini L, Colucci S, Zambonin G. Synthesis, analytical characterization, and osteoblast adhesion properties on RGD-grafted polypyrrole coatings on titanium substrates. J Biomater Sci Polym Ed 2000, 11(10), 1073–1083

[165] Aflori M, Drobota M, Dimitriu DG, Stolca I, Simionescu B, Harabagiu V. Collagen immobilization on polyethylene terephthalate surface after helium plasma treatment. Mater Sci Eng B 2013, 178(19), 1303–1310.

[166] Sun L, Dai J, Baker GL, Bruening ML. High-capacity, protein-binding membranes based on polymer brushes grown in porous substrates. Chem Mater 2006, 18, 4033–4039.

[167] Cullen SP, Liu X, Mandel IC, Himpsel FJ, Gopalan P. Polymeric brushes as functional templates for immobilizing ribonuclease A: Study of binding kinetics and activity. Langmuir 2008, 24, 913–920.

[168] Iwata R, Satoh R, Iwasaki Y, Akiyoshi K. Covalent immobilization of antibody fragments on well-defined polymer brushes via sitedirected method. Colloids Surf B 2008, 62, 288–298.

[169] Salvador-Morales C, Zhang L, Langer R, Farokhzad OC. Immunocompatibility properties of lipid–polymer hybrid nanoparticles with heterogeneous surface functional groups. Biomaterials 2009, 30, 2231–2240.

[170] Ham HO, Liu Z, Lau KHA, Lee H, Messersmith FB. Facile DNA Immobilization on Surfaces through a Catecholamine Polymer. Angew Chem 2011, 50(3), 732–736.

[171] De Giglio E, Cometa S, Ricci MA, Zizzi A, Cafagna D, Manzotti S, Sabbatini L, Mattioli-Belmonte M. Development and characterization of rhVEGF-loaded Poly(HEMA-MOEP) coatings electrosynthesised on titanium to enhance bone mineralization and angiogenesis. Acta Biomaterialia 2010, 6, 282–290.

[172] De Giglio E, Cometa S, Ricci MA, Cafagna D, Savino AM, Sabbatini L, Orciani M, Ceci E, Novello L, Tantillo GM, Mattioli Belmonte M. Ciprofloxacin-modified electrosynthesised hydrogel coatings to prevent titanium implant-associated infections. Acta Biomaterialia 2011, 7, 882–891.

[173] Halliday AJ, Moulton SE, Wallace GG, Cook MJ. Novel methods of antiepileptic drug delivery – Polymer-based implants. Adv Drug Deliv Rev 2012, 64(10), 953–994.

[174] Bonifacio MA, Cerqueni G, Cometa S, Licini C, Sabbatini L, Mattioli-Belmonte M, De Giglio E. Insights into arbutin effects on bone cells: Towards the development of antioxidant titanium implants. Antioxidants 2020, 9, 579.

[175] Rahman CV, Ben-David D, Dhillon A, Kuhn G, Gould TW, Müller R, Rose FR, Shakesheff KM, Livne E. Controlled release of BMP-2 from a sintered polymer scaffold enhances bone repair in a mouse calvarial defect model. J Tissue Eng Regen Med 2014, 8(1), 59–66.

[176] Mundargi R, Babu V, Rangaswamy V, Patel P, Aminabhavi T. Nano/micro technologies for delivering macromolecular therapeutics using poly(D,L-lactide-co-glycolide) and its derivatives. J Control Release 2008, 125, 193–209.

[177] Zhang Y, Huang Y, Li S. Polymeric micelles: Nanocarriers for cancer-targeted drug delivery. AAPS Pharm Sci Tech 2014, 15, 862–871.

[178] RameshKumar S, Shaiju P, O'Connor KE, Babu R. Bio-based and biodegradable polymers – State-of-the-art, challenges and emerging trends. Curr Opin Green Sust Chem 2020, 21, 75–81.

[179] Prajapati SK, Jain A, Jain A, Jain S. Biodegradable polymers and constructs: A novel approach in drug delivery. Europ Polym J 2019, 120, 109191.

[180] Sivakumar PM, Cometa S, Alderighi M, Prabhawathi V, Doble M, Chiellini F. Chalcone embedded polyurethanes as a biomaterial: Synthesis, characterization and antibacterial adhesion. Carb Pol 2012, 87, 353–360.

[181] Dikici S, Claeyssens F, MacNeil S. Pre-seeding of simple electrospun scaffolds with a combination of endothelial cells and fibroblasts strongly promotes angiogenesis. Tissue Eng Regen Med 2020, 17, 445–458.

[182] Naarmann H. Polymers, Electrically Conducting. Ullmann's Encyclopedia of Industrial Chemistry, Weinheim, Germany, Wiley WCH, 2002.

[183] Maziz A, Özgür E, Bergaud C, Uzun L. Progress in conducting polymers for biointerfacing and biorecognition applications. Sens Actuators Rep 2021, 3, 100035.

[184] Ranathunge TA, Ngo DT, Karunarathilaka D, Attanayake NH, Chandrasiri I, Brogdon P, Delcamp JH, Rajapakse RMG, Watkins DL. Designing hierarchical structures of complex electronically conducting organic polymers *via* one-step electro-polymerization. J Mater Chem C 2020, 17(8), 5934–5940.

[185] Thomas SW, Khan RR, Puttananjegowda K, Serrano-Garcia W. Conductive polymers and metal oxide polymeric composites for nanostructures and nanodevices. In: Guarino V, Focarete ML, Pisignano D, Eds, Advanced Nanomaterials, Advances in Nanostructured Materials and Nanopatterning Technologies, Netherlands, Elsevier, Amsterdam, 2020, 243–271.

[186] Cotrone S, Cafagna D, Cometa S, De Giglio E, Magliulo M, Torsi L, Sabbatini L. Microcantilevers and organic transistors: Two promising classes of label-free biosensing devices which can be integrated in electronic circuits. Anal Bioanal Chem 2012, 402, 1799–1811.

[187] Zhang X, Cheng J, Wu L, Mei Y, Jaffrezic-Renault N, Guo Z. An overview of an artificial nose system. Talanta 2018, 184, 93–102.

[188] Torsi L, Tanese MC, Cioffi N, Gallazzi MC, Sabbatini L, Zambonin PG, Raos G, Meille SV, Giangregorio MM. Side-chain role in chemically sensing conducting polymer field-effect transistors. J Phys Chem B 2003, 107(31), 7589–7594.

[189] Henderson RD, Guijt RM, Andrewartha L, Lewis TW, Rodemann T, Henderson A, Hilder EF, Haddad PR, Breadmore MC. Lab-on-a-Chip device with laser-patterned polymer electrodes for high voltage application and contactless conductivity detection. Chem Comm 2012, 48, 9287–9289.

[190] Lee S, Ozlu B, Eom T, Martin DC, Shim BS. Electrically conducting polymers for bio-interfacing electronics: From neural and cardiac interfaces to bone and artificial tissue biomaterials. Biosens Bioelectron 2020, 170, 112620.

[191] Fu X, Cheong Y-H, Zhou AAC, Robert C, Krikstolaityte V, Gordon KC, Lisak G. Diagnostics of skin features through 3D skin mapping based on electro-controlled deposition of conducting polymers onto metal-sebum modified surfaces and their possible applications in skin treatment. Anal Chim Acta 2021, 1142, 84–98.

[192] Murad AR, Iraqi A, Aziz SB, Abdullah SN, Brza MA. Conducting polymers for optoelectronic devices and organic solar cells: A review. Polymers 2020, 12, 2627.

[193] Barsan MM, Pinto EM, Brett CMA. Electrosynthesis and electrochemical characterisation of phenazine polymers for application in biosensors. Electrochim Acta 2008, 53(11), 3973–3982.

[194] O'Neil KD, Semenikhin OA. AFM phase imaging of thin films of electronically conducting polymer polybithiophene prepared by electrochemical potentiodynamic deposition. Russ J Electrochem 2010, 46(12), 1345–1352.

[195] Meana-Esteban B, Lete C, Kvarnström C, Ivaska A. Raman and in situ FTIR-ATR characterization of polyazulene films and its derivate. J Phys Chem B 2006, 110(46), 23343–23350.

[196] Koo J. Environmental and health impacts for nanomaterials and polymer nanocomposites. In: Jk K, Ed, Fundamentals, Properties, and Applications of Polymer Nanocomposites, Cambridge UK, Cambridge University Press, 2016, 605–647.

[197] Krishnaswamy K, Orsat V. Sustainable delivery systems through green nanotechnology. In: Am G, Ed, Nano- and Microscale Drug Delivery Systems, Netherlands, Elsevier, Amsterdam,, 2017, 17–32.

[198] Anstey A, Chang E, Kim ES, Rizvi A, Ramezani Kakroodi A, Park KD, Loe PC Nanofibrillated polymer systems: Design, application, and current state of the art. Progr Polym Sci 2021, 113, 101346.

[199] Sardoiwala MN, Kaundal B, Choudhury SR. Development of Engineered Nanoparticles Expediting Diagnostic and Therapeutic Applications across Blood–Brain Barrier.

In: Hussain CM, Ed, Micro and Nano Technologies, Handbook of Nanomaterials for Industrial Applications, Netherlands, Elsevier, Amsterdam, 2018, 696–709.

[200] ISO/TS 80004-2. Nanotechnologies-vocabulary-part 2: nano-objects 2015, 10.

[201] Sun J, Ma Q, Xue D, Shan W, Liu R, Dong B, Zhang J, Wang Z, Shao B. Polymer/Inorganic nanohybrids: an attractive materials for analysis and sensing. Trends Anal Chem 2021, 140, 116273.

[202] Gao N, Yu J, Chen S, Xin X, Zang L. Interfacial polymerization for controllable fabrication of nanostructured conducting polymers and their composites. Synth Met 2021, 273, 116693.

[203] De Alvarenga G, Hryniewicz BM, Jasper I, Silva RJ, Klobukoski V, Costa FS, Cervantes TNM, Amaral CDB, Schneider JT, Bach-Toledo L, Peralta-Zamora P, Valerio TL, Soares F, Silva BGJ, Vidotti M. Recent trends of micro and nanostructured conducting polymers n health and environmental applications. J Electroanal Chem 2020, 879, 114754.

[204] Fadillah G, Saputra OA, Saleh TA. Trends in polymers functionalized nanostructures for analysis of environmental pollutants. Trends Environ Anal Chem 2020, 26, e00084.

[205] Ramachandraiah K, Hong G-P. Polymeric nanomaterials for the development of sustainable plant food value chains. Food Bioscience 2021, 41, 100978.

[206] Lu Y, Shah KW, Xu J. Synthesis, morphologies and building applications of nanostructured polymers. Polymers 2017, 9(10), 506.

[207] Ghosh B, Biswas S. Polymeric micelles in cancer therapy: State of the art. J Control Rel 2021, 332, 127–147.

[208] Wibowo D, Jorritsma SHT, Gonzaga ZJ, Evert B, Chen S, Rehm BHA. Polymeric nanoparticle vaccines to combat emerging and pandemic threats. Biomaterials 2021, 268, 120597.

[209] Fuentes CA, Zhang Y, Guo H, Oigk W, Masania K, Dransfeld C, De Coninck J, Dupont-Gillain C, Seveno D, Van Vuure AW. Predicting the adhesion strength of thermoplastic/glass interfaces from wetting measurements. Colloids Surf A Physicochem Eng Asp 2018, 558, 280–290.

[210] Lingaraju D, Ramji K, Mohan Rao NBR, Rajya Lakshmi U. Characterization and prediction of some engineering properties of polymer – Clay/Silica hybrid nanocomposites through ANN and regression models. Proc Eng 2011, 10, 9–18.

[211] Chiellini E, Corti A, D'Antone S, Baciu R. Oxo-biodegradable carbon backbone polymers – Oxidative degradation of polyethylene under accelerated test conditions. Polym Degr Stab 2006, 27, 2739–2747.

Elvira De Giglio, Nicoletta Ditaranto, Luigia Sabbatini

2 Polymer surface chemistry: characterization by XPS

2.1 Introduction

We have seen in Chapter 1 that when a polymer interacts with another material or with its environment, the chemical and physical structures of the polymer surface determine the nature of the interaction and can undergo dramatic changes. Therefore, the understanding of the chemical behavior needs surface characterization with a high degree of chemical specificity in terms of composition and structure. This has been the driving force for the evolution of traditional analytical methods as well as for the development of new techniques. The application of X-ray photoelectron spectroscopy (XPS) to polymer surface analysis since its beginning largely contributed to its widespread use. Indeed, the requirements for polymer surface analysis have driven the development of peculiar instrumental facilities because of the importance of this materials field. The XPS perspectives in this sector are widely recognized, and there are plenty of literature that cover, in more or less details, the application of this technique to polymer-based materials. It is clear, however, that most polymer surface studies devoted to chemical characterization benefit from the combined use of XPS with other techniques such as secondary ions mass spectrometry (SIMS), ion scattering spectroscopy, and Fourier transform infrared spectroscopy in attenuated total reflectance mode (ATR/FTIR), which provide information from varying surface thicknesses, ranging from the top few angstroms (SIMS), to 2–10 nm (XPS), to a few micrometers (ATR/FTIR).

Some reviews describe many types of information available from XPS measurements of polymers [1–3]. XPS is now routinely used to obtain surface composition of homopolymers, copolymers, polymer blends, hybrid (i.e., organic/inorganic) systems, composites, and to follow processing steps, application-oriented surface modifications, such as chemical derivatization, grafting, or physical treatments (plasma, ion or electron beam, laser irradiation, etc.), as well as polymer degradation chemistry [4–6].

In this chapter, we will see how XPS is able to answer to important questions such as:

- Which elements are present on the surface? Which is their chemical state?
- Which is the atomic/mass percent surface composition?
- If the material is present as a thin film at the surface, how thick is the film? How uniform is the thickness and/or the chemical composition of the film?
- Which is the spatial distribution of elements (chemical mapping)?

https://doi.org/10.1515/9783110701098-002

Some fundamentals of the technique will be provided as well as basic information on the instrumental apparatus; applications to polymer surface analysis, both routine and specialized, will be illustrated.

2.2 Photoelectron spectroscopy: a brief history

The origin of XPS can be traced back to the end of the nineteenth century with the discovery of the photoelectric effect by Hertz in 1887. After that, several groups analyzed the energies of electrons emitted from metals bombarded by hard X-rays, the most active in the field being Rutherford's group in Manchester. By 1914, Rutherford provided the early basic XPS equation:

$$KE = hv - BE \tag{2.1}$$

where KE is the photoelectron kinetic energy, hv is the exciting photon energy, and BE is the electron binding energy in the solid.

Different groups performed research in the field, but the turning point was the construction by Kai Siegbahn [7] of an instrument capable of measuring electron kinetic energy with a resolving power of 10^5. This allowed KE, and hence BE, to be measured accurately for the first time. Siegbahn had full consciousness of the potential of XPS in chemical analysis and coined the acronym ESCA, for electron spectroscopy for chemical analysis, which is still largely used to address the technique [7]. A further spread was driven by the development of suitable commercial instruments that started to appear around 1969–1970 with the construction of the first UHV system in 1972.

2.2.1 Basic principles

XPS is a spectroscopic technique capable of providing atomic and molecular information about the surface of a solid material. In XPS, the sample surface is irradiated with X-rays providing surface atoms to emit electrons after the entire photon energy is transferred to them.

The relationship among the parameters involved in an XPS experiment is very similar to eq. (2.1), implemented with a further term:

$$KE = hv - BE - W \tag{2.2}$$

where W is the spectrometer work function.

The electron kinetic energy is an experimental quantity that can be measured by the spectrometer, but this is dependent on the photon energy of the X-rays used; the BE of the electron is the parameter that identifies the electron specifically, both in terms of its parent element and atomic energy level. Because no two elements possess the same set of BE, the measured kinetic energy provides for the elemental analysis (except for hydrogen and helium). From eq. (2.2), it is a simple matter to calculate the BE of the electron because all the other three terms are known or, at least, measurable. In practice, this task will be performed by the control electronics or data system associated with the spectrometer and the operator merely selects a binding or kinetic energy scale, whichever is considered the more appropriate.

A scheme of the photoemission process is reported in Fig. 2.1.
The scheme represents an incident X-ray impinging the material and causing an electron from an internal shell to be ejected from the atom (Fig. 2.1a). The photo-electron spectrum will reproduce the electronic structure of an element quite accu-rately because all electrons with a BE less than the photon energy (hv) will feature in the spectrum. Because of the high energy of X-ray radiation, the photoelectrons lines coming from the core levels are the most intense in an XP spectrum; unre-solved lines of low intensity also occur in the low-binding-energy region, produced by photoelectron emission from molecular orbitals (valence band). Those electrons that are excited and escape without energy loss contribute to the characteristic peaks in the spectrum; those that undergo inelastic scattering contribute to the background of the spectrum. The former comes from a depth that depends on the inelastic mean free path (IMFP, λ) of the electrons through the sample and on geo-metric factors (see Fig. 2.6 for details) and is called "sampling depth." This varies within the range of few nanometers and determines the surface-sensitive nature of XPS. After the photoemission, the ionized atom returns to its ground state through two main processes: emission of an X-ray photon, known as X-ray fluorescence, or ejection of an Auger electron (Fig. 2.1b) with a KE given by the relationship:

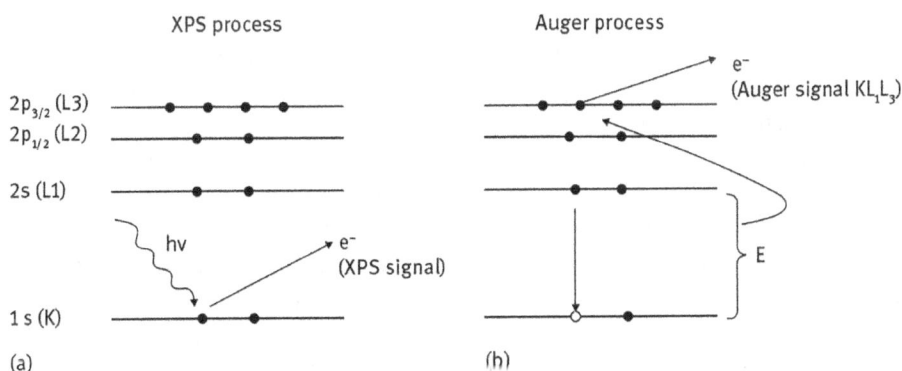

Fig. 2.1: (a) Schematic diagram of the XPS process and (b) of the Auger process.

$$KE = E_K - E_{L_1} - E_{L_3} - W \tag{2.3}$$

where suffixes K, L_1, and L_3 are related to the different levels depicted in Fig. 2.1. Thus, Auger electrons are produced as a consequence of the XPS process, and this kind of relaxation is not competitive at all with the photoemission process. Moreover, the energy of an Auger electron is independent of the incident $h\nu$, as evident from eq. (2.3). Ultimately, Auger signals yield valuable chemical information usually complementary to that obtained from XPS signals. Other important features that can be observed in an XP spectrum are X-ray satellites and shake-up lines. The former comes from a nonmonochromatic source (see Section 2.3.2 for details) and are related to minor X-ray components at higher photon energies. The latter comes from a complex photoelectric process because there is a finite probability that an ion (after photoionization) will be left in an excited energy state a few electron volts above the ground state. When this happens, the kinetic energy of the emitted photoelectron is reduced, and this will result in a "shake-up" peak at a higher BE than the main line.

XP spectra will then appear with these features, deriving from both photoelectron and Auger emission. To better understand the "language" of the spectra, it can be useful to learn some nomenclature rules.

2.2.2 Spectroscopic and X-ray notations

The formalism used to describe which electrons are involved with each of the observed transitions refers to two different notations: XPS uses the spectroscopic notation, whereas Auger electrons are identified by the X-ray notation. They are substantially equivalent, and the differences arise from historical reasons. Basically, both are based on the j–j coupling scheme describing the electron motion around the nucleus and whose total angular momentum is the vectorial sum of electron spin and angular momenta. Referring to n, l, and s, quantum numbers, it is possible to derive the total angular momentum for each electron, which is given by $j = l + s$. The spectroscopic notation provides that states with $l = 0$, 1, 2, 3, etc. are, respectively, named s, p, d, f, etc., preceded by the number n. It should be noted that with $l = 0$, j can assume only one value, and the corresponding signals deriving from these orbitals are single signals. For each value of l different from zero, the quantum number j assumes two values; this is in accordance with doublet signals originating from orbitals with $l > 0$. In X-ray notation, states with $n = 3$, 4, etc. are designated with the capital letters K, L, M, N, etc. with a number as suffix, depending on the different j values. Table 2.1 better explains the notations and their relationship.

Tab. 2.1: The "language".

Quantum numbers				Spectroscopic notation	X-ray notation
n	l	s	$j = l + s$		
1	0	+1/2	1/2	1s	K
2	0	+1/2	1/2	2s	L1
2	1	−1/2	1/2	$2p_{1/2}$	L2
2	1	+1/2	3/2	$2p_{3/2}$	L3
3	0	+1/2	1/2	3s	M1
3	1	−1/2	1/2	$3p_{1/2}$	M2
3	1	+1/2	3/2	$3p_{3/2}$	M3
3	2	−1/2	3/2	$3d_{3/2}$	M4
3	2	+1/2	5/2	$3d_{5/2}$	M5
...	...	...	...	...	...

2.3 Photoelectron spectroscopy: a brief history

An XPS apparatus substantially consists of a high-vacuum environment, a source of fixed energy radiation, an electron energy analyzer coupled with a lens system to disperse the emitted electrons according to their kinetic energy, an electron detector to measure the flux of emitted electrons of a particular energy, and a flood gun for charge compensation. Typically, the apparatus is then connected to a computer with a proper software interface, which registers the signal coming from the detector, and a spectrum of counts rate *vs* kinetic energy will be displayed in real time. After acquisition, data processing tools are generally provided to extract all the required information. Figure 2.2 depicts a block diagram of a typical XPS spectrometer.

2.3.1 Vacuum system

Electron spectroscopic techniques require strict vacuum conditions, that is, ultra-high vacuum (UHV) range of 10^{-8} to 10^{-10} mbar. The reasons for this are itemized in the following:
- The analytical signal of low-energy electrons is easily scattered by the residual gas molecules in the environment, thus leading to the collapse of the signal.
- More importantly, the UHV environment is necessary because of the surface sensitivity of the techniques themselves. Indeed, at 10^{-6} mbar, it is possible for a

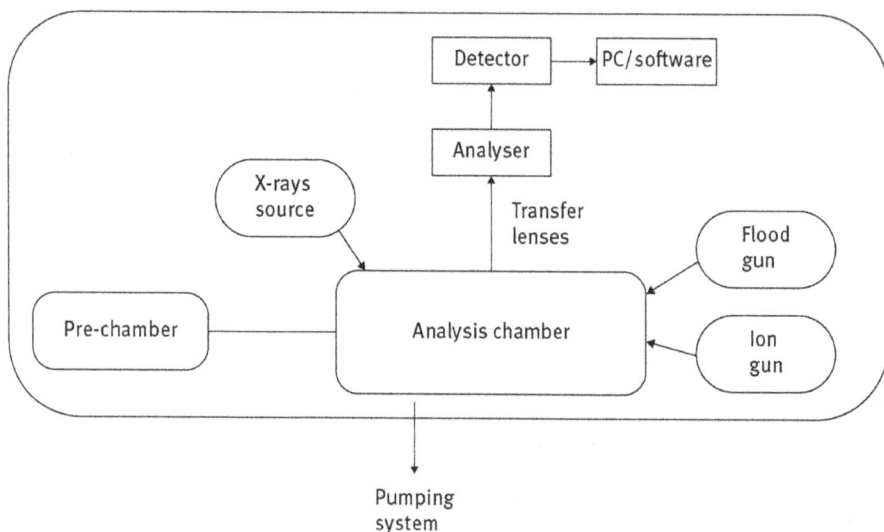

Fig. 2.2: Schematic diagram of an XPS spectrometer.

monolayer of gas to be adsorbed onto a solid surface in about 1 s, too short compared with the time usually required for a typical spectral acquisition. Therefore,
the need for a UHV environment during analysis is mandatory.

How to establish such a vacuum will depend on different manufacturer preferences.
Typically, the chambers are made of stainless steel, and joints are made using
flanges that are tightened onto copper gaskets. UHV conditions are then usually obtained and maintained by dynamic pumping down of the whole spectrometer
through a proper pumping system. Turbomolecular pumps along with a titanium
sublimation pump to assist the primary pumping are generally used to achieve the
desired vacuum level. All UHV systems need baking from time to time to remove
adsorbed layers from the chamber walls, and the baking temperature is usually in
the range of 100–160 °C for routine use.

2.3.2 X-ray sources

2.3.2.1 Dual Mg/Al anode X-ray tube

The photon beam energy for XPS source is usually obtained from an X-ray tube with
a hot filament at high voltage ranging from 10 to 15 kV and a current of 10–15 mA,
which emits electrons. These electrons are accelerated toward a grounded potential
anode. The maximum anode current is set by the efficiency with which the heat can
be dissipated; therefore, X-ray anodes are usually water-cooled. The choice of anode

material determines the energy of the X-ray transition generated, and it is usually chosen to be high enough to excite an intense photoelectron peak from all elements of the periodic table. It must also possess a natural X-ray line width, narrow enough to achieve a good resolution. Based on these considerations, the most popular anode materials that satisfy the requirements are aluminum and magnesium. These are usually supplied in a single X-ray gun with a twin-anode configuration providing AlKα or MgKα photons of nearly mono-energetic (monochromatic) X-rays with energy 1,486.6 and 1,253.6 eV, respectively. An XP spectrum is often rich of many signals coming from both main and secondary photoelectron peaks and Auger peaks. Therefore, very often, the use of two different X-ray energies allows the elimination of ambiguities/interferences in the analysis because the kinetic energy of XP signals is *hv* dependent, whereas the kinetic energy of Auger peaks is only related to the electronic structure of the emitting element. For these reasons, many XPS instruments have dual anodes (Al and Mg) that can be excited separately.

2.3.2.2 Monochromatic source

There are a number of reasons for choosing to use an X-ray monochromator on an XPS spectrometer: removal of satellites, elimination of the Bremsstrahlung background, and the reduction in X-ray line width. A monochromator works by filtering a narrower band of X-rays from the resonance peak by X-ray diffraction from a quartz crystal or from bent SiO_2 crystal. In this way, only certain wavelengths can be reinforced into a spot so that the monochromatic X-rays can be directed at the sample. The crystal spacing is such that X-ray wavelengths, which are multiples of the AlKα X-ray resonance, are reinforced following the Bragg condition for X-ray diffraction.

2.3.2.3 Synchrotron radiation source

The synchrotron radiation (SR) is produced when charged particles, for example, electrons, are forced to move in a circular orbit. Continuous Bremsstrahlung X-radiations of different energies are thus emitted; when relativistic velocities are reached, these photons are emitted in a narrow cone in the forward direction, at a tangent to the orbit. Different possibilities are available using SR source because it possesses a number of unique properties: (1) high brightness: the SR is so intense; it is hundreds of thousands of times higher than conventional X-ray anodes; (2) wide energy spectrum: SR is emitted with a wide range of energies, allowing a beam of any energy to be produced; (3) SR is highly polarized. The major advantage of SR is essentially a surface sensitivity better than the sensitivity obtained using AlKα or MgKα. Meanwhile, damage on the sample surface during the measurement may easily occur, probably due to sample heating when SR excitation is used.

2.3.3 Energy analyzers

In XPS, the analysis of the ejected photoelectron energies is the key issue. Therefore, the analyzer is the most important component of an XPS spectrometer. Before discussing about the energy analyzers, few words on the energy resolution are needed. The latter is defined in two ways: the absolute resolution, that is, the full width at half maximum (FWHM) of a chosen peak (ΔE) and the relative resolution, defined as the ratio of ΔE to the KE of the peak position. The absolute resolution is independent of peak position, whereas the relative one should be referenced to a specific KE. When running an XPS experiment, we need the same absolute resolution at all the energies.

There are two types of electron energy analyzer for XPS: the cylindrical mirror analyzer (CMA) and the hemispherical sector analyzer (HSA). The CMA is used when it is not important that the highest resolution is achieved. Meanwhile, spectral resolution plays a key role in XPS, and this led to the development of the HSA as a design of analyzer with sufficiently good resolution. The addition of a transfer lens to the HSA and multichannel detection increases its sensitivity to the point where both high transmission and high resolution are possible, and this type of analyzer may now be used with excellent results.

The CMA consists of two concentric cylinders: the inner cylinder is held at earth potential, whereas the outer is ramped at a negative potential. An electron gun is often mounted coaxially within the analyzer.

The HSA, also known as a concentric hemispherical analyzer (CHA), uses an electric field between two hemispherical surfaces to disperse the electrons according to their kinetic energy. Between the sample and the analyzer, there is usually a lens or a series of lenses. The hemispheres with radii R_1 (inner) and R_2 (outer) are positioned concentrically, and potentials V_1 and V_2 are applied to these spheres, respectively, with V_2 greater than V_1. If electrons arrive at the entrance of the analyzer with $E = eV_0$, they will be focused on the exit of the analyzer only if the following relationship holds:

$$V_0 = \frac{V_1 R_1 + V_2 R_2}{2R_0} \tag{2.4}$$

which defines the potential of mean free path analyzer through radius R_0 (medium equipotential surface between R_1 and R_2). Because R_1 and R_2 are fixed, changing V_1 and V_2 will allow, in principle, scanning of electron kinetic energy following mean free path through hemispheres.

The resolution of CHA is given by $R = \Delta E / KE$. Because the resolution R is energy dependent, it will not remain uniform across the entire XP spectrum. This can be achieved by retarding the electrons entering energy analyzer to fixed kinetic energy, called the pass energy E_0, so that fixed resolution applies across entire spectrum. For this purpose, either the pass energy can be decreased or R_0 can be increased. The increased resolution for an analyzer is typically ~0.1–1.0 eV.

The CHA is typically operated in two different modes: constant analyzer energy (CAE), also known as fixed analyzer transmission, and constant retard ratio (CRR), also known as fixed retard ratio. In the CAE mode, electrons are accelerated or retarded according to a user-defined pass energy. In the CRR mode, electrons are retarded to some user-defined fraction of their original kinetic energy as they pass through the analyzer. In practice, in the CRR mode, the pass energy is proportional to the kinetic energy, and so the relative resolution is constant throughout the energy range. In the CAE mode of operation, the absolute resolution remains constant throughout the energy range of the scan, ensuring that XPS quantification is more reliable and accentuates the XPS peaks at the low kinetic energy (high BE) end of the spectrum.

2.3.4 Detectors

Once the photoelectrons have been analyzed and transmitted, it is necessary to count the individual electrons arriving at the detector. To achieve this, electron multipliers are used. Although there are many types of electron multipliers, only two types are commonly used: channel electron multipliers (CEMs, or channeltrons) and channel plates.

A CEM is an electron detector able to multiply each electron up to 108 times, thus providing a pulse output suitable for further amplification by conventional electronic circuits. CEM is a bent tube coated with a photoelectric material with a high secondary electron coefficient. The tube is kept at a potential of about 2.5 kV. When an electron strikes the mouth of the tube, a number of secondaries are produced that are accelerated in the channeltron. A single photon is capable of creating an output of 10 million electrons. This process continues down the length of the CEM, creating more and more secondaries along the way. This amplified "cascade" passes through the preamplifier and is thus detected. This pulse is then shaped and ultimately registered by the acquisition electronics. In a similar way, pulses of electrons are produced in the microchannels of a microchannel plate. When this device is used in combination with a position-sensitive anode, electrons can be detected with a lateral resolution of some tens of a micrometer.

A channel plate consists of a disc having an array of small holes. Each of these holes behaves as a small channeltron. The gain of an individual channel is much lower than that of a channeltron; thus, it is common to use a pair of channel plates in tandem. Channel plates are used when it is necessary to detect data in two dimensions; spectrometers have been designed using channel plates to measure signals for parallel acquisition of photoelectron images, XPS line scans, and the parallel acquisition of angle-resolved electrons.

2.3.5 Charge compensation

The electrical properties of the solid materials analyzed in an XPS experiment can vary from the conductive metal to the insulating piece of plastic. Photoemission from an insulating sample causes electrostatic charging to occur in the positive direction. This results in a shift of the peak position in the direction of higher BE, making equivocal the peak attribution. When the photoemission is excited by a non-monochromatic X-ray source, there are usually a sufficient number of low-energy electrons available in the neighborhood of the sample to effectively neutralize the sample surface. When monochromatic X-ray sources are used, these low-energy electrons are not produced in such a large number near the sample, and thus, neutralization does not take place. When analyzing polymers, this effect should be taken into account because they are completely insulating materials. Surface charge compensation is generally accomplished by flooding the surface with low-energy electrons. From the instrumental point of view, there is an additional gun, the flood gun, producing a uniform negative charge of known magnitude at the surface of the sample. This technique minimizes the risk of differential or nonuniform charging. The electron beam used for charge compensation should be of low energy to avoid the risk of damage to the surface of the sample and at the same time supply a high enough flux to adequately compensate for the charging of the sample surface. Nowadays, in modern flood guns, the energy of the electrons is as low as about 1 eV.

2.3.6 Small-area XPS: imaging and mapping

It is often highly desirable to analyze small features of a surface, and to do this instrumental helps are needed. Two ways are possible: (1) limiting the area from which the photoelectrons are collected using the transfer lens and (2) focusing a monochromated beam of X-rays into a small spot on the sample. The first option is referred to as lens-defined analysis: it consists in operating the transfer lens to produce a photoelectron image at some point in the electron optical column where a small aperture is placed: therefore, only electrons emitted from a defined area of the sample can pass through the aperture and reach the analyzer. If the magnification of the lens is M and the diameter of the aperture is d then the diameter of the analyzed area is d/M. Using this technique, commercial instruments can provide high-quality small-area analysis down to about 15 μm, but with the main disadvantage of reducing the detected flux per unit area of the sample. This means longer analysis time, hence a longer exposure of the sample to X-rays, potentially resulting in radiation damage.

The second possibility is the so-called source-defined small-area analysis, when a monochromatic X-ray source is available. The monochromatic X-rays diffracted from a quartz crystal can be focused on the sample using a magnification of unity, which means that the size of the X-ray spot on the sample is approximately equal to the size of the electron spot on the X-ray anode. Analysis areas down to about 10 μm can be achieved in many recent commercial instruments using this method. Because the analysis area is source-defined, the lens can be operated at its maximum transmission, thus dramatically increasing the sensitivity of a spectrometer. This reduces sample damage during analysis and eliminates radiation damage to the surrounding area of the sample(s).

An important practical consequence of small-area XPS is the possibility to register an image or chemical map of the surface. Such an image or map shows the distribution of an element or a chemical state on the surface of the sample. There are two distinct approaches, used by manufacturers, to obtaining XPS maps: serial acquisition, in which each pixel of the image is collected in turn (by X-ray spot, sample stage, or lens scanning all over the defined area to be mapped); parallel acquisition in which data from the whole of the analysis area are collected simultaneously (by additional lenses and a two-dimensional detector).

2.3.7 Ambient-pressure photoelectron spectroscopy

The reasons for strict vacuum conditions for XPS have been itemized in Section 2.3.1. Nevertheless, techniques that provide spectroscopic information of surfaces while exposed to gases or liquids in the atmospheric pressure range are highly desirable. This is the field where ambient-pressure photoelectron spectroscopy (AP-XPS or APPES) has been developed, producing dramatic changes in the study of liquid and solid surfaces, particularly in areas such as atmospheric, environment, and catalysis sciences [8, 9]. The state of the art approach to perform XPS at the mbar range, defined as near ambient pressure XPS (NAP-XPS), consists in confining the high pressure only at the sample region, shortening as much as possible the path of photoelectrons in the high pressure region [10]. With NAP-XPS, XPS can probe moderately volatile liquids, biological samples, porous materials, and/or polymeric materials that outgas significantly.

Recently, Jain et al. investigated polytetrafluoroethylene (PTFE) samples showing NAP-XPS spectra of this insulator polymer [11]. Using different background gas (air) pressures the authors observed that peaks decrease in width, shift toward literature values, and improve in shape with increasing background gas pressure.

However, the application of AP-XPS in polymers science is not so widespread and the potentiality of this approach will be only briefly discussed in this chapter.

One characteristic of surfaces in ambient-pressure environments is that they are covered by dense layers of molecules: water, for example, is known to form layers several molecules thick at room temperature in humid environments; metals readily form oxide films several layers thick in oxygen atmospheres; layers of adsorbed molecules can also be produced in UHV, simply cooling the sample to cryogenic temperatures. Therefore, AP-XPS should rely on instrumental facilities able to circumvent the strong interaction of electrons and ions with the gas. They all involve capturing the particles at distances comparable with their mean free path, which is in the millimeter range for electrons of a few hundred electron volts at millibar-type pressures. This is accomplished by small apertures near the sample and by differential pumping to remove gases leaking through it. All instruments developed to make possible electron spectroscopy under environmental gases are based on this idea, which is accomplished by bringing the sample close to a small aperture at a distance equal or less than λ.

Differential pumping is the key to lower the pressure at the other side of the aperture to decrease the collision rate.

In lab-based AP-XPS systems, the basic designs of coupling an electron energy analyzer to a sample analysis chamber are summarized in Fig. 2.3 [12].

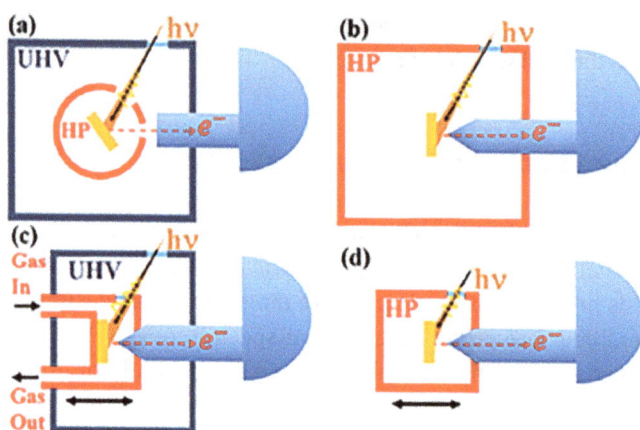

Fig. 2.3: Design schemes of lab-based AP-XPS systems. (a) Early design of a differentially pumped pressurized cell (red) in a UHV chamber which is connected to an electron energy analyzer (blue). (b) Analysis chamber (red) connected to a differentially pumped analyzer terminated by an aperture sitting above a sample surface. (c) Internal gas cell in a UHV chamber. (d) Removable vacuum chamber. Reprinted with permission from Arble C, Jia M, Newberg JT, Surface Science Reports 73, 37–57, 2018. Copyright 2018 Elsevier [12].

The principles of an early pressurized cell contained within a larger UHV chamber are displayed in Fig. 2.3a. The larger UHV chamber differentially pumps the pressurized cell and lowers the pressure near the entrance to the electron energy analyzer (blue). In Fig. 2.3b, an aperture is placed near the sample surface, behind which are differential pumping stages within the analyzer. The sample itself sits within a vacuum chamber (red) which is usually backfilled with gas pressure during analysis. The system in Fig. 2.3c is an internal gas flow cell within a UHV chamber, where the gas cell is formed from mounting the cell to an o-ring on the differential pumping aperture to create a vacuum seal. Figure 2.3d is similar to Fig. 2.3b, except the analysis chamber (red) has a much smaller volume and is intentionally removable.

As might be expected, all the components discussed above will be present in an ambient-pressure spectrometer, but they are especially designed to operate in a different pressure range.

2.4 Chemical information from XPS

The first information that can be derived in characterizing the surface chemistry of the specimen under investigation is the identification of the elements present. To achieve this, it is usual to record a survey, or wide-scan, spectrum over an energy region that will provide peaks for all elements in the periodic table. The individual peaks may be identified thanks to a database generally included in the software. Figure 2.4 shows a typical wide scan along with the proper element attribution.

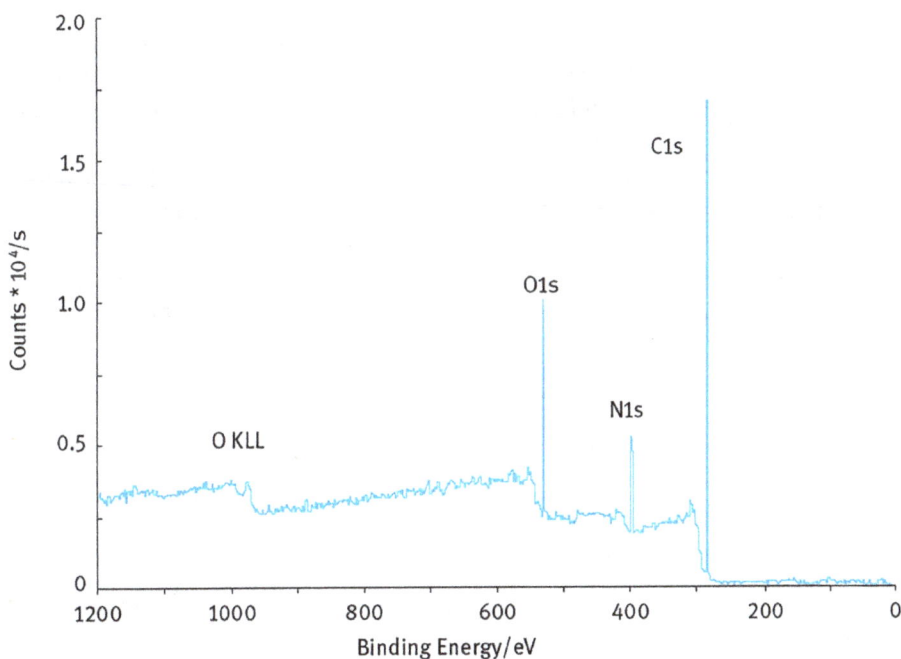

Fig. 2.4: XP wide-scan spectrum with element identification labels.

The labels in the Fig. 2.4 indicate the symbol of the element and the core level that generated the peak (O1s, C1s, and N1s). When the symbol is followed by three letters, then the label is referred to an Auger peak (see Fig. 2.1b for details). Hence, the surface compositional (elemental) analysis can be performed by identifying the peaks with the specific atoms. Quantitative information can be also derived through the measurement of the relative areas of the photoelectron peaks, thus allowing the composition of the sample to be determined. Before calculating the peak areas, however, the background subtraction is needed. Three approaches to defining the spectral background are commonly described in the literature: linear subtraction, Shirley polynomial subtraction, and Tougaard method (see Fig. 2.5); the theory underlying these approaches is beyond the scope of this textbook and can be found elsewhere [13]. However, it is worth noting that for quantitative analysis of polymer spectra the use of the straight-line background or Shirley one is generally recommended.

After background subtraction, the quantification of peak intensity data is performed using peak areas. In particular, the factors affecting the peak intensity are well described in the following equation:

$$I_{i,c} = fN_i\sigma_{i,c}\lambda\cos\alpha \; FTDA \tag{2.5}$$

where $I_{i,c}$ is the photoelectron intensity for core level c of element i, f is X-ray flux in photons per unit area per unit time, N_i is the number of atoms of element i per unit volume, $\sigma_{i,c}$ represents the photoelectric cross section for core level c of element i, λ is the IMFP of the photoelectron in the sample matrix, and α is the angle between the direction of photoelectron and the sample normal. Other parameters are also the analyzer solid angle of acceptance (F), the analyzer transmission function (T), the detector efficiency (D), and the area of sample from which photoelectrons are detected (A). As can be easily argued, this equation is too complex for usual analyses, and empirical sensitivity factors whose values are specific for different elements have been introduced [14]. Now, peak intensity can be described by the simple form: $I = N$ x S,

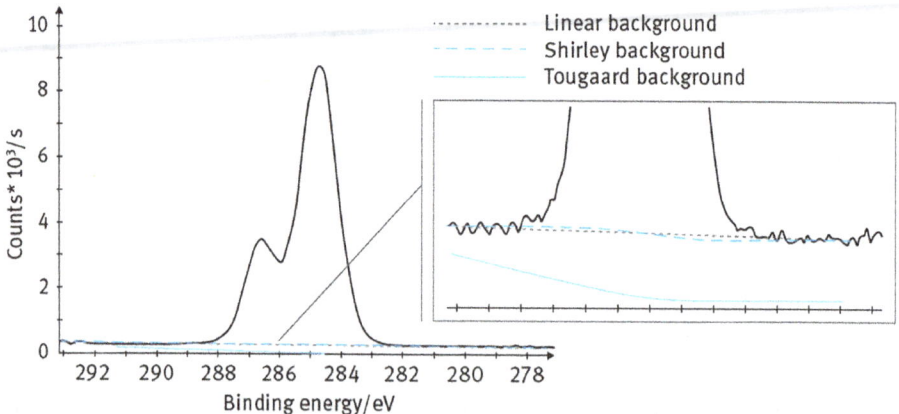

Fig. 2.5: Comparison of experimental background subtraction with the three different methods.

where S is the sensitivity factor. The result is a semiquantitative information in the form of relative atomic percentages:

$$\text{Atomic \%} = \frac{\frac{I_i}{S_i}}{\sum_j \frac{I_j}{S_j}} \times 100\% \tag{2.6}$$

Because the photoelectrons are strongly attenuated by passage through the sample material itself, the information obtained comes from the sample surface, with a sampling depth of the order of 7–10 nm.

In detail, photoelectrons are created at a depth z below the surface; in straight-line trajectory, only inelastic scattering leads to electron attenuation, which is as-sumed to follow the Beer–Lambert (BL) law, as does light attenuation in an absorbing medium. Therefore,

$$I_z = I_0 \exp(-z/\lambda \sin\theta), \tag{2.7}$$

where I_z is the intensity from the atoms at depth z, I_0 is the intensity from the surface atoms, and θ is the electron take-off angle to the surface ($\theta = 90° - \alpha$). In this case, λ is the IMFP of the measured electrons, defined as the average distance that an elec-tron with a given energy travels between successive inelastic collisions. In the case of exponential decay, the escape depth d is $\lambda\sin\theta$. Commonly, a surface layer as thick as $3d$ is taken to be the "sampling depth."

Greater sampling depths can be achieved by removing the sample material layer by layer through ion bombardment (down to hundreds of nanometers); vari-able sampling depth, comprised between near zero (very topmost layers, up to less than 1 nm) and a maximum of $3d$, can be obtained by changing θ. Both modes are defined as depth profiling because they provide the distribution of chemical compo-sition to depths. Ion bombardment is indeed more destructive and is instrumentally performed by an ion gun (positive argon ions accelerated on the sample surface). The application of destructive depth profiling in the polymer analysis has been lim-ited for a long time because of the possible degradation upon ion bombardment along with the loss of chemical information. Recently, this drawback was success-fully faced by the development of guns using gas cluster ions instead of monoa-tomic ions: the bombardment is gentler and the chemical information is preserved.

Surface sensitivity enhancement by variation of θ (take-off angle) is called angle-resolved XPS (ARXPS). Despite the achievable depth regime for this opera-tional mode is limited to $3d < 10$ nm, it is a particularly important technique for polymer surface studies because many important aspects of polymer surface behav-ior are governed by major compositional or structural variations within this layer thickness. The geometry involved is schematically depicted in Fig. 2.6.

From the instrumental point of view, the take-off angle variation is traditionally achieved tilting the sample, that is, the sample holder. A second possibility has been recently developed: the angular acceptance of the transfer lens is set to provide

angular resolution using a two-dimensional detector. In this instrumental arrangement at the output plane of the analyzer, the photoelectron energy is dispersed in
one direction (as with a conventional lens analyzer arrangement) and the angular
distribution dispersed in the other direction. Such an arrangement provides a parallel angle-resolved collection of spectra, with a number of advantages over the
conventional method.

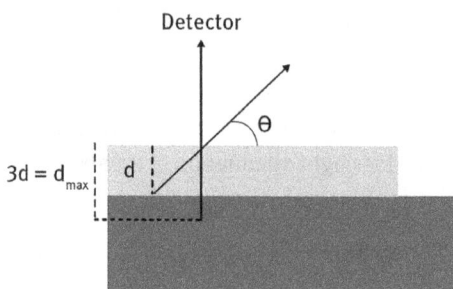

Fig. 2.6: Scheme of geometric parameters
involved in angle-resolved XPS experiments.

2.5 Chemical shift and its significance
in the analysis of polymers

XPS can be also used to derive information about the atomic oxidation state from
the sample surface. As discussed in Section 2.4, core-level binding energies and relative intensities provide atomic identification and relative concentrations. In addition, the exact core-level BE varies slightly depending on the nature of chemical
bonding and the presence of neighboring atoms affecting the BE and producing the
so-called chemical shift. The BE is determined by the difference between the total
energies of the initial-state atom and the final-state ion: chemical bonding will
clearly have an influence on both the initial state energy of the atom and the final
state energy of the ion created by emission of the photoelectron. These changes can
be calculated by quantum chemical methods and are basically due to the redistribution of electrons depending upon the electronegativities of the atoms involved.

Unfortunately, the range of these shifts is small – few electron volts also in the
case of C1s – and therefore the energy resolution as well as the peak width are critical in the functional group recognition. Table 2.2 reports a series of core-level BE for
C1s in different common functional groups.

Consequently, when the molecule structure involves the same atom in more
than one chemical environment, the resulting chemically shifted peaks will probably overlap to produce a complex envelope. Of course, this problem is a crucial
issue in the analysis of polymers where many information can be deduced mainly
from the C1s core-level spectra due to the different functional groups the carbon is

involved in. Extraction of structural information from such spectra requires the component peaks to be resolved and requires knowledge of individual component lineshapes [15]. The approach to do this is the curve fitting technique, whose mathematical basis are well discussed in reference [16].

Before curve fitting, an appropriate background function should be introduced. Afterward, in curve fitting itself a series of components is generated and placed within the measured envelope at appropriate BE positions. The software calculates the sum of the components, and from comparison with the experimental data, it obtains a measure of the "goodness of fit" by the least squares method. This process cannot produce a unique solution, and a great deal of operator inputs are usually required to ensure that the solution is reasonable both from a spectroscopic and chemical point of view. Parameters such as peak lineshape, peak full width at half maximum, peak position, and intensity as well as background position during iteration should be correctly indicated. This way, the fitting process can be "guided" toward a solution that includes the maximum amount of operator understanding of the system.

Two examples of C1s curve fitting are reported in Fig. 2.7: (a) C1s spectrum of a polylactic acid (PLA) where three components are clearly evident; (b) a more complex case, where the different chemical components constitute an envelope still solvable by accurate curve fitting. In the latter spectrum, it can also be observed the substantial chemical shift of carbon when bound to a very electronegative atom such as fluorine.

Tab. 2.2: C1s typical chemical shifts range for different functional groups.

Functional group	BE range (eV)	Functional group	BE range (eV)	Functional group	BE range (eV)
C—C	285.00	C—NO₂	285.76	⬡—X (ring)	284.44–284.80
C=C, C≡C	284.69–284.76	C—N	285.56–286.41	C—Si	284.22–284.39
C—OH, C—O—C	286.13–286.75	C—N⁺—	285.99–286.22	C—S	285.21–285.52
C=O, O—C—O	287.81–288.06	C—C≡N	286.35–286.46	C—SO₂	285.31–285.64
O=C—C=O	288.64–289.23	—C≡N	286.73	C—SO₃⁻	285.16
HO—C(=O)	289.18–289.33	C—ONO₂	287.62	C—Br	285.74
O=C—O⁻	289.30–289.34	N—C—O (ring)	287.78	⬡—Cl (ring)	285.99–286.07
C—O—C (anhydride)	289.36–289.46	N—C=O (ring)	287.97–288.59	C—Cl	287.00–287.03
O—C—O⁻ (carbonate)	290.35–290.44	C—N—C (amide)	288.49–288.61	—CCl₂	288.56
		N—C, N—	288.84	C—F	287.91
		N—C, O—	289.60	—CF₂	290.90
				—CF₃	292.65–292.75

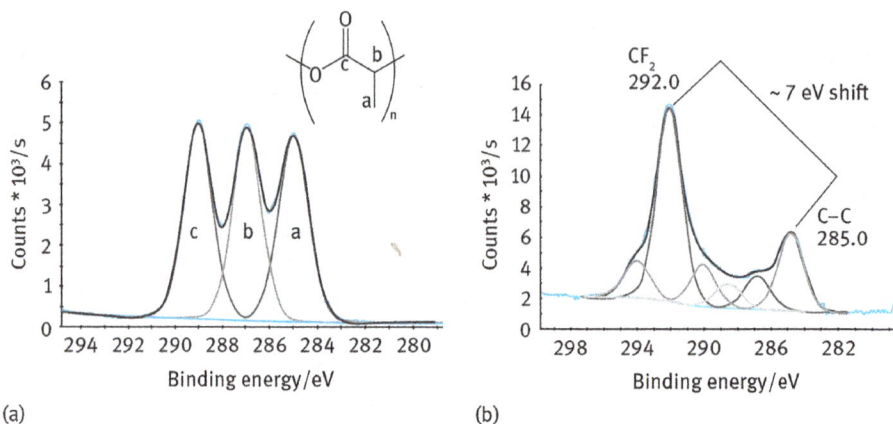

Fig. 2.7: Curve fitting of (a) C1s XP spectrum of PLA polymer and (b) C1s XP spectrum of a fluorinated polymer.

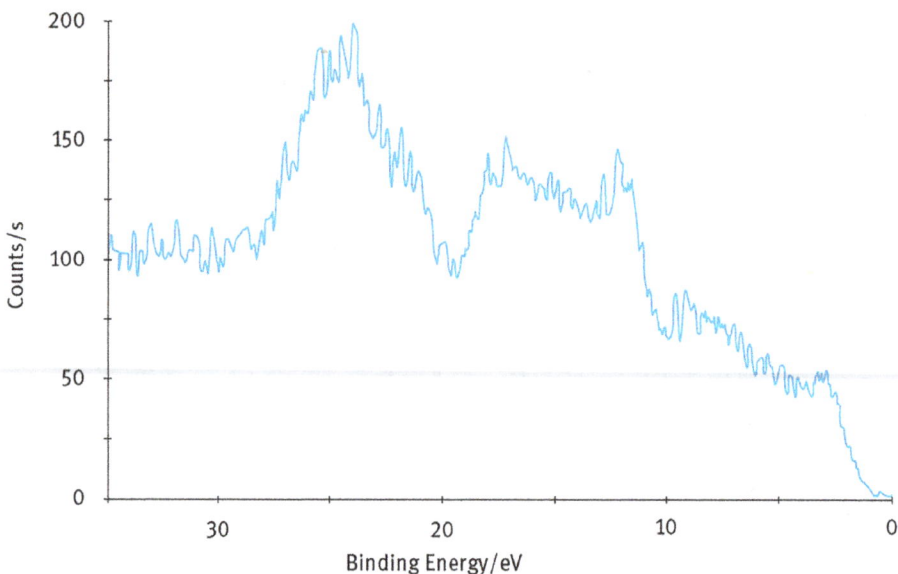

Fig. 2.8: Valence band of a polymer.

In some cases, the chemical shifts observed in core-level XP spectra are not sufficient to discriminate the surface chemistry of a sample. Especially for XPS analysis of polymers, changes in carbon spectra can be so subtle that other spectral region are necessary to be investigated. Owing to the use of monochromators and high counting rates in recent spectrometers, the study of valence bands has become very useful for identification of materials. Rather than specific molecular orbitals identification, it is used as fingerprints for a sample. In particular, the fingerprints of valence bands may be used in the identification of polymers and/or quantification of polymers mixtures by linear least squares fitting [17] (Fig. 2.8).

Additionally, in the case of aromatic polymer compounds, further features can be very useful: the characteristic shake-up lines. They originate from the photoelectric process described in Section 2.2.1 and involving the energy of the $\pi–\pi^*$ transition.

2.6 Chemical derivatization techniques in conjunction with XPS

We have seen in the previous paragraph that polymer surfaces are not always amenable to direct XPS analysis because BE differences between some functional groups may be too small to be distinguishable to spectroscopic probes such as XPS. Curve-fitting procedures are a valuable tool in this case: however, when the BE values of different species overlap, curve-fitting results of core-level photoemission lines may be ambiguous. In such cases, it would be more convenient to have a labeling method to clearly extract the presence of certain reactive functional groups from the spectral envelope. This has led to the development of chemical derivatization techniques: herein, the functional group of interest is labeled with an elemental or molecular tag not present on the original surface [2, 18], thus offering a valid support to peak-fitting analysis.

Chemical derivatization techniques in conjunction with XPS (CD-XPS) can be used both to confirm and exclude the presence of specific functional groups on the surface of polymers as well as to quantify their concentration when they are present and react selectively and stoichiometrically with a specific derivatizing reagent that contains a unique chemical tag.

To provide an accurate estimate of the relative concentration of a particular functional group, the derivatization reaction must be highly selective, particularly when real-world samples, frequently multifunctional (e.g., chemically modified or plasma-modified polymer surfaces) [19], are to be analyzed.

A very simple example is the identification of carbon–carbon double bonds on the surface of a treated polymer when in the presence of fully saturated bonds. Reaction with bromine in carbon tetrachloride is well known from undergraduate chemistry laboratory experiments, and it yields two bromines for every double bond initially present. For surface analysis, however, there are at least two general considerations: (1) solution reactions frequently fail to occur on surfaces; (2) XPS analysis depth is about 10 nm, and any labeling would require penetration and quantitative reaction to at least three times that depth.

The requirements [20] for an effective labeling analysis have been summarized:
- The reaction should preferably be quantitative or should at least always proceed to the same degree.
- Because XPS analyzes samples on the surface to a depth of approximately 10 nm, the concentration distribution of the chemical derivative with depth should be explored. It is preferable if the reaction occurs uniformly throughout the sampling depth.
- The reaction should be rapid and the reaction kinetics should be understood.
- The reaction product should be stable under vacuum and to electron and X-ray sources.
- The reaction product should contain a unique marker atom or group, preferably one that has a high photoemission cross section.
- Cross-reactions of a specific derivatizing reagent with functional groups not under investigation should not occur or should be well understood.
- The chemical reaction should not lead to a rearrangement of the surface structure that is being probed.

Some vapor-phase derivatization reactions meet these criteria. Figure 2.9 reports some of these reactions and the derivatization method.

For CD-XPS, fluorine atoms are often used as the chemical tags when not present on polymers surfaces; furthermore, the F(1s) core-level transition has a high XP cross section. Some examples of fluorine-based derivatization reactions on model polymers surfaces are reported in the following:
- TFAA derivatization reaction: the vapor-phase trifluoroacetic anhydride (TFAA) derivatization reaction has been widely used to tag hydroxyl groups on surfaces [21, 22]. Carbonyl and carboxyl groups showed low reactivity toward TFAA, evidenced by the low intensity of F1s signals on TFAA-derivatized PVMK and PAA. In particular, the fluorine concentration was equal to 4 at% of the acid groups in PAA and less than 6 at% of the ketone groups in PVMK. Meanwhile, TFAA-derivatized PVA showed, within the XPS sampling depth, the complete reactivity of the polymer hydroxyl groups (33 at%). In Fig. 2.10a, the high-resolution C1s spectra of PVA, before and after the reaction with TFAA, have been reported [23]. The new peaks that appear in the C1s region of a TFAA-derivatized PVA can be attributed to the ester carbon (at 289.5 eV) and to the CF_3 carbon atoms (at 292.6 eV). The experimentally determined $CHx:COC:CO_2:CF_3$ peak area ratio of 1.2:1.0:1.0:1.0 for these peaks is close to the 1:1:1:1 ratio estimated for a stoichiometric reaction. TFAA vapor is not selective toward surface hydroxyl groups in polymeric systems containing epoxide groups. In particular, the reactivity of TFAA vapor with epoxides is comparable to its reactivity with hydroxyls.

Thus, the vapor-phase reaction with TFAA and subsequent quantitation with XPS cannot be used to assay the relative surface concentration of hydroxyl groups in the

$$\boxed{}\text{-HCO} + \text{NH}_2\text{NH}_2 \longrightarrow \boxed{}\text{-HC=N-NH}_2 + \text{H}_2\text{O}$$

$$\boxed{}\text{-OH} + (\text{CF}_3\text{CO})_2\text{O} \longrightarrow \boxed{}\text{-O-CO-CF}_3 + \text{CF}_3\text{-COOH}$$

$$\boxed{}\text{-COOH} + \text{CF}_3\text{-CH}_2\text{-OH} + (\text{CH}_3)_3\text{C-N=C=N-C(CH}_3)_3 \longrightarrow \boxed{}\text{-CO}_2\text{-CH}_2\text{-CF}_3 + (\text{CH}_3)_3\text{C-NH-}\overset{\displaystyle O}{\overset{\displaystyle \|}{\text{C}}}\text{-NH-C(CH}_3)_3$$

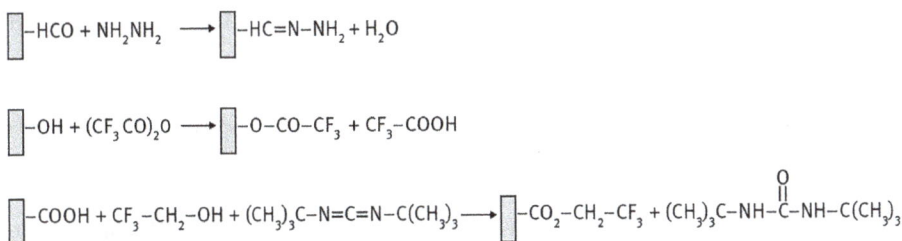

Fig. 2.9: Some vapor-phase derivatization reactions used in CD-XPS.

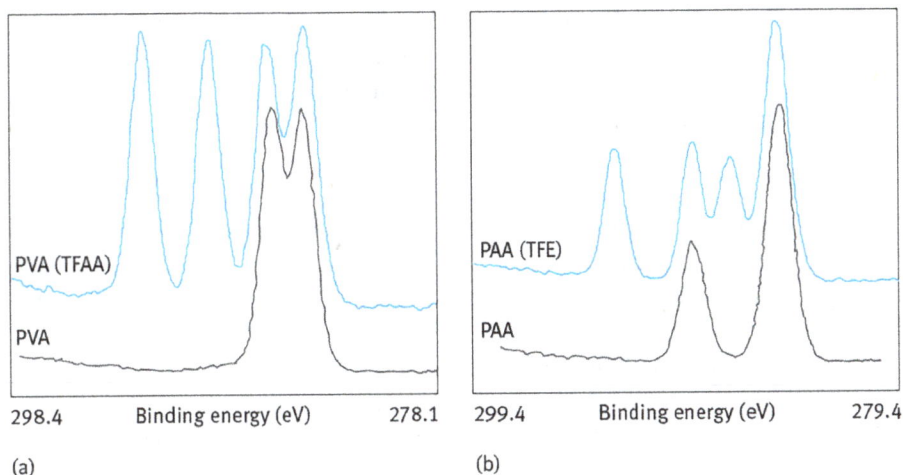

(a)

(b)

Fig. 2.10: High-resolution C1s spectra of (a) PVA (bottom spectrum) and TFAA-derivatized PVA (top spectrum). (b) High-resolution C1s spectra of (a) PAA (bottom spectrum) and TFE-derivatized PAA (top spectrum). Reprinted with permission from Chilkoti A, Ratner BD, Briggs D, *Chem Mater*, 3, 51–61, 1991. Copyright 1991 American Chemical Society [23].

presence of epoxide groups [24, 25]. TFAA was also used as derivatizing agent to detect secondary amines or imines [26], although some authors reported a low yield of the derivatization reaction between TFAA and NH$_2$ groups [27].

Recently, Duchoslav et al. [28] developed and tested a new gas-phase derivatization protocol on different model polymers containing NH$_2$, OH, and a combination of both functional groups, for the determination of amine and hydroxyl functional groups. Besides the derivatization agent, TFAA, the new protocol includes for gas-phase derivatization a catalyst, pyridine, and tetrahydrofurane as a solvent. Besides the formation of the desired reaction products on the treated surface (amide from NH$_2$ and ester from OH group), the new protocol also dramatically increases the yield of the derivatization reaction for NH$_2$, reaching over 80%, maintaining at the same time the high yield of up to 90% for OH groups. Thanks to the derived BE differences in the C1s spectra between the CF$_3$ and amide or ester groups, an alternative way of qualitatively and quantitatively differentiating between the original NH_2 and OH groups on unknown surfaces is now at hand.

– Trifluoroethanol (TFE) derivatization reaction: The derivatization reaction of carboxyl groups with TFE and diterbutylcarbodiimide (Di-tBuC) [23, 29] was performed in vapor-phase thanks to the high volatility of Di-tBuC that minimizes nonspecific adsorption interactions previously observed using the liquid phase reaction of TFE and dicyclohexylcarbodiimide (DCC). Figure 2.10b shows the XPS C1s spectra of PAA before and after TFE derivatization [23]. Upon derivatization, the presence of peaks at 287.2 and 292.6 eV can be ascribed to COC and CF_3 species, respectively. Moreover, the $COC:CO_2:CF_3$ peak area ratio equal to 1.1:1.1:1.0 suggests the complete reactivity of the PAA acid groups. This finding was confirmed by the fluorine amount detected that was equal to 30 at%. Investigating a variety of polymers containing different oxygen functional groups, the selectivity of this reaction toward carboxyl groups has been found to be excellent.

Derivatization methods have been widely used on plasma-treated polymer surfaces to determine surface chemistry changes as a result of different plasma experimental conditions [30]. Indeed, the complete knowledge of the surface chemistry of the plasma treated films is particularly difficult due to their complex chemical nature, which is often multifunctional and displays a range of unsaturation degrees and crosslinking. In this respect, Manakhova et al. observed that the applicability of CD-XPS, usually reported for conventional polymers, can be more problematic for the analysis of highly crosslinked plasma polymers [31]. The authors highlighted the strong variation of fluorine content with the analyzed depth of trifluoromethyl benzaldehyde (TFBA) derivatized amine layers on plasma polymers. This inhomogeneous fluorine depth profile is most probably related to the low permeation of TFBA molecules inside the derivatized plasma polymers due to the large size of this probe molecule. Hence, the use of TFBA derivatization can result in underestimated densities of primary amines. On the other hand, the derivatization of anhydride groups with trifluoroethylamine (TFEA) seems to be more reliable due to the small size of the probe molecule that facilitates its permeation inside the layers. Therefore, in order to avoid heterogeneous and underestimated densities of functional groups, the CD-XPS on plasma polymers should be performed with smaller probe molecules such as TFEA. Another solution can be the utilization of angularly resolved XPS analysis combined with more detailed modeling of the fluorine depth profile in order to monitor the heterogeneity of the elemental composition at least at the outermost surface.

Thiol (SH)-terminated surfaces have been progressively gaining interest over the past years as a consequence of their widespread potential applications [32]. To quantify thiol concentrations, the chemical derivatization reaction with N-ethylmaleimide was exploited. N-ethylmaleimide reacts selectively with SH-groups via a nucleophilic addition between the S atom and the double bond in the maleimide structure, forming a stable thioether. The method was qualitatively and quantitatively validated on self-assembled monolayers of 3-mercaptopropyltrimethoxysilane exhibiting a SH-terminated group used as "model" surface [33].

Malitesta et al. [34] reported an interesting application of CD-XPS to the characterization of complex matrices such as polypyrroles (PPYs) electrosynthesized from aqueous solution. In these polymers, oxygen–carbon and nitrogen–carbon functionalities are contemporaneously present, as well as acidic, basic, charged, and reducible groups. Indeed, CD-XPS allowed to shed light about PPY structure, involving species either not fully resolved or present at low concentration levels, in particular C–OH, COOH, and C=O groups. The derivatizing agents titanium di-isopropoxide-bis (2,4-pentanedionate), TFE, and pentafluorophenylhydrazine were used to label the C–OH, COOH, and C=O groups, respectively. This study represents the first attempt to enlighten unknown structures of electrosynthesized polymers and a significant test for the applicability of CD-XPS to complex samples.

Recently, derivatization reactions have been also developed to quantify functional groups on different carbonaceous surfaces such as carbon nanotubes or carbon fibers [35, 36]. CD-XPS provides to gain information about the identity and concentration of oxygen-containing functional groups formed as a result of an oxidation process. In particular, the quantification of three major oxygen-containing functionalities, hydroxyl, carbonyl, and carboxyl groups can be performed. In three separate reactions, hydroxyl groups are labeled by reaction with TFAA, carbonyl by trifluoroethylhydrazine, and carboxyls via a coupling reaction of TFE and a carbodiimide. For a more sensitive reaction, carbonyl groups can be labeled by pentafluorophenyl hydrazine, which introduces five fluorine atoms per carbonyl group [36].

An efficient method for surface functional groups recognition involves labeling carboxyl, hydroxyl, and carbonyl species with fluorescent tags [37]. In many ways, fluorescent labeling is complementary to CD-XPS; the fluorescent tags can enable lower concentrations of oxygen functional groups to be detected. However, at higher oxygen functional group concentrations, the bulky fluorescent chromophores are more likely to encounter problems associated with steric hindrance, making quantitative analysis more difficult. Under these circumstances, CD-XPS approach is more effective.

2.7 Polymers surface segregation

As described in Chapter 1, surface dynamics of polymeric materials can be much more complex than those of other rigid materials (i.e., metals, ceramics, etc.), due to the high mobility of macromolecules on the surface. Indeed, polymer blends, copolymers, or additives in polymer systems can show the tendency to rearrange in the nano- or microdomains, and this feature is particularly evident on surface as a consequence of interfacial interactions between polymer and the external environment. In particular, for copolymers [38, 39], if there are two or more covalently linked components having different polarity and configuration, certain components could migrate to the surface to satisfy the thermodynamic requirement of minimizing the surface energy of the whole system. In general, components having lower surface energy always aim at enrich the surface: for instance, amphiphilic multiblock copolymers tend to rearrange so that the hydrophilic portion will dominate the interface in a polar hydrophilic medium, whereas the hydrophobic portion tend to dominate in air. Moreover, in going from a dry environment to a wet one, a structural rearrangement of the exposed surface occurs due to the different polarity of the components. In a few cases, if the segmental self-assembly is thermodynamically impervious to restructuring, a locking of the surface can be achieved [40, 41]. The extent to which phase separation occurs in copolymers to form supramolecular structures depends on these inherent critical features (1) compositional dissimilarity between blocks, i.e., difference in solubility parameters or chain interactions; (2) molecular weight of the segments, i.e. block length; (3) differences in crystallinity degree of the segments [42, 43]. The rearrangement of block copolymers can also depend on the way in which the materials are formed. For example, copolymer films obtained by casting using different solvents showed a dependence in segregation according to the solvent used.

Generally, the domain size for block copolymers is smaller than that of polymer blends, because the different blocks in a chain of block copolymers are connected by covalent bonds, whereas the polymer blends are a mixture of at least two polymers or copolymers. Polymer blends can exhibit miscibility or phase separation as well as various levels of mixing in between these extremes (i.e., partial miscibility). Therefore, a larger size scale of phase-separated domains can occur in polymer blends with respect to block copolymers. The domain dimensions of block copolymers could be in the nanometric scale. The periodicity of these morphological structures is determined by molecular weight and chemical composition in the block copolymers and typically is in the range of 10 to 100 nm. These nanoheterogeneous structures might generate many of the novel and useful mechanical and surface properties. In this respect, surface segregation of copolymers has been successfully exploited to design novel materials that reconstruct their surface in response to a change in the environment.

XPS technique can allow to experimentally reveal such effects in different copolymer systems: in detailed studies of some di- and tri-block copolymer systems [44–46], supplying depth profiles of block copolymers [43, 47–49], or in random copolymers [50, 51]. A typical example of surface segregation investigation by XPS concerns cast films in chloroform solution of polystyrene (PS)-poly(dimethylsiloxane) (PDMS)-based block copolymers, in AB di-block and ABA and BAB tri-block structures. PDMS surface segregation changes are also observed for some of the block copolymers after annealing the as-cast films. Because of the large differences in surface energy and the large mixing enthalpy of the two different blocks of PS and PDMS, PDMS blocks prefer to migrate to the surface, forming a surface PDMS microdomain and leaving PS blocks in the outermost layer [52]. Surface composition of these amorphous-semicrystalline block copolymers depends also on the way these films are prepared. Mixed solvents can be used to control the competition between crystallization and microphase separation. Blending a small amount of a block copolymer (AB), where the B blocks have a substantially lower surface energy than the A blocks, into a homopolymer (A) has also been demonstrated as an effective way to change the surface composition and lower the surface tension of the polymer blends [53]. The block copolymer added in small amounts acts as a surfactant and segregates to the surface region. The interesting feature is that only the surface properties are changed, whereas the bulk ones remain the same as the homopolymer.

In this respect, a fundamental approach to monitor how these block copolymers change the surface of the homopolymers is represented by AR-XPS, which supplies the composition profile of the blends within the technique sampling depth. Govers et al. [49] investigated block copolymers containing polydimethylsiloxane (PDMS) and poly(ethylene oxide) (PEO) or poly(propylene oxide) (PPO) with varying molar masses. The functional homopolymer blocks and final diblock copolymers were characterized using proton nuclear magnetic resonance (1 H NMR) and matrix assisted laser desorption/ionization time of flight mass spectroscopy (MALDI-ToF-MS). These polymers were then incorporated in an industrially relevant solvent-borne coating formulation. XPS and a combination of angle-resolved and depth profiling measurements were used to obtain concentration profiles of the block copolymer in the top few nanometers of the coating. These amphiphilic molecules were found to be extremely surface active, and high levels of PDMS enrichment of the coating surface were observed at only minimal concentrations. The extent of segregation is sensitive to the exact mass of both the siloxane and polyether block, where an increase in the size of either part resulted in an overall decrease in surface enrichment. PDMS-PPO was found to be more compatible with the coating network than PDMS-PEO, as evidenced by the substantial lower surface enrichment of the former.

Another class of polymers that can undergo surface rearrangement is represented by polyurethane (PU)-based copolymers because they usually contain hard moieties, consisting of a diisocyanate chain extended with a low-molecular-weight diol or diamine, and soft moieties, which usually are long-chain macrodiols. The hard segments have a high density of urethane groups of high polarity, and for this reason, they are rigid at room temperature, whereas soft segments are flexible at room temperature. The polarity of hard segments produces a strong attraction between them, which causes a high degree of aggregation, and order in this phase, forming crystalline or pseudo-crystalline areas located in a soft and flexible matrix.

They can be synthesized as di-, tri-, or multi-block copolymers, i.e., copolymers where the different macrodiols are polymerized, giving homopolymer subunits and successively linked by junction units (diisocyanate), or random copolymers, i.e., co-polymers where the macrodiols were polymerized together, so the sequence of mac-rodiols residues follows a statistical rule. In both cases, the presence of surface phase separation of the hard and soft segments in PU is very common.

Different authors studied the surface segregation phenomenon of various PU-based copolymers using XPS because this technique, both in conventional and AR modes, can also predict the in vivo performance of multiblock copolymer-based bio-materials. In particular, Cometa et al. [50] carried out XPS measurements to charac-terize the surface chemistry of polycaprolactone (PCL)-polyethyleneglycol (PEG) hydro-biodegradable copolymers having different PEG/PCL ratio and different molec-ular weights, evidencing surface segregation in these common multiphase amphi-philic polymer systems, as also described by different authors [54–56]. Moreover, using three different casting solvents (in terms of polarity and boiling point), XPS analysis evidenced that surface composition was enriched by PCL when the cast-ing solvent was more hydrophobic and/or it evaporated slowly. Therefore, the air-polymer interface was found to be different from the glass-polymer one: when the solution was cast on a Petri dish, the copolymer systems assembled according to the high hydrophilic glass properties (Fig. 2.11).

Finally, the surface segregation of PCL-PEG copolymers, substantiated by a deep XPS characterization, provided a satisfactory explanation of the biological re-sponse to the investigated systems: in vitro cell adhesion tests have indeed shown no dependence on the PCL/PEG nominal ratio, whereas different molecular weights elicited different cell adhesion. Similarly, Martinelli et al. [57] studied (meth)acrylic terpolymers carrying siloxane (Si), fluoroalkyl (F), and ethoxylated (EG) side chains with comparable molar compositions and different lengths of the Si and EG side chains, while the length of the fluorinated side chain was kept constant. Such ter-polymers were used as surface-active modifiers of polydimethylsiloxane (PDMS)-based films. The surface chemical compositions were determined by AR-XPS. The terpolymer was effectively segregated to the polymer–air interface of the films inde-pendent of the length of the constituent side chains. The exceptionally high enrich-ment in F chains at the surface caused the accumulation of EG chains at the surface as well. The response of the films to the water environment was also proven to strictly depend on the type of terpolymer contained. While terpolymers with shorter EG chains appeared not to be affected by immersion in water for seven days, those containing longer EG chains underwent a massive surface reconstruction.

Another example of XPS investigation on the surface segregation phenomenon in PU is that of Vaidya and Chaudhury [58]. Segmented polyurethanes (SPU), contain-ing perfluoropolyether (PFPE), polydimethylsiloxane (PDMS), and PEG segments, ex-hibit different responses to the polarity of the contacting medium, having oleophobic, hydrophobic, and hydrophilic domains. The oleophobic and hydrophobic properties of the SPU are due to the segregation of PFPE segments at the polymer-air interface. In

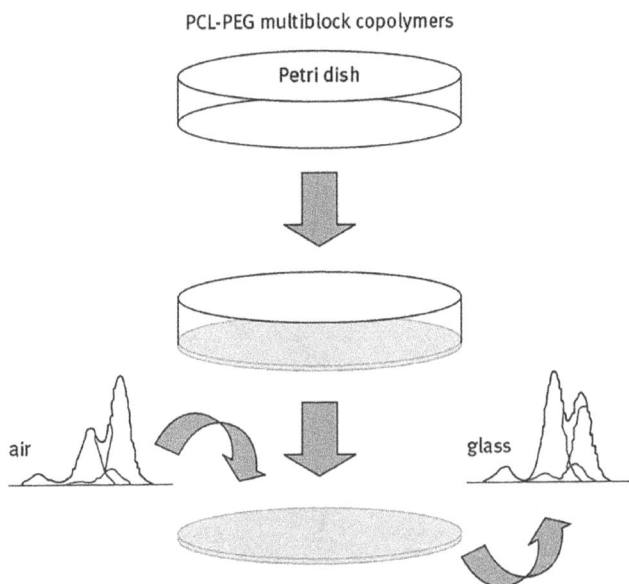

Fig. 2.11: XPS analysis of both air- and glass-exposed sides of PCL-PEG multiblock copolymer films obtained by solvent casting on Petri dishes. The reported spectra refer to C1s regions. Reprinted with permission from Cometa S, Chiellini F, Bartolozzi I, Chiellini E, De Giglio E, Sabbatini L, *Macromol Biosci*, 10, 317–27, 2010. Copyright 2010 Wiley-VCH Verlag GmbH & Co. KGaA, Weinheim [50].

this respect, XPS provides the ability to rewrite the block positioning in the air–polymer interface, as depicted in Fig. 2.12.

In Fig. 2.12a, the C1s region at a 90° take-off angle in SPU containing 0.86% w/w PFPE film is reported. In this figure, the contributions relevant to the C–H groups, due to the aromatic ring and PDMS segments, and to the C–O bonding present in PEG segments can be recognized, as well as the three peaks corresponding to carbon atoms bonded to fluorine (labeled as CF). To maximize the signal obtained from the surface-enriched segments, detailed spectra were also taken at a 15° take-off angle (Fig. 2.12b), showing that the intensity of CF peaks increases and the relative intensity of the CO peak decreases, compared with those taken at a 90° take-off angle. This indicates that the SPU surface is enriched with PFPE segments, whereas the PEG segments are present close to the surface. These findings suggested that the PFPE segments readily segregate at the polymer-air interface, thus reducing the surface energy of the polymer to very low values. During its segregation, the PFPE groups drag the hydrophilic PEG groups close to the surface (most likely in the subsurface region).

Copolymers containing fluorinated moieties are a typical class where surface segregation occurs. For instance, Wu et al. [59] investigated a series of amphiphilic diblock polymers of poly(hydroxyethylacrylamide)-b-poly(1H,1H-pentafluoropropyl

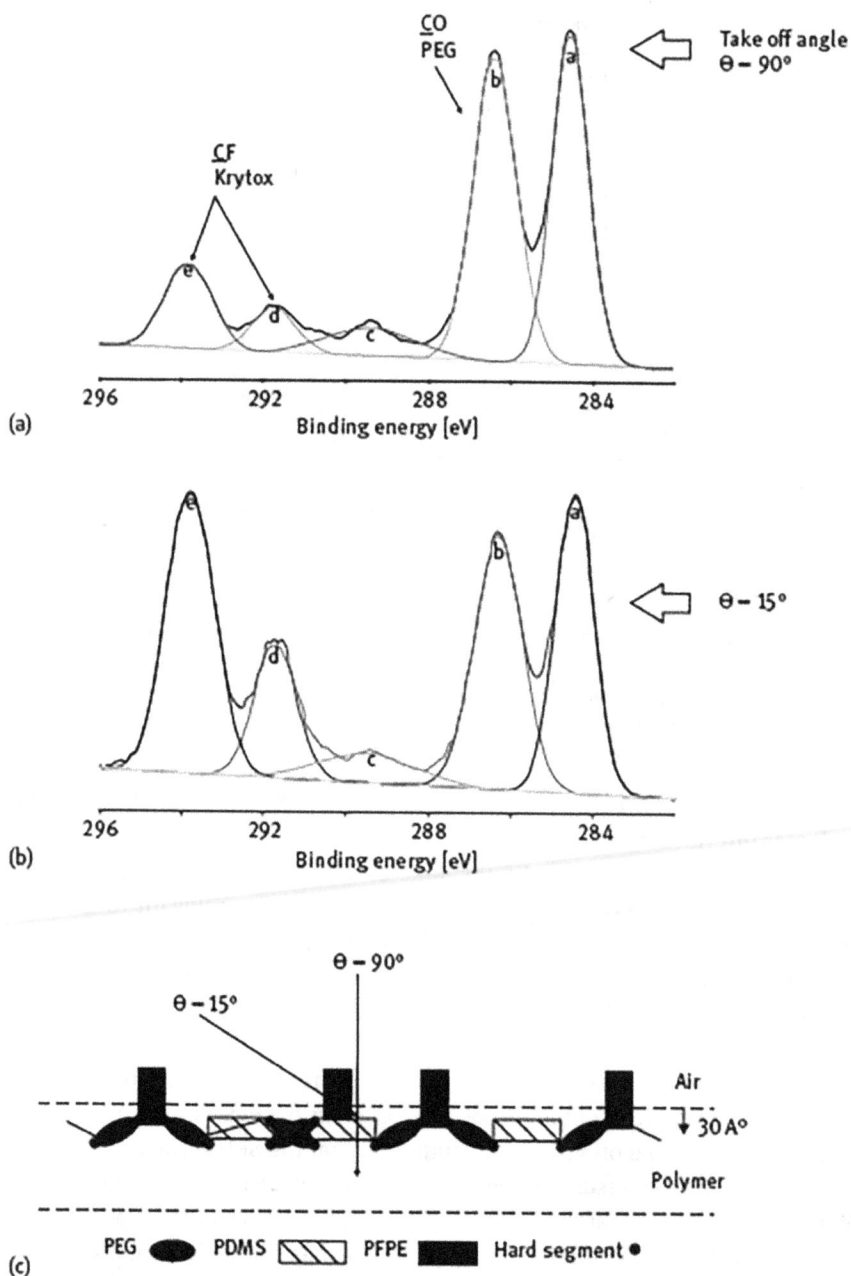

Fig. 2.12: High-resolution C(1s) regions at take-off angles (a) 90° and (b) 15° in the case of PFPE-modified SPU films. (c) Schematics of the structure of PFPE-modified SPU at polymer-air interface. Reprinted from Vaidya A, Chaudhury MK, *J Colloid Interface Sci*, 249, 235–45, 2002. Copyright 2002, with permission from Elsevier [58].

methacrylate) (PHEAA-b-PFMA) grafted from silicon wafer via surface-initiated atom transfer radical polymerization (SI-ATRP). The results show that only when the grafting density and thickness of PHEAA brush were in the range of 0.9–1.3 (chain/nm^2) and 6.6–15.1 nm, respectively, and the ratio of PFMA/PHEAA varied from 89/42 to 89/94, could the diblock copolymer phase separate into nanostructures. Moreover, the study of the antiprotein adsorption performance of the modified surfaces against BSA, fibrinogen, and lysozyme provide further evidence that surface composition and microphase segregation of fluorinated moieties of block copolymer brushes significantly impact protein adsorption behaviors.

Surface segregation is a common phenomenon deeply studied for the development of membranes with different applications [46, 60–62]. In this respect, Xix-Rodriguez et al. obtained hemocompatible membranes by blending polylactic acid (PLA) with three different amphiphilic block copolymers [60]. The copolymers segregation to the membrane surface was confirmed by XPS. All membranes exhibited hemocompatibility and non-hemolyzing properties and the proper combination of amphiphilic copolymer with PLA will allow fine-tuning the middle-molecule toxin separation with high retention of large desirable proteins through the membranes. Jiang et al. provided a new method for antifouling modification of gravity-driven membranes by a novel tailored amphiphilic multi-arms polymer poly(propylene glycol)-silane-poly(ethylene glycol) (PPG-Si-PEG) synthesized and blended into polyvinylidene fluoride (PVDF) matrix [61]. The incorporation of multi-arms polymers affected phase separation and surface segregation behaviors, resulting in the formation of more porous structure. Moreover, the hydrophilic PEG arms in polymers were enriched onto membrane surface through surface segregation and utilized to improve the hydrophilicity of membranes, as confirmed by XPS, FTIR and contact angle measurements. Sun et al. studied a novel mussel-inspired sticky catechol-functionalized poly (ethylene glycol) (Cate-PEG) synthesized and deployed as an additive to modify the hydrophobic polyvinylidene fluoride (PVDF) ultrafiltration membrane for reducing the leakage of PEG from membrane matrix toward practical water treatment applications [62]. By the interesting surface segregation, the sticky Cate-PEG polymer could migrate from matrix onto the membrane surface and internal pores, endowing the modified membrane with excellent hydrophilicity.

2.8 Polymers physical treatments/grafting

Polymer treatment by plasma represents one of the most common strategies to improve polymer performances [63]. Plasmas are environmentally friendly and fast, contrary to traditional wet chemistry. Plasmas are very complex media full of highly reactive particles [radicals, ions, electrons, ultraviolet (UV) photons], all of which reach the polymer surface. Because of this complexity, the mechanisms involved in the surface modification of polymer are not yet fully understood. In view of this, CD-XPS provides a successful approach to elucidate the surface structure of plasma-treated polymers [64]. Low- and high-pressure plasma have been used empirically for the treatment of surfaces for many decades, but the plasma treatment of surfaces really started to gain importance with the development of commercially available surface analysis techniques; in particular, XPS and water contact angle (WCA). Plasma treatment of polymers results mainly in two mechanisms: ablation of the polymer and/or grafting of new species on its surface. The two reactions are taking place at the same time, but depending on the experimental conditions, one can prevail on the other. It is worth noting that plasmas modify only the surface, while maintaining unaltered the bulk properties of the polymers. The most common use is the grafting of new functionalities on the sample surfaces. In terms of industrial applications, the most important reason for the use of plasma is undoubtedly the increase of the adhesion properties of a polymer, i.e., to obtain composite fibers with enhanced tensile strength and stability [65–69]. Indeed, plasma treatments improve the very poor adhesion properties of a polymer, increasing its surface energy by grafting polar species on its surface. In a similar way, plasma treatments are efficiently used to properly modify other critical surface properties such as wettability and biocompatibility. It is worth mentioning that plasma technology can be used for direct polymerization, which allows to deposit polymers on a very large number of substrates, even another polymer. In this case, a precursor is sent into the plasma, and the reactive species from the plasma will activate the precursor molecules that will react with each other and polymerize. A very large variety of polymers can be created by this technique, ranging from polyacrylates [70–72] to diamond-like carbon [73], polyaniline [74] to PS [75] and polytetrafluoroethylene (PTFE) [76] films. They present several advantages over traditional polymers. For instance, these films usually exhibiting a higher degree of cross-linking, therefore, have superior mechanical properties that allow to deposit ultrathin films that present good adhesion on various substrates. XPS analysis gives not only information about the atomic composition of the films but also about the degradation of the monomer in the plasma and therefore provides information about the mechanism of the reaction. Weibel et al. [77] reports a study on the surface chemical changes produced on PU membranes after acrylic acid (AA) vapor plasma treatment XPS. In all the experimental conditions studied, AA plasma treatment increased the concentrations of carbonyl and ester groups at the PU surface compared with traditional oxygen plasma modification. XPS C1s spectra, reported in Fig. 2.13, showed that a thin film with similar chemical properties to the poly(acrylic acid) (PAA) was formed on top of the PU membrane after 5-W AA plasma treatments.

The UV-ozone (UVO) treatment relies upon the combined effects of UV light and ozone, produced in situ from a gas-phase photodissociation of molecular oxygen. UV and ozone photooxidation were first recognized as a potential polymer surface treatment in the early 1980s. This treatment, initially considered as a surface

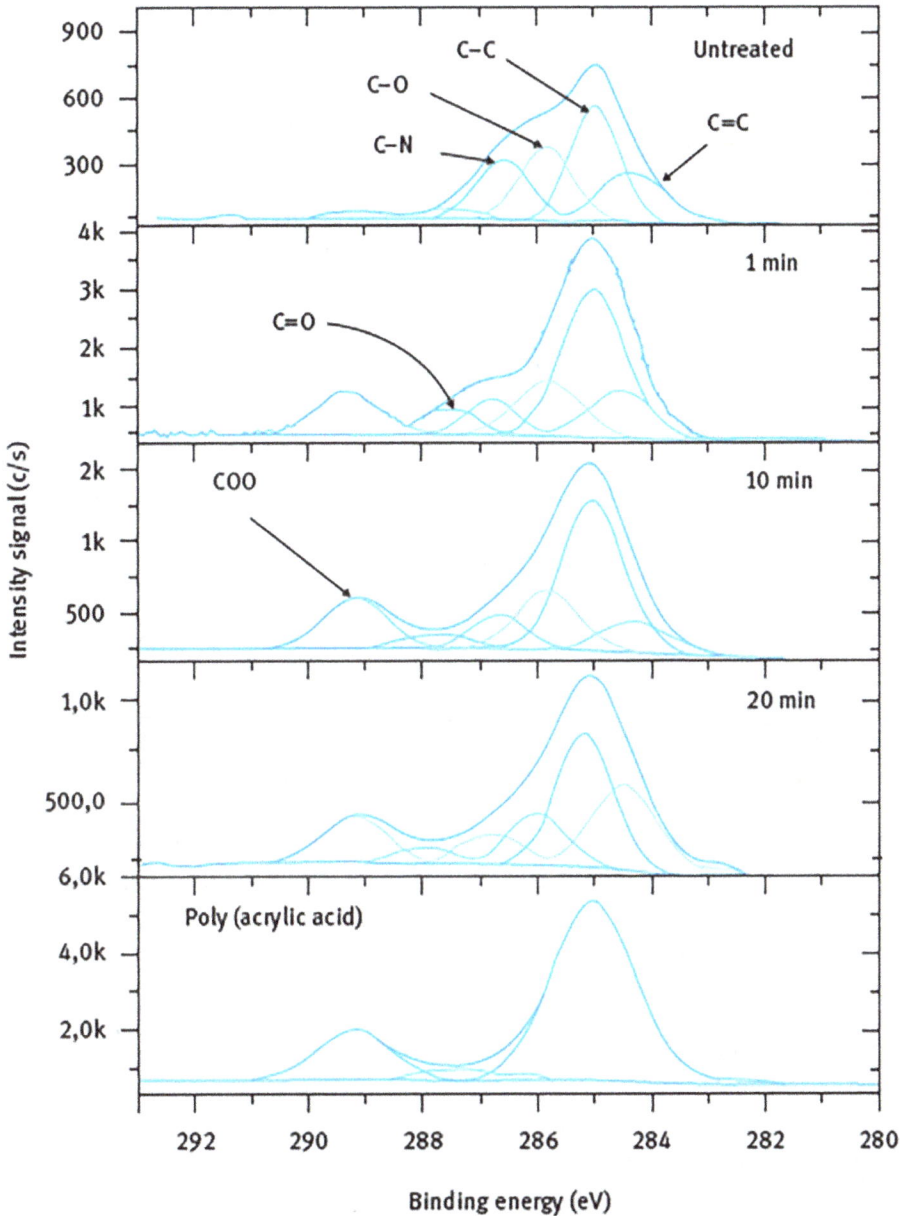

Fig. 2.13: High-resolution XPS spectra of the C1s peak of an untreated PU membrane, AA plasma-treated PUs at several plasma treatment times, and synthesized PAA. Plasma power: 5 W. The $C = O$ and COO groups are shown in different spectra for better presentation. Reprinted from Weibel DE, Vilani C, Habert AC, Achete CA, *J Membr Sci*, 293, 124–32, 2007. Copyright 2007, with permission from Elsevier [77].

cleaning method [78], has been recently demonstrated to modify the surface chemistry and improve the wetting characteristics of natural and synthetic polymers. UVO treatment causes the surface energy of the polymers to increase through the breaking of the polymer chain by the insertion of oxygen-containing functional groups. Tsui and co-workers exploited UVO treatment to alter the viscosity of PS films supported on silica [79]. XPS analyses revealed that surface oxygen levels on UVO-treated polymer surfaces increase quite markedly after relatively short exposure times (<5 min) [80]. In particular, such treatment can be useful in a range of technological applications to increase the surface energy of many organic polymers for the substrate to perform its desired task (i.e., printability, wettability, biocompatibility, etc.). As an example, wood-plastic composites (WPC), widely exploited in building and furniture industry for their high outdoor resistance, are often difficult to coat and paint. In this respect, Yáñez-Pacios et al. exposed WPC composites to UVO treatment, enhancing their surface energy, polarity, roughness and adhesion properties [81]. A principal advantage of the UVO method is that it can be applied under ambient conditions as a continuous treatment method, on three-dimensional substrates and in small batches with a very high degree of control. The individual and combined effects of UV light and ozone have been studied for a variety of surfaces, including polyethylene, polypropylene, PET, poly(ether ether ketone), poly(PEGMA):polysulfone, poly (3,4-ethylenedioxythiophene)polystyrene sulfonate (PEDOT:PSS) and PS. Rafique and co-workers proposed an UVO-treated graphene oxide/polymeric system (GO/PEDOT:PSS) as a hole transport layer for organic solar cells [82]. The UVO-treatment enhanced the contact conditions between the hole transport layer and the photoactive layer, leading to highly efficient and stable devices. Optical, morphological and physicochemical features were studied by means of UV-vis spectroscopy, AFM and XPS. The latter highlighted the surface chemical changes occurred after UVO treatment. A substantial increase in hydroxyl and epoxy groups, as well as of carboxylated moieties was detected. Similarly, Lee et al. observed an increase of surface oxygen content on polysulfone substrates, UVO-treated to graft poly(PEGMA) brushes with high selectivity for CO_2 [83]. Indeed, XPS analyses of the poly(PEGMA)/polysulfone membranes revealed an enhancement of oxygen content, from 13.12% of polysulfone substrate to 29.08% of the membrane. Furthermore, O1s spectrum deconvolution was performed to check the UVO-induced grafting. Two types of functional groups, i.e., S=O/C=O (530.6 eV) and C–O (531.6 eV) were present on the bare polysulfone substrate, while the relative intensity of the C–O moiety became remarkable after UVO exposure, indicating the grafting of poly(PEGMA).

The ion beam co-sputtering (IBS) technique can be successfully used to deposit different polymeric and inorganic materials in an extremely controlled manner by selecting appropriate targets and adjusting sputtering deposition conditions. In this context, the incorporation of metal nanoclusters into polymeric matrices is a research topic of considerable interest because the metallic species can dramatically influence the chemical and physical properties of the resultant composite material. The high research and industrial interest in metal nanoparticle (NP)/fluoropolymer composites (metal-CF_x) is due to their wide range of applications, for example, in microelectronics, nonlinear optics, optical recording, and in the development of biocompatible coatings. Metal-fluoropolymer ($CF_x(M)$)-nanostructured composites have been successfully fabricated using technologies such as ion co-sputtering [84, 85]. The latter is particularly interesting, being a cost-effective and environment-friendly technology, also allowing a high degree of control over important deposition parameters such as the sputtering energy and the metal loading. In this respect, Sachdev et al. described a ferromagnetic nanocomposite film prepared by IBS, embedding cobalt nanoparticles in a PVA substrate [86]. The polymeric matrix acted as seeding agent for the growth of hexagonal closed pack metallic nanoparticles, while preventing their environmental degradation. XPS analyses showed a single-peak ascribable to metallic cobalt, demonstrating that the latter was successfully embedded inside PVA. Cioffi et al. [87] reported a systematic spectroscopic and morphological characterization of gold-fluoropolymer ($CF_x(Au)$) nanocomposites, deposited by ion beam co-sputtering. These composites are technologically relevant because they exhibit vapor-sensing properties based on the swelling phenomenon. XPS analysis assessed that gold codeposition induces a progressive defluorination of the polymeric chains, leading to a significant increase in polymer chain branching and in the concentration of unsaturated fluorinated carbons. Meanwhile, angle-resolved XPS indicated a nonhomogeneous in-depth distribution of the gold particles, with the outer surface being less rich in NPs. Typical C1s spectra relative to gold/fluoropolymer composite are reported in Fig. 2.14 as a function of the metal loading; six photoelectron components and a shake-up weak shoulder have been used to fit the signal. The attribution of the different species has been reported in the following. Starting from the left side of the figure, a weak shake-up signal (BE, ~ 296.0 ± 0.3 eV) can be seen, due to fluorinated-unsaturated groups, whose intensity slightly increases with Φ. The peak falling at 293.8 ± 0.2 eV (293.3 ± 0.1 eV for $\Phi = 0.25$) is attributed to CF3 chain terminations; CF2 groups are responsible for the third peak (BE = 291.8 ± 0.3 eV for $\Phi < 0.15$; BE = 291.4 ± 0.2 eV for $\Phi = 0.25$); the relative intensity of this chemical environment decreases as Φ increases. The spectral region falling at lower BE values comprises three peaks whose attribution is not unequivocal because of an overlap of signals from fluorinated and oxygenated species. A general inspection of the C1s spectra indicates that at higher Φ values ($\Phi = 0.15$ and $\Phi = 0.30$), the structure of the fluoropolymer matrix itself dramatically changes: there is a very high degree of branching and the reduced abundance of strongly electron-attractive fluorine substituents leads to an appreciable decrease of the chemical shift of some photoelectron peaks. At low metal content ($\Phi < 0.10$), the typical C1s spectrum of a fluoropolymer with a low degree of branching and low abundance of oxygenated species can be recorded.

In literature, **annealing treatments** of polymers have been suggested to produce more uniform and less porous films [88, 89]. This phenomenon likely originates from a viscous flow in the polymer chains leading to smoother and more compact polymer

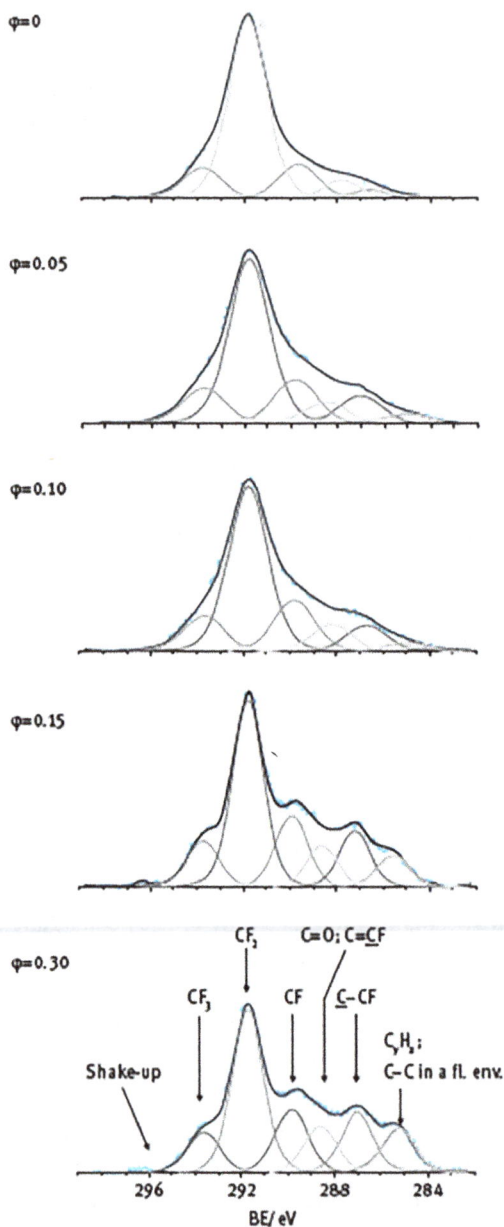

Fig. 2.14: C1s XPS region of Au–CF$_x$ nanocomposites, as a function of the different gold loading Φ. Reprinted with permission from Cioffi N, Losito I, Torsi L, *et al.*, *Chem Mater*, 14, 804–11, 2002. Copyright 2002 American Chemical Society [87].

films. This finding also clarifies why the polymer thickness decreases after annealing. This process was recently exploited to prepare n-type silicon transistors with enhanced electron mobility and reduced voltage threshold [90]. First, a povidone-silica nanocomposite was deposited on n-type Si (100) substrate and then it afforded thermal treatment at 423 K. Deconvolution of Si2p and C1s photoemission spectra demonstrated that the annealing process led to strong siloxane bonds, replacing silanol hydrogen bonds observed on the unannealed substrates. Usually, annealing is performed at a temperature greater than the polymer glass transition temperature (Tg). De Giglio et al. [91] reported a study in which an annealing treatment was applied to polymethyl methacrylate (PMMA) to improve the performance of the polymer coating as barrier films against corrosion of titanium based orthopedic implants. PMMA coatings were annealed at 200 °C in air for 10 min, and XPS was exploited to ascertain possible changes in polymer surface chemistry after the annealing treatment. A comparison between C1s spectra of annealed and unannealed PMMA films showed that no alteration of surface chemical composition was introduced by the annealing treatment. Moreover, SEM investigation showed that the morphology of the annealed films was improved with respect to uniformity and porosity. The final results of this investigation indicated that the annealing process produces coatings with considerable anticorrosion performances.

The **covalent grafting** of target molecules on polymer surfaces is a widely used strategy to address the polymer response toward specific applications. In this context, XPS technique represents a useful tool to characterize step by step the polymer grafting reactions involved. For example, the grafting of bioactive molecules on a polymer surface is one of the most used strategies in the biomaterials research field to improve the biological response of the investigated system. In this context, the covalent grafting of peptides containing the Arg-Gly-Asp (RGD) minimal recognition sequence has been attempted onto different polymer substrates to promote cell adhesion, thanks to the formation of cell membrane integrin-RGD bonds. In these investigations, XPS proved to be an efficient tool to assess the presence and availability of the peptide on the polymer surface [92–94]. Another intriguing example of covalent grafting of active moieties has been reported by Huang et al., who described the functionalization of silica microspheres combining mussel-inspired chemistry and Kabachnik–Fields (KF) reaction [95]. The mussel-inspired layer enabled to achieve a dopamine self-polymerized coating, rich in catechols, amines and aromatic groups. The latter were exploited as platforms to further graft phosphate groups onto the SiO_2 microspheres by KF reaction. TGA, FTIR, and XPS indicated that the polymeric coating was successfully attached on the surface of the microspheres. Furthermore, interesting results have been reported in a study in which an annealing treatment and a grafting reaction have been both applied to improve the biological performances of a PAA coating electrosynthesized on titanium substrates [96]. The annealing treatment provided optimal anti-corrosion performances to the PAA coating while preserving the surface polymer composition, as demonstrated by XPS analyses performed before and after the annealing treatment.

Meanwhile, the presence of carboxylic moieties on the polymer film surface allows the grafting of biomolecules promoting positive interface reactions with the surrounding biological system. As an example, a simple model compound, the amino acid residue 4-fluorophenylalanine, was grafted onto the annealed PAA: the fluorine target atom enabled a simple monitoring of the reaction effectiveness as well as of the reaction yield by estimating the peak area of the F1s XPS signal.

2.9 Polymers aging

As it happens to all materials, interaction with the environment induces, with time, chemical modifications on polymer surfaces, which are referred to as the aging phenomenon.

It must be emphasized, however, that aging is particularly prominent and often remarkably accelerated in those polymers or polymer coatings obtained by plasma treatments. One commonly observed phenomenon, for example, is the deterioration of the beneficial surface properties (such as hydrophilicity) upon storage in air. The problem is complex because of the variety of processes that may be involved, e.g. the diffusion and reaction of free radicals created by exposure to plasma, the adsorption of polar and nonpolar contaminants, and the reorientation of polar groups away from the surface [97–101]. As a matter of fact, although the use of plasmas for the modification of polymer surfaces currently represents a mature technology, research is still needed to forecast the evolution of the treated surfaces with time.

Vandencasteele et al. [63] demonstrated, by contact angle measurements, that the storage in water of a PTFE surface, treated by a CO_2-plasma, causes a significant improvement in wettability. Meanwhile, when the samples are stored in a hydrophobic environment such as air, a hydrophobic recovery occurs, which is not happening when the samples are stored in a hydrophilic environment such as water [102].

Touzin et al. [103] carried out a study, during a period of 4 weeks, on the stability of a plasma-polymerized coating deposited on 316 L stainless steel cardiovascular stents. The deposition of an ultrathin (from 10 to 100 nm) plasma-polymerized fluorocarbon coating (Teflon-like) has been proposed to improve the long-term performance and safety of these devices (i.e., corrosion prevention). The stability of the plasma-polymerized coating has been investigated using deionized (DI) water as a simple aging medium. After different periods of immersion in DI water, the characterization of the morphology and chemistry of the coating was performed. In particular, a constant decrease in the fluorine content and an increase in the oxygen and carbon contents were observed. The XPS F1s/C1s area ratio decreases gradually, and this trend was in agreement with the decreasing of contact angle values. In addition, an increase in the O/C ratio is observed during aging. The loss in fluorine is then compensated by an increase in the concentration of oxygen. A rapid incorporation of oxygen between the second and the third week of aging, associated with an oxidation of the coating surface, was detected. In Fig. 2.15, the high-resolution XPS spectra of C(1s) regions relevant to plasma polymer coatings before and after their immersion in DI water at different periods are reported. Before immersion, negligible oxygen amount was detected on the coating surface; therefore, the C1s spectrum of the coating was curve fitted with five spectral components assigned to C–H or –C–C– (285 eV), –C–CF (286.5 eV), –C–CF (288.8 eV), –CF$_2$ (291.4 eV), and –CF$_3$ (293.8 eV) groups.

After immersion, the C1s spectrum was still fitted using five components, but the increase of the signal intensity in the low binding energies (and the concomitant increase of oxygen peak intensity) suggests that oxygen components such as C–O (286.4–287 eV) and C=O (288–288.4 eV), overlapping the –C–CF and –C–CF contributions, should be considered. Before immersion, the coating mainly consists of CF$_x$ groups with a significant percentage of CF$_3$ end groups, leading to a hydrophobic

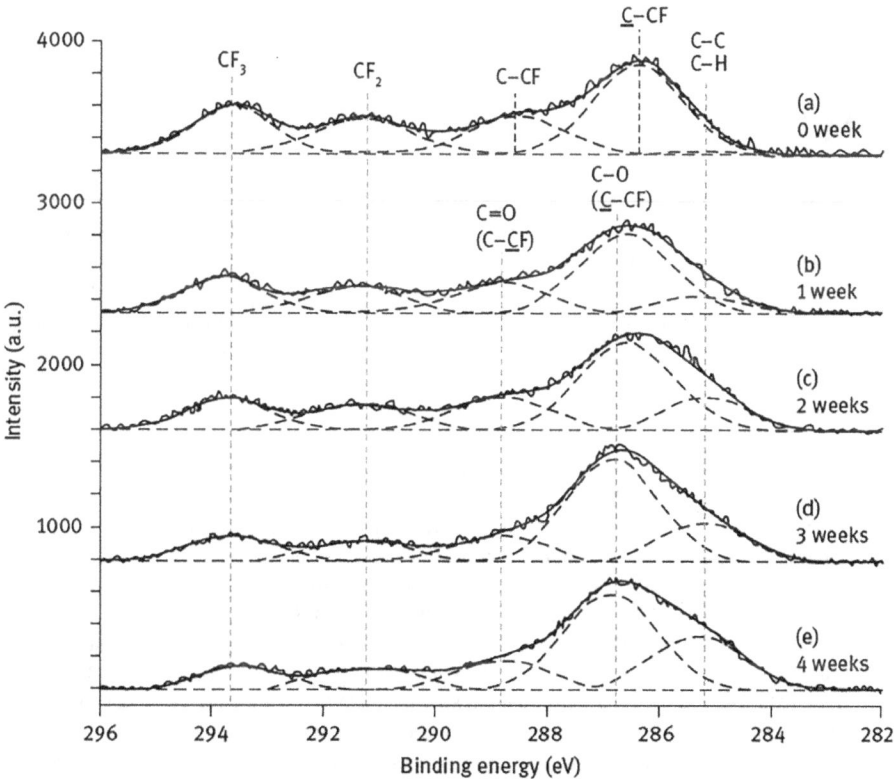

Fig. 2.15: High-resolution C1s X-ray photoelectron spectra of plasma polymer coatings: (a) before, (b) after 1 week, (c) after 2 weeks, (d) after 3 weeks, and (e) after 4 weeks of immersion in DI water. Reprinted from Touzin M, Chevallier P, Lewis F, et al., *Surf Coat Technol*, 202, 4884–91, 2008. Copyright 2008, with permission from Elsevier [103].

coating; defluorination and oxygen incorporation upon aging causes a loss of hydrophobicity.

An aging phenomenon was also observed by Tran et al. [104] when a plasma polymerized fluoropolymer coating was exposed under humid and UV conditions for up to 105 h to test its stability.

Yun et al. investigated air stability of polymeric systems employed in electronic industry [105]. Despite their excellent electrical and mechanical properties, practical application of poly(3,4-ethylenedioxythiophene) polymerized with poly(4-styrenesulfonate) (PEDOT:PSS) face the challenge of ensuring air stability in electronic industry. Degradation mechanism of PEDOT:PSS-based films in air is clearly demonstrated through X-ray/ultraviolet photoelectron spectroscopy (XPS/UPS) and its depth-profiling technique. As the duration of air exposure increases, the PEDOT:PSS-based films alter molecular structures with the formation of $S-O_x$ bond in PEDOT and $C-N_x$ bond growth, changing ratios of insulating part (PSS–, PSSH, oxidized PEDOT) to conducting part and deteriorating their electrical conductivities.

References

[1]   Briggs D. New developments in polymer surface analysis. Polymer 1984, 25, 1379–1391.
[2]   Batich CD. Chemical derivatization and surface analysis. Appl Surf Sci 1988, 32, 57–73.
[3]   Merrett K, Cornelius RM, McClung WG, Unsworth LD, Sheardown H. Surface analysis methods for characterizing polymeric biomaterials. J Biomater Sci-Polym E 2002, 13, 593–621.
[4]   Ha C-S, Gardella JA Jr.. X-ray photoelectron spectroscopy studies on the surface segregation in poly(dimethylsiloxane) containing block copolymers. J Macromol Sci, Part C – Polym Rev 2005, 45, 1–18.
[5]   Vandencasteele N, Reniers F. Plasma-modified polymer surfaces: Characterization using XPS. J Electron Spectrosc 2010, 178–9, 394–408.
[6]   Briggs D, Seah MP. Practical Surface Analysis, Auger and X-ray Photoelectron Spectroscopy, Chichester (UK), Wiley, 1990.
[7]   Siegbahn K, Nordling C, Fahlman A et al. ESCA: Atomic Molecular and Solid State Structure Studied by Means of Electron Spectroscopy, Uppsala (Sweden), Almquist & Wiksells, 1967.
[8]   Salmeron M, Schlögl R. Ambient pressure photoelectron spectroscopy: A new tool for surface science and nanotechnology. Surf Sci Rep 2008, 63, 169–199.
[9]   Zhong L, Chen D, Zafeiratos S. A mini review of in situ near-ambient pressure XPS studies on non-noble, late transition metal catalysts. Catalysis Sci & Tech 2019, 9(15), 3851–3867.
[10] Amati M, Gregoratti L, Zeller P, Greiner M, Scardamaglia M, Junker B, Ruß T, Weimar U, Barsan N, Favaro M, Alharbi A, Jensen IJT, Ali A, Belle BD. Near ambient pressure photoelectron spectro-microscopy: From gas–solid interface to operando devices. J Phys D: Appl Phys 2021, 54, 204004.
[11] Jain V, Bahr S, Dietrich P, Meyer M, Thißen A, Linford MR. Polytetrafluoroethylene, by near-ambient pressure XPS. Surf Sci Spec 2019, 26, 014028.
[12] Arble C, Jia M, Newberg JT. Lab-based ambient pressure X-ray photoelectron spectroscopy from past to present. Surf Sci Rep 2018, 73, 37–57.

[13] Tougaard S, Jansson C. Comparison of validity and consistency of methods for quantitative XPS peak analysis. Surf Interface Anal 1993, 20, 1013–1046.

[14] Wagner CD, Davis LE, Zeller MV, Taylor JA, Raymond RH, Gale LH. Empirical sensitivity factors for quantitative analysis by electron spectroscopy for chemical analysis. Surf Interface Anal 1981, 3, 211–225.

[15] Briggs D. Surface Analysis of Polymers by XPS and Static SIMS, Cambridge (UK), Cambridge University Press, 1998.

[16] Sherwood PMA. Data analysis in X-ray photoelectron spectroscopy. In: Briggs D, Seah MP, eds. 2nd, Chichester (UK), Wiley, 1990.

[17] Moulder JF, Stickle WF, Sobol PE, Bomben KD. Handbook of X-ray Photoelectron Spectroscopy, Eden Prairie (MN), Perkin Elmer, 1992.

[18] Windawi H, Ho F. Applied Electronic Spectroscopy for Chemical Analysis, New York (NY), Wiley-Interscience, 1982.

[19] Baker AD, Brundle CR. Electron Spectroscopy: Theory, Techniques and Applications, Vol. IV, London (UK), Academic Press, 1981.

[20] Ratner BD, Castner DG. Advances in X-ray photoelectron spectroscopy instrumentation and methodology: Instrument evaluation and new techniques with special reference to biomedical studies. Colloid Surf B 1994, 2, 333–346.

[21] Ruiz JC, Girard-Lauriault PL, Wertheimer MR. Fabrication, characterization, and comparison of oxygen-rich organic films deposited by plasma- and vacuum-ultraviolet (VUV) photo-polymerization. Plasma Processes Polym 2015, 12(3), 225–236.

[22] Pochan JM, Gerenser LJ, Elman JF. An e.s.c.a. study of the gas-phase derivatization of poly (ethylene terephthalate) treated by dry-air and dry-nitrogen corona discharge. Polymer 1986, 27, 1058–1062.

[23] Chilkoti A, Ratner BD, Briggs D. Plasma-deposited polymeric films prepared from carbonyl-containing volatile precursors: XPS chemical derivatization and static SIMS surface characterization. Chem Mater 1991, 3, 51–61.

[24] Chilkoti A, Ratner BD. An X-ray photoelectron spectroscopic investigation of the selectivity of hydroxyl derivatization reactions. Surf Interface Anal 1991, 17, 567–574.

[25] Nietzold C, Dietrich PM, Ivanov-Pankov S, Lippitz A, Gross T, Weigelb W, Ungera WES. Functional group quantification on epoxy surfaces by chemical derivatization (CD)-XPS. Surf Interface Anal 2014, 46, 668–672.

[26] Choukourova A, Kousala J, Slavìnskà D, Biedermana H, Fuocob ER, Tepavcevicb S, Saucedob J, Hanley L. Growth of primary and secondary amine films from polyatomic ion deposition. Vacuum 2004, 75, 195–205.

[27] Kehrer M, Duchoslav J, Hinterreiter A, Cobet M, Mehic A, Stehrer T, Stifter D. XPS investigation on the reactivity of surface imine groups with TFAA. Plasma Process Polym 2019, 16, e1800160.

[28] Duchoslav J, Kehrer M, Hinterreiter A, Duchoslav V, Unterweger C, Fürst C, Steinberger R, Stifter D. Novel protocol for highly efficient gas-phase chemical derivatization of surface amine groups using trifluoroacetic anhydride. Appl Surf Sci 2018, 443, 244–254.

[29] Manakhov A, Michlíček M, Nečas D, Polčák J, Makhneva E, Eliáš M, Zajíčková L. Carboxyl-rich coatings deposited by atmospheric plasma co-polymerization of maleic anhydride and acetylene. Surf Coatings Technol 2016, 295, 37–45.

[30] Povstugar VI, Mikhailova SS, Shakov AA. Chemical derivatization techniques in the determination of functional groups by X-ray photoelectron spectroscopy. J Anal Chem 2000, 55, 405–416.

[31] Manakhova A, Michlíčekb M, Feltend A, Pireauxd JJ, Nečasb D, Zajíčková L. XPS depth profiling of derivatized amine and anhydride plasma polymers: Evidence of limitations of the derivatization approach. Appl Surf Sci 2017, 394, 578–585.

[32] Kasparek E, Tavares JR, Wertheimer MR, Girard-Lauriault PL. Sulfur-rich organic films deposited by plasma- and vacuum-ultraviolet (VUV) photo-polymerization. Plasma Process Polym 2016, 13, 888–899.

[33] Thiry D, Francq R, Cossement D, Guerin D, Vuillaume D, Snyders R. Establishment of a derivatization method to quantify thiol function in sulfur-containing plasma polymer films. Langmuir 2013, 29(43), 13183–13189.

[34] Malitesta C, Losito I, Sabbatini L, Zambonin PG. Applicability of chemical derivatization X-ray photoelectron spectroscopy (CD-XPS) to the characterization of complex matrices: Case of electrosynthesized polypyrroles. J Electron Spectrosc 1998, 97, 199–208.

[35] Wepanick KA, Billy AS, Bitter JL, Fairbrother DH. Chemical and structural characterization of carbon nanotube surfaces. Anal Bioanal Chem 2010, 396, 1003–1014.

[36] Hinterreiter AP, Duchoslav J, Kehrer M, Truglas T, Lumetzberger A, Unterweger C, Fürst C, Stifter D. Determination of the surface chemistry of ozone-treated carbon fibers by highly consistent evaluation of X-ray photoelectron spectra. Carbon 2019, 146, 97e105.

[37] Flores-Rizo J, Esnal I, Osorio-Martínez CA, Gómez-Durán CFA, Bañuelos J, López Arbeloa I, Pannell KH, Metta-Magaña AJ, Peña-Cabrera E. 8 Alkoxy- and 8 aryloxy-BODIPYs: Straightforward fluorescent tagging of alcohols and phenols. J Org Chem 2013, 78, 5867–5877.

[38] Patterson AL, Wenning B, Rizis G, Calabrese DR, Finlay JA, Franco SC, Zuckermann RN, Clare AS, Kramer EJ, Ober CK, Segalman RA. Role of backbone chemistry and monomer sequence in amphiphilic oligopeptide- and oligopeptoid-functionalized PDMS- and PEO-based block copolymers for marine antifouling and fouling release coatings. Macromolecules 2017, 50(7), 2656–2667.

[39] Choi JW, Carter MCD, Wei W, Kanimozi C, Speetjens FW, Mahanthappa MK, Lynn DM, Gopalan P. Self-assembly and post-fabrication functionalization of microphase separated thin films of a reactive azlactone-containing block copolymer. Macromolecules 2016, 49(21), 8177–81868.

[40] Andruzzi L, D'Apollo F, Galli G, Gallot B. Synthesis and structure characterization of liquid crystalline polyacrylates with unconventional fluoroalkylphenyl mesogens. Macromolecules 2001, 34, 7707–7714.

[41] Bertolucci M, Galli G, Chiellini E, Wynne KJ. Wetting behavior of films of new fluorinated styrene-siloxane block copolymers. Macromolecules 2004, 37, 3666–3672.

[42] Losito I, Sabbatini L, Gardella JA Jr.. Electron, ion, and mass spectroscopy. In: Brady RF Jr, ed. Comprehensive Desk Reference of Polymer Characterization and Analysis, New York (NY), Oxford University Press, 2003, 375–407.

[43] Zhou Y, Liu C, Gao J, Chen Y, Yu F, Chen M, Zhang H. A novel hydrophobic coating film of water-borne fluoro-silicon polyacrylate polyurethane with properties governed by surface self-segregation. Prog Org Coatings 2019, 134, 134–144.

[44] Chen X, Ye X, Lu L, Qian Y, Wang L, Bi Y, Wang Z, Cai Z. Preparation of cross-linkable waterborne polyurethane-acrylate coating films with multifunctional properties. Coatings 2020, 10(1), 65.

[45] Guo Y, Zhao E, Zhao X, Zhang C, Yao L, Guo X, Wang X. Synergistic effect of electric field and polymer structures acting on fabricating beads-free robust superhydrophobic electrospun fibers. Polymer 2021, 213, 123208.

[46] Zhao J, Wang Q, Yang J, Li Y, Liu Z, Zhang L, Zhao Y, Zhang S, Chen L. Comb-shaped amphiphilic triblock copolymers blend PVDF membranes overcome the permeability-selectivity trade-off for protein separation. Sep Purif Technol 2020, 239, 116596.

[47] Lundin JG, Giles SL, Fulmer PA, Wynne JH. Distribution of quaternary ammonium salt encapsulated polyoxometalates in polyurethane films. Prog Org Coatings 2017, 105, 320–329.

[48] Mandal J, Arcifa A, Spencer ND. Synthesis of acrylamide-based block-copolymer brushes under flow: Monitoring real-time growth and surface restructuring upon drying. Polym Chem 2020, 11(18), 3209–321614.

[49] Govers SPW, Alexander N, Al-Masri M, Omeis J, Van Der Ven LGJ, De With G, Esteves ACC. Surface segregation of polydimethylsiloxane-polyether block copolymers in coatings driven by molecular architecture. Prog Org Coatings 2021, 150, 105991.

[50] Cometa S, Chiellini F, Bartolozzi I, Chiellini E, De Giglio E, Sabbatini L. Surface segregation assessment in poly(e-caprolactone)-poly(ethylene-glycol) multiblock copolymer films. Macromol Biosci 2010, 10, 317–327.

[51] Weng L-T, Ng K-M, Cheung ZL, Lei Y, Chan C-M. Quantitative analysis of styrene-pentafluorostyrene random copolymers by ToF-SIMS and XPS. Surf Interface Anal 2006, 38, 32–43.

[52] Ha C-S, Gardella JA Jr.. X-ray photoelectron spectroscopy studies on the surface segregation in poly(dimethylsiloxane) containing block copolymers. J Macromol Sci, Part C – Polym Rev 2005, 45, 1–18.

[53] Dwight DW, McGrath JE, Riffle JS, Smith SD, York GA. ADXPS/STEM studies of surface and bulk microphase behavior in block copolymers and graft copolymers and their blends. J Electron Spectrosc 1990, 52, 457–473.

[54] Alvarez Albarran A, Rosenthal-Kim EQ, Kantor J, Liu L, Nikolov Z, Puskas JE. Stimuli-responsive antifouling polyisobutylene-based biomaterials via modular surface functionalization. J Polym Sci A: Polym Chem 2017, 55(10), 1742–1749.

[55] Xie W, Weng L-T, Yeung KL, Chan C-M. Segregation of dioctyl phthalate to the surface of polystyrene films characterized by ToF-SIMS and XPS. Surf Interf Anal 2018, 50(12–13), 1302–1309.

[56] Andruzzi L, Hexemer A, Li X, et al. Control of surface properties using fluorinated polymer brushes produced by surface-initiated controlled radical polymerization. Langmuir 2004, 20, 10498–10506.

[57] Martinelli E, Guazzelli E, Glisenti A, Galli G. Surface segregation of amphiphilic PDMS-based films containing terpolymers with siloxane, fluorinated and ethoxylated side chains. Coatings 2019, 9(3), 153.

[58] Vaidya A, Chaudhury MK. Synthesis and surface properties of environmentally responsive segmented polyurethanes. J Colloid Interface Sci 2002, 249, 235–245.

[59] Wu H-X, Zhang X-H, Huang L, Ma L-F, Liu C-J. Diblock Polymer Brush (PHEAA-b-PFMA): Microphase Separation Behavior and Anti-Protein Adsorption Performance. Langmuir 2018, 34(37), 11101–11109.

[60] Xix-Rodriguez C, Varguez-Catzim P, Alonzo-García A, Rodriguez-Fuentes N, Vázquez-Torres H, González-Diaz A, Aguilar-Vega M, González-Díaz MO. Amphiphilic poly(lactic acid) membranes with low fouling and enhanced hemodiafiltration. Sep Purif Technol 2021, 259, 118124.

[61] Jiang H, Zhao Q, Wang P, Ma J, Zhai X. Improved separation and antifouling properties of PVDF gravity-driven membranes by blending with amphiphilic multi-arms polymer PPG-Si-PEG. J Membran Sci 2019, 588, 117148.

[62] Sun H, Yang X, Zhang Y, Cheng X, Xu Y, Bai Y, Shao L. Segregation-induced in situ hydrophilic modification of poly (vinylidene fluoride) ultrafiltration membranes via sticky poly (ethylene glycol) blending. J Membran Sci 2018, 563, 22–30.

[63] Vandencasteele N, Reniers F. Plasma-modified polymer surfaces: Characterization using XPS. J Electron Spectrosc 2010, 178–9, 394–408.

[64] Chilkoti A, Ratner BD, Briggs D. Plasma-deposited polymeric films prepared from carbonyl-containing volatile precursors: XPS chemical derivatization and static SIMS surface characterization. Chem Mater 1991, 3, 51–61.

[65] Fazeli M, Florez JP, Simão RA. Improvement in adhesion of cellulose fibers to the thermoplastic starch matrix by plasma treatment modification. Compos B Eng 2019, 163, 207–216.

[66] Shen T, Liu Y, Zhu Y, Yang DQ, Sacher E. Improved adhesion of Ag NPs to the polyethylene terephthalate surface via atmospheric plasma treatment and surface functionalization. Appl Surf Sci 2017, 411, 411–418.

[67] Giorcelli M, Guastella S, Mandracci P, Liang Y, Li X, Tagliaferro A. Carbon fibre functionalization by plasma treatment for adhesion enhancement on polymers. AIP Conf Proc 2018, 1981, 020142.

[68] Kim MT, Kim MH, Rhee KY, Park SJ. Study on an oxygen plasma treatment of a basalt fiber and its effect on the interlaminar fracture property of basalt/epoxy woven composites. Compos B Eng 2011, 42, 499–504.

[69] Enciso B, Abenojar JA, Martínez MA. Influence of plasma treatment on the adhesion between a polymeric matrix and natural fibres. Cellulose 2017, 24, 1791–1801.

[70] Cools P, Van Vrekhem S, De Geyter N, Morent R. The use of DBD plasma treatment and polymerization for the enhancement of biomedical UHMWPE. Thin Solid Films 2014, 572, 251–259.

[71] Bitar R, Cools P, De Geyter N, Morent R. Acrylic acid plasma polymerization for biomedical use. App Surf Sci 2018, 448, 168–185.

[72] Ramkumar MC, Pandiyaraj KN, Kumar AA, Padmanabhan PVA, Kumar SU, Gopinath P, Bendavid A, Cools P, De Geyter N, Morent R, Deshmukh RR. Evaluation of mechanism of cold atmospheric pressure plasma assisted polymerization of acrylic acid on low density polyethylene (LDPE) film surfaces: Influence of various gaseous plasma pretreatment. Appl Surf Sci 2018, 439, 991–998.

[73] Ozeki K, Hirakuri KK. The effect of nitrogen and oxygen plasma on the wear properties and adhesion strength of the diamond-like carbon film coated on PTFE. Appl Surf Sci 2008, 254, 1614–1621.

[74] Wu S, Su F, Dong X, Ma C, Pang L, Peng D, Wang M, He L, Zhang Z. Development of glucose biosensors based on plasma polymerization-assisted nanocomposites of polyaniline, tin oxide, and three-dimensional reduced graphene oxide. Appl Surf Sci 2017, 401, 262–270.

[75] Merche D, Poleunis C, Bertrand P, Sferrazza M, Reniers F. Synthesis of polystyrene thin films by means of an atmospheric-pressure plasma torch and a dielectric barrier discharge. IEEE T Plasma Sci 2009, 37, 951–960.

[76] Reniers F, Vandencasteele N, Burry O. Procédé pour déposer une couche fluorée à partir d'un monomère précurseur. EP 2 098 305 A1 (2009) Bruxelles (BE).

[77] Weibel DE, Vilani C, Habert AC, Achete CA. Surface modification of polyurethane membranes using acrylic acid vapour plasma and its effects on the pervaporation processes. J Membr Sci 2007, 293, 124–132.

[78] Vig JR. UV/ozone cleaning of surfaces. J Vac Sci Technol A 1985, 3, 1027–1034.

[79] Yu X, Beharaj A, Grinstaff MW, Tsui OK. Modulation of the effective viscosity of polymer films by ultraviolet ozone treatment. Polymer 2017, 116, 498–505.

[80] Mitchell SA, Poulsson AHC, Davidson MR, Emmison N, Shard AG, Bradley RH. Cellular attachment, spatial control of cells using micro-patterned ultra-violet/ozone treatment in serum enriched media. Biomaterials 2004, 25, 4079–4086.

[81] Yáñez-Pacios AJ, Martín-Martínez JM. Surface modification and adhesion of wood-plastic composite (WPC) treated with UV/ozone. Compos Interfaces 2018, 25, 127–149.

[82] Rafique S, Roslan NA, Abdullah SM, Li L, Supangat A, Jilani A, Iwamoto M. UV-ozone treated graphene oxide/PEDOT: PSS bilayer as a novel hole transport layer in highly efficient and stable organic solar cells. Org Electron 2019, 66, 32–42.

[83] Lee JY, Park CY, Moon SY, Choi JH, Chang BJ, Kim JH. Surface-attached brush-type CO2-philic poly (PEGMA)/PSf composite membranes by UV/ozone-induced graft polymerization: Fabrication, characterization, and gas separation properties. J Membran Sci 2019, 589, 117214.

[84] Atta A, Abdeltwab E, Bek A. Surface physical properties of ion beam sputtered copper thin films on poly tetrafluoroethylene. Surf Topogr Metrol Prop 2021, 9, 025013.

[85] Satulu V, Mitu B, Ion V, Marascu V, Matei E, Stancu C, Dinescu G. Combining fluorinated polymers with Ag nanoparticles as a route to enhance optical properties of composite materials. Polymers 1640, 2020(12).

[86] Sachdev P, Banerjee M, Mukherjee GS. Magnetic and Microstructural Studies on PVA/Co Nanocomposite Prepared by Ion Beam Sputtering Technique. Def Sci J 2014, 64, 290–294.

[87] Cioffi N, Losito I, Torsi L, et al. Analysis of the surface chemical composition and morphological structure of vapor-sensing gold-fluoropolymer nanocomposites. Chem Mater 2002, 14, 804–811.

[88] Saffar A, Ajji A, Carreau PJ, Kamal MR. The impact of new crystalline lamellae formation during annealing on the properties of polypropylene based films and membranes. Polymer 2014, 55, 3156–3167.

[89] Malgas GF, Motaung DE, Arendse CJ. Temperature-dependence on the optical properties and the phase separation of polymer-fullerene thin films. J Mater Sci 2012, 47, 4282–4289.

[90] Hashemi A, Bahari A, Ghasemi S. The low threshold voltage n-type silicon transistors based on a polymer/silica nanocomposite gate dielectric: The effect of annealing temperatures on their operation. Appl Surf Sci 2017, 416, 234–240.

[91] De Giglio E, Cometa S, Sabbatini L, Zambonin PG, Spoto G. Electrosynthesis and analytical characterization of PMMA coatings on titanium substrates as barriers against ion release. Anal Bioanal Chem 2005, 381, 626–633.

[92] Hao H, Huang J, Liu P, Xue Y, Wang J, Ren K, Jin Q, Ji J, Greiner A, Agarwal S. Rapid build-up of high-throughput screening microarrays with biochemistry gradients via light-induced thiolene "click" chemistry. J Mater Chem B 2021, 9, 3032–3037.

[93] Yildirim E, Choi H, Schulte A, Schönherr H. Synthesis of end group-functionalized PGMA-peptide brush platforms for specific cell attachment by interface-mediated dissociative electron transfer reversible addition-fragmentation chain transfer radical (DET-RAFT) polymerization. Eur Polym J 2021, 148, 110370.

[94] Dong X, Cheng Q, Long Y, Xu C, Fang H, Chen Y, Dai H. A chitosan based scaffold with enhanced mechanical and biocompatible performance for biomedical applications. Polym Degrad Stab 2020, 181, 109322.

[95] Huang Q, Liu M, Chen J, Wan Q, Tian J, Huang L, Jiang R, Deng F, Wen Y, Zhang X, Wei Y. arrying the mussel inspired chemistry and Kabachnik–Fields reaction for preparation of SiO2 polymer composites and enhancement removal of methylene blue. Appl Surf Sci 2017, 422, 17–27.

[96] De Giglio E, Cometa S, Cioffi N, Torsi L, Sabbatini L. Analytical investigations of poly(acrylic acid) coatings electrodeposited on titanium-based implants: A versatile approach to biocompatibility enhancement. Anal Bioanal Chem 2007, 389, 2055–2063.

[97] Esbah Tabaei PS, Ghobeira R, Cools P, Rezaei F, Nikiforov A, Morent R, De Geyter N. Comparative study between in-plasma and post-plasma chemical processes occurring at the surface of UHMWPE subjected to medium pressure Ar and N2 plasma activation. Polymer 2020, 193, 122383.

[98] Amorim MKM, Rangel EC, Landers R, Durrant SF. Effects of cold SF6 plasma treatment on a-C: H, polypropylene and polystyrene. Surf Coat Technol 2020, 385, 125398.

[99] Rafique S, Abdullah SM, Badiei N, McGettrick J, Lai KT, Roslan NA, Lee HKH, Tsoi WC, Li L. An insight into the air stability of the benchmark polymer: Fullerene photovoltaic films and devices: A comparative study. Org Electron 2020, 76, 105456.

[100] Thompson R, Austin D, Wang C, Neville A, Lin L. Low-frequency plasma activation of nylon 6. Appl Surf Sci 2021, 544, 148929.

[101] Chen TF, Siow KS, Ng PY, Nai MH, Lim CT, Yeop Majlis B. Ageing properties of polyurethane methacrylate and off-stoichiometry thiolene polymers after nitrogen and argon plasma treatment. J Appl Polym Sci 2016, 133(42), 44107.

[102] König U, Nitschke M, Pilz M, Simon F, Arnhold C, Werner C. Stability and ageing of plasma treated poly(tetrafluoroethylene) surfaces. Colloid Surf B 2002, 25, 313–324.

[103] Touzin M, Chevallier P, Lewis F, et al. Study on the stability of plasma-polymerized fluorocarbon ultra-thin coatings on stainless steel in water. Surf Coat Technol 2008, 202, 4884–4891.

[104] Tran ND, Dutta NK, Choudhury NR. Weatherability and wear resistance characteristics of plasma fluoropolymer coatings deposited on an elastomer substrate. Polym Degrad Stab 2006, 91, 1052–1063.

[105] Yun D-J, Jung J, Sung YM, Ra H, Kim J-M, Chung J, Kim SY, Kim Y-S, Heo S, Kim K-H, Jeong YJ, Jang J. In-situ photoelectron spectroscopy study on the air degradation of PEDOT:PSS in terms of electrical and thermoelectric properties. Adv Electron Mater 2020, 6(11), 2000620.

Adam P. Hitchcock

3 Polymer surface characterization by near-edge X-ray absorption fine structure spectroscopy

3.1 Introduction

3.1.1 Scope of chapter

There are many challenges to understanding polymer surfaces and how a specific surface interacts with its surroundings. Such information may be of fundamental interest – *why are some polymer surfaces more hydrophilic than others*? In many cases the properties of a polymer surface are critical for a specific application – *how can the surface of a polymer be made biocompatible and thus suitable for medical applications*? X-ray absorption spectroscopy, and in particular, the near-edge X-ray absorption fine structure (NEXAFS) signal, has much to offer for analyzing polymers and polymer surfaces. In addition to identifying elemental composition, NEXAFS can identify the molecular repeat unit structure, the spatial orientation of functional groups at the surface, the presence and lateral spatial distribution of adsorbates; and how those properties might change under the influence of changes to the physical (T, P) or chemical (e.g., solvent) environment of the surface. This chapter provides an introduction to the principles and instrumentation of NEXAFS, with emphasis on its application to surface-sensitive studies of polymers. NEXAFS microscopy is described. Applications are presented in the areas of organic electronics and biomaterials.

3.1.2 Why NEXAFS?

Tunable monochromatic X-rays are required for NEXAFS spectroscopy. The vast majority of NEXAFS measurements are performed using synchrotron light sources [1], which are large, accelerator-based sources of broad-band electromagnetic radiation (microwave to gamma ray). These facilities have significant barriers to use due to long delays to access, need to travel, remote location, etc. Given the inconvenience relative to home-lab based instrumentation, it is very reasonable to ask, *"why use NEXAFS?"* The short answer is that NEXAFS provides information that is either unique (e.g., polymer orientation at surfaces), or better obtained by NEXAFS than by competing methods. The longer answer can be obtained by reading this chapter.

https://doi.org/10.1515/9783110701098-003

3.1.3 History

NEXAFS is a type of electronic spectroscopy based on the excitation of an inner-shell electron to an energy level or band which is partially of completely unoccupied in the initial state of the sample (typically the ground electronic state). The earliest *soft X-ray* inner-shell spectroscopy to my knowledge was recorded photographically on a number of solids and gases by Hanawalt in 1931 [2]. Hard X-ray (>5 keV) absorption spectroscopy was developed shortly after the discovery of X-rays by Roentgen [3]. However, it was only many years later, after development of improved vacuum, electronics, and optics that systematic studies of soft X-ray absorption spectroscopy began, first using electron energy loss, then monochromated X-rays produced by a synchrotron radiation source. The modern era of inner-shell spectroscopy dates to the mid-1970s when the group of Chris Brion at the University of British Columbia carried out systematic studies of the inner-shell electron energy loss spectra (ISEELS) of many small molecules in the gas phase [4]. Small molecule spectra are of great use in interpreting the NEXAFS spectra of polymers since the ISEEL spectrum of a molecule that is structurally similar to the repeat unit of a polymer is often very close to that of the NEXAFS spectrum of the polymer. My group has continued gas-phase molecular inner-shell spectroscopy studies over the past 40 years. The results have been reported in many papers. A bibliography of atomic and molecular inner-shell spectra and downloadable experimental ISEELS data for over 400 molecules are available at http://unicorn.mcmaster.ca/corex/cedb-title.html.

Synchrotron-based X-ray absorption spectroscopy in the soft X-ray region, which includes the C 1s, N 1s, and O 1s inner-shell spectra that are most relevant to polymer science, also started in the late 1970s [5] but really only emerged as an effective technique with the development of high-resolution monochromators such as the "Dragon"-type spherical grating monochromator [6]. The transition from fundamental studies of atoms and small molecules to studies of solids, surfaces, organic adsorbates on surfaces, and bulk polymers occurred during the late 1970s and early 1980s. That history, along with an excellent exposition of the principles and early applications of NEXAFS, is provided by Joachim Stöhr's excellent book, *NEXAFS Spectroscopy* [7], which, despite being published in 1992, still remains the best monograph dedicated to NEXAFS spectroscopy. An area of application that really brought attention to the power of NEXAFS was studies by the Stöhr group and others of the structure and orientation of small molecules on single-crystal metal surfaces [8]. During the 1980s the first studies of bulk polymers, and then of polymer surfaces were carried out [9–11]. In 1992, Ade et al. reported the first NEXAFS imaging study of a heterogeneous polymer system using scanning transmission x-ray microscopy (STXM) [12].

In addition to Stöhr' book [7], there are a number of excellent reviews of NEXAFS, including ones by Hähner [13] focusing on thin organic films and liquids; DeLongchamp et al. [14] dealing with NEXAFS of molecules and polymers for organic electronics; and Watts et al. [15] providing an excellent summary of physical principles

and experimental methods. Other review articles dealing with applications of NEX-
AFS to polymers and polymer surfaces include those by Unger et al. [16] comparing
NEXAFS and X-ray photoelectron spectroscopy (XPS) as applied to polymer surfaces;
Ade and Urquhart [17] outlining the C 1s spectroscopy and microscopy of natural and
synthetic polymers; Ade and Hitchcock [18] focusing on NEXAFS microscopy and resonant X-ray scattering of polymers; Son et al. [19] comparing the capabilities of NEX-
AFS to other spectroscopic techniques for characterizing polymer nanocomposite
materials. Fingerprint methods are frequently used in interpreting NEXAFS spectra
and thus databases of the NEXAFS spectra of molecules (analogs of polymer repeat
units) and polymers are very useful. Dhez et al. [20] reported the spectra of 24 common polymers, organized by functional groups. These spectra can be accessed at
https://www.physics.ncsu.edu/stxm/polymerspectro/. Urquhart et al. [21–24] reported the NEXAFS spectra of a number of polyurethanes, polyureas, and polyethers.
The review articles by Hähner [13] and Smith et al. [25] contain spectra of many polymers. Graf et al. [26] reported NEXAFS of aliphatic and aromatic amine-terminated
polymers in their study of functionalized surfaces. Watts et al. [27] reported calibrated
NEXAFS spectra of common conjugated polymers used in organic electronics. Finally, the database of elemental X-ray absorption cross-sections (50 eV to 30 keV),
originally published by Henke et al. [28], is an important resource for fundamental
properties of X-ray interactions with matter across the whole periodic table. This information is continuously updated and available at the website of the Centre for X-ray
Optics (https://henke.lbl.gov/optical_constants/).

3.2 Principles of NEXAFS

3.2.1 Physical processes

When a photon is absorbed by a sample, the photon disappears and its energy is
taken up by the internal states of the sample, which, in the X-ray energy regime, involves promotion of an inner-shell (core) electron to an unoccupied level of the target
(see Fig. 3.1). Energy is conserved so the absorption will only occur when the energy
of the X-ray photon matches the difference in energy of the inner-shell excited state
and the ground state. The upper state can be symbolized by (IS^{-1}, U^1) where IS is the
inner-shell level, U is the unoccupied level and the superscripts indicate that, relative
to the ground state, there is one less electron in an IS orbital (e.g., C 1s, the most
important inner-shell level in the context of NEXAFS of organic polymers) and there
is one more electron in one of the unoccupied orbitals of the ground state (e.g., π^* in
a closed shell, unsaturated organic molecule) or partly occupied (e.g., 3d in an open-
shell transition metal species). The high-energy excited state created by an X-ray absorption event is very short lived (femtosecond or shorter lifetime) because it rapidly

decays by either X-ray fluorescence radiation or by a non-radiative Auger electron decay (Fig. 3.1). In the radiative X-ray fluorescence process, an electron in one of the occupied, less tightly bound inner-shell or valence (V) orbitals fills the inner-shell hole and emits an X-ray at the same time to conserve energy. In non-radiative Auger decay, two electrons in less tightly bound orbitals are involved, with one filling the inner-shell hole and the other being ejected from the sample. In the soft X-ray region (<1000 eV), the probability of X-ray fluorescence is very low (<1%) and the Auger process dominates. NEXAFS spectra can be acquired in many different modes, which allows adjustment of the spatial sensitivity and other aspects of the method. Thus, if transmission or X-ray fluorescence (FY) detection is used, the spectra reflect the properties of the bulk of the material because X-rays penetrate far into matter (50–1000 nm in the 50 eV to 1 keV range). Various signals involving emitted electrons – drain current (total electron yield, TEY), photoelectron yield (PE-EY), Auger electron yield (AEY), or secondary electron yield (SEY) with a defined energy bandwidth (PEY) – provide variable extents of sensitivity to the sample surface, with the sampling depth dependent on which type of electrons detected. Other channels, such as partial or total ion yield can provide even higher surface sensitivity, while others, such as luminescence or thermal detection, can enhance bulk sensitivity. Lateral sensitivity (NEXAFS microscopy) [29] can be achieved by use of a focused X-ray beam (STXM or scanning photoelectron

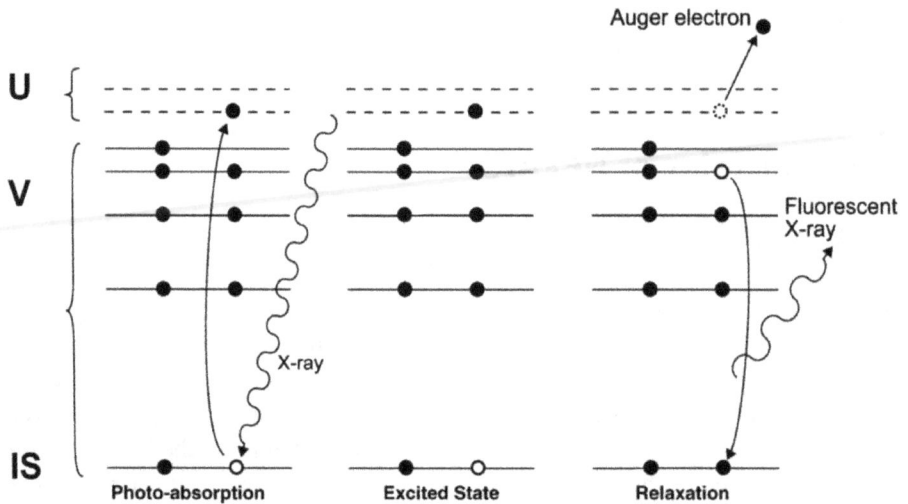

Fig. 3.1: Schematic of the photoabsorption and subsequent relaxation processes involved in near-edge X-ray absorption fine structure (NEXAFS) spectroscopy. A monochromatic X-ray photon is absorbed by the system in a resonant process that excites an electron from an inner-shell (IS) level to an unoccupied (U) level. This creates a highly excited state which rapidly relaxes by an electron in a less tightly bound occupied valence (V) or IS level filling the IS hole, simultaneously ejecting an Auger electron or emitting an X-ray fluorescent photon (adapted from fig. 1 of [15], used with permission).

microscopy, SPEM) or by imaging the spatial distribution of transmitted photons (transmission X-ray microscopy, TXM) or ejected electrons (X-ray photoemission electron microscopy, XPEEM). More details on NEXAFS microscopy are provided at the end of this chapter.

3.2.2 Theory

X-ray absorption and X-ray fluorescence processes are governed by electric dipole selection rules, which in the simplest form, strictly applicable only to atoms, requires a unit change in the orbital angular momentum quantum number, ℓ:

$$\Delta \ell = \ell \pm 1 \tag{3.1}$$

In the context of NEXAFS of polymers, where the C 1s, N 1s, and O 1s edges are most commonly measured, the strongest inner-shell excitations are those which involve promotion of 1s electrons to unoccupied orbitals with a large 2p component on the same atom where the inner-shell excitation took place. The intensity of an X-ray absorption process involving the transition from an initial, $|i\rangle$, to a final state, $|f\rangle$ is governed by the transition dipole moment (TDM):

$$TDM = <f|e.r|i> \tag{3.2}$$

Since the TDM contains a spatial overlap component and inner-shell (core) orbitals are very compact and centered about a single atom, NEXAFS transitions are spatially localized at the atom whose inner-shell electron is excited. This is true even in the case of species where there are symmetry equivalent atoms, such as ethane, since creating the inner-shell hole breaks the symmetry [30]. Thus, a simplistic but often effective way of estimating 1s excitation intensities is the contribution of 2p atomic orbitals on the inner-shell excited atom to the upper molecular orbital of the transition (i.e. the square of the coefficient for the 2p atomic orbital on the core excited atom to the final molecular orbital). This approach has been used to approximate inner-shell excitation spectra from simple, semi-empirical molecular orbital calculations [31, 32]. More correctly, the inner-shell spectrum can be predicted by quantum chemical calculations in which the energies and wavefunctions of the initial ground state and the final inner-shell excited states are calculated, and used to compute the transition dipole matrix element (eq. (3.2)). Ideally, such calculations should take into account all relevant interactions including relaxation of all energy levels in the presence of the core hole, electron correlation, and the interaction of isolated molecules with their local environment (e.g., an underlying substrate in the context of surface adsorbed species). It is only in recent years that the computational power and sophistication of quantum chemical calculations, particularly density functional theory (DFT) codes, have developed to the level where calculations can accurately predict inner-shell spectra of modestly complex systems [33–35]. The

GSCF3 method [36], an intermediate level of quantum calculation which includes a localized core hole but not electron correlation, has had much success at reproducing energies and intensities of discrete core excitations below the inner-shell ionization limit in many molecular systems [21–24, 37, 38].

3.3 Techniques

3.3.1 Experimental aspects

To date almost all NEXAFS spectra of organic molecules and polymers have been measured using synchrotron radiation. Lab-based spectrometers exist [39] but only recently have they begun to approach the performance of synchrotron systems [40]. Lab systems have not yet had much impact on polymer surface science so they will not be discussed further. Figure 3.2a is a cartoon of a soft X-ray (50–3000 eV) synchrotron radiation beamline that is used for NEXAFS. The X-ray source may be a bend magnet (BM) or (increasingly) an undulator [41]. BM, which guides the electron beam around the storage ring, provides a continuum of X-rays with strong linear polarization in the horizontal plane. Undulators (Fig. 3.2b), a type of insertion device [41], consist of a periodic array of magnets above and below a straight section of the storage ring. These magnet arrays are mounted on moveable rails, which allow tuning the undulator emission. The undulating magnetic field modifies the electron trajectory, and converts the intrinsically broad BM spectrum into a series of narrow, intense peaks of radiation (Fig. 3.2c) [42, 43]. By displacing the magnets vertically (closing or opening the gap), the photon energy is changed. If one of the magnet arrays is shifted along the straight relative to the other array, the relative phasing of the magnets is changed which modifies the direction of linear polarization. If the top and bottom arrays are split so there are four magnet arrays, motion of three of these arrays relative to the fourth allows the user to fully control the polarization of the X-rays: fully circular; fully linear with the electric vector positioned over an angular range of ±180°; and, arbitrary elliptical polarization[42]. Such elliptically polarizing undulators (EPU) provide the largest flexibility and are particularly useful in NEXAFS microscopy [44]. Mirrors are used to focus and redirect X-rays. A grazing incidence diffraction grating is used to disperse the polychromatic X-rays in space. The exit slit selects a specific energy of monochromatic light. By precisely rotating the diffraction grating, a range of photon energies can be scanned while collecting one or more signals sensitive to the X-ray absorption. All parts of the system are in ultra-high ($<10^{-9}$ mbar) vacuum.

A typical synchrotron beamline used for NEXAFS is the soft X-ray beam line and Brookhaven National Laboratory, operated by scientists from the National Institute for Science and Technology (NIST). Originally this was on the UV ring of NSLS-1 [11],

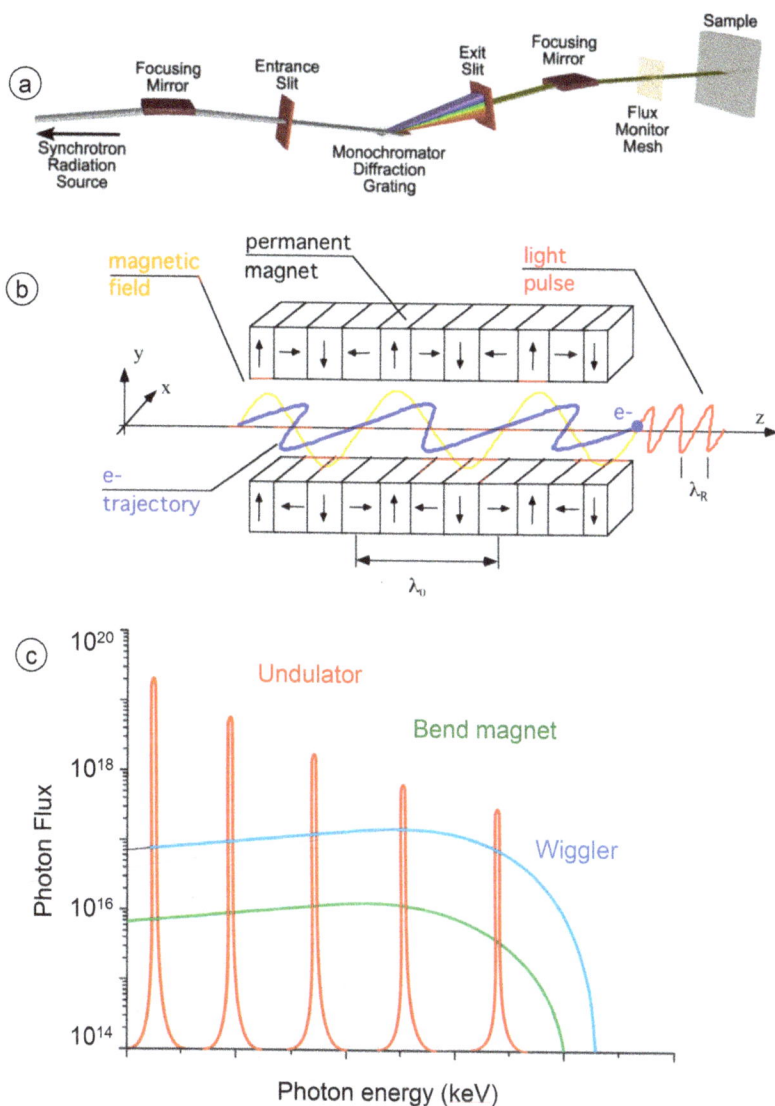

Fig. 3.2: (a) Schematic of a soft X-ray (50–3000 eV) synchrotron radiation beamline. Mirrors are used to focus and redirect X-rays. A grazing incidence diffraction grating is used to spatially disperse the polychromatic X-rays. An exit slit selects a specific energy of monochromatic light. All parts of the system are in ultra-high vacuum (<10^{-9} mbar). (b) Cartoon of an undulator. A periodic array of strong, permanent magnets imposes a sinusoidal motion (wavelength λ_u) of the electron beam in a straight section of the storage ring. The light emitted at each of the N turns in the trajectory reinforces each other in a laser-like fashion, producing a 2N-fold increase in flux and 1/N reduction in the angular spread of the X-ray beam. (c) Multi-harmonic undulator spectrum compared to that of a wiggler and a bend magnet source. (adapted from [43], used with permission).

but it has recently been replaced with a higher performance system, BL 7-ID-1 at the fourth generation source, NSLS-II [45]. Typically synchrotron beamlines are used for a variety of experiments, such as X-ray photoemission (XPS) which has been described in Chapter 2, resonant inelastic X-ray scattering (RIXS), etc., and the portion of beamtime used for polymer studies is limited. In an earlier era, users might even bring their own experimental chamber to synchrotron facilities. These days it is overwhelming the case that the facility builds, commissions, maintains, and evolves the capabilities of both the source (storage ring and beamline) and the experimental end stations. The users (a government, industrial, or academic scientist, often accompanied by postdoctoral fellows and graduate students) are responsible for defining the scientific goals, fabricating samples, bringing them to the beamline, verifying their integrity (e.g., off-line surface conditioning and characterization), operating the system (under supervision of the beamline scientist), analyzing the results and preparing publications. Access to synchrotron facilities is administered through a peer review process which involves proposal submission, review by experts in the field, scoring and beamtime allocation based on those scores. At competitive beamlines, success rates for proposals can be less than 25%, in some cases less than 10% for NEXAFS microscopes. Typically there is a 4 month interval between proposal submission and the period when beam might be allocated. Many facilities provide access for up to 2 years (four 6-month periods) on the same proposal, although some require submission of a new proposal for each 6-month period.

3.3.2 Detection schemes

Figure 3.3a summarizes the most common NEXAFS detection schemes. In general the incident flux intensity will vary with photon energy, and with time, for light sources that operate in decay mode. Increasingly, the light sources are operated in top-up mode, in which the storage ring current is kept constant by frequent small refill procedures. In order to get correct NEXAFS spectra it is necessary to measure the incident intensity (I_o), ideally with the same detector (which is realistically only possible in transmission geometry) or one with a similar spectral response in the energy regime measured [15]. For transmission detection the transmitted intensity (I) and I_o are input to the Beer–Lambert law (see eq. (3.3)) to obtain a valid NEXAFS spectrum. For all yield measurements (TEY, PEY, AEY, FY) the yield signal is divided by the I_o signal.

3.3.2.1 Transmission

When transmission detection is used, the incident (I_o) and transmitted (I) signals are measured, and the NEXAFS spectrum is obtained in the absorbance or optical density (OD) form using the Beer–Lambert law:

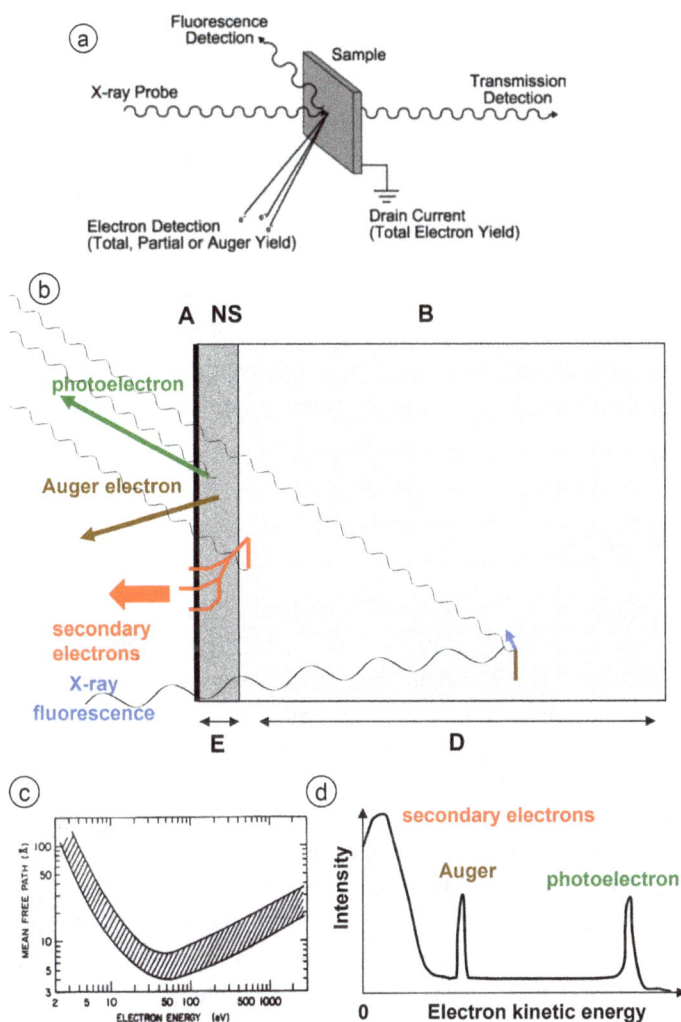

Fig. 3.3: (a) Signals are used to detect NEXAFS. Transmission or X-ray fluorescence if the sample is very thin; various types of energy-resolved electron yield (secondary, photoelectron, Auger); and total electron yield via sample current. (b) Schematic of X-ray–sample interactions. Soft X-ray photons penetrate far into samples – from several hundred nanometers up to a few microns, depending on the X-ray energy and the material. X-ray fluorescence (XRF) photons have a similar range, such that XRF detection will give the NEXAFS of the bulk (B) of the sample. Photoelectrons or Auger electrons can only escape from the near surface (NS), < 2 nm into the sample, and thus are best for detecting absorbates (A). Secondary electrons, usually 5–50 eV kinetic energy, generated by inelastic scattering of photoelectrons or Auger electrons, can escape from zone E, up to 10 nm into the surface, depending on photon energy. (b) Universal inelastic mean free path for electron scattering. (c) Dependence of the inelastic mean free path in solids (distance an electron can travel before an inelastic collision) on the electron kinetic energy. (d) Sketch of a typical kinetic energy spectrum of electrons ejected from an X-ray irradiated sample.

$$OD = -\ell n\left(\frac{I}{I_o}\right) = \sigma\rho t \qquad (3.3)$$

where σ is the photon energy dependent cross-section, ρ is the sample density, and t is the thickness. In order to use transmitted X-rays to measure a NEXAFS spectrum, the sample must have an appropriate thickness, which is optimally that which gives an OD of 1 at the most intense absorption peak. If the sample is too thick, such that the peak OD is above 3, the resulting absorption saturation can greatly distort the shape of NEXAFS spectra. Absorption saturation occurs in thick samples due to contributions from electronic and other sources of backgrounds, and higher order spectral contributions when most of the incident photons are absorbed by the sample [46]. If the sample is too thin, such that the peak OD is less than ~0.1, it is difficult to obtain good statistics.

3.3.2.2 Electron yield methods

In general electron detection methods provide much higher surface sensitivity than photon detection methods (transmission, fluorescence yield). The simplest way to measure the total electron yield is to electrically isolate the sample and measure the drain current between the sample and electrical ground. The drain current is directly related to the number of electrons ejected when the incident X-ray beam hits the sample. The ejected electron current is the sum of electrons from the primary photoionization and Auger decay processes, along with secondary electrons generated indirectly from ionizing inelastic scattering events as the photo-electron and Auger electron travel to the surface (see Fig. 3.3b). The X-ray photon penetration depth (roughly 1 µm per keV) is much larger than the electron escape depth (1–10 nm). The sampling depth (region of the sample contributing to an electron yield NEXAFS spectrum) depends on the kinetic energy of the electrons detected and their inelastic mean free path (Fig. 3.3c) [47, 48]. The primary photoelectron and Auger signals, which have well-defined electron kinetic energies, can only be detected if their kinetic energy is unchanged and thus only come from very near to the surface. The secondary electron signal, especially the low kinetic energy electrons (<20 eV), will come from a much larger range of depths, including several tens of nanometers. In order to measure the primary electrons with well-defined kinetic energy, an electron energy analyzer and a high, or ultrahigh vacuum system are required. TEY and PEY methods can be done in a low vacuum (<10^{-3} mbar) using retarding grids and a suitable electron detector such as a photodiode, a channeltron, or a channelplate, combined with appropriate electronics. For all types of electron yield detection, sample charging in the case of insulating samples can be an issue. In such cases, a very thin coating of a conducting material (e.g., Pt) can be used, although that may compromise the polymer surface chemistry of interest. Charging is worst for the primary electrons since PE-EY and AEY detection modes require stable, well-defined kinetic energies. PEY and SEY are relatively tolerant to some charging, while transmission and XRF modes are immune to sample charging.

3.3.2.3 Photon yield methods

X-ray absorption can be measured by detecting X-ray fluorescence photons, as either a total fluorescence yield or partial fluorescence yield (PFY). While FY is very important for studies of the bulk of solids, especially at higher X-ray energies where the fluorescence yield is high, it is rarely used for surface studies in the soft X-ray regime, in part due to the very low fluorescence yield from core edges below 1000 eV. Situations where FY detection has been used to advantage is in the study of organic polymers that are part of another material that did not contain carbon, with the polymers either as a thin coating on the surface, or as a thin layer that is buried at a depth less then the escape depth of the C Kα X-ray fluorescence photon (260 eV), about 300 nm [13, 49]. Simultaneous acquisition of the polarization dependent electron yield (TEY or SEY) and FY spectra are often used to compare properties of the surface and bulk of a polymeric material, in terms of both composition and molecular orientation [50, 51]. If the fluorescence X-rays are detected with an energy resolving detector such as a silicon drift detector, further options become available, including selective detection of minority elements in a polymer (e.g., S in thiophene-containing organic photovoltaic materials [52]) and use of inverse PFY which reduces the distortions from absorption saturation of fluorescent X-rays by too thick material [53]. In general, although the C Kα fluorescence yield is very low (<0.1% of the decay of C 1s core holes), the ever increasing brightness and stability of modern synchrotron light sources tends to compensate for the low fluorescence yield, although radiation damage can be a limiting factor on high flux beamlines, and especially in NEXAFS microscopy where the focused X-ray beam can deliver a very large radiation dose [29, 48].

3.3.3 Surface-sensitive detection

If the element of interest is localized at the surface and the underlying material does not contain that element, then any NEXAFS detection method may suffice since inner-shell edges are widely spaced. However, in order to be sensitive to the very surface – defined as one monolayer – a detection method that has a very shallow escape depth is needed, especially if the goal is to differentiate the structure/composition of the very surface from underlying layers which have the same or a similar composition. Among the methods described above, PE-EY and AEY are the most surface sensitive. It is also possible to enhance surface sensitivity by using ions ejected from the surface as the detection channel, either as a total ion current (TIY), or with mass spectrometric detection of individual mass-resolved ions (PIY) [54]. However the ion signal is weak, due to recapture of ions at the surface. Also, since soft X-rays penetrate hundreds of nanometers into the sample, energetic electrons (photoelectrons, Auger, secondary) can ionize atoms in the near-surface region, thereby mixing and often overwhelming the surface-specific signal with a spectral signature characteristic of regions much deeper into the sample [55]. Another way to enhance surface sensitivity is to tilt the sample and use a grazing incidence X-ray beam. This will reduce the penetration depth of the photons, and increase the contribution from X-ray absorption in the surface region. However, given that a 300 eV photon penetrates on the order of 500 nm, an angle of ~1° (~0.1°) is needed to reduce the penetration to ~10 (1) nm. If the width of the X-ray beam is 100 μm (a typical beam size for spectroscopy synchrotron beamlines) the lateral spread of the beam on the surface will be 5 mm at 1° incidence and 50 mm at 0.1° incidence.

3.4 Interpreting NEXAFS spectra

3.4.1 Common features of a NEXAFS spectrum

The energy of an inner-shell spectral feature will depend on the energies of both the core level and the ground state unoccupied level to which the electron is excited. The core-level energies, which are essentially those detected in X-ray photoelectron spectroscopy (XPS) (see Chapter 2) are determined by the local environment, with the main factor being the electronegativity of the atoms directly attached to the atom where the core excitation takes place. Thus the C 1s level in aliphatic polymers (only C–C and C–H single bonds) are ~ 285 eV. When carbon is bound to electronegative atoms the C 1s core-level binding energy (BE) shifts to lower energy. Thus the C 1s BE of $\underline{C}$–O carbons is ~ 286 eV, that of $\underline{C}$ = O carbons is ~ 289 eV, that of $\underline{C}$–F is ~ 289 eV, that of $\underline{C}F_2$ is ~ 292 eV and that of $\underline{C}F_3$ is ~ 294 eV [56]. The unoccupied levels of the ground state, which are the upper level of the transition, are quite sensitive to the chemical environment and thus the fine details of a NEXAFS spectrum reflect the molecular structure. Typically the strongest NEXAFS features are those involving excitations to compact valence-type levels which have large 2p density on the core excited atom. For organic polymers these are π^* levels for unsaturated species, and σ^* levels for both unsaturated and saturated species. The 1s → π^* transitions tend to be sharp and at lower energy; they are usually the spectral features used for qualitative identification (see Fig. 3.4). The 1s → σ^* transitions occur at somewhat higher energy. For σ^* features associated with bonds between second row atoms (B – F), there is a correlation between term values (difference in energy of the 1s → σ^* transition and the 1s binding energy) and bond lengths [57], which can be used for spectral assignment and compound identification.

Detailed assignment and understanding of all the details of the spectrum of a specific molecule or polymer requires accurate quantum mechanical calculations of the electronic states of the molecule (see Section 3.2.1). However in many cases the spectra can be adequately understood by comparison to small molecules with a structure similar to that of a polymer – for example, ethane as a model for polyethylene. In general additivity principles apply – for example the C 1s spectrum of phenylalanine is quite accurately simulated by the sum of the C 1s spectra of benzene and alanine [58]. In addition, when there are characteristic functional groups, such as pendant phenyl groups, nitrile or carbonyl groups, these tend to have transitions at similar energies in all molecules. Finally, by recording the NEXAFS spectra of reference materials, fingerprinting identification methods can be used.

3.4.2 Polymer NEXAFS – fingerprinting and functional group identification

Figure 3.4 presents a montage of C 1s NEXAFS spectra of a range of polymers. While a detailed analysis of each spectrum can and has been done for most of these spectra, the main message of this figure is to show that the C 1s spectra of even chemically closely related polymers – e.g., polystyrene (PS) and brominated polystyrene (PBrS) – are readily differentiated. This means that NEXAFS can be used in a fingerprint fashion by comparing the spectrum of an unknown polymer to a data base of polymer spectra (e.g., https://www.physics.ncsu.edu/stxm/polymerspectro/). While NEXAFS of closely related species can be differentiated, there are many similarities for polymers containing the same functional groups. For instance, all polymers containing a phenyl group have a sharp intense C 1s $\rightarrow \pi^*$ peak around 285 eV, ester groups have a prominent C 1s $\rightarrow \pi^*_{C=O}$ peak at 285.4–285.6 eV, etc. Figure 3.5 summarizes the energy ranges where different functional groups are typically found. Within the class of carbonyl species (aldehydes, ketones, amides, esters, ureas, urethanes, carbonates, etc.), the sensitivity of NEXAFS to the type of carbonyl species rivals that of infrared spectroscopy [59].

3.4.3 Linear dichroism – theory

This is one of the most important and widely used properties in synchrotron-based NEXAFS studies of polymer surfaces. In the following, the X-rays are assumed to be 100% linearly polarized, as can achieved with undulators [43]. Bend magnet beamlines typically have only 85% linearly polarized light [7, 11]. That reduced degree of polarization must be taken into account in quantitative analyses of angle-dependent

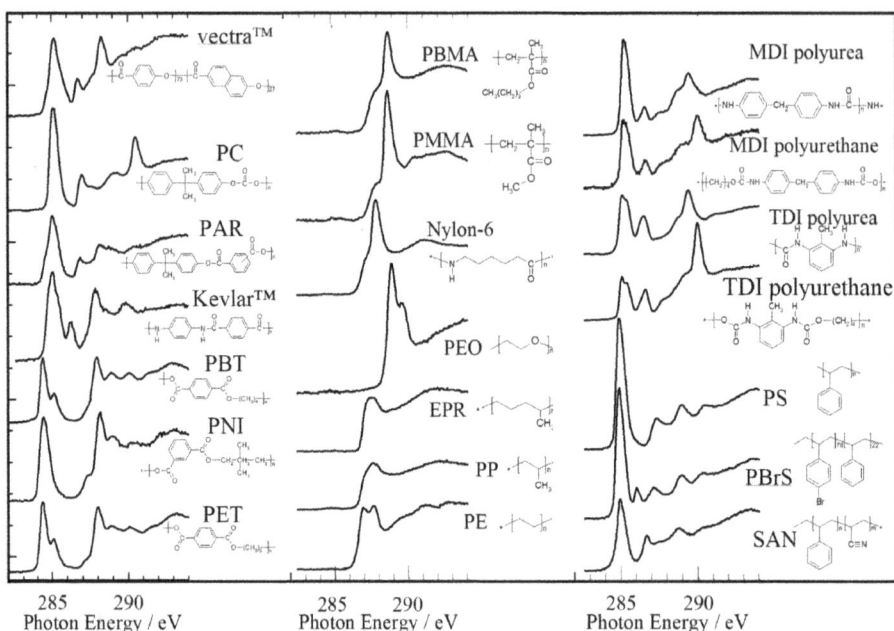

Fig. 3.4: C 1s NEXAFS spectra of 21 different polymers showing the capability of NEXAFS for identification and chemical analysis of polymers (composite of data reported in references [20–25]).

NEXAFS studies. The intensity of a NEXAFS transition is related to the square of the electric dipole transition matrix element (TDM, eq. (3.2)). The angular distribution of the intensity of each transition is governed by the symmetry properties of the TDM and its orientation relative to the electric vector (E) of the incident X-rays [7, 60]:

$$I \alpha |E.\langle f|\mu|i\rangle|^2 \alpha \cos^2 \theta \tag{3.4}$$

where θ is the angle between the electric vector (E) and the direction of the TDM. The following treatment is restricted to excitations of electrons from 1s orbitals, since C 1s, N 1s, and O 1s are the most important edges for organic polymers. For a 1s initial orbital and a directional final orbital, O, the matrix element $|E. <f|\mu|$ $1s >^2$ points in the direction of the final orbital O and the transition intensity becomes

$$I \alpha |E.\langle f|\mu|i\rangle|^2 \alpha |E.O|^2 \alpha \cos^2 \theta \tag{3.5}$$

In general, 1s excitations have the highest intensity when the 2p component of the unoccupied orbital to which the 1s electron is excited is aligned along the electric vector of the linearly polarized X-rays. Thus, for a monolayer of benzene that is adsorbed on a Ag(110) surface, where the plane of the molecule is parallel to the surface, the intensity of the C 1s → π* transitions are greatest when measured with

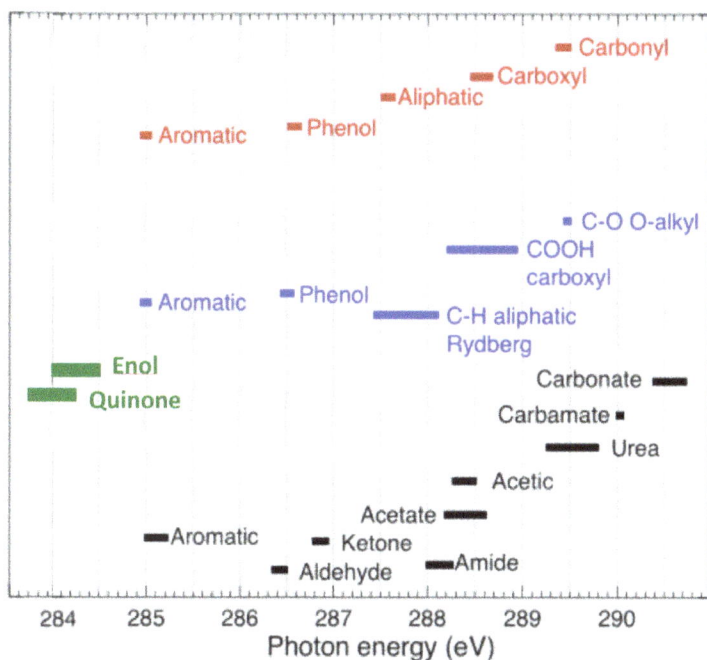

Fig. 3.5: Energy ranges for characteristic, mostly C 1s → π*, transitions of different functional groups, including carbonyl compounds (black) [59], enol and quinone compounds (green) [102, 103], and organic components of soils (blue [104, 105] and red [106]).

grazing incidence X-rays (where the $\underline{E}$-vector is nearly perpendicular to the surface) and weakest when measured with normal incidence X-rays (where the $\underline{E}$-vector is parallel to the surface) [61, 62] (see Fig. 3.6). In contrast C 1s → σ* transitions are strongest at normal incidence (where the E-vector is in the plane of the benzene ring, as are the σ* orbitals), and weakest at grazing incidence [61, 62]. While this example illustrates the concept of using NEXAFS linear dichroism to qualitatively determine molecular orientation at a surface, the real power of NEXAFS linear dichroism comes from a quantitative analysis of how the intensity of a specific NEXAFS transition varies with systematic changes in the angle between the $\underline{E}$-vector and the sample [63]. Both polar and azimuthal tilting can be used. A detailed geometric approach describing 1s excitation to specific unoccupied orbital shapes (linear, vector, and planar) and involving four separate angles was developed by Stöhr [7]. A symmetry analysis [63–65] is conceptually simpler, and also treats the most general cases, including excitation from p, d, and f inner-shell levels.

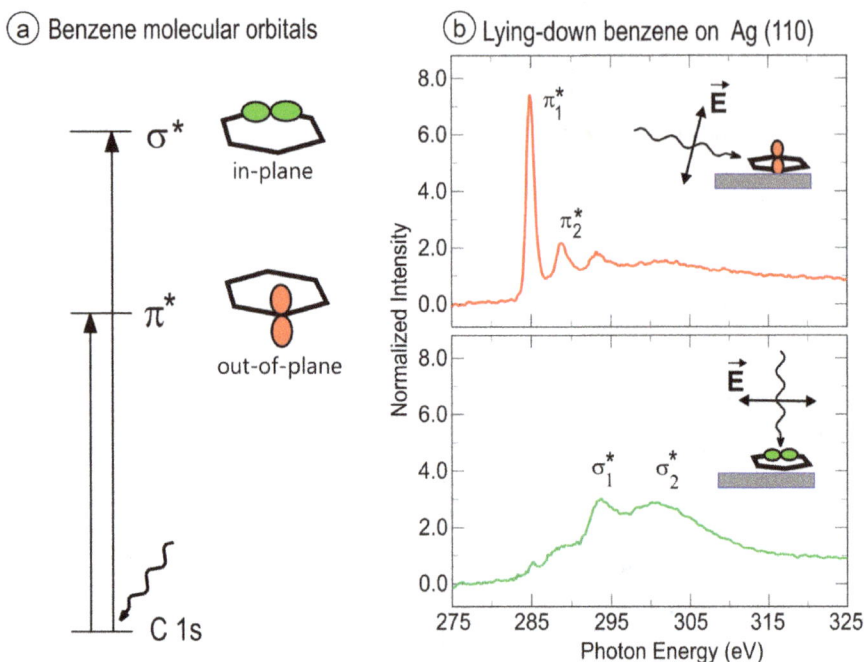

(a) Benzene molecular orbitals

(b) Lying-down benzene on Ag (110)

Fig. 3.6: (a) Schematic of inner-shell excitations to the unoccupied π* and σ* energy levels that dominate the NEXAFS spectra of benzene. (b) Polarization dependence of the C 1s NEXAFS spectrum of a monolayer of benzene on a Ag(110) surface where the molecule adsorbs with the molecular plane parallel to the surface [62]. C 1s → π* excitations are most intense when the electric vector of the X-rays ($\vec{E}$) is perpendicular to the plane of the ring. C 1s → σ* excitations are most intense when $\vec{E}$ lies in the plane of the ring (used with permission, private communication).

3.4.4 Linear dichroism – examples

Figure 3.7 shows an example of a linear dichroism study of the C 1s NEXAFS of a thin film of polytetrafluoroethylene (PTFE) on a silicon wafer, aligned parallel to the surface by gently rubbing a piece of PTFE across the Si surface which had been heated to 150oC [66]. These PEY NEXAFS spectra were recorded using X-rays at normal incidence (E parallel to the surface) and at grazing (20°) incidence (E almost perpendicular to the surface). The C 1s → σ*$_{C–C}$ transition at 286 eV is most intense in the normal incidence spectrum, which is consistent with the –(CF_2–CF_2)– chains lying parallel to the surface. The C 1s → σ*$_{C–F}$ transitions at 292 and 299 eV are most intense at glancing incidence but have some intensity at normal incidence. Detailed analysis of the polarization dependence, assisted by calculations, showed these polarization-dependent NEXAFS spectra were consistent with a 13/6 helical conformation (6 turns in 13 repeat units), and not consistent with other helical structures or a zigzag conformation known to exist at high pressure [66].

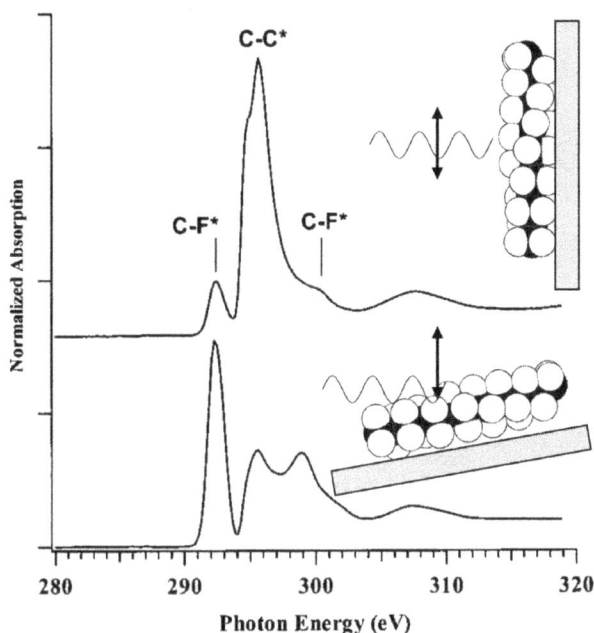

Fig. 3.7: (a) C 1s NEXAFS of an oriented thin film of polytetrafluoroethylene (PTFE) on a Au-coated Si wafer, recorded with the X-rays at normal incidence ($\vec{E}$ parallel to the surface) and grazing (20°) incidence ($\vec{E}$ almost perpendicular to the surface [59]. The C 1s → σ^*_{C-C} transition (286 eV) is most intense in the normal incidence spectrum which is consistent with the $-(CF_2-CF_2)-$ chains lying parallel to the surface in the indicated 13/6 helical conformation. The C 1s → σ^*_{C-F} transitions (292 and 299 eV) are most intense at glancing incidence but have intensity at normal incidence. Detailed analysis of the polarization dependence, assisted by calculations, showed the NEXAFS data was consistent with a 13/6 helical conformation (13 advances for 6 turns of the helix), and not consistent with tighter (7/6) or looser (19/6) coiled configurations, or with a zigzag conformation known to exist at high pressure (adapted with permission from fig. 1 of [66]. Copyright 2002, American Chemical Society).

Figure 3.8 presents results of a NEXAFS polarization study of an annealed 50 nm thick pentacene thin film which is a material used in organic thin film transistors (OTFT) [11, 14]. The results display a very systematic variation of the intensities of the NEXAFS features with the angle between the E-vector and the sample. Notably, the low lying C 1s → π^* spectral features at 284.2 and 285.8 eV increase in intensity when the orientation of the X-ray beam relative to the surface approaches normal incidence while the σ^* features at 294.9 and 301 eV increase in intensity when the incident X-ray direction approaches glancing incidence. Thus, the π^* orbitals are preferentially oriented in the substrate plane, and the pentacene molecules are oriented edge on in the thin film. The π^* variation indicates only the orientation of the conjugated plane. It cannot be used to determine the orientation of the long axis of the pentacene (see insert to Fig. 3.8b). For that, the dichroism of the σ^*

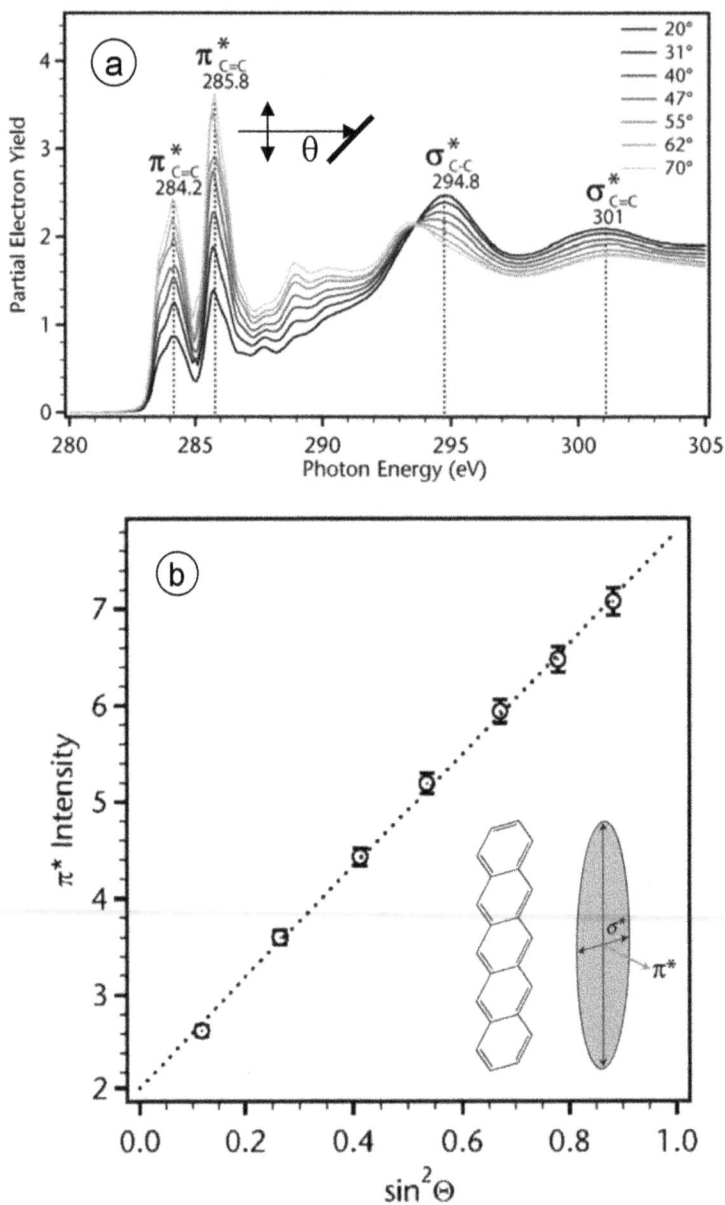

Fig. 3.8: Pentacene C 1s NEXAFS dichroism. (a) C 1s spectra of a free-standing 50 nm thick pentacene film, that was highly oriented by deposition on a Si wafer followed by annealing [11]. (b) Plot of the integrated C 1s → π* intensity (282–286.5 eV) versus $\sin^2 \theta$, where θ is the angle between the surface and the X-ray beam. The inset figure indicates the orientational vectors of the in-plane σ* and out-of-plane π* orbitals which are the terminus of the C 1s excitations (adapted from [11], used with permission).

resonances is used. The σ* intensity is most intense near glancing incidence, where the electric field vector is normal to the substrate plane. Thus, the long axis of the σ* ellipse is preferentially normal to the substrate, and the long axis of pentacene is normal to the substrate. The edge-on orientation deducted from the π* dichroism is consistent with the standing up orientation deduced from the σ* dichroism. The exact orientation of the pentacene molecule in this particular film was quantified by integrating the π* intensity from 283 eV to 286.5 eV. Figure 3.8b plots that intensity versus $\sin^2 \theta$. A detailed analysis [11] quantified the extent of edge-on ordering in this sample by deriving a dichroic ratio, R given by

$$R = \frac{I(90°) - I(0°)}{I(90°) + I(0°)} \tag{3.6}$$

with the 0° and 90° values obtained by extrapolation from the intensities at the seven angles measured. In this approach to quantifying NEXAFS linear dichroism, $R = +1$ if the final orbital lies in the plane of the substrate, $R = 0$ when the final orbital is oriented at the magic angle ($\theta = 54.7°$), and $R = -1$ for an orbital aligned perpendicular to the substrate [11]. The analysis yielded an R value of 0.59(1) which indicates the pentacene molecules are strongly oriented in an edge-on fashion and the distribution of the molecular orientation is narrow. This was actually a result of significance to the goal of the study. When pentacene films of this type are used in OTFT, an edge-on orientation enhances π interactions in the source-drain plane, explaining the excellent high hole mobility when the pentacene molecules are properly oriented relative to the interface. In order to further optimize such devices NEXAFS was used to measure pentacene orientation as a function of film thickness, substrate chemistry, and substrate temperature. In other studies of pentacene, C 1s NEXAFS microscopy (STXM) has been used to map in-plane grain orientation based on the C 1s NEXAFS polarization dependence [67].

3.5 Examples

3.5.1 Pure polymeric materials

3.5.1.1 Fused aromatic rings

In the case of polymers with an aliphatic CH–CH backbone and pendant groups (e.g., polystyrene) bonds, there is little electronic interaction between repeat units, and the spectra are independent of chain length, except for very short oligomers. However, when there is extensive delocalization across repeat units (as is important in conducting polymers) the NEXAFS spectra can be strongly dependent on electronic interactions among adjacent units. Figure 3.9a shows how the C 1s → π* region (283–287 eV) of the NEXAFS spectra of a series of fused phenyl rings evolves as the length of the fused ring system grows [68]. The delocalization of the π^* orbitals across the full structure (Fig. 3.9b) leads to splittings which push the π^* bands as low as 283.4 eV and as high as 286.2 eV, which is a much larger energy range than found for noninteracting phenyl groups. The complex pattern in these spectra, detected with a very high-resolution spectrometer ($E/\Delta E > 14{,}000$), is due to a combination of electronic level spitting and C–H and C–C vibrational excitation accompanying the C 1s excitation. While the details are beyond the scope of this chapter, it is noteworthy that

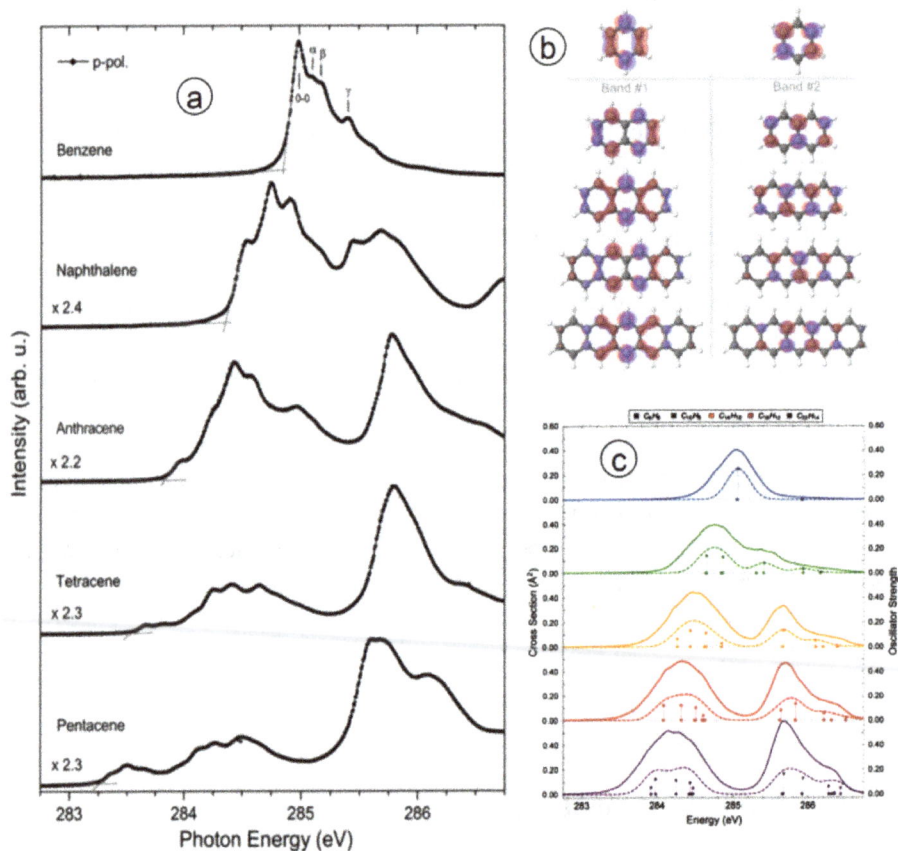

Fig. 3.9: Development of π^* band structure of fused ring polyacenes [68]. (a) High-resolution C 1s NEXAFS of benzene, naphthalene, anthracene, tetracene and pentacene in the region of the C 1s → π^* transitions. The fine structure is a combination of electronic and vibrational excitations. (b) Dominant band 1 and band $2\pi^*$ molecular orbitals for naphthalene, anthracene, tetracene, and pentacene, related to the $1\pi^*$ and $2\pi^*$ of benzene. (c) Results of high-level density functional theory (DFT) calculations of these spectra. The vertical lines indicate positions of the π^*electronic transitions (fixed ground state geometry), the dashed curves consider only electronic processes, while the full curves include consideration of vibrational excitations (adapted with permission from figs. 1–3 of [68]. Copyright 2018, American Chemical Society).

many of the spectral features arising from electronic interactions, nuclear motion and core hole effects were predicted with reasonable accuracy using DFT (Fig. 3.9b, c) [68]. The major deficiency of the computational aspect of this study, the over prediction of the intensity of the first band (283–285 eV) was attributed to insufficient treatment of the relaxation in the presence of the core hole. The authors speculated that using other functionals or using independent parameters for short- and long-range electronic interactions could remedy this deficiency [68].

3.5.1.2 Polysaccharides

Polysaccharides are biopolymers important in many aspects of biology and biotechnology. A NEXAFS study of eight different saccharides and polysaccharides [69] showed that fine details of the C 1s spectra could be used to differentiate each species (Fig. 3.10a). Further, with the support of FEFF8 multiple scattering calculations

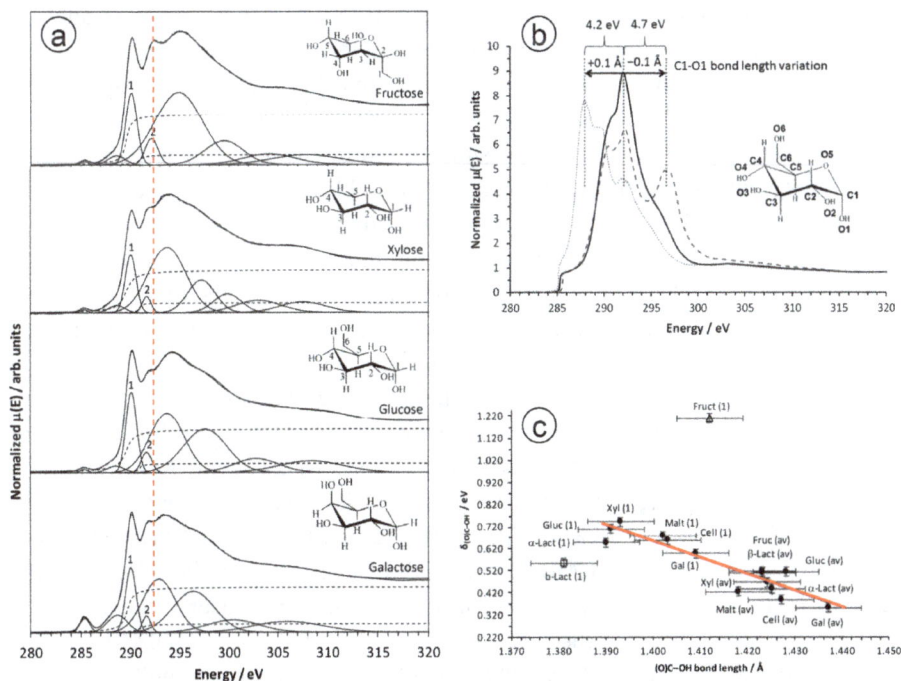

Fig. 3.10: C 1s NEXAFS of saccharides and polysaccharides[69]. (a) C 1s spectra of fructose, xylose, glucose and galactose, curve fit using Athena. Peak 1 is assigned to C 1s(C_1) → σ^*_{C-OH} and peak 2 is assigned to C 1s(C_1) → $\sigma^*_{(O)C-OH}$ transitions. (b) Feff8 simulation of the spectrum of the C_1 carbon of glucose, at the correct structure, and with +10 pm and −10 pm shifts of the C_1–OH bond length. (c) Linear correlation between the term value of the C 1s(C_1) → $\sigma^*_{(O)C-OH}$ transition and the C_1–OH bond lengths (adapted from figs. 1, 4, and 5 of [69], Creative Commons).

[70] of equilibrium and distorted structures of glucose (Fig. 3.10b), it was shown that the term value of the C 1s $(C_1) \rightarrow \sigma^*_{(O)C-OH}$ transition – the difference in energy of the ionization potential for the C_1 carbon and the energy of peak 2 around 292 eV (Fig. 3.10a) – correlated with the length of the C–OH bond at the hemiacetal/hemiketal position (Fig. 3.10c). This type of bond length – term value correlation, first reported in 1984 [57], works well for bond lengths in small molecules constructed from second row atoms. This is one of the most subtle examples of this type of structure–spectral correlation to my knowledge. The high quality of the spectral peak fitting using Athena [71, 72] software was critical to this analysis. Athena is an important processing tool for NEXAFS data.

3.5.2 Surface versus bulk chain orientation

NEXAFS has been used to study the orientation of pendant groups at the surface of homopolymers. For example, Lenhart et al. [51] showed that there was a net alignment of phenyl groups normal to the exposed surface of various types of polystyrene, but that there is a wide range of tilt angles, particularly in more bulky substituted phenyl polystyrenes. A nice example of using NEXAFS to study molecular orientation of the bulk and surface of a complex polymer has been reported by Schuettfort et al. [73] These results are summarized in Fig. 3.11. The polymer studied was poly(N,N-bis-2-octyldodecylnaphthalene-1,4,5,8-bis-dicarboximide-2,6-diyl-alt5,5–2,2-bithiophene) [p(NDI2OD-T2)], a high-mobility conjugated polymer used in organic field-effect transistors (OFET). It was spun cast from a dilute aqueous solution on to a Si wafer (which established the chain orientation), and then floated off to form a free standing ~ 55 nm thin film. High surface-sensitive analysis is critical in this application since charge transport in the OFET occurs along a ~1 nm thick layer at the semiconductor/ polymer interface. Figure 3.11a defines the set of angles needed to describe the orientation of the polymer chains at the surface while Fig. 3.11b shows how the Auger electron yield (AEY), total electron yield (TEY) and transmission C 1s NEXAFS spectra were measured simultaneously. By comparing the polar angle dependence of the AEY, TEY and transmission C 1s NEXAFS spectra (Fig. 3.11c), they showed that the surface layer of the solvent cast p(NDI2OD-T2) film had a significantly smaller tilt angle (41°) than the bulk of the polymer (55°) (Fig. 3.11d). Further they measured the azimuthal angle dependence of the C 1s spectra of as-cast and melt-annealed p(NDI2OD-T2) films and showed that the polar and azimuthal alignment of the molecular units at the surface was the same in both preparations. In contrast, the molecular orientation in the bulk differed significantly (see Fig. 3.11e). The NEXAFS results [73] provided a consistent microstructural explanation of bulk and surface charge transport phenomena previously observed in p(NDI2OD-T2) films. In particular, the observation of a distinct edge-on surface orientation in spin-coated P(NDI2OD-T2) accounts for the high-charge transport mobilities observed in top-gate transistors. As part of this study, the same

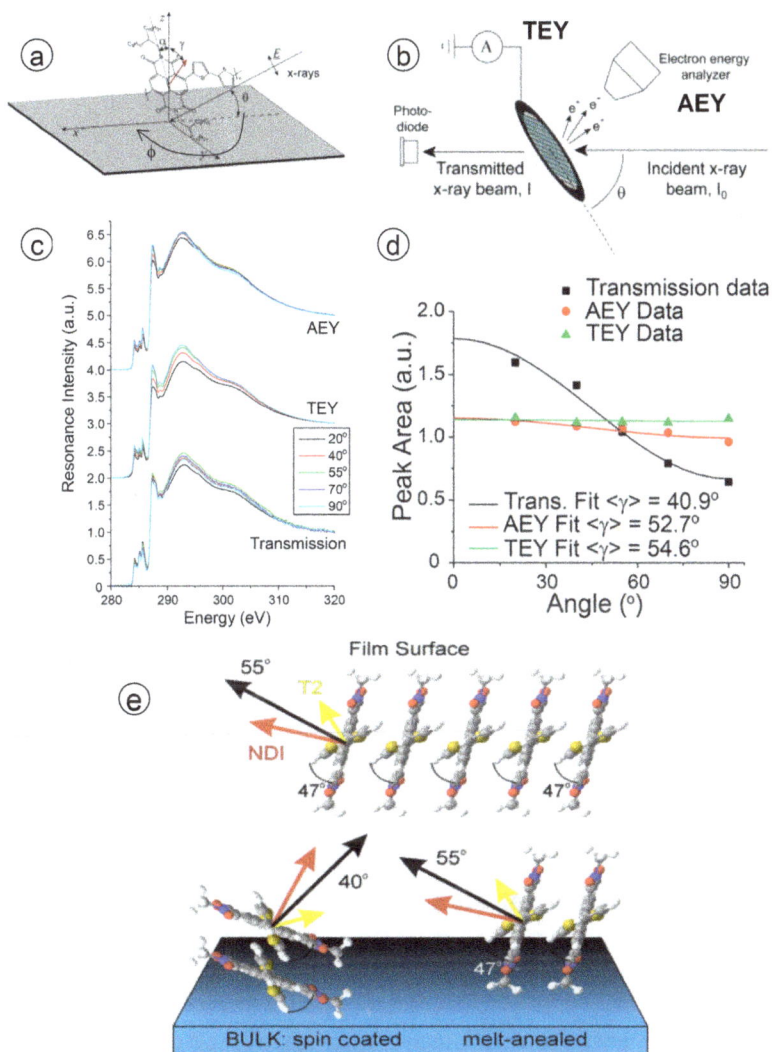

Fig. 3.11: Using angle-resolved NEXAFS to deduce surface versus bulk orientation of P(NDI2OD-T2), an organic field-effect transistor material [73]. (a) Schematic defining the experimental geometry and the polar (θ) and azimuthal (ϕ) angles, for which the polarization dependence was measured. The arrow perpendicular to the molecular plane indicates the dipole transition dipole moment (TDM). (b) Experimental geometry used to simultaneously measure total electron yield (TEY), Auger electron yield (AEY), and transmission signals. (c) Polar angle-resolved AEY, TEY, and transmission NEXAFS spectra of the as-cast P(NDI2OD-T2) film. (d) Polar angle dependence of the 1s → π* peak area for the three different types of spectra. (e) Cartoon of molecular orientation at the surface and bulk based on the measured values of the overall TDM (black arrow). The deduced orientation of the individual naphthalene diimide (NDI, red) and thiophene (T2, yellow) units relative to the TDM are indicated for a dihedral angle of 47°. Only the NDI and T2 cores are shown (adapted with permission from [73] . Copyright 2013, American Chemical Society).

samples were also measured by grazing incidence wide-angle X-ray scattering (GI-WAXS) [73]. However, GIWAXS was unable to differentiate the surface and bulk structure, due to its much larger sampling depth than AEY- or TEY-NEXAFS.

3.6 Spectromicroscopy of polymers

C 1s NEXAFS spectromicroscopy has proven to be of tremendous value in many studies of both the bulk and surface of homogeneous and heterogeneous polymers since the capability was first demonstrated by Ade et al. [12]. Because polymers such as Kevlar exhibit strong molecular alignment, one can use the difference in x-ray absorption between spectra recorded with linear polarization in two (ideally orthogonal) directions to characterize the polymer orientation in the plane of the thin section [74]. A wide variety of multicomponent polymer materials have been studied using implementations of NEXAFS in soft X-ray microscopes. Several reviews of soft X-ray NEXAFS microscopy have been published [17, 18, 75, 76]. C 1s NEXAFS microscopy has been applied to a large number of polymer types [20, 21, 25]. A bibliography of the soft X-ray microscopy literature can be accessed at http://unicorn.mcmaster.ca/xrm-biblio/xrm_bib.html. Two major soft X-ray spectromicroscopy techniques have been applied to polymer surfaces: X-ray photoemission electron microscopy (X-PEEM) and scanning transmission X-ray microscopy (STXM). Here these two NEXAFS microscopies are described, and examples of their application to adsorption of proteins to polymeric biomaterial surfaces are given.

3.6.1 NEXAFS microscopy methods

3.6.1.1 X-ray photoemission electron microscopy (XPEEM)

XPEEM [77] is a full-field microscopy in which a lateral area (2–60 μm) of a surface is illuminated by monochromatic X-rays (Fig. 3.12a). Electrons created near the surface are ejected and accelerated by an electric field (<1 kV/mm) into an electrostatic or magnetic electron microscope column to make a magnified image. The image is detected by a suitable imaging system, typically a channel plate to amplify the electron signal, a phosphor to convert electron pulses to visible light, followed by a digital camera (Fig. 3.12a). XPEEM instruments are housed in ultrahigh vacuum (UHV) chambers ($P < 10^{-9}$ mbar) so polymer films which outgas cannot be studied. NEXAFS spectroscopy and NEXAFS imaging are obtained by measuring sequences of XPEEM images over a range of X-ray energies. Due to the high field at the surface and the high sensitivity of the XPEEM analyzer to secondary electrons, the sampling depth is somewhat larger than typical TEY sampling depths, of the order of 10 nm

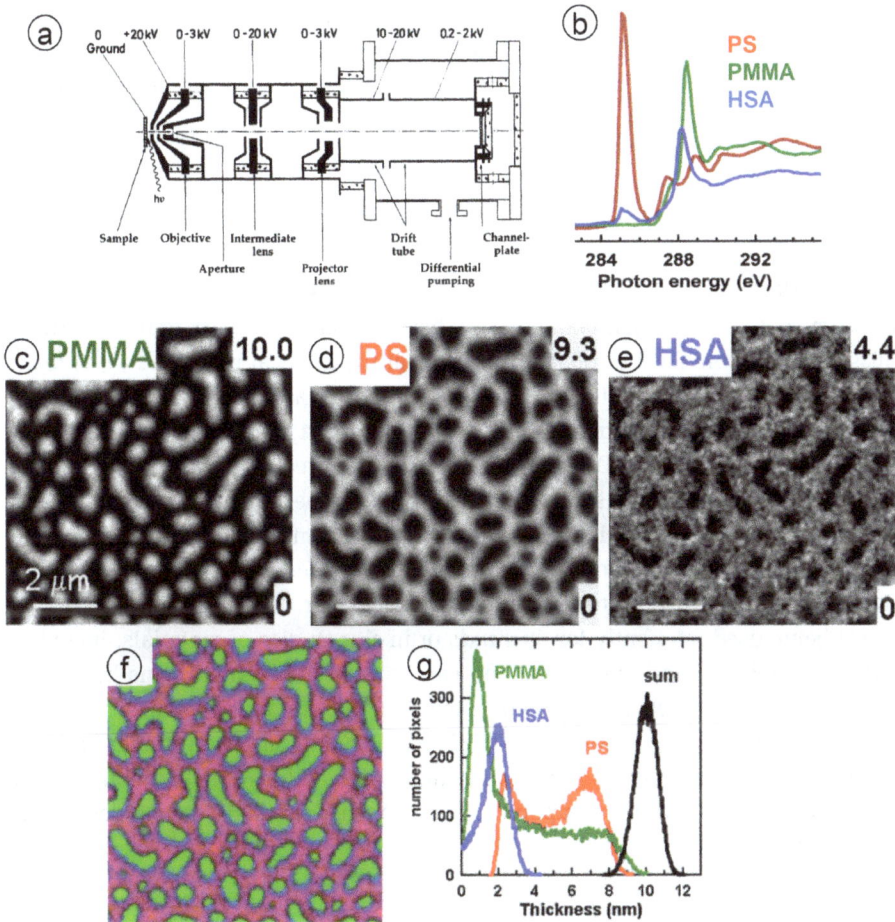

Fig. 3.12: XPEEM visualization of the adsorption of human serum albumin (HSA) on a spun-cast phase-segregated polystyrene/polymethylmethacrylate (PS/PMMA) surface [7, 80]. (a) Schematic of an X-ray photoemission electron microscope (XPEEM).(b) C 1s spectra of PS, PMMA and HSA. Maps of (c) PMMA, (d) PS, and (c) HSA in the single domain thick film. (f) Rescaled color-coded composite of the PMMA (green), PS (red), and HSA (blue) component maps. (g) Histograms of the thickness distributions of the three components. The sum of the gray scales of the three components has been normalized to 10 nm, a sampling depth determined experimentally (from [80], used with permission).

[42]. There are significant challenges to applying XPEEM to polymer surfaces, including sample charging, UHV vacuum incompatibility, radiation damage [48], and sample damage by arcing caused by the high field at the surface. However, with thin samples (<50 nm), rapid shuttering, short exposures, and careful control of total exposure times, XPEEEM has been applied successfully to a number of polymer surfaces, including environmentally responsive polymer brush surfaces [78],

and candidate biomaterials [79]. An example from a study of protein adsorption on a polymer blend is given below. In other studies, XPEEM has been used to investigate competitive adsorption of peptides versus proteins [80] and pH-responsive surfaces [81].

3.6.1.2 Scanning transmission X-ray microscopy (STXM)

In STXM [18, 29, 75, 76] a Fresnel zone plate (ZP) is used to create a finely focused (<30 nm spot size) monochromated X-ray beam. A sample with partial X-ray transmission at the core level of interest is placed at the ZP focal point and (x, y) raster scanned while recording the transmitted X-ray intensity (see Fig. 3.13a). Images constructed from the transmitted intensity are recorded over a range of photon energies, and converted to OD using an I_o signal measured by STXM through a path that includes all except the sample. Since STXM is a type of X-ray in, X-ray out technique, the vacuum requirements are not very stringent, such that measurements can be made at 1 bar He, reduced pressure with saturated water vapor, and on fully hydrated samples with up to 3 μm thick aqueous layer. This flexibility has lead to STXM being used frequently for *operando* or in situ studies of materials, including polymers [82]. Electron yield detection can be used and comparisons have been made of bulk properties, probed by transmission, and surface properties, probed by TEY [83]. Here, in order to compare XPEEM and STXM applied to a similar problem, Section 3.7.3 presents results from visualization of the selective adsorption of a protein at the complex surface of a polyurethane polymer containing two types of reinforcing particles [77].

3.6.2 XPEEM of protein adsorption on PS/PMMA

Many polymer materials are used in medical technology. Frequently, it is desirable to optimize the innate or modified surface chemistry and morphology for specific medical applications. Characterization of the surfaces of such biomaterials and quantitative evaluation of their interaction with relevant proteins, peptides, and other biologically active species is an important part of biomaterials optimization. The goal of imaging polymeric biomaterial surfaces is to determine and understand the chemical and morphological factors which improve biocompatibility. In general, proteins or peptides are the first species to adsorb to biomaterials and thus much of biomaterials optimization involves controlling protein–surface interactions. In this context control may refer to complete prevention or minimization of adsorption (protein resistance and antifouling), or it may refer to the selective promotion of adsorption of one specific protein from the complex mix of species present in a biological tissue or fluid with which the biomaterial is in contact. Thus, studies in which the competition for adsorption by

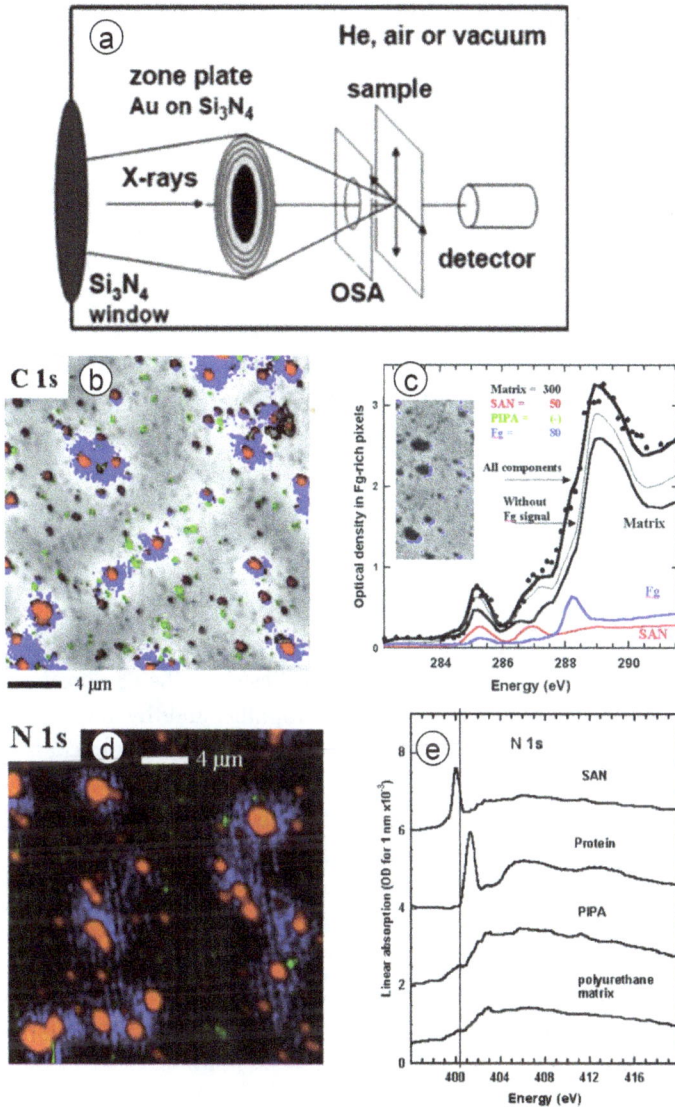

Fig. 3.13: STXM mapping of the adsorption of fibrinogen (Fg) on a polyurethane (matrix) with reinforcing styrene acrylonitrile (SAN) and poly-isocyanate polyaddition product (PIPA) particles [84]. (a) Schematic of a scanning transmission X-ray microscope (STXM). (b) Rescaled color-coded composite of the four-component maps derived from a C 1s image sequence – polyurethane matrix (gray scale); SAN (red); PIPA (green); Fg (blue). (c) Curve fit analysis of the C 1s spectrum of the pixels identified as having large Fg content. The thick line, which is a good match to the data (points), includes the Fg component in the fit, while the thin line, excludes Fg from the fit. The insert shows the pixels selected (Fg in blue on the grayscale polyurethane). (d) Rescaled color-coded composite of SAN (red), PIPA (green), and Fg (blue) distributions derived by fitting an N 1s image sequence. (e) N 1s spectra of the SAN, Fg, PIPA, and polyurethane spectra, used in the fit (from [84], used with permission).

several species to the same surface is studied by chemically sensitive surface imaging are of interest.

Figure 3.12 presents results from an XPEEM study of a 0.005 mg/mL aqueous solution of human serum albumin (HSA), interacted for 20 min with the surface of a phase segregated 30 wt% polystyrene/70 wt% polymethyl methacrylate (PS/PMMA) blend [80]. The surface was thoroughly rinsed and then placed directly into the load lock chamber, and pumped for ~20 min to dry it. Figure 3.12b presents the C 1s NEXAFS spectra of the three species on a comparable absolute intensity scale, showing they can be easily differentiated. These reference spectra were used to fit a C 1s image stack measured by XPEEM and thus derive maps of the surface distribution of the three species. The sampling depth was measured to be 10 nm [80]. The ~ 50 nm thick polymer blend film was prepared by spin-coating dilute toluene solutions of the indicated composition on to native oxide silicon wafers. This polymer blend surface, which has been studied extensively by both XPEEM [79–81] and STXM [84–87] naturally phase segregates to produce isolated 0.5–2 µm sized PMMA domains (Fig. 3.12c) in a continuous PS domain (Fig. 3.12d). The fact that the minority species is the continuous domain is surprising. This is believed to be due to preferential attraction of the PMMA to the hydrophilic oxidized silicon substrate [88], although other factors, such as the higher surface activity of PS [89], may also play a role. This blend surface is an interesting candidate for competitive adsorption studies since the PS domain is strongly hydrophobic, the PMMA domains are quite hydrophilic, and the interphase, a region of 10–30 nm at the boundary of the PS and PMMA domains has both characteristics. Figure 3.12e presents a map of the signal from HSA on the PS/PMMA surface, derived from a three-component fit to a C 1s image sequence [80]. Comparison of the maps of HSA and PS suggests that the protein prefers to adsorb to the hydrophobic PS domain. However, the color-coded composition of the three maps (Fig. 3.12f) shows there is a clear preference to the PS–PMMA interphase. HSA is an amphiphilic protein, with both hydrophobic and hydrophilic regions at its surface, so the interphase offers more opportunities for strong binding. In addition to mapping of HSA on PS/PMMA, competitive adsorption of HSA and SUB-6, an antimicrobial peptide, to the PS/PMMA surface was also measured [80].

3.6.3 STXM of protein (Fg) adsorption on a reinforced polyurethane

Polyurethanes (PU) are commonly used in medical applications due to their favorable mechanical and chemical properties. The adsorption of fibrinogen (Fg) to a complex multicomponent polyurethane thin film in which the polyether-rich toluene-di-isocyante (TDI) polyurethane matrix was reinforced with styrene-b-acrylonitrile (SAN) particles and poly-isocyanate poly-addition product (PIPA, a methylene diphenyl diisocyanate (MDI)-based hard segment-like material) particles. A STXM study of the chemistry and morphology of the polyurethane [90, 91] showed the SAN and PIPA particles could be readily differentiated from their NEXAFS signals. The lateral distribution of the Fg protein on the TDI-PU_SAN_PIPA surface was determined by STXM [84] for three types of samples: (i) Fg exposure from phosphate buffer solution, followed by rigorous rinsing with buffer, then dried. (ii) the same protocol but using exposure to a pure aqueous overlayer, followed by rinsing and drying. (iii) an in situ study in which the polyurethane film covered by ~ 1 µm of dilute protein solution trapped between two silicon nitride windows. Figure 3.13 presents results from a C 1s and N 1s STXM study of the in situ sample [84]. Both the C 1s (Fig. 3.13b,c) and N 1s (Fig. 3.13d,e) results show that Fg prefers to adsorb to regions of the matrix at the edges of the larger SAN particles. While the C 1s analysis was quite subtle, and required careful peak fitting analysis (Fig. 3.13c), the N 1s edge was particularly informative, since protein has many N atoms and a characteristic spectrum. While the SAN particles also have significant amounts of nitrogen, the N 1s $\rightarrow \pi^*_{nitrile}$ transition (399.8 eV) is easily distinguished from the N 1s $\rightarrow \pi^*_{amide}$ transition (401.2 eV) (see Fig. 3.13e). Figure 3.13d shows a color-coded composite derived from a N 1s image sequence. A similar conclusion as to preferred adsorption site was reached for the ex situ Fg exposed sample measured after rinsing and drying. While the detailed mechanism of Fg adsorption to the SAN–polyurethane interface is not clear, surface topography may be playing a role since the SAN particles protrude up to 50 nm from the surface.

3.7 Summary

From a purely spectroscopy perspective, there is ample flux at existing synchrotron facilities to apply NEXAFS spectroscopy to most polymer science problems where the sample is unchanging or only changing on a second or slower timescale. Radiation damage is a significant limitation for NEXAFS microscopy of many polymers, particularly fluorinated species such as perfluorosulfonic acid (PFSA) ionomer [48, 92, 93]. However, if an unfocused or intentionally defocused beam is used, the dose can be reduced to the point where radiation damage is not a problem, of course, at the expense of spatial resolution. At this time, a revolution happening in synchrotron sources improves the flux and brightness of the monochromated X-rays [94]. These two to three orders of magnitude improvement will make NEXAFS spectroscopy and microscopy significantly more powerful. For purely spectroscopic applications, the new fourth-generation light sources will be able to follow the kinetics and dynamics of chemical and physical transformations of polymer surfaces [33, 82, 95–97]. As an example, the surface structure of hydrated polymers used as bioscaffolds for artificial tissue is known to be important to the success of tissue growth [98]. Static, ex situ approaches are giving insights [79] but the ability to follow the surface re-organization and watch the tissue formation in real time is probably within the capabilities of NEX-AFS spectroscopy at fourth-generation light sources. In addition to improving "conventional" STXM, the much higher brightness will make dramatic improvements to the newly emerging method of soft X-ray ptychography which already is providing spatial resolutions below 10 nm [99, 100]. It is expected, once systems are fully optimized, the spatial resolution will routinely be limited only by wavelength of the X-rays (~4 nm at the C 1s edge). This is almost the case for the dedicated COSMIC ptychography system at the Advanced Light source (Berkeley, CA, USA) [100]. As an example, F 1s ptychography has been used to map PFSA ionomer in the membrane electrode assembly of polymer electrolyte membrane fuel cells (PEM-FC) with a resolution better than 15 nm in 2D and 30 nm in 3D [101]. In general, in situ and *operando* studies of polymer surfaces are emerging as an increasingly important application of NEXAFS. The future of NEXAFS, as a tool to study both bulk and surface properties of polymers, is very bright indeed!

References

[1]   Rubensson J-E. Synchrotron Radiation: An everyday application of special relativity. IOP Concise Phys 2016.
[2]   Hanawalt JD. The dependence of x-ray absorption spectra upon chemical and physical state. J D Phys Rev 1931, 37, 715–726.
[3]   Röntgen WC. Uber eine neue art von strahlen.I Mitteilung. Sitzungs-Berichtedel Physikalisch-medicinischen Gesellshaft Zu Würzburg 1895, 137, 132–141.
[4]   Brion CE, Daviel S, Sodhi R, Hitchcock AP Recent advances in inner-shell excitation of free molecules by electron energy loss spectroscopy, AIP Conference Proceedings 1982;92:429–446.
[5]   Koch EE, Sonntag BF. Molecular spectroscopy. In: Kunz C, ed. Synchrotron Radiation Techniques and Applications, Berlin, Springer, Topics Curr Phys, 1979, 10.269-355.
[6]   Chen CT, Sette F. Performance of the Dragon soft x-ray beamline. Rev Sci Instr 1998, 60, 1616–1621.
[7]   Stöhr J. NEXAFS Spectroscopy, Springer Ser Surf Sci, Vol. 25, Springer,, 1992.
[8]   Stöhr J, Outka DA. Determination of molecular orientations on surfaces from the angular dependence of near-edge x-ray-absorption fine-structure spectra. Phys Rev B 1987, 36, 7891–7905.
[9]   Jordan-Sweet JL, Kovac CA, Goldberg MJ, Morar JF. Polymer/metal interfaces studied by carbon near-edge x-ray absorption fine structure spectroscopy. J Chem Phys 1988, 89, 2482–2489.
[10]   Outka DA, Stöhr J, Rabe JP, Swalen JD. The orientation of Langmuir–Blodgett monolayers using NEXAFS. J Chem Phys 1988, 88, 4076–4087.
[11]   DeLongchamp D, Lin EK, Fischer DA. Organic semiconductor structure and chemistry from near-edge x-ray absorption fine structure (NEXAFS) Spectroscopy. Proc SPIE Organic Field-Effect Transistors IV 2005, 5940:59400A.
[12]   Ade H, Zhang X, Cameron S, Costello C, Kirz J, Williams S. Chemical contrast in X-ray microscopy and spatially resolved XANES spectroscopy of organic specimens. Science 1992, 258, 972–975.

[13]   1Hähner G. Near edge X-ray absorption fine structure spectroscopy as a tool to probe electronic and structural properties of thin organic films and liquids. Chem Soc Rev 2006, 35, 1244–1255.

[14]   DeLongchamp DM, Kline RJ, Fischer DA, Richter LJ, Toney MF. Molecular characterization of organic electronic films. Adv Mater 2011, 23, 319–337.

[15]   Watts B, Thomsen L, Dastoor PC. Methods in carbon K-edge NEXAFS: Experiment and analysis. J Electron Spectrosc Rel Phenom 2006, 151, 105–120.

[16]   Unger WES, Lippitz A, Wöll C, Heckmann W. X-ray absorption spectroscopy (NEXAFS) of polymer surfaces. Fresenius J Anal Chem 1997, 358, 89–92.

[17]   Ade H, Urquhart SG. NEXAFS spectroscopy and microscopy of natural and synthetic polymers, Chapter 4 of "Chemical applications of synchrotron radiation" (T.K. Sham, ed.) Part I: Dynamics and VUV Spectroscopy. In: Advanced Series in Physical Chemistry Vol 12A, Singapore, World Scientific, 2002, 154–227.

[18]   Ade H, Hitchcock AP. NEXAFS microscopy and resonant scattering: Composition and orientation probed in real and reciprocal space. Polymer 2008, 49, 643–675.

[19]   Son D, Cho S, Nam J, Lee H, Kim M. X-ray-Based spectroscopic techniques for characterization of polymer nanocomposite materials at a molecular level. Polymers 2020, 12, 1053.

[20]   Dhez O, Ade H, Urquhart SG. Calibrated NEXAFS spectra of some common polymers. J El Spec Rel Phen 2003, 128, 85–96.

[21]   Urquhart SG, Hitchcock AP, Smith AP, et al. NEXAFS spectromicroscopy of polymers: Overview and quantitative analysis of polyurethane polymers. J Electron Spectrosc Rel Phenom 1999, 100, 119–135.

[22]   Urquhart SG, Smith AP, Ade HW, Hitchcock AP, Rightor EG, Lidy W. Near-edge X-ray absorption fine structure spectroscopy of MDI and TDI polyurethane polymers. J Phys Chem B 1999, 103, 4603–4610.

[23]   Urquhart SG, Hitchcock AP, Leapman RD, Priester RD, Rightor EG. Analysis of polyurethanes using core excitation spectroscopy. Part I: Model polyurethane foam polymers. J Polym Sci Part B: Polym Phys 1995, 33, 1593–1602.

[24]   Urquhart SG, Hitchcock AP, Priester RD, Rightor EG. Analysis of polyurethanes using core excitation spectroscopy. Part II: Inner shell spectra of ether, urea and carbamate model compounds. J Polym Sci Part B: Polym Phys 1995, 33, 1603–1620.

[25]   Smith AP, Urquhart SG, Winesett DA, Mitchell G, Ade H. Use of near edge x-ray absorption fine structure spectromicroscopy to characterize multicomponent polymeric systems. Appl Spectrosc 2001, 55, 1676–1681.

[26]   Graf N, Yegen E, Gross T, et al. XPS and NEXAFS studies of aliphatic and aromatic amine species on functionalized surfaces. Surf Sci 2009, 603, 2849–2860.

[27]   Watts B, Swaraj S, Nordlund D, Lüning J, Ade H. Calibrated NEXAFS spectra of common conjugated polymers. J Chem Phys 2011, 134, 024702.

[28]   Henke BL, Gullikson EM, Davis JC. X-ray interactions: Photoabsorption, scattering, transmission, and reflection at E = 50-30,000 eV, Z = 1-92. At Data Nucl Data Tables 1993, 54, 181–342.

[29]   Jacobsen C. X-ray Microscopy, Cambridge University Press, 2019.

[30]   Guillemin R, Decleva P, Stener M, et al. Selecting core-hole localization or delocalization in CS_2 by photofragmentation dynamics. Nat Commun 2015, 6, 6166.

[31]   Ishii I, Hitchcock AP. A Quantitative experimental study of the core excited electronic states in formamide, formic acid and formyl fluoride. J Chem Phys 1987, 87, 830–839.

[32]   Wen AT, Ruehl E, Hitchcock AP. Inner-shell excitation of organoiron compounds by electron impact. Organometallics 1992, 11, 2559–2569.

[33] Su GM, Cordova IA, Brady MA, Prendergast D, Wang C. Combining theory and experiment for X-ray absorption spectroscopy and resonant X-ray scattering characterization of polymers. Polymer 2016, 99, 782–796.

[34] Su GM, Patel SN, Pemmaraju CD, Prendergast D, Chabinyc ML. First-principles predictions of near-edge x-ray absorption fine structure spectra of semiconducting polymers. J Phys Chem C 2017, 121, 9142–9152.

[35] Michelitsch GS, Reuter K. Efficient simulation of near-edge x-ray absorption fine structure (NEXAFS) in density-functional theory: Comparison of core-level constraining approaches. J Chem Phys 2019, 150, 074104.

[36] Kosugi N. Strategies to vectorize conventional SCF-CI algorithms. Theoret Chim Acta 1987, 72, 149–173.

[37] Schöll A, Fink R, Umbach E, Mitchell GE, Urquhart SG, Ade H. Towards a detailed understanding of the NEXAFS spectra of bulk polyethylene copolymers and related alkanes. Chem Phys Lett 2003, 370, 834–841.

[38] Nagasaka M, Yuzawa H, Kosugi N. Intermolecular interactions of pyridine in liquid phase and aqueous solution studied by soft x-ray absorption spectroscopy. Zeitschrift Für Physikalische Chemie 2018, 232, 705–722.

[39] Zimmermann P, Peredkov S, Abdala PM, et al. Modern X-ray spectroscopy: XAS and XES in the laboratory. Coord Chem Rev 2020, 423, 213466.

[40] Lewis S, Gelb J, Sh L, Yun W, Vine D, Stripe B, Seshadri S. Recent Developments in laboratory x-ray microanalytical techniques for electronic structure, chemical composition, and microstructure of metals and materials. Microsc Microanal 2002, 26(S2), 516–516.

[41] Paroli B, Potenza MAC. Radiation emission processes and properties: Synchrotron, undulator and betatron radiation. Adv Phys X 2017, 2, 978–1004.

[42] Wolfgang Grünert and Konstantin Klementiev, X-ray absorption spectroscopy principles and practical use in materials analysis, De Gruyter, (accessed: March 5, 2020 at https://doi.org/10.1515/psr-2017-0181)

[43] Dohlus M, Rossbach J, Bethge KHW, et al. Application of accelerators and storage rings. In: Myers S, Schopper H, eds. Particle Physics Reference Library : Volume 3: Accelerators and Colliders, Springer International Publishing, 2020, 661–795.

[44] Kaznacheyev KV, Karunakaran C, He F, Sigrist M, Summers T, Obst M, Hitchcock AP. CLS ID-10 chicane configuration: From "simple" sharing to extended performance with high speed polarization switching. Nucl Inst Meth A 2007, 582, 103–106.

[45] Gann E, Crofts T, Holland G, et al A NIST facility for resonant soft x-ray scattering measuring nano-scale soft matter structure at NSLS-II. J Phys: Condens Matter 2021, 33, 164001.

[46] Hanhan S, Smith AM, Obst M, Hitchcock AP. Optimization of analysis of Ca 2p soft X-ray spectromicroscopy. J Electron Spectrosc Rel Phenom 2009, 173, 44–49.

[47] Gergely G, Menyhard M, Sulyok A, et al. Evaluation of the inelastic mean free path (IMFP) of electrons in polyaniline and polyacetylene samples obtained from elastic peak electron spectroscopy (EPES). Central Eur J Phys 2007, 5, 188–200.

[48] Wang J, Morin C, Li L, Hitchcock AP, Scholl A, Doran A. Radiation damage in soft X-ray microscopy. J El Spec Rel Phen 2009, 170, 25–36.

[49] Aygül U, Batchelor D, Dettinger U, et al. Molecular orientation in polymer films for organic solar cells studied by NEXAFS. J Phys Chem C 2012, 116, 4870–4874.

[50] Wu W, Sambasivan S, Wang C-Y, Wallace WE, Genzer J, Fischer DA. A direct comparison of surface and bulk chain-relaxation in polystyrene. Eur Phys J E 2003, 12, 127–132.

[51] Lenhart JL, Fischer DA, Chantawansri TL, Andzelm JW. Surface orientation of polystyrene based polymers: Steric effects from pendant groups on the phenyl ring. Langmuir 2012, 28, 15713–15724.

[52] Tamenori Y, Morita M, Nakamura T. Two-dimensional approach to fluorescence yield XANES measurement using a silicon drift detector. J Syn Rad 2011, 18, 747–752.

[53] Asakura D, Hosono E, Nanba Y, et al. Material/element-dependent fluorescence-yield modes on soft X-ray absorption spectroscopy of cathode materials for Li-ion batteries. AIP Adv 2016, 6, 035105.

[54] Chiang Y-J, Huang W-C, Ni C-K, Liu C-L, Tsai -C-C, Hu W-P. NEXAFS spectra and specific dissociation of oligo-peptide model molecules. AIP Adv 2019, 9, 085023.

[55] Ikeura-Sekiguchi H, Sekiguchi T, Baba Y, Imamura M, Matsubayashi N, Shimada H. Direct and indirect processes in photon-stimulated ion desorption from condensed formamide. Surf Sci 2005, 593, 303–309.

[56] Tardio S, Cumpson PJ. Practical estimation of XPS binding energies using widely available quantum chemistry software. Surface & Interface Analy 2018, 50, 5–12.

[57] Sette F, Stöhr J, Hitchcock AP. Determination of intramolecular bond lengths in gas phase molecules from K shell shape resonances. J Chem Phys 1984, 81, 4906–4914.

[58] Cooper G, Gordon M, Tulumello D, Turci C, Kaznatcheev K, Hitchcock AP. Inner shell excitation of glycine, glycyl-glycine, alanine and phenylalanine. J Electron Spectros Relat Phenomena 2004, 137–140, 795–799.

[59] Urquhart S, Ade H. Trends in the carbonyl core (C 1s, O 1s) → $\pi^*_{C=O}$ transition in the near-edge x-ray absorption fine structure spectra of organic molecules. J Phys Chem B 2002, 106, 8531–8538.

[60] Fu J, Urquhart SG. Linear dichroism in the X-ray absorption spectra of linear n-alkanes. J Phys Chem A 2005, 109, 11724–11732.

[61] Horsley JA, Stöhr J, Hitchcock AP, Newbury DC, Johnson AL, Sette F. Resonances in the K shell excitation spectra of benzene and pyridine: Gas phase, solid, and chemisorbed states. J Chem Phys 1998, 83, 6099.

[62] Liu AC, Stöhr J, Friend CM, Madix RJ. A critical interpretation of the near-edge X-ray absorption fine structure of chemisorbed benzene. Surf Sci 1990, 235, 107–115.

[63] Urquhart SG, Lanke U, Fu J. Characterization of molecular orientation in organic nanomaterials by X-ray linear dichroism. Int J Nanotech 2008, 5, 1138–1170.

[64] Nordén B. Applications of linear dichroism spectroscopy. Appl Spectrosc Rev 1978, 14, 157–248.

[65] Rodger A, Norden B. Circular Dichroism and Linear Dichroism, Oxford, U.K, Oxford University Press, 1997.

[66] Gamble LJ, Ravel B, Fischer DA, Castner DG. Surface structure and orientation of PTFE films determined by experimental and FEFF8-calculated NEXAFS spectra. Langmuir 2002, 18, 2183–2189.

[67] Bräuer B, Virkar A, Mannsfeld SCB, et al. X-ray microscopy imaging of the grain orientation in a pentacene field-effect transistor. Chem Mater 2010, 22, 3693–3697.

[68] Rocco MLM, Häming M, De Moura CEV, et al. High-resolution near-edge x-ray absorption fine structure study of condensed polyacenes. J Phys Chem C 2018, 122, 28692–28701.

[69] Gainar A, Stevens JS, Jaye C, Fischer DA, Schroeder SLM. NEXAFS sensitivity to bond lengths in complex molecular materials: A study of crystalline saccharides. J Phys Chem B 2015, 119, 14373–14381.

[70] Ankudinov AL, Ravel B, Rehr JJ, Conradson SD. Real-space multiple-scattering calculation and interpretation of x-ray-absorption near-edge structure. Phys Rev B 1998, 58, 7565–7576.

[71] Ravel B, Newville M. Athena, Artemis, Hephaestus: Data analysis for X-ray absorption spectroscopy using Ifeffit. J Synchrotron Radiat 2005, 12(Pt 4), 537–541.

[72] Ravel B, Newville M. Athena and Artemis: Interactive graphical data analysis using Ifeffit. Phys Scr 2005, T115, 1007.

[73] Schuettfort T, Thomsen L, McNeill CR. Observation of a distinct surface molecular orientation in films of a high mobility conjugated polymer. J Am Chem Soc 2013, 135, 1092–1101.

[74] Ade H, Hsiao B. X-ray linear dichroism microscopy. Science 1993, 262, 1427–1429.

[75] Hitchcock AP, Stöver HDH, Croll LM, Childs RF. Chemical mapping of polymer microstructure using soft x-ray spectromicroscopy. Aus J Chem 2005, 58, 423.

[76] Hitchcock AP. Soft X-ray imaging and spectromicroscopy Chapter 22. In: Tendeloo GV, Dyck DV, Pennycook SJ, eds. Handbook on Nanoscopy, Volume II, Wiley, 2012, 745–791.

[77] Anders S, Padmore HA, Duarte RM, et al. Photoemission electron microscope for the study of magnetic materials. Rev Sci Inst 1999, 70, 3973–3981.

[78] Li M, Pester CW. Mixed polymer brushes for "smart" surfaces. Polymers 2020, 12, 1553.

[79] Leung BO, Brash JL, Hitchcock AP. Characterization of biomaterials by soft x-ray spectromicroscopy. Materials (Basel) 2010, 3, 3911–3938.

[80] Leung BO, Hitchcock AP, Cornelius RM, Brash JL, Scholl A, Doran A. Using X-PEEM to study biomaterials: Protein and peptide adsorption to a polystyrene–poly(methyl methacrylate)-b-polyacrylic acid blend. J El Spec Rel Phen 2012, 185, 406–416.

[81] Hitchcock AP, Leung BO, Brash JL, Scholl A, Doran A. Soft X-ray spectromicroscopy of protein interactions with phase-segregated polymer surfaces. In: Proteins at Interfaces III State of the Art. Vol 1120. ACS Symposium Series.; 2012:731–760.

[82] Fink RH, Rosner B, Du X, et al. In-operando soft X-ray microspectroscopy of organic electronics devices. Microsc Microanal 2018, 24(S2), 424–425.

[83] Watts B, McNeill CR. Simultaneous surface and bulk imaging of polymer blends with X-ray spectromicroscopy. Macromol Rapid Commun 2010, 31, 1706–1712.

[84] Hitchcock AP, Morin C, Heng YM, Cornelius RM, Brash JL. Towards practical soft X-ray spectromicroscopy of biomaterials. J Biomater Sci, Polym Ed 2002, 13, 919–937.

[85] Morin C, Ikeura-Sekiguchi H, Tyliszczak T, et al. X-ray spectromicroscopy of immiscible polymer blends: Polystyrene–poly(methyl methacrylate). J Electron Spectros Relat Phenomena 2001, 121, 203–224.

[86] Li L, Hitchcock AP, Robar N, et al. X-ray microscopy studies of protein adsorption on a phase-segregated polystyrene/polymethyl methacrylate surface. 1. Concentration and exposure-time dependence for albumin adsorption. J Phys Chem B 2006, 110, 16763–16773.

[87] Li HAP, Cornelius R, Brash JL, Scholl A, Doran A. X-ray microscopy studies of protein adsorption on a phase segregated polystyrene/ polymethylmethacrylate surface. 2. Effect of pH on site preference. J Phys Chem B 2008, 112, 2150–2158.

[88] Harris M, Appel G, Ade H. Surface morphology of annealed polystyrene and poly(methyl methacrylate) thin film blends and bilayers. Macromolecules 2003, 36, 3307–3314.

[89] Winesett DA, Story S, Luning J, Ade H. Tuning substrate surface energies for blends of polystyrene and poly(methyl methacrylate). Langmuir 2003, 19, 8526–8535.

[90] Hitchcock AP, Koprinarov I, Tyliszczak T, et al. Optimization of scanning transmission X-ray microscopy for the identification and quantitation of reinforcing particles in polyurethanes. Ultramicroscopy 2001, 88, 33–49.

[91] Rightor EG, Urquhart SG, Hitchcock AP, et al. Identification and quantitation of urea precipitates in flexible polyurethane foam formulations by X-ray spectromicroscopy. Macromolecules 2002, 35, 5873–5882.

[92] Wu J, Melo LGA, Zhu X, et al. 4D imaging of polymer electrolyte membrane fuel cell catalyst layers by soft X-ray spectro-tomography. J Power Sources 2018, 381, 72–83.

[93]   Martens I, Melo LGA, Wilkinson D, Bizzotto D, Hitchcock AP. Characterization of X-ray damage to perfluorosulfonic acid using correlative microscopy. J Phys Chem C 2019, 123, 16023–16033.

[94]   Hitchcock AP, Toney MF. Spectromicroscopy and coherent diffraction imaging: Focus on energy materials applications. J Synchrotron Rad 2014, 21, 1019–1030.

[95]   Liu Y, Russell TP, Samant MG, et al. Surface relaxations in polymers. Macromolecules 1997, 30, 7768–7771.

[96]   Winter AD, Larios E, Alamgir FM, et al. Thermo-active behavior of ethylene-vinyl acetate / multiwall carbon nanotube composites examined by in situ near-edge x-ray absorption fine-structure spectroscopy. J Phys Chem C 2014, 118, 3733–3741.

[97]   Winter AD, Czaniková K, Larios E, et al. Interface Dynamics in strained polymer nanocomposites: Stick–slip wrapping as a prelude to mechanical backbone twisting derived from sonication-induced amorphization. J Phys Chem C 2015, 119, 20091–20099.

[98]   Dhandayuthapani B, Yoshida Y, Maekawa T, Kumar DS. Polymeric scaffolds in tissue engineering application: A review. Int J Polym Sci 2011, 2011, e290602.

[99]   Stampanoni M, Menzel A, Watts B, Mader KS, Bunk O. Coherent x-ray imaging: Bridging the gap between atomic and micro-scale investigations. Chimia Int J Chem 2014, 6, 66–72.

[100] Shapiro DA, Babin S, Celestre RS, et al. An ultrahigh-resolution soft x-ray microscope for quantitative analysis of chemically heterogeneous nanomaterials. In: Science Advances, Vol. 6, 2020, eabc4904.

[101] Wu J, Zhu XH, West MM, et al. High resolution imaging of polymer electrolyte membrane fuel cell cathode layers by soft X-ray spectro-ptychography. J Phys Chem C 2018, 122, 11709–11719.

[102] Francis JT, Hitchcock AP. Inner-shell spectroscopy of p-benzoquinone, hydroquinone, and phenol: Distinguishing quinoid and benzenoid structures. J Phys Chem 1992, 96, 6598–6610.

[103] Francis JT, Hitchcock AP. Distinguishing keto and enol structures by inner-shell spectroscopy. J Phys Chem 1994, 98, 3650–3657.

[104] Scheinost AC, Kretzschmar R, Christ I, Jacobsen C. Carbon group chemistry of humic and fulvic acid: A comparison of C-1s NEXAFS and ^{13}C-NMR spectroscopies. Chapter 9 in: Humic Substances: Structures. In:: Ghabbour EA, Davies G, eds. Models and Fnuctions, Cambridge, RSC, 2001, 39–48.

[105] Schumacher M, Christl I, Scheinost AC, Jacobsen C, Kretzschmar R. Chemical heterogeneity of organic soil colloids investigated by scanning transmission x-ray microscopy and C-1s NEXAFS microspectroscopy. Environ Sci Technol 2005, 39, 9094–9100.

[106] Schäfer T, Hertkorn N, Artinger R, Claret F, Bauer A. Functional group analysis of natural organic colloids and clay association kinetics using C(1s) spectromicroscopy. J Phys IV France 2003, 104, 409–412.

Beat A. Keller, Marco Consumi, Gemma Leone, Agnese Magnani

4 Investigation of polymer surfaces by time-of-flight secondary ion mass spectrometry

4.1 Introduction

4.1.1 Analysis of surfaces

The surface properties of polymers are relevant in many applications such as adhesion, friction and wear, biomaterials, composites, microelectronic devices, protective coatings, and thin-film technology. As surface properties of any system are governed by the molecular structure of its outermost layers, surface analysis is an indispensable tool for the design and development of polymeric systems.

Several surface-sensitive analytical techniques are known today that provide the scientist, who is working in the field of polymer surface characterization, with a set of tools to address the basic questions of analytical science. Among the numerous questions to be answered, the most frequent are, generally, relative to:
1. which compounds are found on the surface,
2. how they are laterally distributed,
3. how much of the material is found on the surface.

It is common knowledge that no single technique provides high-quality answers to all three of the basic analytical challenges. At the same time, it means mastering the field by choosing the most powerful analytical tool to address the problem at hand. ToF-SIMS is acknowledged as one of the major techniques for surface characterization of solids, owing to its unique features, including high molecular specificity, high surface sensitivity, and sub-micrometer imaging resolution [1–3]. In the last two decades, many advances have been made in the SIMS analysis of polymeric materials. In particular, a better understanding of the sputtering mechanisms of polymeric materials has been gained using molecular dynamics simulations [4], multivariate analysis methods have been employed to interpret data [5], and new techniques such as cluster ion beams and metal-assisted SIMS have been developed [6, 7]. These advances not only have improved the SIMS capabilities for polymer analysis, but they have also opened up new application perspectives such as molecular depth profiling, which allows polymeric materials to be analyzed along the depths as wells as at the surface [8–11].

https://doi.org/10.1515/9783110701098-004

This chapter will provide basic insight about the possibilities and limitation of time-of-flight secondary ion mass spectrometry (ToF-SIMS) to obtain useful information about the chemistry of polymer surfaces [1, 12–18]. Experiments concerning measurements about polymer characterization, end-group chemistry, chemically or physically modified polymers, molecular weight determination, polymer blends and additive chemistry have been reported. In recent years, ToF-SIMS has become important also for the characterization and utilization of biopolymers [19]. However, this book chapter does not cover this topic. In principle, the technique is straightforward. A high energy focused and pulsed primary ion beam is directed at the location of interest on the surface, inducing a more or less controlled sputter process. As a consequence, atomic and molecular secondary ions, characteristic of the surface chemistry, are released, collected, and directed through a mass analyzer system to a counting detector.

Mass and identity can then be estimated according to simple kinetic velocity-to-mass relations given by classical charged particle mechanics: $t_{ToF} \propto U_{acc}^{-1} \propto (m/z)^{1/2}$ [20, 21]. The detection of co-emitted secondary neutral species and secondary electrons requires additional detection schemes [22–25].

The first observation of the ion impact-induced emission of positive and neutral ions from a metal plate target was observed in 1910 by J. J. Thomson [26] during his famous "Kanalstrahlen" experiment. After World War II and during the moon exploration period, the first commercial instruments were constructed for the isotopic composition analysis of space-borne materials. At the same time, the analysis of semiconductor and thin film samples has become more important. In fact, the semiconductor industry today is still one of the most important users of applications for SIMS analysis. The generic outline of a ToF-SIMS experiment is schematically shown in Fig. 4.1.

Data can be represented as a spectrum or image. Access to subsurface or three-dimensional (3D) analysis via sputtering is also important but is not within the scope of this work. Raw data obtained from a two-dimensional (2D) imaging experiment can be stored and reloaded for retrospective data analysis. This analytical method is based on selecting areas of interest on the image and subsequently reconstructing a ToF-SIMS spectrum of it. Inversely, signals from a spectrum can be used to reconstruct an image in which the intensity scale of colors assigned to individual mass fragments is proportional to the number of counts in each pixel and therefore represents the 2D distribution of peaks found in the ToF-SIMS spectrum of investigated area. No preselection of signals for ToF-SIMS mapping is necessary because all data are available for inspection long after finishing the experiment.

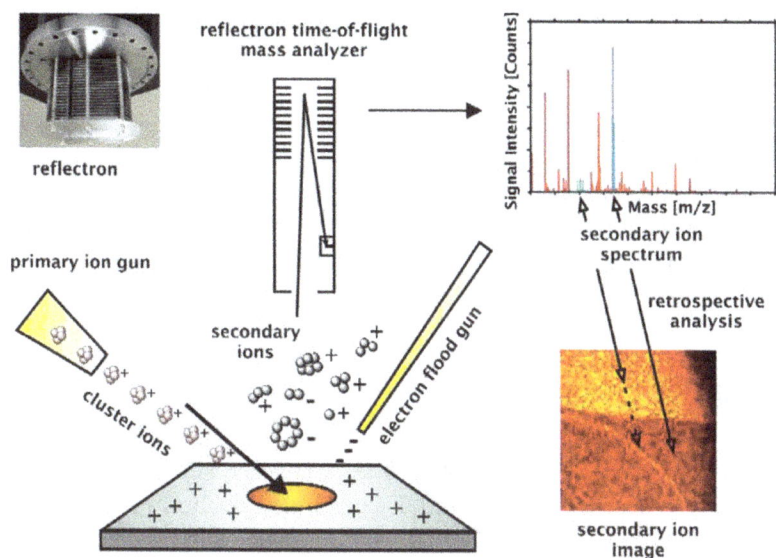

Fig. 4.1: Schematic drawing of a generic ToF–SIMS experiment. Gold- or bismuth primary ion sources have replaced classical gallium, indium or noble gas guns almost completely. The use of a reflectron ion energy focusing element is standard equipment for high-mass-resolution ToF-SIMS applications. Data can be represented as mass spectrum or mass–resolved ion image. The electron flood gun is operating between measurement cycles to compensate excessive surface charging. ©Empa (2012).

4.1.2 The SIMS process: a detailed approach of theory and instruments

The Secondary Ion Mass Spectrometry, SIMS, is the mass spectrometry of surfaces. In particular, it is the mass spectrometry of ionized particles emitted from a surface bombarded by energetic primary particles that are usually ions [1].

True surface sensitivity of the SIMS technique was achieved when Benninghoven and colleagues [27–30] introduced static SIMS (SSIMS) using a high-resolution ToF mass analyzer that allowed limiting the ion current from the primary source to a value of less than approximately 10^{-10} to 10^{-9} A/cm^2. A fraction as little as 1% of the outermost surface layers of a material is typically consumed to obtain a meaningful ToF-SIMS spectrum. This mode of operation was called SSIMS technology. The quality of information depends, of course, on signal strength and particle yield. Based on a number of experiments, it is widely accepted that through SSIMS experiments, a good description of polymer surface chemistry can be achieved [31, 32].

The impact of energetic charged projectiles is always locally destructive. The degree of disruption of the affected area around the site of collision is dependent on the nature of both the primary ion used for analysis and investigated material. Most of the particles that are released as a consequence of primary ion impact are neutral (90%). Only few emitted particles are ions. Thus, a very sensitive analyzer is required to reveal secondary ions. The percentage of secondary ionized particles can be increased *via* post-ionization using a temporally correlated pulsed laser system [22–25, 33]. The surface sensitivity of the SIMS process is not related to the primary ions (source) used to bombard the surface, but it is explained by the fact that the secondary species emitted from the surface are originating only from the topmost monolayers (1–2 nm) of a material. All the species originated within deeper layers are not emitted from the surface because they are stopped during the collisional process.

4.1.2.1 Sputter process and SSIMS approach

The process of secondary ions emission in static SIMS is conceptually simple: when a high-energy (1–25 keV) beam of ions or neutrals bombards a surface, the energy of the impinging particles is transferred to the atoms of the solid matrix in many cases via a collision cascade introduced by Sigmund [34, 35].

The interaction of energetic ions with a solid can be described by the following mechanism (stopping power describes the energy loss inside the solid as function of distance from the surface):

- Nuclear stopping power: Direct interaction between ion and target nuclei (elastic Coulomb interaction). This process is dominant for primary energies $E \leq 100$ keV/z.
- Electronic stopping power: Inelastic coupling between ion and target atom electrons (primary ions in the mega-electron-volt range).
- Exchange stopping power: Exchange of electric charges between ion and target (small).

The typical diameter of a collision cascade is about 10 nm around the ion impact site. The linear cascade theory has been successful in describing sputtering in many materials. However, it does not include extreme thermal processes and therefore situations with high damage. Amorphization, site damage, or ion-induced reaction are taking place within a time span of a few femtoseconds after impact. In complex materials such as polymers or biological samples, the cascade theory, which is based on elastic collisions between point charge particles fails. In cluster ion experiments, the average energy of individual particles is low, and interactions take place *via* large distance and directional energy transfer [36]. It has been realized by Benninghoven and colleagues [27, 28] that an ion current density of 10^{-10} up to 10^{-9} A/cm^2 (ion dose $\leq 10^{13}$ ions/cm^2) is low enough that each impinging projectile finds pristine surface conditions in most materials, and therefore, only *ca.* 1–10% of the surface is influenced during measurement. The kinetic energy of the secondary particles is low (≤ 100 eV) and shows a Thompson distribution for atomic particles [37] and a Maxwell-Boltzmann distribution for molecular fragments. The former is obviously broader than the latter. The angular distribution of atomic secondary ion particles exhibits roughly a cosine dependence of the form [37]

$$\frac{\partial N}{\partial \Omega} = \pi^{-1} \cos \theta \tag{4.1}$$

However, in particular, when ions such as sodium or silver as contaminants (or by intention) are present at the surface, ion-induced cationization can give rise to different, non-classical situations. The most important factors determining the efficiency of the sputtering process are the mass of the primary ion, its angle of incidence, and the energy. After ion impact, all subsurface atoms within the collision zone are moved (species redistribution). This is called "collisional mixing" and limits the depth resolution of the eroding process for monoatomic primary ions (e.g. in depth profiling studies).

For Cs^+ ions, the following formula can be applied [38]:

$$\lambda = 1.84\, E^{0.68} \cos(\theta) \tag{4.2}$$

where λ is the penetration depth (nm), E is the impact energy (keV), and θ is the angle of incidence.

Numerous other material-related effects influencing the emission of secondary particles have been discussed in the literature [13, 38].

In reality, the situation is often more complex because materials exhibit individual character in terms of composition, contamination, crystal structure, and density. This individualism is frequently called matrix effect and limits the possibility of SIMS investigations requiring precise quantification. In addition, insulating materials such as ceramics or polymers accumulate positive electric charges deposited during a ToF- SIMS experiment. Unless properly compensated by a low-energy electron gun (as shown in Fig. 4.1), surface charging will change the acceptance properties of the analyzing system and degrade mass resolution as well as signal intensity (SI). Because of matrix effects, ToF-SIMS is usually not suitable for precise quantitative analysis. Although studies in this direction have been published [39], X-ray photoelectron spectroscopy (XPS), which provides much more reliable quantitative data, is often used as complementary analytical technique to ToF-SIMS. Quantification is one of the most important limitations of the technique. However, when compared with ToF-SIMS, XPS is less surface sensitive and has a lower molecular specificity [17, 40, 41].

Sputtered particles of molecular fragments generally show vibrational and rotational excitation that sometimes can lead to dissociation if a metastable doorway state is excited. If this takes place in the field free drift tube of a ToF analyzer, such particles can only be detected using a straight-line detection geometry.

The high sensitivity of SIMS also allows for trace analysis of many atomic species in the parts-per millions (10^{-6}) and in some cases in the parts-per-billion (10^{-9})

range, which is particularly important in the investigation of electronic materials. In this case, the mass resolution is more important than the mass range. However, the rather low-duty cycle can yield to long measurements, e.g. for deeply located quantum wells. When investigating surfaces of polymeric samples, the mass resolution is decreasing because of the complexity of the matrix. However, the analytical accessibility of high mass species or molecular fragments provides the experimenter with detailed information about differences in composition and end-group chemistry [42].

The sputter process can be modeled by molecular dynamics (MD) computer simulations to better understand its complexity. MD simulations in which atomic and molecular solids are bombarded by Ar_n (n = 60–2953) clusters, have been used to explain the physics that underlie the "universal relation" of the sputtering yield Y per cluster atom versus incident energy E per cluster atom (Y/n vs E/n) [43, 44]. This study showed that a better representation to unify the results was $Y/(E/U_0)$ versus $(E/U_0)/n$, where U_0 is the sample cohesive energy per atom or molecular equivalent, and the yield Y is given in the units of atoms or molecular equivalents for atomistic and molecular solids, respectively. A synergistic cluster effect was also identified that manifests itself in the way that a larger cluster induces a greater yield than the yield that would increase just proportionately to the cluster total energy E or, equivalently, the cluster size n. This synergistic effect can be described in the high $(E/U_0)/n$ regime as scaling of Y with $(E/U_0)\alpha$, where $\alpha > 1$.

MD simulations were used to model cluster bombardment of pure hydrocarbon [polyethylene (PE) and polystyrene (PS)] and oxygen-containing [paraformaldehyde (PFA) and polylactic acid (PLA)] polymers by 20 keV C_{60} projectiles at a 45° impact angle to investigate the chemical effect of oxygen in the substrate material on the sputtering process [45]. The simulations demonstrated that the presence of oxygen enhances the formation of small molecules such as carbon monoxide, carbon dioxide, water, and various molecules containing the $C=O$ moiety. The explanation for the enhanced small molecule formation is the higher stability of carbon–oxygen multiple bonds with respect to the carbon–carbon multiple bonds. This chemistry is reflected in the fraction of the ejected material that has a mass not higher than 104 amu. For PFA and PLA, the fraction is approximately 90% of the total mass, whereas for PE and PS, it is less than half.

MD computer simulations have been employed to investigate processes leading to particle ejection from free-standing two-layered graphene irradiated by keV argon gas cluster projectiles with different kinetic energy and size [46]. It was observed that both the effect of the primary kinetic energy and projectile size significantly affect the ejection process. A significant portion of the primary kinetic energy is deposited into the sample. Part of this energy is used for particle emission, which is substantial. As a result, circular nanopores of various dimensions are created depending on the bombardment conditions. A great part of the deposited energy is also dispersed as acoustic waves. Different mechanisms leading to particle ejection and defect formation were identified depending on the projectile energy per atom. Ar cluster projectiles and ultrathin graphene substrates can be applied for SIMS analysis of organic overlayers in transmission geometry. Studies with C_{60} projectiles showed that intact organic molecules are emitted by the unfolding of the topmost graphene layer. In this case, the graphene sheet acts as a catapult that can gently hurl molecules into the vacuum. There is a considerable amount of energy associated with this movement, which means that even very large molecules can be uplifted. During argon cluster bombardment, there is more energy in this motion, and the movement extends to a much higher lateral distance from the point of impact. Consequently, a larger number of adsorbed molecules could be ejected by a single projectile impact, making analysis of small amounts of organic material even more viable.

MD computer simulations were also used to model ejection of particles from β-carotene samples bombarded by 15 keV Ar_{2000} [47]. The effect of the incidence angle on the angular and kinetic energy distributions was investigated and it was found that both of these distributions are sensitive to the variation of the incidence angle, particularly near the normal incidence.

Recently, MD simulations of ion impact and secondary ion emission has been published to model erosion and chemical alteration of organic samples during ToF-SIMS depth profiling analysis with cluster projectile ion beams [48]. The integration of the results of MD simulations in a transport/reaction model was used for the prediction of the SIMS depth profile. In particular, the model allowed the simulation of the SIMS depth profiles of polymeric samples under cluster primary beam irradiation and was able to take into account ion-beam-induced reactions and the effect of reactive gas dosing. The model was used to describe phenomena taking place during depth profiling of polystyrene samples by 20 keV C_{60}, Ar_{872}, and Ar_{1000} projectiles, as well as to determine the overall efficiency of nitrogen monoxide molecules in eliminating the radicals responsible for polystyrene cross-linking induced by analyzing ion beams. This study evidenced a good agreement between theoretical and experimental results.

4.1.2.2 Mass analyzer systems

Three standard types of mass analyzers are used in SIMS technology:
– quadrupole analyzer;
– magnetic sector analyzer;
– ToF analyzer.

The theory of ToF analyzer systems has been discussed in detail in the literature [20, 21, 49]. Compared with quadrupole and magnetic sector analyzers, ToF systems show up to 10 times better transmission efficiency. Combined with the parallel (multiplex) mass transfer, the overall sensitivity can outnumber quadrupole as well as magnetic sector analyzers by a factor 10^3 to 10^4. The attainable mass range limit depends on the length of the field free drift tube, the data rate, and also the performance of the detector and counting electronics. Detected secondary particles can be assigned to specific mass with parts-per-million (ppm) accurateness with respect to the exact mass calculated from isotopic masses of the fragment constituents. Parts per million is defined as

$$\text{ppm} = \frac{M_{\text{obs}} - M_{\text{exact}}}{M_{\text{exact}}} \times 10^6$$

where Mobs is the observed m/z mass and Mexact is the exact m/z mass.

The mass accuracy that can finally be achieved by the analyzer ionoptics is important for later analysis of peak identity and also for successfully applying methods like PCA. Modern systems provide numbers close to 10–15 ppm. Other figures of merit which are important for the performance of the mass analyzer are the transmission and the mass resolution. Usually there is a trade-off between the two, but in timely instruments an elaborate design (patented by ION-TOF GmbH, Münster, Germany) will allow high mass resolution (up to a ratio of m/Δm around 20,000 at mass $m/z = 29$) and high transmission in a parallel optimized experimental setup. As discussed earlier, the acceptance geometry of an analyzer system of any design is responsible for the sensitivity of the ToF-SIMS instrument, because it finally accounts for the number of ions that can be counted at the detector. This high duty cycle is important in dual-beam sputter and imaging experiments (see Section 4.1.2.8) [50].

4.1.2.3 Ion sources and primary ions

Inorganic and metal oxide materials show restricted breakup into fragment secondary ions and generate most of the signals in the low-mass range ("fingerprint domain"), which is restricted to signals with mass $m/z \leq 250$. Many organic species do not yield useful characteristic signals at masses m/Δm ≤ 150, particularly in positive secondary ion spectra. Since the generation of high-mass secondary ions is strongly dependent on the type, mass, energy, and impact angle of the primary ion beam, fragmentation is not uniform from all regions of the area influenced by the impinging ion. Local heating is highest in the immediate vicinity of the impact site, and therefore, fragmentation is excessive at this location. With increasing distance, quasi-molecules remain intact. Obviously, signals collected from the periphery of the impact crater of the primary ion reflect the initial composition and structural features more realistically. Gentle SIMS (G-SIMS) (Section 4.2.2.7) is based on this concept.

The first commercially available ion sources were operating with Ga^+, Cs^+, or In^+, which all provided relatively small focused beams (*ca.* 100 μm) with sufficiently high ion densities. However, in the high mass range, signal intensity is decreasing rapidly for monoatomic projectiles because of excessive fragmentation and low yield of intact material. Property-specific molecular fragments are of low abundance. Therefore, the majority of molecular fragments induced by monoatomic ion sources are detected in the fingerprint range of the SIMS spectrum, which leaves the experimenter with complex spectra and unspecific mass signals. When dealing with organic materials such as polymers, the impact of a high-energy charged projectile not only induces the emission of secondary particles but also disrupts the molecular identity of the solid. The consequence is a crater of damage extending over several nanometers around the impact site. With the introduction of cluster ion sources to ToF-SIMS spectrometry in the mid to late 1980s, this effect could be minimized at least for surface spectrometry. Access to subsurface domains by using depth-profiling experiments became realistic with the use of gas cluster ion beams (GCIBs) as sputter source.

Routine use of large data libraries, the application of the G-SIMS theory and the introduction of principal component analysis (PCA) to ToF-SIMS were all developed to facilitate the interpretation of SSIMS experiments. Some of these concepts are discussed in Section 4.2.2.7.

4.1.2.4 Surface ionization ion guns

Surface ionization ion guns have first been developed for use in ion implant technology [51]. For some time, monoatomic alkali metal sources such as Cs^+ or In^+ were in use to acquire surface spectra. With the development of cluster primary ion guns, the application of monoatomic ion sources for surface analytical investigations was replaced completely, and this type of primary ions are almost entirely used to perform depth profiling studies of electronegative species in semiconductor samples. Ion emission is stimulated by heating an adsorbed layer of Cs (the most commonly used element) on a low-work-function material. The ionization potential of cesium is low, and therefore highly mobile electrons can move to the substrate (e.g. a porous tungsten or iridium plug). Under vacuum conditions, using a suitable extraction potential, ion pulses with both sufficient intensity and energy spread can be generated. Although beam brightness often succeeds 5×10^{-5} A cm^{-2} sr^{-1}, the focus diameter of Cs^+ beams is limited to values of $ca.$ 100 μm and therefore the gun is not suitable for high lateral resolution applications (Section 4.1.2.8).

4.1.2.5 Duoplasmatron and gas ion sources

Gaseous elements are ionized in duoplasmatron ion sources by electron impact. High and stable beams of Ar^+, Xe^+, O^-, and SF_5^+ can be generated. Oxygen and noble gas primary ion beams are used to obtain depth profile data from nonorganic materials.

Gillen and Roberson [52] demonstrated, for the first time, true depth profiling through poly(methyl-methacrylate) (PMMA) using an SF_5^+ cluster beam source for sputtering. SF_5^+ clusters are produced by leaking SF_6 gas into an electron impact chamber. Other smaller clusters are formed in parallel inside the ionization plume, but SF_5^+ is the most abundant species. Later, depth profiling of polymers and biological samples [53] became a vital field using cluster ion beams. For some time, C_{60}^+ ion beams were popular [54, 55] due to the high mass of the primary ion and energy splitting among a large number of carbon atoms. In a thermal effusive source, in which a powder of C_{60} (or other carbon compounds, e.g. C_{70} or coronene) is heated to 700 K. The carbon gas is then ionized by electron impact. Although providing a sufficiently high ion current (up to 500 nA, continuous), the source suffers from long-term stability and ease of operation. GCIBs (Section 4.1.2.7) show superior performance for this application and are nowadays the best choice for depth profiling of organic compounds.

4.1.2.6 Liquid metal ion guns

Liquid metal ion guns (LMIGs) [56] have become the most popular type for day-to-day operation of ToF-SIMS instruments. They are easy to operate, reliable, and provide a stable ion current. For a long period, gallium has been used as source material because of its low melting point (302.9 K) and excellent focusing properties, which proved ideal for imaging applications. Ga beams can routinely be focused down to 60 nm and are therefore also widely used in focused ion beam (FIB) ion microscopy. However, secondary ion efficiency is low, and high mass fragments usually are not accessible with sufficient intensity for small-area mapping. Because gallium does not occur as a pristine element in nature, it has to be purified to avoid split signals in high-resolution SIMS.

As sources in SSIMS technology and surface analysis, gold and bismuth cluster ion sources provide up to typically 1,000 times higher signal levels above the fingerprint mass region and most often also show less fragmentation because of primary ion energy distribution over several beam particles [7, 57]. The principal evidence for the superior performance of cluster ions as projectiles came, in fact, from the observations that "output" (secondary ions emission) per unit "input" (number of atoms) was unusually or unexpectedly high. The advantage of using polyatomic projectiles is the ability to produce what are termed "enhanced" ion yields [58].

Walker and Winograd [59] described the use of a LMIG ion source generating Au^+, Au^{2+}, and Au_2^+ primary ion particles. Eutectic alloys with Si and Be, or Ge were experimentally tested. The melting point of a SiBeAu tip lies around 693 K. For the investigated materials, Au^{2+} projectiles induced the least damage to the surface. However, all gold particles produced considerably stronger secondary ion yields than achieved by Ga-LMIG sources. Commercially available gold sources were introduced in 2003 [60] and produced routinely Au_n^+ clusters (n = 1–5).

The use of Bi_n^{m+} clusters (n denoting the cluster size and m the number of charge) has replaced most other types of primary ions for surface characterization experiments because of the high secondary ion yield at high mass range. The cluster size can vary between 1 and 7, with Bi_1^+, Bi_3^+, and Bi_3^{2+} the most often used species. The metal or metal alloy compound deposited on a fine needle with a reservoir is liquid at ca. 540 K. The most recent designs [61] can be focused almost down to the ultimate physical limit (ca. 15–30 nm) as defined by Orloff et al. [62]. Cluster ions break up after impact into component atoms, which carry only a small fraction of the initial kinetic energy and therefore induce less damage on the surface and subsurface layers. The penetration depth of primary ions is strongly dependent on impact energy and angle [63].

Numerous other, more exotic, clusters consisting of ionic species that were produced by ^{252}Cf fission fragment-induced desorption of source material deposited as thin layer on an Aluminum-Mylar foil (CsI^+, Cs_2I^+, and $Cs_3I_2^+$ [64, 65] or (Bi_2O_3) BiO^+ ions [65]) were experimentally tested. LMIG sources are capable of profiling through thin polymer films but show increased damage accumulation as compared with gas cluster and C_{60}^+ sources [59].

4.1.2.7 Gas cluster ion sources

GCIBs are presently the source of choice for organic depth profiling, including investigation of polymers [66]. Although many different gases have been studied, the most often used system is Argon.

In a first step, the noble gas is condensed into a neutral cluster by supersonic cooling. This effect dates back to early experiments performed for gas phase reaction studies [67]. The still neutral clusters of different sizes are ionized by electron impact and are afterwards mass selected. Modern commercially available ion sources produce argon clusters with component numbers ranging from 50 up to more than 10,000.

A study comparing the emission of secondary ions from PMMA generated by a single atom and by large, size selected argon clusters (m_{av} = 700, 1,000 and 1,350) was published by Hashinokuchi et al. [68] in 2008. The SI for fragment ions with masses m/z = 100, 125, 300, 500, and 800, respectively, was found to be strongly enhanced with increasing cluster size for a PMMA film deposited on silver when compared with single atom experiments. Later experiments on PMMA, polystyrene (PS), polycarbonate (PC), and indium tin oxide (ITO)/glass samples bombarded with 5.5-keV Ar_{700}^+ ions under SSIMS conditions were presented by Ninomiya et al. [8]. The signals detected were compared with the literature and found in good agreement for PMMA [69] (m/z = 69, $C_4H_5O^+$, and m/z = 115, $C_6H_{11}O^+$), PS [70] (m/z = 91, $C_7H_7^+$ and m/z = 115, $C_9H_7^+$), and PC [71] (m/z = 77, $C_6H_5^+$ and m/z = 135, $C_9H_{11}O^+$), respectively.

However, to make the equipment useful as pulsed beam in commercially available high-mass-resolution ToF-SIMS instruments, a small fraction of the cluster beam is selected by special 90° beam geometry. This allows for the generation of short, well- defined ion pulses, enabling the acquisition of high-resolution surface mass spectra. Kayser et al. [72] investigated the differences in negative ion ToF-SIMS spectra of PC using a Bi_3^+ and for analysis an Ar_{1500}^+ primary ion beam. While the former exhibited average particle energy of 6.6 keV/atom, the GCIB source only deposited 2.6 eV/atom onto the surface. The bismuth beam generated a considerable amount of fragmentation. In contrast, the spectrum obtained from the argon cluster source was dominated by characteristic molecular fragments. As proven in earlier investigations for other cluster guns [73, 74], the degree of surface charging was considerably reduced in several ToF-SIMS experiments using GCIB cluster guns.

4.1.2.8 High-resolution ion images obtained from cluster LMIG sources

Highly focused ion beams are also a prerequisite for high-resolution imaging ToF-SIMS by raster scanning the beam over a selected surface area. Collecting mass spectra at each pixel of the image as a function of position (usually 128 × 128 or 256 × 256) and retrospectively analyzing the signal intensities provides the analyst with a powerful tool for two-dimensional analysis of surface chemistry. All ions are detected in parallel. Secondary electrons co-produced after each primary ion impact yield secondary electron images of the analyzed area [75]. Charles Evans and Associates developed in the early 1990s, a ToF-SIMS design based on coherent ion transfer through a TRIFT mass analyzer and subsequent position-sensitive ion detection (resistive anode encoder, or RAE) capable of mass-resolved imaging. Modes of detection for ToF-SIMS imaging are represented in Fig. 4.2.

The lateral resolution of a ToF-SIMS image is instrumentally limited by the degree of beam focus and the detection efficiency of a specific characteristic fragment ion from the area under investigation. The number of secondary ions that are emitted from a defined area is limited by the damage cross section of the material. Therefore, the achievable lateral resolution is not only limited by the primary ion focus but rather by the material-dependent number of ions that can be detected from a small area in an SSIMS experiment.

The efficiency, E, is defined as

$$E(M\pm) = \frac{Y(M\pm)}{\sigma(M\pm)}$$

where Y is the secondary ion yield and σ is the disappearance cross section.

A sample-dependent useful lateral resolution Δ_l from an area A_D can be defined [70] as $\Delta_l = (A_D)^{1/2} = (N/E(M\pm)^{1/2}$. The useful lateral resolution is related to the secondary ion emission (E) and is sample dependent. It should not be confused with

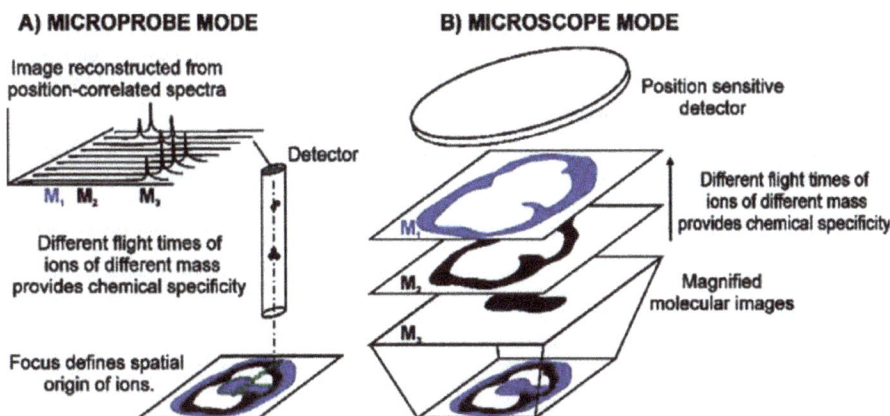

A) MICROPROBE MODE

Image reconstructed from position-correlated spectra

M_1 M_2 M_3

Different flight times of ions of different mass provides chemical specificity

Focus defines spatial origin of ions.

Detector

B) MICROSCOPE MODE

Position sensitive detector

Different flight times of ions of different mass provides chemical specificity

Magnified molecular images

M_1

M_2

M_3

Fig. 4.2: Two different modes of detection are possible for ToF–SIMS imaging. In microprobe configuration In Figure 4.2: A) mass spectra a collected from a pre-selected pixel array and images are rebuilt retrospectively from an extended raw data set. In the microscope mode In Figure 4.2: B) images are transferred coherently through a time-of-flight analyzer system and image distribution is magnified before detected at a position sensitive resistive anode encoder (RAE) detector. Retrospective analysis is also possible. Reprinted with permission from [75]. Copyright (2004) American Chemical Society.

lateral resolution, which is dependent on the spot size of the primary ion beam. N is the number of secondary ions that is detected from the area A_D before the uppermost monolayer of the material is removed by the primary ion beam:

$$N(t) = N(0)e^{(\sigma It/Ae)}$$

where $N(t)$ depends on the disappearance cross section, I is the primary ion current, t is the time, A is the bombarded area, and e is the electric charge, respectively [76].

The issue of lateral resolution in SSIMS experiments is of increasing importance for detecting small, nanosized features in imaging mode. Touboul et al. [76] estimated values of D_l for the cholesterol $[M+H]^+$ signal at $m/z = 385$ after Bi_n^{q+} bombardment (Tab. 4.1).

A full imaging data set is composed of the complete set of time (i.e. mass) dependent counts arising from each pixel. However, in cases where very small features are investigated, the dimension can approach the minimum size of a pixel. In such cases, it is more realistic to compare neighboring pixel intensities than observe average signal intensities over a number of pixels (binning). Several authors defined a number $N = 4$ as adequate pixel counts. However, it remains speculative to which extent this number really describes small-scale features on a sample surface. Counting statistics of ToF-SIMS imaging of complex surfaces is not yet fully described in all details.

The simultaneous possibility of high transmission and high mass resolution as discussed in Section 4.1.2.2 allows for the detection of extremely low signal numbers

Tab. 4.1: Ion bombardment efficiency and useful lateral resolution for $Bi_n{}^{q+}$ primary ions measured for the cholesterol $[M+H]^+$ signal at $m/z = 385$ in rat brain sections. Data and discussion found in: Touboul D, Kollmer F, Niehuis E, Brunelle A, Laprévote O, *J Am Soc Mass Spectrosc* 16, 1608–18, 2005. Copyright (2005) American Society for Mass Spectrometry [76].

Primary ion	Primary ion current [pA] (at ~10 kHz repetition rate)	Ion bombardment efficiency [cm^{-2}]	Useful lateral resolution Δ [nm]
$Bi_1{}^+$	1.0	$3.04 \times 10^{+8}$	1148
$Bi_3{}^+$	0.34	$1.71 \times 10^{+9}$	484
$Bi_3{}^{2+}$	0.185	$1.29 \times 10^{+9}$	557
$Bi_5{}^+$	0.045	$7.14 \times 10^{+8}$	749
$Bi_5{}^{2+}$	0.044	$2.81 \times 10^{+9}$	377
$Bi_7{}^+$	0.004	$2.12 \times 10^{+8}$	1374

per pixel. This is a prerequisite if it's the beam diameter of the ion source is focused to the near-theoretical limit and, at the same time, the data rate remains high enough to obtain a reasonable low measurement time. The higher the achievable pulsed primary ion current, the better. However, stability of the source is necessary for stable measurement conditions. Depending on the experimental condition and the nature of the sample, a field of view of 10×10 µm^2 seems realistic.

Under the brand name Nanoprobe 50 an ion source (Iontof GmbH, Münster, Germany) capable to operate with 50 kHz repletion frequency and a primary ion current of 40 pA ($Bi_3{}^{2+}$) has been presented. With this equipment, a routine resolution of less than 50 nm could be achieved.

In combination with large Ar-cluster ions, as presented in Section 4.1.2.7 (90°-geometry), 3-dimensional image of polymers or biological samples are possible. Since the total beam energy of 20 kev per Ar_{10000}-cluster primary ion breaks up to 2 eV per Argon atom, a very soft eroding process takes place on the sample surface, and therefore high-resolution depth-profiling is possible [77].

The use of gas cluster ion beams (GCIBs) in SIMS produces three main benefits: reduced damage accumulation, improved depth resolution, and decreased fragmentation in the mass spectrum. The decreased damage accumulation has immediate benefits for 3D imaging. If the steady state signal of a species in the mass spectrum can be increased then the increase in sensitivity can be used to improve useful lateral, and potentially depth, resolution. Organic depth profiling with GCIBs can routinely produce depth resolution better than 10 nm on well prepared reference materials. Improved molecular signal levels compared with C_{60} bombardment for lipid species in cell and tumor samples have been reported [78] and most recently Bich and coworkers showed persistent lipid signal when a depth profile was performed through a rat brain section that was sliced at 14 µm thickness and analyzed after freeze drying when the tissue thickness had been reduced to approximately 1.2 μm [79]. The implication is that these new beams will offer improved performance in a wide range of 3D imaging experiments on organic and biological samples. The majority of reports related to the use of GCIBs in SIMS itemize dual beam analysis where the GCIB is used at low energy to perform sample etching while a second beam, usually a liquid metal ion beam such as Bi_3^+, is used for analysis. This is due to the GCIBs being difficult to pulse quickly for use in conventional ToF-SIMS instruments, being difficult to focus and having low ionization efficiency. Angerer et al. have recently reported the application of a high-energy (40 keV) GCIB for organic and biological analysis where a spatial resolution < 3 μm was achieved and signal levels for intact lipid species on mouse brain were 10–30× higher than those observed on the same tissue analyzed with a 40 keV C_{60}^+ ion beam [80].

4.2 TOF-SIMS investigations of polymer materials

4.2.1 General remarks

The physical and chemical characterization of materials surfaces requires, in most cases, the application of several complementary techniques. The most important are discussed in other chapters of this textbook. However, the use of mass spectrometric methods was identified as powerful analytical approach to obtain information about the chemical composition of surfaces and subsurface domains.

Several very elaborate and detailed review articles and other book chapters on the application of mass spectrometry and ToF-SIMS spectrometry to the analysis of polymeric materials and their surface have been published [3, 15, 39, 81–88].

4.2.2 Polymers

Bombardment of polymers with energetic charged particles can induce different phenomena such as cross-linking, chain scission, loss of chemical groups, depolymerization, loss of hydrogen or oxygen, monomer units, and even the formation of new single or double bonds.

Polymers that tend to favor cross-linking [e.g. PS, PC, polypropylene (PP), polyethylene (PE)] show excessive signal decay with increasing primary ion fluence. On the other hand, materials that are known to degrade *via* a random main-chain scission process (e.g. polymethacrylates, polyesters) tend to retain characteristic molecular signals with progressive beam intensity [9].

Depending on the type of information that is required, the experimental setup and sample handling have to be chosen or adapted. Thick layer or bulk polymer surfaces need little or no sample preparation and are usually analyzed by inserting a small piece of the material in the instrument. Certainly, in the case of highly insoluble materials, this is a good experimental approach. Because of the sometimes rather high surface charging [89], which can be difficult to fully compensate using a low energy electron gun, the mass spectra of bulk polymers are often restricted to the fingerprint domain ($m/z \leq 150$). This is most evident in experiments using a monoatomic primary ion source. Surface charging has been described as less distinct in cluster ion ToF-SIMS experiments. However, data generated of monoatomic ions provide information about the repeat unit of the polymer, possible surface contaminants, and also the end-group chemistry. Leggett [90] divided the formation of molecular fragments in SSIMS spectra of poly(styrene) into three distinct processes:

- High energy fragmentation, close to the impact site, characterized by high energy transfer to the polymer. This leads to the formation of low mass fragments, often containing unspecific chemical information.
- Unzipping region at a more distant area surrounding the impact site. Energy density is decreasing, and larger molecular fragments are formed. Often, information about monomer interactions, cross-linking behavior, and structural rearrangements is found. Laser post-ionization can be used to obtain further information from the neutral particle-rich domain. Mahoney [9] published a compilation of polymeric structures according to their cross-linking behavior as type I or type II under irradiation of charged particles.
- Simple low energy fragmentation characterized by formation of large fragments. The energy (in the range of electron volts) available for fragmentation is low, and therefore, bond breaking in certain chemical structures is impossible. The ejection of monomer fragments is dominating. The information about subtle molecular rearrangements is often very useful to obtain structural information.

As outlined earlier, some of these results are questioning the application of the simple linear Sigmund cascade theory [34, 35] for the description of fragment ejection from polymers.

Gilmore and Seah [91] developed a model based on the dependence of SI upon ion dose to characterize the probability of bond breaking and fragmentation pattern of linear polymers at the end and in the middle. This effect was confirmed by investigating the fragmentation of poly(terahydrofluoroethylene) (PTFE) and poly(ethyleneterephthalate) (PET). Whereas PTFE is characterized by multiple bond breaking decay, PET shows a fast exponential behavior for the molecular fragments at masses $m/z = 149$ and $m/z = 193$, but slow decay was observed at $m/z = 93$. These early investigations later became the foundations of G-SIMS (Section 4.2.2.7).

However, if dealing with structure determination and molecular weight assignment, elaborate sample preparation techniques are sometimes required. The most common is the dissolution of the polymer in a suitable solvent and subsequent spreading on a solid support in a concentration of about 1 mg/mL. Thoroughly cleaned silicon wafers or acid etched thin silver sheets (dilute HNO_3) are most often used for this purpose. The formation of secondary ions in this case is not generated *via* the collision cascade mechanism but by cationization of the parent ion or molecular fragments. Etched silver provides characteristic ^{107}Ag/^{109}Ag doublets after cationization. In many cases, the cation is added to the polymer solution (e.g. Na$^+$) in advance. Keller and Hug [92] investigated the use of ion exchanged Nafion117 as source for selective silver [93] or rubidium cationization. In this way, ToF-SIMS has many similarities with matrix-assisted laser desorption ionization mass spectrometry (MALDI-MS). Cationization becomes more probable with increasing polarizability of a molecule. Therefore, some uncertainty remains if independence from end-group chemistry is a realistic approach of measured SI of cationized species [94].

Molecular weight information from ToF-SIMS experiments is obtained by investigation of cationized fragment pattern. Lee et al. [95] reported a reasonable good correlation with gel permeation chromatography and MALDI-ToF for data from the derivatives poly(α-methylstyrene, poly(4-methylstyrene), poly(4-*tert*-butyl styrene) investigated as thin films of different average mass on silver substrates in the mass range 0–5000. For the aromatic residue-substituted poly(4-vinyl biphenyl) and poly (vinylnaphthalene), however, experimental values were slightly lower than the manufacturer specifications. Galuska [96] analyzed the low-mass fingerprint region of polyisoprene, 1,4-poly(butadiene), polyisobutylene, PE, PS, PP, and poly(1-butene) (P1-B) with respect to possible correlation between the average molecular weight and apparent monomer signal intensities in the ToF-SIMS spectra. In this experiment, the protonated (M + H$^+$) fragment was observed. Signal intensities correlated well with molecular weight values higher than 20,000 but lacked a general linear relationship at lower molecular masses.

A general linear dependence was reported for a wide variety of polymers:

$$\text{Relative}(M+H)SI = M(MW/1,000)E + B \tag{4.3}$$

where B describes the ion ratio intercept and E is a polymer-type-dependent exponent with values between −0.5 and −0.6 [96]. The same type of molecular weight estimation was applied by Coullerez et al. [97] to study a series of 2,2-bis(hydroxymethyl) propionic acid with different molecular weights. Signals in the fingerprint range of positive and negative ToF-SIMS mode were assigned to the fragmentation of the (bis- MPA) hyperbranched aliphatic polyester and used to calibrate the molecular weight.

4.2.2.1 Polydimethylsiloxane

Polydimethylsiloxane (PDMS) is one of the most extensively studied polymers from the polysiloxane (often termed silicones) family. The polymer has found widespread application in the industry as anti-foaming component in shampoos and household cleaners, as lubricating agent, water repellant, and anti-fouling protective coat. PDMS is frequently found on surfaces responsible for adhesive or printing failure and is also found as contaminant on extruded parts because of its use as low-temperature mold-release agent (e.g. baking paper or chocolate process casting mold). Its low toxicity, clear appearance, and safe handling make PDMS an ideal food-processing additive.

The use of PDMS as a component for microfluidic devices has been reviewed by Ng et al. [98]. The authors demonstrated the versatility of the material for fabrication of pneumatically actuated switches, passive structures, and chaotic mixers. Because of the straightforward surface modification by oxidative processing [99], PDMS can be easily made hydrophilic by exposure to an oxidizing plasma (e.g. air and oxygen). This process destroys the methyl (Si-CH$_3$) groups and introduces silanol functional groups (Si-OH) to the surface.

However, as most siloxanes, PDMS is migrating to materials surfaces, and due to its high mobility, it can form a dense overlayer that completely masks chemical entities of interest. In such cases, the fingerprint regions of ToF-SIMS spectra are dominated by the characteristic signals of PDMS [83] identified as Si$^+$ ($m/z = 29.79$), CH$_3$Si$^+$ ($m/z = 42.99$), CH$_3$SiO$^+$ ($m/z = 58.98$), C$_3$H$_7$O$^+$ ($m/z = 59.03$), and (CH$_3$)$_3$Si$^+$ ($m/z = 73.06$), Si$_2$OC$_5$H$_{15}$$^+$ ($m/z = 147$), Si$_3$O$_3$C$_5$H$_{15}$$^+$ ($m/z = 207$), respectively. Cyclic fragments are observed at mass $m/z = 221$ (Si$_3$O$_2$C$_7$H$_{21}$$^+$) [100]. Figure 4.3 shows the typical positive-ion PDMS ToF-SIMS spectrum in the fingerprint range below $m/z = 300$. Molecular weight distributions were obtained by Dong et al. [101, 102] and later Inoue et al. [103] by analyzing the (M+Ag)$^+$ fragments of different silver-deposited PDMS modifications. Molecular weight distribution with C$_2$H$_6$SiO ($m/z = 74$) monomeric repeat unit, was detected in the high mass range ($m/z \geq 1,000$) of the ToF-SIMS spectra.

4.2.2.2 Polystyrene

PS rapidly degrades under ion irradiation. In contrast to polyacrylates, no difference in fragmentation was observed after irradiation of PS with different cluster ions. The inability to sputter intact polymer chain fragments because of entanglement of the rather bulky benzyl group might be responsible for this effect. Chain scission is the dominant degradation process with the formation of multiple aromatic fragment ions.

4.2.2.3 Polyacrylates

The application of ToF-SIMS to the surface analysis of polymethacrylates was described by Belu et al. [104]. The authors investigated a series of well-defined polyacrylate polymers, which were identical in their structural identity but had incremental additions of C$_2$H$_4$ groups to the backbone and side chain. Characteristic signals were

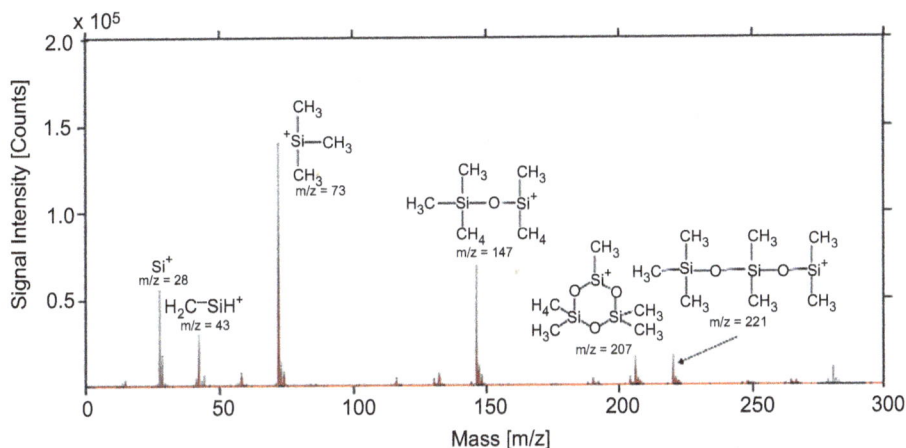

Fig. 4.3: Static ToF–SIMS spectrum of poly(dimethylsiloxane) (PDMS) with characteristic fragmentation signals in the fingerprint spectrum region. The formation of hexa–cyclic fragment ions is energetically favored over linear species. Adapted from [83] with copyright © 1999, Elsevier.

identified for all investigated polyacrylates. In addition, the specific backbone and side-chain chemistry revealed unique mass signals in the ToF-SIMS spectra, which could be assigned to individual well-controlled polymer materials. Leeson et al. [105] showed that molecular weight had a large effect on signal intensities obtained from solvent cast films produced of monodispersed PMMA standards with $M_W = 2965$–$1,200,000$. Negative secondary ion intensities were found to be very sensitive to trace amounts of solvent.

The degradation of methacrylate polymers under irradiation of a 5-keV SF_5^+ primary ion beam and an ion dose of approximately 2×10^{14} ions/cm^2 was described by Wagner in a series of publications [106–108]. Clearly, such extended bombardment exceeds the SSIMS limit considerably, but certain trends of degradation behavior also occur in SSIMS-fragmentation. Modifications in the main chain and end-group chemistry of poly(acrylates) revealed improved degradation of poly (methyl acrylate) and poly(methacryl acrylate) as compared with PMMA. These results were suggesting that main-chain or pendant methyl groups clearly reduce depolymerization and increase the rate of intermolecular or intramolecular cross-linking in the polymer matrix.

Mahoney [109] published data on PMMA thin films with variable tacticity. Damage cross-sections of the stereospecific polymer were estimated in a temperature-dependent study between −198 and −123 K under SF_5^+ bombardment. The isotactic samples exhibited lower damage accumulation compared with atactic and syndiotactic PMMA, which was explained by increased main-chain scission due to increased strain along the polymer chain.

Combination of static and dynamic ToF-SIMS has been successfully used to investigate surface and bulk composition and structure of some PMMA copolymer blends. The SIMS data evidenced an enrichment of the surface area in the most hydrophobic component, although H-bonding interaction occurred between the different blend components [110].

4.2.2.4 Fluorinated polymers

Positive and negative secondary ion spectra of neat Teflon surfaces were published by Léonard et al. [111]. Characteristic signals in the fingerprint mass range included C^+ ($m/z = 12$), CF^+ ($m/z = 31$), CF_3^+ ($m/z = 69$), $C_2F_5^+$ ($m/z = 119$), $C_3F_5^+$ ($m/z = 131$), and $C_3F_7^+$ ($m/z = 169$), respectively. The negative ion spectra were dominated by F^- ($m/z = 19$), CF^- ($m/z = 31$), F_2^- ($m/z = 38$), CF_3^- ($m/z = 69$), and $C_2F_5^-$ ($m/z = 119$).

In an early study, Fowler et al. [94] analyzed perfluorinated polyether (PFPE) on Ag and Si substrates. The use of PFPE (Krytox™, DuPont) as surface lubricant stimulated the investigation. Characteristic molecular fragments and the molecular weight distribution were detected by ToF-SIMS. The complete chemical structure could be reconstructed using integrated intensities of a series of molecular ion signals extracted from investigated compounds. Information of end-group chemistry was obtained also from the spectra, however, with an uncertainty of fragment assignment.

Nafion® (DuPont) is a block copolymer consisting of a fluoropolymer backbone and several other fluoro-homopolymer units. The polymer is often poorly defined, but the most popular representative, Nafion117®, with only one extended fluoropolymer group and otherwise unity numbered subunits, has the well-defined structure shown in Fig. 4.4a. The super acidic end-group proton of the sulfonic acid in the α-position is very mobile, and the polymer has found widespread application as membrane material in fuel cell technology [112]. The polymer is structurally very cavernous and exchange reactions with water and small cations take place with almost 100% yield [92, 93]. As in most ionomers, proton exchange with various cations can be induced in suitable media [113].

Liu et al. [114] studied Aquivion (Hyflon), which is a perfluorosulfonate polymer built of a Teflon backbone and a shorter sulfonic acid side chain than that found in Nafion. The chemical structure of Hyflon is shown in Fig. 4.4b. An electroactive actuator of Hyflon with 1-butyl-2,3-dimethylimidazolium chloride (BMMI-Cl) was analyzed using a C_{60}^+ ion gun. Ions originating from BMMI (BMMI$^+$, $m/z = 153$) and chlorine ($^{35}Cl^-$ and $^{37}Cl^-$) were detected with high intensity. Fragments characteristic to the backbone included CF^+ ($m/z = 31$) and F^- ($m/z = 19$). A signal assigned to SO_3^- was observed at mass $m/z = 80$. Depth profiles were recorded with these signals as markers for the individual components ≈ 7 µm inside the polymer.

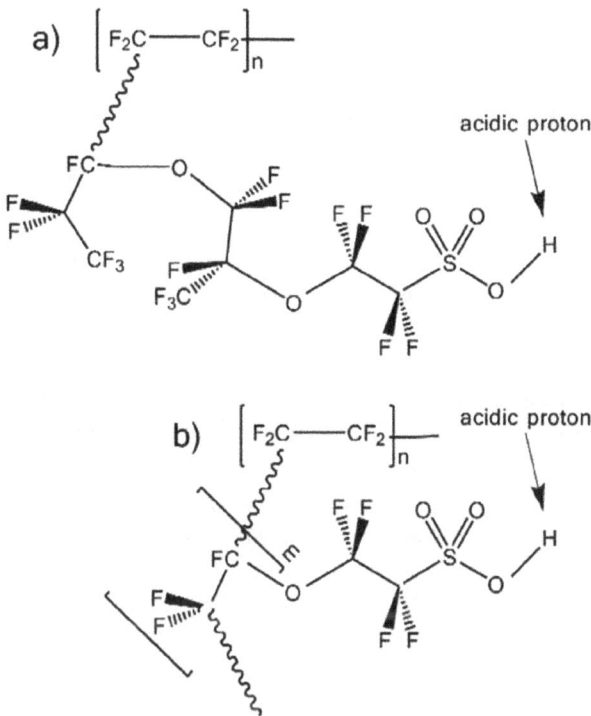

Fig. 4.4: Chemical structures of a) Nafion and b) Aquivion (Hyflon). The shorter side chain of Hyflon promotes a more pronounced formation of subsurface located ionic clusters because of increased chain mobility. ©Empa (2011).

Structures of partially fluorinated bottlebrush polymers in coating films were obtained by combining MD simulation methods with ToF-SIMS measurements. These studies allowed to understand the influence of the bottlebrush architecture on the spatial distribution of fluorinated moieties and confirmed that fluorine atoms in the bottlebrush are preferentially located at the air–film interfaces. Combining with contact angle measurements this study confirmed the tenability of surface energy with polymeric architecture [115].

Both Feng et al. [116] and Zhon et al. [117] demonstrated that the chain sequence structure in fluorinated polymers differently affect the fragmentation patterns of positive and negative ion ToF-SIMS spectra. Alternating ethylene-tetrafluoroethylene (ETFE) copolymers and poly(vinylidene fluoride) (PVDF) can be distinguished by their positive ions ToF-SIMS spectra, especially in the high mass region. The chain sequence structure does not have much effect on the negative spectra of the two polymers that are both dominated by the F^- (m/z =19), since F^- is very stable, due to its large electronegativity value, and fluorine concentration is relatively high in both ETFE and PVDF.

4.2.2.5 Poly(ethyleneterephthalate)

Because its often very inhomogeneous surface, the analysis of PET has attracted much interest to use ToF-SIMS for characterizing the chemical groups present. Nysten et al. [118] studied surface inhomogeneity induced into a sample obtained by compression molding of the two incompatible polymers PET and PP. ToF-SIMS data in combination with scanning electron microscopy (SEM) and lateral force microscopy (LFM) revealed a surface structure consisting of a continuous phase with inclusions of nodules whose size were between 1 and 100 µm. The continuous phase could be assigned to PET and the nodules to PP.

High lateral resolution tandem secondary ion MS imaging was employed to determine the composition of surface features on PET precipitated during heat treatment with a lateral resolving power of <200 nm. These features that are found to be comprised of ethylene terephthalate trimer at greater abundance than is observed in the surrounding polymer matrix, represent the first chemical identification of PET surface precipitates without the use of either an extraction step or a reference material [119].

4.2.2.6 Polyethylene glycol

Keller and Hug [83] published a positive-ion ToF-SIMS spectrum of polyethylene glycol (PEG) in the mass range of $m/z = 0-800$, recorded with a 8-kV Cs^+ primary ion beam. As shown in Fig. 4.5, the fingerprint range of the spectrum is dominated by the monomer fragment signal $(CH_2\text{-}CH_2\text{-}OH)^+$ at $m/z = 45$. Protonated fragments with the general chemical composition $(CH_2CH_2\text{-}O)_nH^+$ were also detected in the mass range $\leq m/z = 150$. For larger oligomers (index $n \geq 5$), a second series with repeating units $(CH_2\text{-}CH_2\text{-}O)_nOH_3^+$ showing an intensity maximum around $m/z = 415$ $(CH_2\text{-}CH_2\text{-}O)_9OH_3^+$ was reported. The insert in Fig. 4.5 indicates higher signal intensities for oxygen terminated oligomer series. Fragmentation process of PEG in ToF-SIMS spectral analysis was also simulated by Quantum mechanical dynamics (QMD) method, assuming the fragmentation process as thermal decomposition. The study demonstrated that the peaks position in the experimental ToF-SIMS spectra depends on the energy deposited to the PEG molecule [120].

4.2.2.7 Spectra libraries and G-SIMS approach

In the early days of ToF-SIMS analytical experiments, the interpretation of spectra has largely benefited from the compilation of spectra libraries, mainly covering the fingerprint region. Characteristic masses from often-analyzed materials were compared with stored signals from the library. This approach has a long tradition in analytical

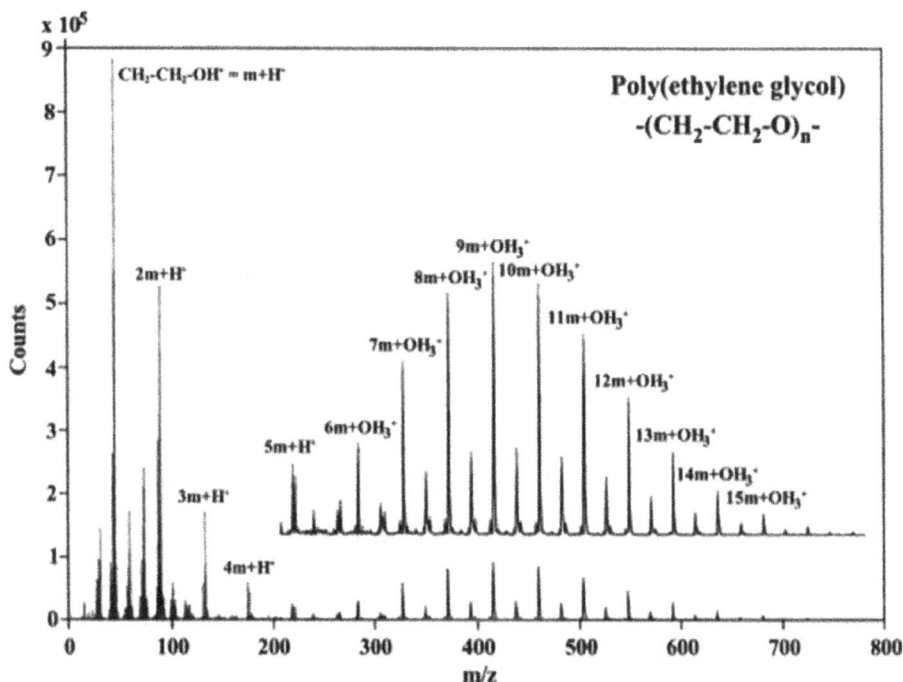

Fig. 4.5: Positive-ion ToF-SIMS spectrum of PEG obtained after bombardment with an 8-keV Cs+ primary ion beam. The sample was prepared as thin film deposited from a 1-mg/mL ethanol solution. With permission from Keller BA, Hug P, *Anal Chim Acta* 393, 201–12, 1999. Copyright 1999, Elsevier [83].

chemistry and dates back to the developing of mass spectrometry as tool for the organic chemist. However, in contrast to traditional electron impact ionization, which can easily be standardized, the impact of charged projectiles onto a polymer surface with several kiloelectron volts (keV) energy induces fragmentation and, most often, also matrix-dependent structural changes in secondary ions. Of these, only a very small fraction is usually related to the original chemical structure, and therefore the spectra of complex surfaces are very difficult and time consuming to interpret. Although the concept of library matching works rather well in some cases for monoatomic primary ions (e.g. Ga^+, Cs^+, In^+, Ar^+) and serves often as starting point for more advanced interpretation, it completely fails for the use of cluster ion guns.

Green et al. [121] estimated that the contents of the popular public chemical database PubChem have grown more than two orders of magnitude containing a very large number of complex molecules, all accessible within ToF-SIMS mass range. To obtain a more detailed picture of fragmentation mechanism in polymers, Gilmore and colleagues [69, 122, 123] introduced the concept of gentle SIMS (G-SIMS). Although a complete description of G-SIMS is not within the scope of this book, a short outline is given. Briefly, the concept is based on the fact that secondary ions

released after ion impact from areas close to the impinging point exhibit more fragments than those more distant from it, due to different energy density. Therefore, a
set of secondary ions is generated with signals arising from structure-related ions
together with fragmented secondary particles. If intelligently steering the energy of
impact, the amount of the former species can be enhanced and the latter decreased.
A G-SIMS theory [69] was developed based on a partition function for a population
of intact molecules with a characteristic temperature T_p. The theory requires two
spectra with signal intensities S_{1x} and S_{2x} and masses M_x acquired under different
conditions and therefore with different surface temperatures T_1 and T_2 ($T_1 < T_2$). A
G-SIMS spectrum can be computed as

$$G_x = M_x S_{1x}(S_{1x}/S_{2x})^g \qquad (4.4)$$

where g is the so-called g-index. For atomic primary ions, this number is typically
about 13, but it has a considerably lower value for cluster primary ions, e.g. Bi_n^{m+}.
In this latter situation, the fragmentation is low, and the nonlinear intensity enhancement of fragments in the high mass range largely simplifies peak specificity.
Because the kinetic energy is distributed over several atoms in a cluster beam, modern LMIG sources operating with gold or bismuth reservoirs are providing very effective low energy conditions [124]. To correlate spatial alignment of the two independent
beams, a Bi/Mn emitter, known as the "G-tip," has been developed [125] for use in
commercial LMIG ion sources. G-SIMS has become accessible to many analysts and
the software to compute G-SIMS spectra has been incorporated directly into some
vendor's software. G-SIMS has been successfully applied to the analysis of polymers
[124, 126, 127].

4.2.2.8 Multivariate analysis and principal component analysis

Usually, materials collected from the real world or modified by advanced laboratory
processing exhibit a complex surface chemistry [126]. Reducing the dimension of
this complexity by multivariate statistical analysis (MVA) allows the extraction of
meaningful information about individual compounds and their relationship. MVA
is concerned with data acquired from multiple independent measurements of samples
[128]. A fundamental relationship between variables is obtained from measurements
and subsequent data analysis. A typical ToF-SIMS spectrum contains hundreds of
peaks, the intensity of which can vary due to composition, structure, order, and orientation of the surface species. ToF-SIMS data are inherently multivariate since the relative intensities of many of the peaks within a given spectrum are related, due to the
fact that they often originate from the same surface species. The challenge is to determine which peaks are related to each other, and how they relate to chemical differences present on the surface. Yet knowing the information buried within the complexity

of acquired mass spectra or ToF-SIMS images, it is possible to extract sample chemical composition, orientation of surface adsorptions, contaminants, or compound homogeneity. Technically, a list of fragments is selected from a series of 10–15 individual spectra. After preprocessing (e.g. peak normalization to the sum of intensities, mean centering), a matrix is generated with one row describing the samples and the other describing the signal intensities (normalized). A software is computing a scores matrix, describing the relationship between the samples (i.e. comparing the chemical identity), and the loadings matrix is illustrating the relationship between the original variables and the principal components. Timely computer equipment allows for running data processing algorithm to reduce complexity in SIMS spectra and identify important molecular signals. MVA methods such as PCA and partial least squares (PLS) have been successfully used for data reduction in analytical science and also in ToF-SIMS. However, it remains in many cases difficult to find the proper preprocessing and data selection to obtain useful information about a given analytical question [129–133].

The efficacy of the MVA approach applied to ToF-SIMS spectra of polymers has been demonstrated by a secondary ion mass spectrometry study of a set of standard polymers (PET, PTFE, PMMA, and LD-PE). The polymer types were discriminated to a moderate extent by PCA, but were easily skewed with saturated species or contaminants present in ToF-SIMS data. Artificial neural networks, in the form of self-organizing maps, were introduced, providing a non-linear approach to classifying data and focusing on similarity between samples. The classification outcome achieved was excellent for both the different polymer types and spectra from a single polymer type generated by using different primary ions. This method offered great promise for the investigation of complex systems including polymer classes and blends [134].

Aoyagi et al. [126] applied G-SIMS and multivariate curve resolution (MCR) to the investigation of fragment ion patterns of PEG with several different molecular weight distributions (PEG600, PEG1000, mixtures). The authors prepared thin films of the neat materials and mixtures on silicon substrates, and were able to extract pure spectra of each deposited compound by MCR. The use of G-SIMS allowed for classification of the relationship between secondary ions and proposed fragmentation mechanism. Samples were measured with Mn^+, Bi^+, and Bi_3^+, respectively. A G-SIMS spectrum of PEG showing enhanced specific signals at mass $m/z = 113, 157, 201$, and 245 $(C_2H_4O^+)_n$ $(n = 2–5)$ was discussed. Signals at $m/z = 113$ were assigned to $(C_2HC_2H_4O^+)_2$. As concluded from the experiments, G-SIMS spectra obtained with 25 keV Bi^+ and Mn^+ primary ions (g index = 13) showed the most prominent signal enhancement in the fingerprint mass range ($m/z \leq 300$) and MCR provided pure spectra from the investigated complex polymer mixtures.

4.2.3 Polymer additives

Additives are essential parts of synthetic technical polymers and enhance the application range of the product. Common types of additives are lubricants, plasticizers, antioxidants, UV absorbers, impact modifiers, and fillers. Release agents are usually not
classified as typical additive compounds, but are often found on technical polymer surfaces and notorious as harmful to desired properties of the later product. A broad
range of additives are routinely used in polymer manufacturing, and their relative composition is specifically tailored to the later application. Organic materials include
classes of alkyl phenols or hydroxybenzophenones, inorganic and organometallic
additive, fillers comprised of different oxides, salts, metallocarboxylates, and finally accelerators (Ni, Zn). For an extensive discussion of the topic, the reader is
referred to monographs of Jan Bart [135, 136] and citations therein.

Materials properties such as brittleness, thermal and light stability, or impact resistance are achieved by additive concentrations ranging from the parts per million
(ppm) to the percentage range. A combination of several additives is usually necessary to give the polymer its desired performance. It has been realized [126] that com-
pounding, i.e. the incorporation of the additive into the polymer matrix, is often a
prerequisite to avoid dynamic processes at the surface [136]. If not controlled properly, the dynamic behavior of near-surface- or subsurface-located additives can lead
to phenomena such as blooming, color change or adhesion and printing failure [83].

The surface segregation of Irgafos 168 (Ciba), one of the most used antioxidant
additives in industry, was investigated by Médard et al. [137] on the surface of poly
(ethylene terephthalate-ethylene isophthalate) (PETI). Thin films of PETI, containing various amounts of Irgafos^{TM} 168, were spin casted from chloroform. The film
thickness was estimated to be 1.5 nm at 3 wt% of additive concentration. ToF-SIMS
spectra obtained in the framework of a small round robin test at three independent
laboratories and acquired using different instrument platforms were reported. To
compare signals, a Ga$^+$ primary ion source was used for all investigations. Strong
segregation of Irgafos 168 to the surface of the substrate accompanied by subsequent crystallization for samples containing 5 wt% of additive concentration was
observed. A uniform covering of the surface with additive was detected by imaging
ToF-SIMS and further confirmed by angle-dependent XPS investigations. Surface
segregation (blooming) of mobile polymer additives usually occurs when the concentration overpasses the solubility within the polymer matrix and decreases the
intensity of characteristic substrate signals. The aging effects on the amount of additive at the surface was then studied, using ToF-SIMS by Poleunis et al. [138] The
ToF-SIMS data of thin films of PETI containing variable concentration of antioxidant (Irgafos^{TM} 168) and a UV-stabilizer (Hostavin^{TM} N30) demonstrated in the case
of a single additive (antioxidant), the increase of the additive intensity peaks at the
polymer surface during the first five aging days (exudation phenomenon), followed by an intensity decrease, which was related to the adsorption of hydrocarbon

contaminations at the sample surface. In the case of binary additive system (anti-oxidant and UV-stabilizer) the antioxidant behavior was similar to that in the single additive system, whereas the UV-stabilizer evolution corresponded to an additive depletion, followed by an exudation.

A surface migration study of brominated flame retardants in high-impact PS (HIPS) was published by Holbrook et al. [139]. The authors detected blooming of poly-brominated diphenyl ether (PBDE, BDE-209) on a HIPS surface, using micro-X-ray fluorescence spectroscopy, FIB-SEM, energy dispersive X-ray spectroscopy, and ToF-SIMS. Experiments based on a Bi_3^+ primary ion beam revealed characteristic signals of BDE-209 ($C_{12}OBr_{10}$) on the HIPS surface at masses at $m/z = 485$, 487, 489, and 491, respectively. The effect was also confirmed by imaging ToF-SIMS investigations (Fig. 4.6).

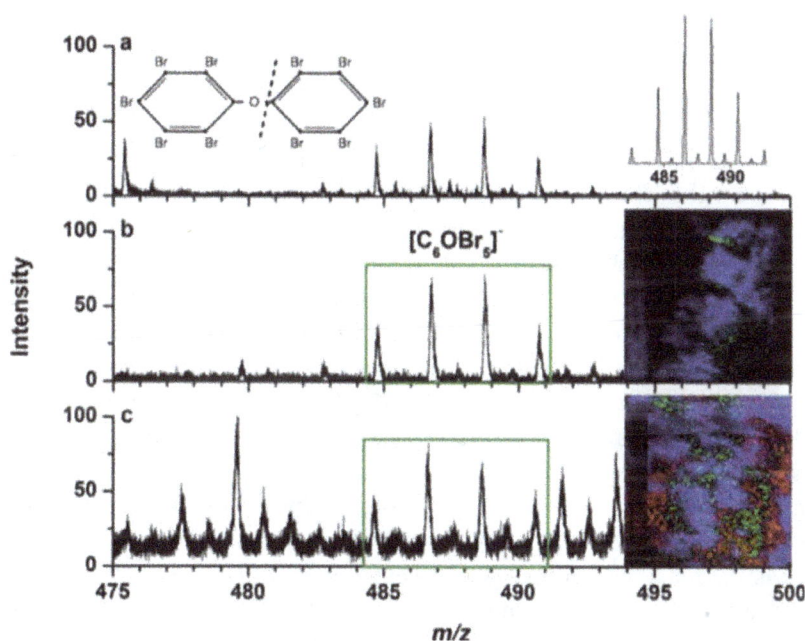

Fig. 4.6: Negative ion secondary mass spectra and images of BDE–209 obtained from standard (a), on silicon) and HIPS samples. The intense high-mass signal $C_6OBr_5^-$ is generated from symmetric bond breaking at the central ether group and reflects the calculated isotopic distribution of bromine at $m/z = 485$, 487, 489 and 491, respectively, as shown in the insert at the upper right hand corner. ToF–SIMS spectra b and c were obtained from the grey (b) and black (c) side of the HIPS surface. Overlay images (500 × 500 µm) show the green (BDE–209), blue (b, $m/z = 183$, HIPS fragment) and (c, $m/z = 293$ blue and $m/z = 463$, red, HIPS fragment). Reprinted from [139] with permission of John Wiley & Sons Ltd.

A novel method for the quantification of organic additives on polymer surfaces by ToF-SIMS using gold deposition was proposed by Murase et al. [140]. The method made it possible to compare the amount of two organic additives among three different polymer matrices (PS, PMMA and PET). Gold deposition on polymer film surfaces containing various concentrations of additives made ToF-SIMS analysis possible for quantitative evaluation of the additives from negative ion mass spectra among the different matrices.

4.2.4 Copolymers

Unlike homopolymers that are built of identical monomer subunits, copolymers consist of two or more different species. Classes of copolymers are defined, based on the alternating arrangement of their subunits, as:
- periodic copolymers with a well-defined sequence such as ABAABBABAABB;
- alternating copolymers, which are a special case with sequence ABABABABAB;
- block copolymers where homopolymer subunits are linked by a covalent bond;
- statistical copolymers with a completely irregular order of monomer sequence.

Kita-Tokarczyk et al. [141] published a study discussing the interaction between 1,2-dipalmitoyl-sn-glycero-3-phosphocholine (DPPC) and 1,2-dioleoyl-sn-glycero-3-phosphocholine with amphiphilic poly(2-methyloxazoline)-block-poly(dimethylsiloxane)- block-poly(2-methyloxazoline) block copolymers. For simplicity, the polymer consisting of a symmetric structure of two terminally poly(2-methyloxazoline) $(PMOXA)_{15}$ units and one $PDMS_{110}$ polymer, located at the center ($PMOXA_{15}$-$PDMS_{110}$-$PMOXA_{15}$) was called $A_{15}B_{110}A_{15}$ (mass 10,700 g/mol). Among other characterizing techniques, ToF-SIMS imaging was used to observe the formation of phase irregularities occurring from fluorescent dye attachment to the lipids under investigation. ToF-SIMS spectra of a LB (Langmuir-Blodgett) film consisting of neat $A_{15}B_{110}A_{15}$ copolymer, and a similarly prepared film with co-deposited lipids were recorded to identify characteristic signals. Peaks assigned to the polymer included CN^- ($m/z = 26$) and CNO^- ($m/z = 42$), whereas DPPC was identified by PO_2^- ($m/z = 62.97$), PO_3^- ($m/z = 78.92$), and a peak attributed to the alkyl chain with composition $C_{16}H_{31}O_2^-$ ($m/z = 255.25$), respectively. As shown in Fig. 4.7, the formation of star-like structures consisting predominantly of DPPC was detected in the ToF-SIMS images of the lipid-covered LB films. This phenomenon was observed, in particular, in films with high surface pressure where part of the lipid is excluded from the polymer film (collapse pressure) and condensation into the star-like structure is energetically favored.

Keller and Mayerhofer [142] detected vesicles of an $A_xB_yA_x$ copolymer in rat liver samples. The spherical objects with an average diameter of 200 nm were identified by mass resolved imaging of $(CH_3)_3Si^+$ ($m/z = 73.047$) as characteristic signal of PDMS. Signal assignment, however required mass discrimination from conflicting fragments of $C_3H_5O_2^+$ ($m/z = 73.029$), $C_3H_7NO^+$ ($m/z = 73.053$), and $C_4H_{11}N^+$ ($m/z = 73.089$) originating from surrounding tissue. Because the sample was measured below 148 K, condensation of water was one of the problems arising with transfer of the material into the UHV measuring chamber of the ToF-SIMS instrument. A water

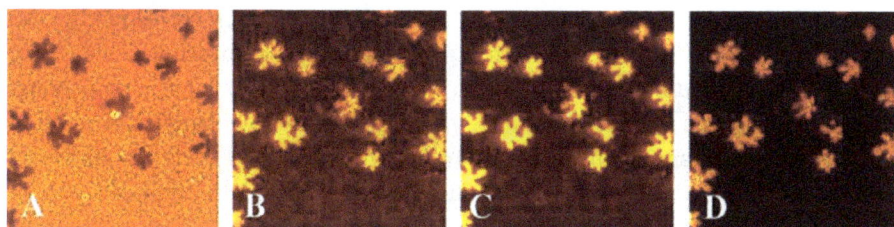

Fig. 4.7: ToF–SIMS images (100 × 100 μm) of Langmuir–Blodgett monolayers from $A_{15}B_{110}A_{15}$–DPPC with molar ratio 0.3–0.7 transferred on a silicon surface at 25 mN/m. Brighter colors indicate higher signal intensity. The images were recorded using a 25 keV Bi_3^{++} ion beam. Fragments were assigned to (a) CN^-, $A_{15}B_{110}A_{15}$ ($m/z = 26.00$), (b) PO_2^-, DPPC ($m/z = 62.97$), (c) PO_3^-, DPPC ($m/z = 78.92$) and (d) $C_{16}H_{31}O_2^-$, alkyl chain ($m/z = 255.25$). Reprinted with permission from [141]. Copyright (2009) American Chemical Society.

cluster with composition $H_9O_4^+$ ($m/z = 73.050$) was detected at the surface of the tissue sample.

The possibility to perform imaging of secondary ions with high mass and high lateral resolution at the same time is therefore required to distinguish between fragments arising from different origin. Tight focusing of temporally narrow ion pulses is not possible in conventional ion guns. The development of primary ion beam technology has met this challenge by operating the LMIG source in a so-called burst mode. The temporally broad ion pulse is split into several sub-pulses (patented technology of IonTof GmbH, Münster, Germany), of which one can be selected for high-mass-resolution imaging. A mass resolution m/Δm ≥ 2'500 is achieved for signals in the fingerprint region of the mass spectrum. However, the mass interference between $H_9O_4^+$ and $(CH_3)_3Si^+$ requires a mass resolution of m/Δm > 24'000 to be clearly resolved. This figure was beyond the performance of LMIG technology and was only attainable because in reality the $(CH_3)_3Si^+$ signal was shifted by +200 ppm due to incomplete charge compensation during measurement.

An electric potential-dependent adsorption/desorption process of the triblock copolymer poly(propylene sulfide-*bl*-ethylene glycol) (PPS-PEG) from a conductive ITO substrate was studied by Tang et al. [143]. Characteristic signals of PEG were identified at masses out of a series with $C_aH_bO_c^+$, eg. $C_3H_7O^+$ ($m/z = 59$). As most intense fragments of the PPS backbone, the authors assigned $C_2H_3S^+$ ($m/z = 58$) and $C_3H_7S^+$ ($m/z = 75$). ToF-SIMS spectra from PPS-PEG adlayers contained signals at masses Na^+ ($m/z = 22.992$), CH_3O^+ ($m/z = 31.019$), S^+ ($m/z = 32.98$), $C_2H_5O^+$ ($m/z = 45.037$), $C_4H_9^+$ ($m/z = 57.071$), $C_2H_3S^+$ ($m/z = 58.995$), $C_3H_7O^+$ ($m/z = 59.054$), $C_3H_7S^+$ ($m/z = 75.027$), $C_6H_9^+$ ($m/z = 81.067$), In^+ ($m/z = 114.905$), and $In-S^-$ ($m/z = 146.866$), respectively.

Electric potential-dependent polymer desorption was observed by comparing signal intensities of $C_3H_7O^+$ ($m/z = 59.05$, PEG) and $C_2H_3S^+$ ($m/z = 58.99$, PPS) after applying potentials between 0 and 2,000 mV to the copolymer-coated ITO substrate. Maximum intensities were measured at 0 mV, which corresponds to an intact surface film. Spectra shown in Fig. 4.8 indicate a significant decrease for both signals at anodic potential ≥ 800 mV as consequence of increasing polymer desorption from the surface. ToF-SIMS spectra recorded from samples exposed to 2'000 mV showed similar peak distribution as the neat ITO surfaces. An intact desorption in molecular form was postulated rather than oxidation of sulfur or chain degradation by XPS investigations of the process.

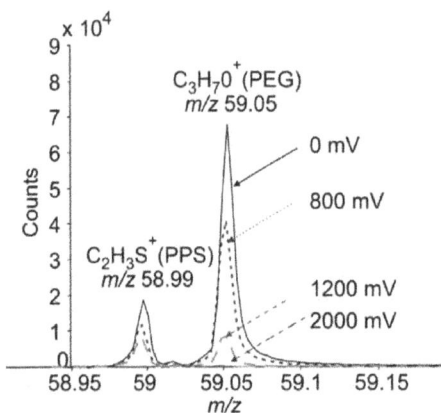

Fig. 4.8: Averaged positive ion high–resolution ToF–SIMS spectra from three independent samples shown around the nominal mass $m/z = 59$. Spectra were recorded as function of applied anodic potential with fragments $C_2H_3S^+$ ($m/z = 58.995$) characteristic of PPS and $C_3H_7O^+$ ($m/z = 59.052$) originating from PEG. Reprinted from [143] with permission from Elsevier.

ToF-SIMS combined with AFM was successfully applied to the investigation of micro-phase separation of a symmetric di-block copolymer: the polystyrene-b-poly(2-ethyl hexyl acrylate), d(PS-PEHA) during annealing processes. The study evidenced that annealing the copolymer films with different thickness for 2 h at 140 °C, produced films with strongly oriented copolymer microdomains parallel to the surface. A depth profile analysis of molecular fragment ions showed consistently regular alternated patterns with the same period as those of elemental ions. It also confirmed that the size of the lamellar structure is a function of the total number of monomers in the copolymer chain [144].

ToF-SIMS molecular depth profiling of PS-b-PMMA block copolymers, performed by ultra-low energy Cs^+, C_{60}^{++} and Ar cluster (Ar_{1500}^+) ion beams, provided data consistent with the theoretical model, emphasizing the ability of ToF-SIMS molecular depth profiling of verifying the film morphology as a function of deposition parameters, such as thickness and annealing time [145]. The study also demonstrated the improved quality of depth profile with the use of polyatomic ion projectiles, in particular the Ar cluster ion beam, with respect to a monomeric one. Weng and Chan [146] demonstrated the ability of ToF-SIMS to perform quantitative analysis in copolymers with a well-defined structure as well as in polymer blends with controllable interaction.

4.2.5 Multicomponent polymers (polymer blends)

Blending of polymers is one of the technologically most important fields of polymer chemistry today because it allows the manufacturing of materials that combine properties of single components. This synthetic concept has been successfully applied to many consumer products or system components with new and superior qualities [147, 148]. However, one of the most remarkable and often cumbersome physico-chemical facts is the often found phase separation of its constituents due to minimal miscibility or partial immiscibility. Depending on the application of the material, the formation of micro-mechanical defect sites is more or less important for the performance of the polymer blend. Feng et al. published a compatibility study of PC and poly(vinylidene fluoride) blends using ToF-SIMS and SEM [149] and a imaging ToF-SIMS study of the interdiffusion between PMMA and PVDF during the blend processing [150]. Yuan and Shoichet [151] investigated the enrichment of poly(trifluorovinyl ether) (PTFVE) in PS blends at low bulk concentration. PTFVE is chemically characterized by a fluorocarbon backbone with oligoether side groups. The surface of the blended material was characterized using ToF-SIMS and XPS and was found to be enriched by PTFVE at concentrations was <1 wt%. Large aggregates with diameters around 200 nm were detected. It was suggested that PTFVE in PS/ PTFVE blends forms core-corona structure elements at the surface with the oligoether groups as corona. The fingerprint mass range of positive-ion ToF-SIMS spectra was investigated with mass $m/z = 91$ ($C_7H_7^+$, tripylium ion) as dominant signal in pristine PS.

In blends of PS with PTFVE, this signal decreased dramatically for only as low as 0.5 wt% of PTFVE, and strong peaks at masses $m/z = 29$ ($C_2H_5^+$), $m/z = 45$ ($C_2H_5O^+$), and $m/z = 73$ ($C_2H_5OC_2H_4^+$), respectively, were detected as characteristic fragments of the oligoether group of PTFVE. Spectra collected from polymer blends with 2.5 wt % PTFVE showed no tripylium ions at mass $m/z = 91$ anymore. Because poly(trifluorovinyl ether) is a viscous liquid at room temperature (Tg = 213 K), this effect is probably due to a surface energy-driven rearrangement process. Polymer films cast from toluene, which is a more potent solvent for PS than e.g. chloroform, required greater PTFVE concentration to show similar surface activity. A surface reconstruction process was also observed by Hug and Keller [113] in modified ethylene-methacrylic acid copolymers.

Combination of ToF-SIMS with AFM were used to study the morphology-driven surface segregation in a blend of PCL and PVC [152, 153]. Imaging ToF-SIMS in combination with XPS and AFM was also used to investigate the development of surface morphology and composition in blends of PS(OH/PVPy) [154].

High-resolution ToF-SIMS imaging of polymer composites containing carbon nanotubes (CNTs) identified CNT clusters inside the admixture. Changes in cluster shape inside the admixture at different concentration evidenced a good correlation with the tensile strength and hardness of the polymer blend as measured by traditional methods, that allows the authors to assess that imaging ToF-SIMS could be used to predict the extent of change in strength of complex polymer mixtures [155].

4.2.6 Plasma modification and deposition

Léonard et al. [111] studied the adhesion properties of metals following H_2-, O_2-, and N_2-plasma modification of Teflon PFA and Teflon AF1600 surfaces. ToF-SIMS measurements confirmed that ablation of fluorine from surfaces exposed to H_2-plasma treatment showed the strongest effect. However, surface incorporated hydrocarbons during the plasma process strongly decreased adhesion efficiency by formation of weak boundary layers. Signals assigned to CN^- on the surface were detected after N_2-plasma exposure and enhanced occurrence of surface oxygen functional groups was characteristic for O_2-plasma treatment.

The introduction of epoxy groups to biologically active surfaces has been recognized to form stable and reliable bonds to peptides. Silanization by solvent-based (3-glycidyloxypropyl)trimethoxysilane has previously found widespread application. However, the practical use is restricted by excessive pre-polymerization in the liquid phase and incomplete formation of a surface layer. Hydrolysis of siloxal bonds is also often observed. Plasma deposition of surface films offers an alternative to solvent based epoxy functionalization. Pulsed plasma polymerization of glycidyl methacrylate [156, 157] produced well-characterized epoxy-rich and almost pinhole-free surfaces. Thierry et al. [158] used a DC plasma deposition process of allyl-glycidylether (AGE) monomer onto various surfaces and demonstrated the homogeneity of the epoxy-carrying surface by XPS and ToF-SIMS experiments. Fingerprint spectra in the range of $m/z = 0$–150 of pulsed and DC-deposited films were compared. Most of the structural information of epoxy-functionalized surfaces was found within this relatively restricted mass span (Fig. 4.9).

The inspection of ToF-SIMS spectra under high-mass-resolution conditions, reveals a split-up of unit mass signals into several distinct peaks. For example, the insert in Fig. 4.9 at mass $m/z = 57$ shows fragment masses at $m/z = 57.03$ ($C_3H_5O^+$) and $m/z = 57.07$ ($C_4H_9^+$). Comparing the intensities of close-spaced signals of different origin allows an estimation of the influence of process parameters. Because $C_3H_5O^+$ is related to the epoxy functionality, the relative enhancement of this signal in spectrum (b) indicates the superior retention of this chemical group under pulsed plasma deposition conditions. PCA was used to detect subtle differences between films produced under distinct conditions. Clustering of related data was observed,

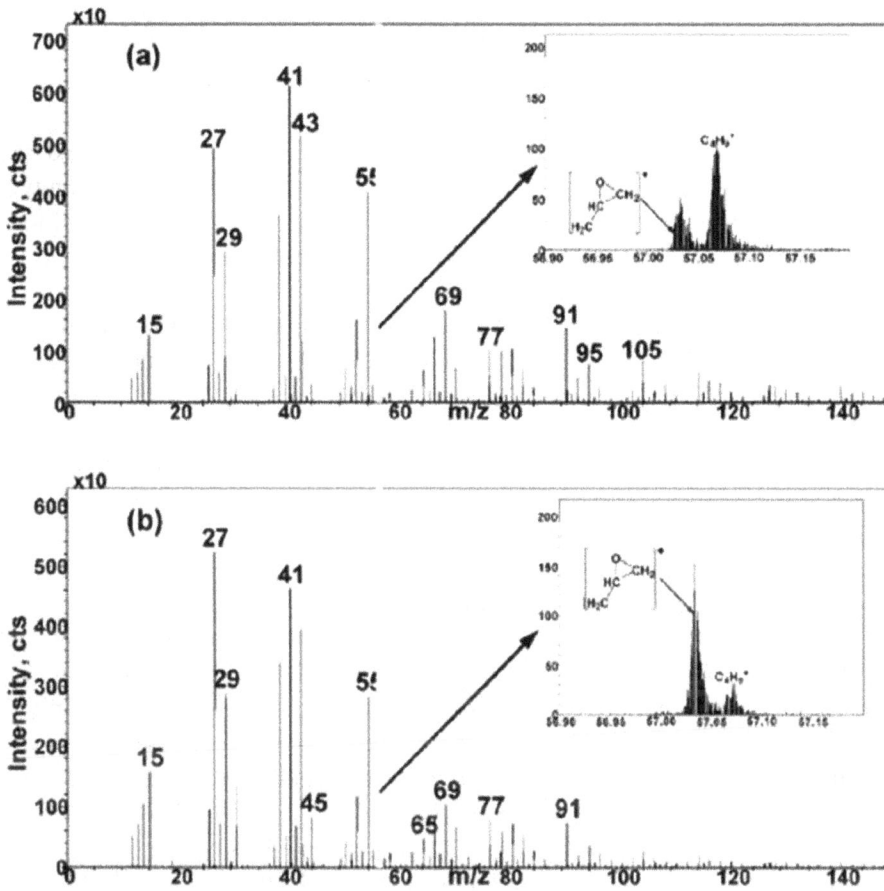

Fig. 4.9: ToF–SIMS spectra of continuous (a) and plasma (b) deposited films of AGE. $C_3H_5^+$ at mass $m/z = 41$ is dominating spectrum (a), while $C_2H_3^+$ at $m/z = 27$ is the most intense signal in spectrum (b). The formation of enhanced low–mass fragments in spectrum (b) is indicating a better retention of the monomer chemical structure under pulsed deposition conditions. Reprinted with permission from [158]. Copyright (2008) American Chemical Society.

which confirmed the chemical homogeneity of the processed surfaces. For details about the peak list and PCA parameters, the reader is referred to the literature [158].

Pulsed plasma polymerization is a very effective and straightforward experimental way to produce uniform surface polymer films of various chemical compositions. It was shown [159] that pulsed plasma polymerization of 2-hydroxyethyl-methacylate leads to the formation of a well-defined poly-HEMA film.

Kinmond et al. [160] detected a high structural retention of perfluoroalkyl groups after the plasma polymerization of 1 H,1 H,2 H-perfluorododecene using

NMR and ToF- SIMS. In contrast, continuous wave conditions lead to a high level of fragmentation and considerable cross-linking in the film.

Static ToF-SIMS was successfully applied to investigate copolymers prepared from ethylene and allyl alcohol in a plasma chemical process. The fact that the OH functional groups in the copolymer vary non-linearly with respect to the feed gas composition was explained in terms of the different chemical reactivity of the two monomers in the plasma. Through a careful selection of the gas composition, plasma copolymerization can be used to control the concentration of surface functional groups in plasma deposited films [161].

Carbon fluorinated films on Silicon and PET at different deposition time were investigated by ToF-SIMS and XPS. Using Multivariate Analysis, important fragmentation in the ToF-SIMS spectra was found to nicely correlate to both the surface structure and plasma parameter (deposition time). In the case of silicon, fluorination took place, forming SiF and CF_x sites on the surface and higher fluorination degree was obtained with increasing deposition time, whereas, in the case of PET substrates, higher crosslinking of the CF_x film occurred [162].

ToF-SIMS depth profiling with large Ar cluster ion beams in combination with PCA was demonstrated to be a useful tool to elucidate the chemical structure of plasma synthetized polystyrene-like coatings. The method provided an accurate picture of the polymer film during growth phase with detailed information at a very high in-depth resolution (few nm) and a lateral micrometric resolution (3D molecular imaging). Reconstruction of the mass spectra from a section of the depth profile, corresponding to the bulk of the film allowed to disregard post-polymerization oxidation triggered by long-lived radicals and other surface phenomena [163].

Ar cluster ion beams have been proposed by Poleunis et al. [164] as a new method to retrieve local physical properties suche as Young modulus and glass transition temperature. These methodological developments made ToF-SIMS a useful tool for both the chemical and physical characterization of complex plasma coatings in three dimensions.

4.2.7 Other applications

The application of ToF-SIMS imaging to small, almost nano-dimensional structures has made great progress in the last decades. The focal diameter of timely LMIG cluster ion sources has been decreased to values close to theoretical limit [61, 62]. The number of secondary ions generated from small areas is now high enough to acquire images with good quality. However, restricted information can be obtained from the surface topography of the material at this lateral dimension. This has classically considered the domain of force microscopy (atomic force microscopy, AFM;

scanning probe microscopy, SPM), where lateral resolution of atomic dimension is possible under suitable experimental conditions. Chemical information gained by SPM or AFM is poor and is usually restricted to phase contrast AFM or Kelvin probe force microscopy.

A recently developed instrument combining a ToF-SIMS instrument equipped with a 10 nm resolution gun with a high-quality, in-line long-range SPM assembly in the same vacuum chamber allowed for the first time true complementary access to high-resolution ToF-SIMS imaging combined with the ultra-high lateral surface morphology information of SPM (3D NanoChemiscope Project) [165]. The instrument was applied to the 3-D analysis of a PMMA/PS blend on silicon via GCIB sputtering.

Although the equipment is intended for use in a depth profiling mode and applications in metrology, the shallow profiles classify the instrument as surface-sensitive technique. A similar experimental setup was presented by Wirtz et al. [166]. Combined SIMS-SPM 3D-mapping of an annealed PS/PMMA blend was measured using a high-precision sample stage mounted inside a NanoSIMS instrument. Secondary ions recorded included $^{12}C^-$, $^{12}C^1H^-$, $^{16}O^-$, and $^{24}C_2^-$, respectively. Images of ^{12}C- and ^{16}O- were presented with a field of view of 10.6×9.8 µm^2. An area of 256×256 pixels was scanned with a 50 nm diameter primary ion beam. Two separate phases were detected, from which PS was identified by the absence of oxygen. Topographical information was simultaneously obtained, revealing a distinct crater-like surface roughness as result of considerably lower sputter yield of PS compared with PMMA.

3D images of organic and biological samples can be obtained using various approaches based on the specific nature of the samples and considering that a general rule for 3D analysis with mass spectrometry is that the resolution in the z dimension is normally better than is routinely achievable in the x and y dimensions. In tissue sections 3D-imaging, for example, the standard approach is to use consecutive sections of the tissue and to image each section. The images in the series are then combined to produce a 3D chemical map. In SIMS the consecutive section method was first demonstrated for the 3D imaging of rat heart by Fornai et al.; 40 individual tissue slices were taken at 100 µm intervals in the z direction through the tissue and imaged using ToF-SIMS [167].

Alternatively, the change in the ion signals from the sample can be monitored with improved mass resolution and mass accuracy as the sample is eroded using polyatomic ion beams (molecular depth profiling) [168]. Critical to a successful molecular depth profile is the sputter yield being high enough to remove any damage that is generated during the primary ion impact. Polyatomic ion beams fulfill this criterion by depositing more energy into the outer surface of the sample, thus enhancing the sputter yield with less subsurface disruption [169]. A high sputtering yield can also facilitate molecular depth profiling as damaged material is more likely to be ejected and not accumulate on the sample surface. Polymers such as polylactic acid (PLA) and polycaprolactone showed high sputter yields providing stable depth profiles [9, 170]. An early example of 3D molecular imaging was published by Gillen et al. using SF_5^+ to generate a 3D image of acetaminophen-doped PLA polymer film [171].

The introduction of C_{60} for molecular depth profiling brought further improvements. Polymeric and biological materials produced reliable depth profiles under C_{60}^+ bombardment. Further successful depth profiles were obtained even up to 120 keV C_{60}^+ energy [172]. Fisher et al. reported advantages of C_{60}^+ versus SF_5^+ for studying the 3D distribution of the sirolimus drug in a poly(lactic-co-glycolic acid) (PLGA) coronary stent coating. The 3D chemical images revealed the distribution of the drug in the surface and subsurface layers and allowed to visualize the molecule distribution as a function of elution time [173].

The Vickerman group at the University of Manchester reconstructed 3D images of a frog egg were from the ToF-SIMS analysis using C_{60}^+ as both the analysis and etching beam [174–176].

A challenge associated with 3D analysis with SIMS is the accurate reconstruction of the 2D data to produce a 3D image. The sample often has inherent, natural, topography whereas the image generated by the instrument is flat. For example, the analysis of cells on a flat substrate results in a 2D image of the cell surrounded by substrate material. Erosion of the cell by the ion beam produces a series of images where the cell appears to be getting smaller and smaller as the thinner extremities of the cell are removed. Visualization of the combined image stack shows a cell that appears to be embedded in the substrate. To compensate for this, the data must be reshaped. For 3D cell imaging, a common approach is to reshape the data, pixel by pixel, based on the appearance of the substrate. The cell/substrate interface can be identified by using a known substrate specific peak or by using multivariate analysis such as principal components analysis (PCA) to identify the change from cellular to substrate signal [177–179]. The data reshaping to obtain 3D images in SIMS can use either multivariate analysis (PCA or others) or topographical approaches (combination of ToF-SIMS with AFM, SPM, DHM) [179–181]. Three-dimensional images reconstructions in SIMS become much more difficult when analyzing complex heterogeneous materials, because of different sputter rates due to the sample chemical and topoghraphical heterogeneity and matrix effect [182].

A further ongoing challenge associated with SIMS imaging of biological samples is the low ionization efficiency of the technique. Several approaches have been demonstrated to enhance ionization during analysis such as metal assisted SIMS and matrix enhanced SIMS. However, the drawback of these approaches is that ionization is only enhanced on the surface of the sample and so the benefits are not applicable to 3D imaging experiments. Hence, the search for a universally applicable method for signal enhancement continues, and recent avenues of research have focused on the possibility of using chemically reactive gas cluster ions. The Winograd group demonstrated enhanced molecular ion generation from trehalose films when 3% CH_4 was incorporated into the Ar gas cluster ions. A 4× increase in the

[M+H]$^+$ signal was observed compared to a 1.5× increase in the [M+Na]$^+$ moiety suggesting proton transfer as a major factor for the enhancement [183]. Another strategy for supplying excess protons to the sample from the primary ion beam has been investigated by researchers at the University of Manchester where a water cluster source has been developed. A 10–20× increase in signal levels has been reported for lipid and pharmaceutical reference materials [184].

A recent application of SIMS to the investigation of organic, polymeric and biological systems is represented by NanoSIMS that, combined with other analytical tools, has been successfully used to investigate the biodegradation in soils of poly (buthylene adipate-co-terephthalate (PBAT), an important polyester used in agriculture, and contributed to assess the key steps involved in the degradation process, such as microbial polymer colonization, enzymatic depolymerization on polymer surface, and microbial uptake and utilization of the released low-molecular weight compounds. In particular, the NanoSIMS-based approach allowed to unambiguously demonstrate incorporation of polyester carbon into soil microbial biomass [185]. The use of antibody functionalized multimodal polymeric nanoparticles, designed to deliver an anti-oxidant (resveratrol) to astrocyte cells, was analyzed by NanoSIMS [186] to measure oxidative stress in stressed cells exposed to the free anti-oxidant compared to the anti-oxidant encapsulated in polymeric nanoparticles (NPs). NanoSIMS data well described the distribution of ^{40}Ca microdomains in the z projection of optic nerve following partial transection injury.

ToF-SIMS, NanoSIMS and Imaging ToF-SIMS (including 3D imaging) in combination with other surface analytical methods and/or Multivariate Analysis have been extensively applied to the characterization of biomaterials [130, 187–192], micro- and nano-structured systems [193–198], surface monolayers and thin films [199, 200], and biological samples (cells and tissues) [201–204].

For the general aspects of ToF-SIMS, Imaging ToF-SIMS, and NanoSIMS applications in biomedical and biological fields, refer to the review articles of A.M. Belu et al. [187], J. Yang and I. Gilmore [192], S. Fearn [190], Fletcher [174], and J. Nunez et al. [205] and references therein.

References

[1] Vickerman JC, Briggs D ed. ToF-SIMS: Surface Analysis by Mass Spectrometry, 1st and 2nd edn, Chichester, UK, IM Publications and Surface Spectra, 2001 and 2013.

[2] Chan CM. Polymer Surface Modification and Characterization. Munich, Vienna, New York, Carl Hansen Verlag, 1994.

[3] Briggs D. Surface Analysis by XPS and Static SIMS, 1st, Cambridge UK, Cambridge University Press, 1998.

[4] Delcorte A, Bertrand P, Garrison BJ. A microscopic view of organic sample sputtering. Appl Surf Sci 2003, 203–204, 166–169.

[5]   Lee JLS, Gilmore I, Seah MP. Quantification and methodology issues in multivariate analysis of ToF-SIMS data for mixed organic systems. Surf Interface Anal 2008, 40, 1–14.

[6]   Delcorte A, Medard N, Bertrand P. Organic secondary ion mass spectrometry: Sensitivity enhancement by gold deposition. Anal Chem 2002, 74, 4955–4968.

[7]   Winograd N. The magic of cluster SIMS. Anal Chem 2005, 77, 142A–9A.

[8]   Ninomiya S, Ichiki K, Yamada H, Nakata Y, Seki T, Aoki T, Matsuo T. Precise and fast secondary ion mass spectrometry depth profiling of polymer materials with large Ar cluster ion beams. Rapid Commun Mass Spectrom 2009, 23, 1601–1606.

[9]   Mahoney CM. Cluster secondary ion mass spectrometry of polymers and related materials. Mass Spectrom Rev 2010, 29, 247–293.

[10]   Wucher A, Winograd N. Molecular sputter depth profiling using carbon cluster beams. Anal Bioanal Chem 2010, 396, 105–114.

[11]   Shard A, Gilmore I, Wucher A. Molecular depth profiling. In: Vickerman JC, Briggs D, eds. ToF-SIMS: Surface Analysis by Mass Spectrometry, Chichester, UK, IM Publications and Surface Spectra, 2013, 331–334.

[12]   Vickerman JC. Molecular surface mass spectrometry by SIMS. In: Vickerman JC, Gilmore IS, eds. Surface Analysis: The Principal Techniques, 2nd, Chichester, UK, Wiley, 2009, 113–205.

[13]   Briggs D, Seah MP eds. Practical Surface Analysis: Ion and Neutral Spectroscopy. Vol 2, Chichester (UK), Wiley, 1983.

[14]   Eccles AJ. SIMS instrumentation. In: Vickerman JC, Brown A, Reed NM, eds. Secondary Ion Mass Spectrometry: Principles and Applications, Oxford (UK), Oxford University Press, 1989, 73–104.

[15]   Grams J. New Trends and Potentialities of ToF-SIMS in Surface Studies, New York (NY), Nova Science Publishers, 2007.

[16]   Smith R, Walls JM. Ion erosion in surface analysis. In: Walls JM. Methods of Surface Analysis: Techniques and Applications, Cambridge, UK, Cambridge University Press, 1989, 20–56.

[17]   Reed NM. Static SIMS for applied surface analysis. In: Vickerman JC, Brown A, Reed NM. Secondary Ion Mass Spectrometry: Principles and Applications, Oxford (UK), Oxford University Press, 1989, 186–243.

[18]   Chan CM, Weng LT, Lau YTR. Polymer surface structure determined using ToF-SIMS. Rev Anal Chem 2014, 33(1), 11–30.

[19]   Thiel V, Sjövall P. Using time-of-flight secondary ion mass spectrometry to study biomarkers. Annu Rev Earth Planet Sci 2011, 39, 125–156.

[20]   Beavis RC. The interpretation of reflectron time-of-flight mass spectra. In: Cotter RJ, ed. ACS Symp Ser, Washington, DC USA, American Chemical Society, 1994, Vol. 549, 49–60.

[21]   Cotter RJ. Time-of-flight Mass Spectrometry: Instrumentation and Applications in Biological Research, Washington, DC, USA, ACS Professional Reference Books, 1997.

[22]   Wood M, Zhou Y, Brummel CL, Winograd N. Imaging with on beams and laser postionization. Anal Chem 1994, 66, 2425–2432.

[23]   Wucher A. Laser post-ionization: Fundamentals. In: Vickerman JC, Briggs D, eds. ToF-SIMS: Surface Analysis by Mass Spectrometry, 1st, Chichester, UK, IM Publications and Surface Spectra, 2001, 347–373.

[24]   Pellin MJ, Calaway WF, Veryovkiu IV. Laser post-ionization for quantitative elemental analysis. In: Vickerman JC, Briggs D, eds. ToF-SIMS: Surface Analysis by Mass Spectrometry, 1st, Chichester, UK, IM Publications and Surface Spectra, 2001, 375–415.

[25]   Lockyer NP. Laser post-ionization for molecular analysis. In: Vickerman JC, Briggs D, eds. ToF-SIMS: Surface Analysis by Mass Spectrometry, 1st, Chichester, UK, IM Publications and Surface Spectra, 2001, 347–373.

[26]   Thomson JJ. Rays of positive electricity. Philos Mag 1910, 20, 752–767.

[27] Benninghoven A. The analysis of monomolecular layers of solids by secondary ion emission. Z Physik 1970, 230, 403–417.

[28] Benninghoven A, Jaspers D, Sichterman W. Secondary ion emission of amino acids. Appl Phys 1976, 11, 35–39.

[29] Hagenhoff B Sekundärionenmassenspektrometrie molekularer Oberflächenstrukturen. Thesis. Münster, Germany, University Münster, 1993 [in German].

[30] Niehuis E, Heller T, Benninghoven A. A new high resolution time-of-flight secondary ion mass spectrometer. J Vac Sci Technol 1987, A5, 1243–1246.

[31] Van Vaeck L, Adriaens A, Gijbels R. Static secondary ion mass spectrometry: (S-SIMS) Part 1: Methodology and structural interpretation. Mass Spectrom Rev 1999, 18, 1–47.

[32] Adriaens A, Van Vaeck L, Adams F. Static secondary ion mass spectrometry (S-SIMS) Part 2: Material science applications. Mass Spectrom Rev 1999, 18, 48–81.

[33] Brummel CL, Willey KF, Vickerman JC, Winograd N. Ion beam induced desorption with postionization using high repetition femtosecond lasers. Int J Mass Spectrosc Ion Process 1995, 143, 257–270.

[34] Sigmund P. Theory of sputtering. I. Sputtering yield of amorphous and polycrystalline targets. Phys Rev 1969, 184, 383–416.

[35] Sigmund P. Sputtering processes: Collision cascade and spikes. In: Tolk NH, Tully JC, Heiland W, White CW, edited by. Inelastic Ion-surface Collisions, New York, Academic Press, Inc., 1977, 121.

[36] Fletcher JS, Conlan XA, Jones EA, Biddulph G, Lockyer NP, Vickerman JC. TOF-SIMS analysis using C_{60}. Effect of impact energy on yield and damage. Anal Chem 2006, 78, 1827–1831.

[37] Garrison BJ. Energy and angular distributions of atoms sputtered from polycrystalline surfaces: Thompson and beyond. Nucl Instrum Method Phys Res 1989, B40/41, 313–316.

[38] Benninghoven A, Ruedenauer FG, Werner HW. Secondary Ion Mass Spectrometry: Basic Concepts, Instrumental Aspects, Applications and Trends, New York (NY), Wiley, 1987.

[39] Van Den Eynde X. Quantitative characterization of polymer surfaces. In: Vickerman JC, Briggs D, eds. TOF-SIMS: Surface Analysis by Mass Spectrometry, 1st, Chichester, UK, IM Publications and Surface Spectra, 2001, 543–573.

[40] Kovac J. Surface characterization of polymers by XPS and SIMS techniques. Mater Technol 2011, 45,191–197.

[41] Hauert R, Keller BA. Chemical surface analysis with nanometer depth resolution. Chimia 2006, 60, 800–804.

[42] Vanden Eynde X, Reihs K K, Bertrand P. Molecular weight dependent fragmentation of selectively deuterated polystyrenes in ToF–SIMS. Macromolecules 1999, 32, 2925–293.

[43] Paruch RJ, Garrison BJ, Mlynek M, Postawa Z. On universality in sputtering yields due to cluster Bombardment. J Phys Chem Lett 2014, 5, 3227–3230.

[44] Paruch RJ, Postawa Z, Garrison BJ. Seduction of finding universality in sputtering yields due to cluster Bombardment of solids. Acc Chem Res 2015, 48, 2529–2536.

[45] Kanski M, Garrison BJ, Postawa Z. Effect of oxygen chemistry in sputtering of polymers. J Phys Chem Lett 2016, 7, 1559–1562.

[46] Gołuński M, Hrabar S, Postawa Z. Mechanisms of particle ejection from free-standing two-layered graphene stimulated by keV argon gas cluster projectile bombardment – molecular dynamics study. Surf Coat Technol 2020, 391, 125683.

[47] Kański M, Postawa Z. Effect of the impact angle on the kinetic energy and angular distributions of β-Carotene sputtered by 15 keV Ar_{2000} projectiles. Anal Chem 2019, 91, 9161–9167.

[48] Tuccitto N, Maciaze D, Postawa Z, Licciardello A. MD-based transport and reaction model for the simulation of SIMS depth profiles of molecular targets. J Phys Chem C 2019, 123, 20188–20194.

164 ⎯ Beat A. Keller et al.

[49] Schueler BW. Time-of-Flight Mass Analyzers. In: Vickerman JC, Briggs D, eds. ToF-SIMS: Surface Analysis by Mass Spectrometry, 1st, Chichester, UK, IM Publications and Surface Spectra, 2001, 75–93.

[50] Finšgar M. The interface characterization of 2-Mercapto-1-methylimidazole corrosion inhibitor on brass. Coatings 2021, 11, 295. DOI: https://doi.org/10.3390/coatings11030295.

[51] Wilson RG. Surface ionization ion sources. IEEE Trans Nucl Sci 1967, 14, 72–74.

[52] Gillen G, Roberson S. Preliminary evaluation of an SF_5+ polyatomic primary ion beam for analysis of organic thin films by secondary ion mass spectrometry. Rapid Commun Mass Spectrom 1998, 12, 1303–1312.

[53] Fletcher JS, Lockyer NP, Vickerman JC. Developments in molecular SIMS depth profiling and 3D imaging of biological systems using polyatomic primary ions. Mass Spectrom Rev 2011, 30, 142–174.

[54] Wong SCC, Hill R, Blenkinsopp P, Lockyer NP, Weibel DE, Vickerman JC. Development of a C_{60}^+ ion gun for static SIMS and chemical imaging. Appl Surf Sci 2003, 203–4, 219–222.

[55] Cheng J, Wucher A, Winograd N. Molecular depth profiling with cluster ion beams. J Phys Chem B 2006, 110, 8329–8336.

[56] Swanson LW. Liquid metal ion sources: Mechanism and applications. Nucl Instrum Methods Phys Res 1983, 218, 347–353.

[57] Appelhans A, Delmore JE. Comparison of polyatomic and atomic primary beams for secondary ion mass spectrometry of organics. Anal Chem 1989, 61, 1087–1093.

[58] Van Stipdouk MJ. Polyatomic Primary Beams. In: Vickerman JC, Briggs D, eds. ToF-SIMS: Surface Analysis by Mass Spectrometry, 1st, Chichester, UK, IM Publications and Surface Spectra, 2001, 309–345.

[59] Walker EV, Winograd N. Prospects for imaging with TOF-SIMS using gold liquid metal ion sources. Appl Surf Sci 2003, 203–4, 198–200.

[60] Davies N, Weibel P, Blenkinsopp N, Lockyer N, Hill R, Vickerman JC. Development and experimental application of a gold liquid metal ion source. Appl Surf Sci 2003, 203–4, 223–227.

[61] Kollmer F, Paul W, Krehl M, Niehuis E. Ultra high spatial resolution SIMS with cluster ions approaching the physical limits. Surf Interface Anal 2013, 45, 312–314.

[62] Orloff J, Utlaut M, Swanson L. High Resolution Focused Ion Beams: FIB and Its Application, New York (NY), Kluwer Academic, 2003.

[63] Kollmer F. Cluster ion bombardment of organic materials. Appl Surf Sci 2004, 231–2, 153–158.

[64] Blain MG, Della-Negra S, Joret H, Le Beyec Y, Schweikert EA. Secondary-ion yields from surfaces bombarded with keV molecular and cluster ions. Phys Rev Lett 1989, 63, 1625–1628.

[65] Harris RD, Baker WS, Van Stipdonk MJ, Crooks RM, Schweikert EA. Secondary ion yields produced by keV atomic and polyatomic ion impacts on a self-assembled monolayer surface. Rapid Commun Mass Spectrom 1999, 13, 1374–1380.

[66] Mahoney CM Ed. Cluster Secondary Ion Mass Spectrometry: Principles and Applications, New Jersey, USA, J. Wiley & Sons, Inc, 2013.

[67] Levine RD. Molecular Reaction Dynamics, Cambridge, UK, Cambridge University Press, 2005.

[68] Hashinokuchi M, Moritani K, Nakagawa J, Kashiwagi T, Toyoda N, Mochiji K. Secondary ion mass spectrometry using size-selected gas cluster ion beam. J Surf Anal 2008, 14, 387–390.

[69] Gilmore IS, Seah MP. Organic molecule characterization -G-SIMS. Appl Surf Sci 2004, 231, 224–229.

[70] Vanden Eynde X, Bertrand P. TOF-SIMS quantification of polystyrene spectra based on principal component analysis (PCA). Surf Interface Anal 1997, 25, 878–888.

[71] Henry M, Dupont-Gillain C, Bertrand P. Conformation change of albumin adsorbed on polycarbonate membranes as revealed by TOF-SIMS. Langmuir 2003, 19, 6271–6276.

[72] Kayser S, Rading D, Moellers R, Kollmer F, Niehuis E. Surface spectrometry using large argon clusters. Surf Interface Anal 2013, 45, 131–133.

[73] Cheng J, Winograd N. Depth profiling of peptide films with TOF-SIMS and a C_{60} probe. Anal Chem 2005, 77, 3651–3659.

[74] Cheng J, Winograd N. Molecular depth profiling of multi-layer systems with cluster ion sources. Appl Surf Sci 2006, 252, 6498–6501.

[75] Luxembourg SL, Mize TH, McDonnell LA, Heeren RMA. High-spatial resolution mass spectrometric imaging of peptide and protein distributions on a surface. Anal Chem 2004, 76, 5339–5344.

[76] Touboul D, Kollmer F, Niehuis E, Brunelle A, Laprévote O. Improvement of biological time-of-flight secondary ion mass spectrometry imaging with a bismuth cluster ion source. J Am Soc Mass Spectrom 2005, 16, 1608–1618.

[77] Holzweber M, Shard AG, Jungnickel H, Luch A, Unger WES. Dual beam organic depth profiling using large argon cluster ion beams. Surf Interface Anal 2014, 46, 936–939.

[78] Fletcher JS, Rabbani S, Barber AM, Lockyer NP, Vickerman JC. Comparison of C_{60} and GCIB primary ion beams for the analysis of cancer cells and tumour sections. Surf Interface Anal 2013, 45, 273–276.

[79] Bich C, Havelund R, Moellers R, Touboul D, Kollmer F, Niehuis E, Gilmore IS, Brunelle A. Argon cluster ion source evaluation on lipid standards and rat brain tissue samples. Anal Chem 2013, 85, 7745–7752.

[80] Angerer TB, Blenkinsopp P, Fletcher JS. High energy gas cluster ions for organic and biological analysis by time-of-flight secondary ion mass spectrometry. Int J Mass Spectrom 2015, 377, 591–598.

[81] Peacock PM, McEwen CN. Mass spectrometry of synthetic polymers. Anal Chem 2006, 78, 3417–3428.

[82] Hanton SD. Mass spectrometry of polymers and polymer surfaces. Chem Rev 2001, 101, 527–569.

[83] Keller BA, Hug P. Time-of-flight secondary ion mass spectrometry of industrial materials. Anal Chim Acta 1999, 393, 201–212.

[84] Reed NM, Vickerman JC. The application of static secondary ion mass spectrometry (SIMS) to the surface analysis of polymeric materials. In: Sabbatini L, Zambonin PG, eds. Surface Characterization of Advanced Polymers, Weinheim, Germany, VCH Verlagsgesellschaft, 1983, 83–162.

[85] Peacock PM, McEwen CN. Mass spectrometry of synthetic polymers. Anal Chem 2002, 74, 2743–2748.

[86] Peacock PM, McEwen CN. Mass spectrometry of synthetic polymers. Anal Chem 2004, 76, 3417–3428.

[87] Hakkarainen M ed. Mass Spectrometry of Polymers New techniques. Adv Polym Sci. Heidelberg, Germany, Springer-Verlag, 2012, 248.

[88] Galuska AA. Analysis of bulk polymers. In: Vickerman JC, Briggs D, eds. ToF-SIMS: Surface Analysis by Mass Spectrometry, 1st, Chichester, UK, IM Publications and Surface Spectra, 2001, 525–541.

[89] Legget GJ, Vickerman JC. Sample charging during static SIMS studies of polymers. Appl Surf Sci 1995, 84, 253–266.

[90] Leggett GJ. Mechanisms of ion production and fragmentation in static SIMS. In: Vickerman JC, Briggs D, Henderson A, eds. The Static SIMS Library, Manchester, UK, Surface Spectra, 1998, 19–38.

[91] Gilmore IS, Seah MP. Static SIMS: A study of damage using polymers. Surf Interface Anal 1996, 24, 746–762.

[92] Keller BA, Hug P. Cationization in TOF-SIMS spectra of biologically active peptides using modified ion-containing polymers as support. In: Benninghoven A, Bertrand P, Migeon HN, Werner HW, eds. Secondary Ion Mass Spectrometry, SIMS XII, Amsterdam, The Netherlands, Elsevier, 2000, 885–888.

[93] Keller BA, Hug P. A new type of support based on ion-containing polymers for selective generation of cationized species in TOF-SIMS spectra. In: Benninghoven A, Bertrand P, Migeon HN, Werner HW, eds. Secondary Ion Mass Spectrometry, SIMS XII, Amsterdam, The Netherlands, Elsevier, 2000, 749–752.

[94] Fowler DE, Johnson RD, Van Leyen D, Benninghoven A. Quantitative time-of-flight secondary ion mass spectrometry of a perfluorinated polyether. Surf Interface Anal 1991, 17, 125–136.

[95] Lee Y, Han S, Yoon J-H, Lim H, Cho J. Studies of polystyrenes by secondary ion mass spectrometry. J Surf Anal 1999, 6, 50–53.

[96] Galuska AA. TOF-SIMS determinations of molecular weights from polymeric surfaces and microscopic phases. Surf Interface Anal 1997, 25, 790–798.

[97] Coullerez G, Lundmark S, Malmström E, Hult A, Mathieu HJ. TOF-SIMS for the characterization of hyperbranched aliphatic polyesters: Probing their molecular weight on surfaces based on principal component analysis (PCA). Surf Interface Anal 2003, 35, 693–708.

[98] Ng JMK, Gitlin I, Stroock AD, Whitesides GM. Components for integrated poly (dimethylsiloxane) microfluidic systems. Electrophoresis 2002, 23, 3461–3473.

[99] Morra M, Occhiello E, Marola R, Garbassi F, Humphrey P, Johnson D. On the aging of oxygen plasma-treated polydimethylsiloxane surfaces. J Colloid Interface Sci 199(137), 11–24.

[100] Delcorte A, Befahy S, Poleunis C, Troosters M, Bertrand P. Improvement of metal adhesion to silicon films: A TOF-SIMS study. In: Mittal KL, ed. Adhesion Aspects of Thin Films, Vol. 2, Boca Raton, FL, USA, CRC Press, 2005, 155–166.

[101] Dong X, Proctor A, Hercules DM. Characterization of poly(dimethylsiloxane)s by time-of-flight secondary ion mass spectrometry. Macromolecules 1997, 30, 63–70.

[102] Dong X, Gusev A, Hercules DM. Characterization of polysiloxanes with different functional groups by time-of-flight secondary ion mass spectrometry. J Am Soc Mass Spectrom 1998, 9, 292–298.

[103] Inoue M, Murase A, Sugiura M. Molecular weight evaluation of poly(dimethylsiloxane) on solid surfaces using silver deposition TOF-SIMS. Anal Sci 2004, 20, 1623–1628.

[104] Belu AM, Brocchini S, Kohn J, Ratner BD. Characterization of combinatorially designed polyacrylates by time-of-flight secondary ion mass spectrometry. Rapid Commun Mass Spectrom 2000, 14, 564–571.

[105] Leeson AM, Alexander MR, Short RD, Briggs D, Hearn MJ. Secondary ion mass spectrometry of polymers: A TOF-SIMS study of monodispersed PMMA standards. Surf Interface Anal 1997, 25, 261–274.

[106] Wagner MS. Degradation of poly(acrylates) under SF_5+ primary ion bombardment studied using time-of-flight secondary ion mass spectrometry. 1. Effect of main chain and pendant methyl groups. Surf Interface Anal 2004, 36, 42–52.

[107] Wagner MS. Degradation of poly(acrylates) under SF_5+ primary ion bombardment studied using time-of-flight secondary ion mass spectrometry. 2. Poly(n-alkyl methacrylates). Surf Interface Anal 2004, 36, 53–61.

[108] Wagner MS. Degradation of poly(acrylates) under SF_5+ primary ion bombardment studied using time-of-flight secondary ion mass spectrometry. 3. Poly(hydroxymethyl methacrylate) with chemical derivatization. Surf Interface Anal 2004, 36, 62–70.

[109] Mahoney CM. Cluster SIMS depth profiling of stereo-specific PMMA thin films on Si. Surf Interface Anal 2010, 42, 1393–401.

[110] Huang HL, Goh SH, Lai DMY, Huan CH. Tof_SIMS studies of poly(methyl methacrylate-co-methacrylic acid), poly(2,2,3,3,3-pentafluoropropyl methacrylate-co-4-vinyl pyridine) and their blends. Appl Surf Sci 2004, 227(1–4), 373–382.

[111] Léonard D, Bertrand P, Shi MK, Sacher E, Martinu L. Plasma surface modification of fluoropolymers studied by ToF-SIMS. Plasmas Polym 1999, 4, 97–111.

[112] Ma ZQ, Chen P, Zhao TS. A palladium-alloy deposited Nafion membrane for direct methanol fuel cells. J Membr Sci 2003, 215, 327–336.

[113] Hug P, Keller BA. Surface reconstruction in modified ethylene-methacrylic acid copolymers probed by TOF-SIMS. In: Olefjord I, Nyborg L, Briggs D, eds. ECASIA'97 (European Conference on Applications of Surface and Interface Analysis), Chichester, UK, Wiley, 1997, 727–730.

[114] Liu Y, Lu C, Twigg S et al. Direct observation of ion distributions near electrodes in ionic polymer actuators containing ionic liquids. Sci Rep 2013, 3, 973.

[115] Chang D, Lorenz M, Burch MJ, Ovchinnikova OS, Hong K, Sumpter BG, Corrillo JMY. Structures of partially fluorinated bottlebrush polymers in thin films. ACS Appl Polym Mater 2020, 2, 209–219.

[116] Feng J, Chan CM, Weng LT. Influence of chain sequence structure of polymers on ToF-SIMS spectra. Polymer 2000, 41(7), 2695–2699.

[117] Zhon H, Chan CM, Weng LT, Ng KM, Li L. Relationship between the structure of polymers with well-defined fluorocarbon segmental lengths and the formation of secondary ions in SIMS. Surf Interface Anal 2002, 33(12), 932–939.

[118] Nysten B, Verfaillie G, Ferain E et al. SEM, TOF-SIMS and LFM morphological study of an heterogeneous polymeric surface. Micros Microanal Microstruct 1994, 5, 373–380.

[119] Fisher GL, Hammond JS, Bryan SR, Larson PE, Herren RMA. The composition of Poly(Ethylene Terephthalate) (PET) surface precipitates determined at high resolving power by tandem mass spectrometry imaging. Micros Microanal 2017, 23(4), 843–848.

[120] Kato N, Kudo M. Simulation of fragmentation of polyethylene glycol by quantum molecular dynamics for ToF-SIMS spectral analysis. Journal Surf Anal 2011, 17(3), 208–211.

[121] Green FM, Gilmore IS, Seah MP. Mass Spectrometry and informatics: Distribution of molecules in the PubChem database and general requirements for mass accuracy. Anal Chem 2011, 83, 3239–3243.

[122] Gilmore IS, Seah MP. Static SIMS: Towards unfragmented mass spectra-the G-SIMS procedure. Appl Surf Sci 2000, 161, 465–480.

[123] Aoyagi S, Gilmore IS, Mihara I, Seah MP, Fletcher IW. Identification and separation of protein, contaminant and substrate peaks using gentle-secondary ion mass spectrometry and the g-ogram. Rapid Commun Mass Spectrom 2012, 2, 2815–2821.

[124] Straif CJ, Hutter H. Investigation of polymer thin films by use of Bi-cluster-ion-supported time-of-flight secondary ion mass spectrometry. Anal Bio Anal Chem 2009, 393, 1889–1898.

[125] Green FM, Kollmer F, Niehuis E, Gilmore IS, Seah MP. Imaging G-SIMS: A novel bismuth-manganese source emitter. Rapid Commun Mass Spectrom 2008, 22, 2602–2608.

[126] Aoyagi S, Mihara I, Kudo M. Mixed polymer sample evaluation using gentle secondary ion mass spectrometry and multivariate curve resolution. Surf Interface Anal 2012, 44, 789–792.

[127] Gilmore IS, Aoyagi S, Fletcher IW, Seah MP. G-SIMS: A powerful method for simplifying and interpretation of complex secondary ion mass spectra. Spectroscopy Europe 2012, 24, 6–13.

[128] Lee JLS, Gilmore IS IS. The application of multivariate data analysis techniques in surface analysis. In: Vickerman JC, Gilmore IS, eds. Surface Analysis: The Principal Techniques, 2nd, Chichester, UK, Wiley, 2009, 563–612.

[129] Graham DJ, Wagner MS, Castner DG. Information from complexity: Challenges of ToF-SIMS data interpretation. Appl Surf Sci 2006, 252, 6860–6868.

[130] Tyler BJ, Rayal G, Castner DG. Multivariate analysis strategies for processing ToF-SIMS Images of biomaterials. Biomaterials 2007, 28(15), 2412–2423.

[131] Wagner MS, Graham DJ, Ratner BD, Castner DG. Maximinzing information obtained from secondary ion mass spectra of organic thin films using multivariate analysis. Surf Sci 2004, 570(1–2), 78–97.

[132] Trinidade GF, Abel ML, Watts JF. simsMVA: A tool for multivariate analysis of ToF-SIMS datasets. Chemom Intell Lab Syst 2018, 182, 180–187.

[133] Trinidade GF, Abel ML, Lowe C, Tshulu R, Watts JF. "A time of flight secondary ion mass spectrometry/ multivariate analysis (ToF-SIMS/MVA) approach to identify phase segregation in blends of incompatible but extremely similar resins. Anal Chem 2018, 90(6), 3936–3941.

[134] Madiona RMT, Welch NG, Russell SB, Winkler DA, Scoble JA, Muir BW, Pigram PJ. Multivariate analysis of ToF-SIMS data using mass segmented peak lists. Surf Interface Anal 2018, 50(7), 713–728.

[135] Bart JCJ. Additives in Polymers: Industrial Analysis and Applications. UK, Chichester, Wiley, 2005.

[136] Bart JCJ. Polymer Additive Analytics: Industrial Practice and Case Studies. Italy, Firenze, Firenze University Press, 2006.

[137] Médard N, Benninghoven A, Rading D et al. Antioxidant segregation and crystallisation at polyester surfaces studied by ToF-SIMS. Appl Surf Sci 2003, 203–4, 571–574.

[138] Poleunis C, Médard N, Bertrand P. Additive quantification on polymer thin films by ToF-SIMS: Aging sample effects. Appl Surf Sci 2004, 231–232, 269–273.

[139] Holbrook RD, Davis JM, Scott KC, Szakal C. Detection and speciation of brominated flame retardants in high-impact polystyrene (HIPS) polymers. J Microsc 2012, 246, 143–152.

[140] Murase A, Kato Y, Sudo E. Quantification of organic additives on polymer surface by time-of-flight secondary ion mass spectrometry with gold deposition. Appl Surf Sci 2020, 509, 144813.

[141] Kita-Tokarczyk K, Itel F, Grzelakowski M, Egli S, Rossbach P, Meier W. Monolayer interactions between lipids and amphiphilic block copolymers. Langmuir 2009, 25, 9847–9856.

[142] Keller BA, Mayerhofer K. The role of TOF-SIMS in nanomedicine. Eur J Nanomed 2009, 2, 37–42.

[143] Tang C, Feller L, Rossbach P et al. Adsorption and electrically stimulated desorption of the triblock copolymer poly(propylene sulfide-bl-ethylene glycol) (PPS-PEG) from indium tin oxide (ITO) surfaces. Surf Sci 2006, 600, 1510–1517.

[144] Lee J, Shin K, Lee KB, Lee Y. Surface analysis of diblock copolymer films by ToF-SIMS in combination with AFM. Surf Interface Anal 2014, 46, 87–91. DOI: 10.1002/sia.5513.

[145] Terlier Tzappalà G, Marie C, Leonard D, Barnes JP, Licciardello A. ToF-SIMS depth profiling of PS-b-PMMA block copolymers using Ar_n^+, C_{60}^{++} and C_s^+ sputtering ions. Anal Chem 2017, 89, 6984–6991.

[146] Weng LT, Chan CM. ToF-SIMS quantitative approaches in copolymers and polymer blends. Appl Surf Sci 2003, 203–204, 532–537.

[147] Mayerhofer KE, Heier J, Maniglio Y, Keller BA. Three dimensional analysis of self-structuring organic films using time-of-flight secondary ion mass spectrometry. Thin Solid Films 2011, 519, 6183–6189.

[148] Yu BY, Lin WB, Ilda SI et al. Effect of fabrication parameters on three-dimensional nanostructures of bulk heterojunctions imaged by high-resolution scanning ToF-SIMS. Nano 2010, 4, 833–840.

[149] Feng J, Weng L-T, Li L, Chan C-M. Compatibilization of polycarbonate and poly(vinylidene fluoride) blends studied by time-of-flight secondary ion mass spectrometry and scanning electron microscopy. Surf Interface Anal 2000, 29, 168–174.

[150] Feng J, Li L, Chan CM, Weng LT. Inter-diffusion between PMMA and PVDF during lamination studied time-of-flight secondary ion mass spectrometry chemical imaging. Surf Interface Anal 2002, 33, 455–458.

[151] Yuan Y, Shoichet MS. Surface enrichment of poly(trifluorovinyl ethers) in polystyrene blends. Macromolecles 2000, 33, 4926–4931.

[152] Chan CM, Weng LT. Surface characterization of polymer blends by XPS and ToF-SIMS. Materials 2016, 9, 655–674.

[153] Muthukumar M. Nucleation Polym Cryst Adv Chem Phys 2004, 128, 1–64.

[154] Lei YG, Cheung ZL, Ng KM, Li L, Weng LT, Chan CM. Surface chemical and morphological properties of a blend containing semi-crystalline and amorphous polymers studied with ToF-SIMS, XPS and AFM. Polymer 2003, 44, 3883–3890.

[155] Karar N, Gupta TK. Study of polymers and their blends using ToF-SIMS ion imaging. Vacuum 2015, 111, 119–123.

[156] Harris LG, Schofield WCE, Badyal JPS. Multi-functional molecular scratchcards. Chem Mater 2007, 19, 1546–1551.

[157] Tarducci C, Kinmond EJ, Badyal JPS, Brewer SA, Willis C. Epoxide-functionalized solid surfaces. Chem Mater 2000, 12, 1884–89.

[158] Thierry B, Jasieniak M, De Smet LCPM, Vasilev K, Griesser HJ. Reactive epoxy-functionalized thin films by a pulsed plasma polymerization process. Langmuir 2008, 24, 10187–10195.

[159] Tarducci C, Schofield WCE, Badyal JPS, Brewer SA, Willis C. Monomolecular functionalization of pulsed plasma deposited poly(2-hydroxyethyl methacrylate) surfaces. Chem Mater 2002, 14, 2541–2545.

[160] Kinmond EJ, Coulson SR, Badyal JPS, Brewer SA, Willis C. High structural retention during plasma polymerization of 1H,1H;2H-perfluorododecene: An NMR and ToF-SIMS study. Polymer 2005, 46, 6829–6835.

[161] Oran U, Swaraj S, Friedrich JF, Unger WES. Static ToF-SIMS analysis of plasma chemically deposited ethylene/allyl alcohol co-polymer films. Appl Surf Sci 2006, 252, 6588–6590.

[162] Von Gradowski M, Jacoby B, Hilghers H, Barz J, Wahl M, Kopnarski M. ToF-SIMS characterization of ultra-thin fluorinated carbon plasma polymer films. Surf Coat Technol 2005, 200, 334–340.

[163] Cristaudo V, Merche D, Poleunis C, Devaux J, Eloy P, Reniers F, Delcorte A. Ex-situ SIMS characterization of plasma-deposited polystyrene near atmosphere pressure. Appl Surf Sci 2019, 481, 1490–1502.

[164] Poleunis C, Cristaudo V, Delcorte A. Temperature dependence of Ar$^+$ cluster back scattering from polymer surfaces: A new method to determine the surface glass transition temperature. J Am Soc Mass Spectr 2018, 29, 4–7.

[165] Moellers R, Bernard L, Dianoux R, et al. Combined SIMS-SFM instrument for the 3-dimensional chemical analysis of nanostructures. Supported by FP7 of the European Commission, Münster (Germany); 2008. (Accessed September 6, 2013, at http://www.3dna nochemiscope.eu/3D_NanoChemiscope/Home.html.

[166] Wirtz T, Fleming Y, Gerard M et al. Design and performance of a combined secondary ion mass spectrometry-scanning probe microscopy instrument for high sensitivity and high-resolution elemental three-dimensional analysis. Rev Sci Instrum 2012, 83, 063702.

[167] Fornai L, Angelini A, Klinkert I, Gisken F et al. Three-dimensional molecular reconstruction of rat heart with mass spectrometry imaging Anal. Bioanal Chem 2012, 404, 2927–2938.

[168] Shon HK, Yoon S, Moon JH, Lee TG. Improved mass resolution and mass accuracy in TOF-SIMS spectra and images using argon gas cluster ion beams. Biointerphases 2016, 11, 02A321.

[169] Garrison BJ, Postawa Z. Computational view of surface based organic mass spectrometry. Mass Spectrom Rev 2008, 27, 289–315.

[170] Mahoney CM, Roberson SV, Gillen G. Depth profiling of 4-acetamindophenol-Doped poly (lactic acid) films using cluster secondary ion mass spectrometry. Anal Chem 2004, 76, 3199–3207.

[171] Gillen G, Fahey A, Wagner M, Mahoney CM. 3D molecular imaging SIMS. Appl Surf Sci 2006, 252, 6537–6541.

[172] Fletcher JS, Conlan XA, Lockyer NP, Vickerman JC. Molecular depth profiling of organic and biological materials Appl. Surf Sci 2006, 252, 6513–6516.

[173] Fisher GL, Belu AM, Mahoney CM, Wormuth K, Sanada N. Three-dimensional time-of-flight secondary ion mass spectrometry imaging of a pharmaceutical in a coronary stent coating as a function of elution time. Anal Chem 2009, 81, 9930–9940.

[174] Fletcher JC. Latest applications of 3D ToF-SIMS bioimaging. Biointerphases 2015, 10, 018902.

[175] Fletcher JS, Lockyer NP, Vaidyanathan S, Vickerman JC. TOF-SIMS 3D biomolecular Imaging of xenopus laevis Oocytes using buckminsterfullerene (C60) primary ions. Anal Chem 2007, 79, 2199–2206.

[176] Fletcher JS, Henderson A, Biddulph GX, Vaidyanathan S, Lockyer NP, Vickerman JC. Uncovering new challenges in bio-analysis with ToF-SIMS. Appl Surf Sci 2008, 255, 1264–1270.

[177] Breitenstein D, Rommel CE, Mollers R, Wegener J,, Hagenhoff B. The chemical composition of animal cells and their intracellular compartments reconstructed from 3D mass spectrometry. Angew Chem Int Ed 2007, 46, 5332–5335.

[178] Fletcher JS, Rabbani S, Henderson A, Lockyer NP, Vickerman JC. Three-dimensional mass spectral imaging of HeLa-M cells–sample preparation, data interpretation and visualisation. Rapid Commun Mass Spectrom 2011, 25, 925–932.

[179] Robinson MA, Graham DJ, Castner DG. ToF-SIMS depth profiling of cells: Z-correction, 3D imaging, and sputter rate of individual NIH/3T3 fibroblasts. Anal Chem 2012, 84, 4880–4885.

[180] Whitby JA, Ostlund F, Horvath P, Gabureac M et al. High spatial resolution time-of-flight secondary ion mass spectrometry for the masses: A novel orthogonal ToF FIB-SIMS Instrument with in situ AFMAdv. Mater Sci Eng 2012, 2012, 180437.

[181] Koch S, Ziegler G, Hutter H. ToF-SIMS measurements with topographic information in combined images. Anal Bioanal Chem 2013, 405, 7161–7167.

[182] Angerer TB, Fletcher JS. 3D imaging of TiO_2 nanoparticle accumulation in Tetrahymena pyriformis. Surf Interface Anal 2014, 46, 198–203.

[183] Wucher A, Tian H, Winograd N. A mixed cluster ion beam to enhance the ionization efficiency in molecular secondary ion mass spectrometry. Rapid Commun Mass Spectrom 2014, 28, 396–400.

[184] Sheraz (Née Rabbani) S, Barber A, Berrueta RI, Fletcher JS, Loycker NP, Vickerman JC. Prospect of increasing secondary ion yields in ToF-SIMS using water cluster primary ion beams. Surf Interface Anal 2014, 46, 51–53.

[185] Zumstein MT, Schintlmeister A, Taylor FN, Baumgartner R, Woebken D, Wagner M, Kohler HPE, McNeill K, Sander M. Biodegradation of synthetic polymers in soils: Tracking carbon into CO_2 and microbial biomass. Sci Adv 2018, 4, eaas9024. DOI: 10.1126/sciadv.aas9024.

[186] Lozic I, Hartz RV, Bartlett CA, Shaw JA, Archer M, Naidu PSR, Smith NM, Dunlop SA, Iyer KS, Kilburn MR, Fitzgerald M. Characterization of polymeric nanoparticles for treatment of partial injury to the central nervous system. Data in Brief 2016, 7, 152–156.

[187] Belu AM, Graham DJ, Castner DG. Time-of-flight secondary ion mass spectrometry: Techniques and application for the characterization of biomaterial surfaces. Biomaterials 2003, 24(21), 3636–3653.

[188] Magnani A, Barbucci R, Lewis KB, Leach-Scampavia D, Ratner BD. Surface properties and restructuring of a crosslinked polyurethane – poly(amido-amine) network. J Mater Chem 1995, 5(9), 1321–1330.

[189] Leone G, Consumi M, Lamponi S, Magnani A. New hyaluroran derivative with prolonged half-life for ophthalmogical formulation. Carbohydrate Polymers 2012, 88, 799–808.

[190] Fearn S. Characterization of biological material with ToF-SIMS: A review. Mater Sci Technol 2015, 31(2), 148–161.

[191] Davis MC, Lynn RAP. A review: Secondary ion mass spectrometry (SIMS) of polymeric biomaterials. Clinical Mater 1990, 5(2–4), 97–114.

[192] Yang J, Gilmore I. Application of secondary ion mass spectrometry to biomaterials, proteins and cells: A concise review. Mater Sci Technol 2015, 31(2), 131–136.

[193] Barbucci R, Magnani A, Lamponi S, Pasqui D, Bryan S. The use of Hyaluronan and its sulphated derivative patterned with micrometric scale on glass substrate in melanocyte cell behaviour. Biomaterials 2003, 24, 915–926.

[194] Magnani A, Priamo A, Pasqui D, Barbucci R. Cell behaviour on chemically microstructured surfaces. Mater Sci and Eng C 2003, 23, 315–328.

[195] Barbucci R, Lamponi S, Magnani A, Piras FM, Rossi A, Weber E. Role of the Hyal-Cu (II) complex on bovine aortic and lymphatic endothelial cells behavior on microstructured surfaces. Biomacromolecules 2005, 6, 212–219.

[196] Leone G, Consumi M, Lamponi S, Magnani A. Combination of static time of flight secondary ion mass spectrometry and infraRed reflection-adsorption spectroscopy for the characterisation of a four steps built-up carbohydrate array. Appl Surf Sci 2012, 258(17), 6302–6315.

[197] Hook AL, Williams PM, Alexander MR, Scurr DJ. Multivariate ToF-SIMS image analysis of polymeric microarrays and protein adsorption. Biointerphases 2015, 10, 019005. DOI: 10.111.

[198] Hook A, Scurr DJ. ToF-SIMS analysis of a polymer microarray composed of poly(meth)acrylates with C6 derivative pendant groups. Surf Interface Anal 2016, 48, 226–236.

[199] Pasquale Totaro P, Favre A, Poggini L, Mannini M, Sainctavit P, Cornia A, Magnani A, Sessoli A. Tetrairon(III)-single-molecule magnet monolayers on gold: Insights from ToF-SIMS and isotopic labelling. Langmuir 2014, 30, 8645–8649.

[200] Leone G, Consumi M, Tognazzi A, Magnani A. Realisation and chemical characterisation of a model system for saccharide-based biosensor. Thin Solid Films 2010, 519, 462–470.

[201] Kern C, Ray S, Gelinsky M, Bellew AT, Pirkl A, Rohnke M. New insights into ToF-SIMS imaging in osteoporotic bone research. Biointerphases 2020, 15, 031005. DOI: 10.1116/6.0000051.

[202] Bonechi C, Consumi M, Matteucci M, Tamasi G, Donati A, Leone G, Rossi C, Menichetti L, Kusmic C, Magnani A. Distribution of gadolinium in rat heart studied by fast-field cycling relaxometry and imaging SIMS. Int J mol sci Special Issue "Advances in Metal Metabolism Research" 2019, 20, 1339–1351.

[203] Volandri G, Menichetti L, Matteucci M, Kusmic C, Consumi M, Magnani A, L'Abbate A, Landini L, Positano V. An image formation model for secondary ion mass spectrometry imaging of biological tissue samples. Appl Surf Sci 2010, 257(4), 1267–1275.

[204] Henss A, Otto SK, Schaepe K, Pauksch L, Lips KS, Rohnke M. High resolution imaging and 3D analysis of Ag nanoparticles in cells with ToF-SIMS and delayed extraction. Biointerphases 2018, 13, 03B410.

[205] Nunez J, Renslow R, Cliff JB, Anderton CR. NanoSIMS for biological applications: Current practice and analyses. Biointerphases 2018, 13, 03B301. DOI: 10.1116/1.4993628.

Filippo Mangolini, Antonella Rossi

5 Advances in attenuated total reflection (ATR) infrared spectroscopy: a powerful tool for investigating polymer surfaces and interfaces

5.1 Principles of infrared (IR) spectroscopy

Infrared (IR) spectroscopy is widely used in research and industry as a simple, reliable, and cost-effective technique for the structural elucidation and chemical analysis of both organic and inorganic compounds. One of the advantages of IR spectroscopy is the possibility of analyzing samples in different states, i.e., liquids, solids, and gases.

IR spectrometers have been available to researchers and scientists since the 1940s. Although numerous technical developments have been made over the years, the most significant advance in IR spectroscopy was the introduction of Fourier-transform IR (FT-IR) spectrometers, which have drastically improved the quality of the spectra, minimizing, at the same time, the spectral acquisition time [1].

The basic principle of IR spectroscopy is the interaction of electromagnetic radiation in the IR region of the spectrum (wavelength range: 0.78–1000 µm; wavenumber range: 10–12,820 cm^{-1}) with molecules. Such an interaction induces transitions between vibrational levels [1–6].

The basic model for describing molecular vibrations is the harmonic oscillator, which assumes the atoms in a molecule to be spheres of mass m_i and the chemical bonds connecting the atoms as massless springs with spring constant k_i [1–6]. Taking into consideration a diatomic molecule (masses of the two atoms m_1 and m_2), the restoring force of the spring (F) is proportional to the atoms' displacement (x) from their equilibrium position (Hooke's law):

$$F = -kx \tag{5.1}$$

where k is spring constant (in N m^{-1}), which is a measure of the bond strength.

The vibrational frequency ($\tilde{v}$ wavenumber in cm^{-1}) of the harmonic oscillator is, then, given by

$$\tilde{v} = \frac{1}{2\pi c}\sqrt{\frac{k}{\tilde{m}}} \tag{5.2}$$

where c is the speed of light and $\tilde{m}$ is the reduced mass ($\tilde{m} = (m_1 m_2/(m_1 + m_2))$).

Acknowledgments: F.M. acknowledges support from the Welch Foundation (grant no. F-2002-20190330), the 2018 Ralph E. Powe Junior Faculty Enhancement Award sponsored by the Oak Ridge Associated Universities (ORAU), and the University of Texas at Austin.

https://doi.org/10.1515/9783110701098-005

In the case of a harmonic oscillator, only changes of ±1 in the vibrational quantum number are allowed ($\Delta n = \pm 1$). However, the molecular potential energy curve can be more properly described using an anharmonic function. A consequence of the anharmonicity of the molecular potential energy curve is to relax the selection rule (i.e., it allows transitions with $|\Delta n| > 1$). Additional peaks, such as overtone and combination bands, can, thus, appear in the mid-IR spectrum together with absorption bands due to fundamental transitions ($\Delta n = \pm 1$).

The vibration of atoms is usually described in terms of normal modes of vibration (q_i), in which all atoms in the molecule under consideration vibrate with the same frequency and simultaneously pass through their equilibrium positions. In order to be IR-active, a change in the molecular dipole (μ_j) associated with the normal mode of vibration (*IR selection rule*) must occur (i.e., $(\partial\mu/\partial q_i)\neq 0$).

As for the direction of the vibrational movement, we can distinguish between stretching and deformation vibrations. The latter can be divided into bending modes, twisting/scissoring modes, rocking modes, and wagging modes. Further subdivisions refer to the vibration symmetry (i.e., symmetric or antisymmetric, in-plane or out-of-plane; Fig. 5.1).

5.2 Theory of attenuated total reflection (ATR) IR spectroscopy

Since the pioneering work of Harrick [7, 8] and Fahrenfort [9], who independently described the theoretical principles and the analytical relevance of attenuated total reflection (ATR) IR spectroscopy, an ever-increasing number of publications, exploiting this technique for a variety of applications, has appeared in the literature [1, 8, 10–14]. Several monographs have already been published on the use of ATR IR spectroscopy in different research fields [8, 11, 12, 14, 15], such as polymer science [16], electrochemistry [17], and the science of biological interfaces [18]. This analytical technique was also demonstrated to have potential for fundamental in situ investigations in other fields of research, where the phenomena occurring at the solid–liquid interface play an important role, such as catalysis [19–22], environmental chemistry [23], and tribology [24–32].

In order to explain the theory of ATR/IR spectroscopy, the physical principles of the propagation of IR radiation through a planar interface dividing two isotropic media will be first outlined.

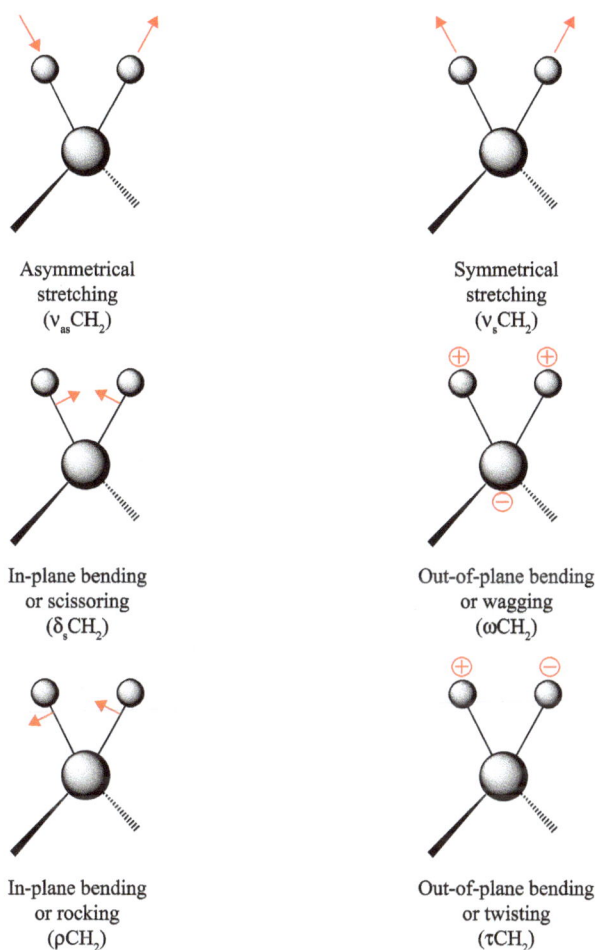

Asymmetrical
stretching
($\nu_{as}CH_2$)

Symmetrical
stretching
(ν_sCH_2)

In-plane bending
or scissoring
(δ_sCH_2)

Out-of-plane bending
or wagging
(ωCH_2)

In-plane bending
or rocking
(ρCH_2)

Out-of-plane bending
or twisting
(τCH_2)

Fig. 5.1: Vibrational modes for a nonlinear group, CH_2 (+ indicates motion from the plane of the page toward the reader; − indicates motion from the plane of the page away the reader).

5.2.1 Propagation of IR radiation through a planar interface between two isotropic media

When an electromagnetic wave propagating in an isotropic medium (medium 1) with refractive index n_1 strikes an interface (assumed to be perfectly planar) with a second isotropic medium (medium 2) with refractive index n_2, the wave splits into two components: the reflected and transmitted (refracted) components (Figs. 5.2) [14, 33].

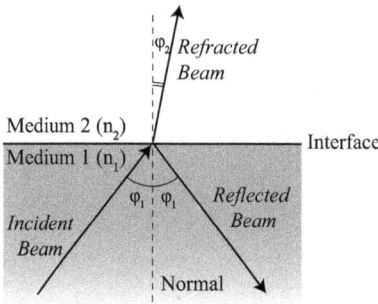

Fig. 5.2: External reflection and refraction of light at a planar interface between two isotropic phases ($n_2 > n_1$).

The refraction angle (φ_2) is related to the incidence angle (φ_1) by Snell's law [33]:

$$n_1 \sin \varphi_1 = n_2 \sin \varphi_2 \tag{5.3}$$

where n_1 and n_2 are the refractive indexes of the input and output media, respectively.

The intensity of the reflected (r) and transmitted (t) beams (for both the s-polarized (in which the electric field is perpendicular to the plane of incidence) and p-polarized components (in which the electric field is parallel to the plane of incidence)) can be calculated using Maxwell's equations and considering the continuity of the magnetic and electric fields across the interface. The coefficients, which interconnect the amplitudes of the waves, are the so-called *Fresnel amplitude coefficients*:

$$r_{12}^p = \frac{\tan(\varphi_1 - \varphi_2)}{\tan(\varphi_1 + \varphi_2)}; \ t_{12}^p = \frac{\frac{2n_1}{\cos \varphi_2}}{\frac{n_2}{\cos \varphi_2} + \frac{n_1}{\cos \varphi_1}} \tag{5.4}$$

$$r_{12}^s = - \frac{\sin(\varphi_1 - \varphi_2)}{\sin(\varphi_1 + \varphi_2)}; \ t_{12}^s = \frac{2 \sin \varphi_2 \cos \varphi_1}{\sin(\varphi_1 + \varphi_2)} \tag{5.5}$$

where the superscripts s and p refer to the polarization of the light.

If the angle of incidence exceeds the critical angle (φ_c), *total internal reflection* occurs:

$$\varphi_c = \arcsin\left(\frac{n_2}{n_1}\right) \tag{5.6}$$

Upon inspecting the optical field near the interface, a tail of the electromagnetic field is found to penetrate into the rarer medium as an inhomogeneous wave:

$$E \propto \exp[i(\omega t - k_x x + k_z z)] \tag{5.7}$$

The analysis of the wavevector, which can be used for describing such a wave, shows that the component along the vertical direction (k_z) is imaginary, implying that the electric field vector exponentially decays in the optically rarer medium, as depicted in Fig. 5.3. Because of that, this wave is usually referred to as *evanescent*, from the Latin *evanescere*, which means "to vanish."

Starting from the Fresnel coefficients (eqs. (5.4) and (5.5)), it can be demonstrated that the radiation energy is totally reflected independently of the polarization state. However, the Fresnel amplitude transmission coefficients ($t^{s,p}$) have non-vanishing values and the wavevectors of the components parallel to the interface are real, indicating the presence of an energy flow parallel to the boundary between the two phases. As a consequence of this energy flow, the reflected wave is laterally displaced (*Goos–Hänchen shift*).

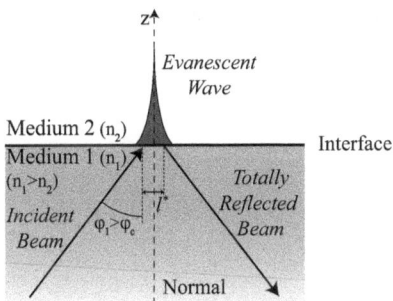

Fig. 5.3: Schematic representation of the physical principle of total internal reflection of light at the planar interface between two media ($n_1 > n_2$). The Goos–Hänchen shift is highlighted (l^*).

However, when the contacting media are absorbing (i.e., they have a non-zero extinction coefficient, $k \neq 0$), the Fresnel coefficients have to be adapted by replacing the real refractive indexes with the complex ones ($\hat{n} = n + ik$):

$$r_{12}^p = \frac{\hat{\varepsilon}_2 \xi_1 - \hat{\varepsilon}_1 \xi_2}{\hat{\varepsilon}_2 \xi_1 + \hat{\varepsilon}_1 \xi_2}; \ t_{12}^p = \frac{2\hat{n}_1 \hat{n}_2 \xi_1}{\hat{n}_2^2 \xi_1 + \hat{n}_1^2 \xi_2} \tag{5.8}$$

$$r_{12}^s = \frac{\xi_1 - \xi_2}{\xi_1 + \xi_2}; \ t_{12}^s = \frac{2\xi_1}{\xi_1 + \xi_2} \tag{5.9}$$

where $\hat{\varepsilon}_1$ and $\hat{\varepsilon}_2$ are the permittivity values of the input and output medium, and ξ_i is the generalized complex index of refraction:

$$\xi_i = \hat{n}_i \cos \varphi_i = \sqrt{\hat{n}_i^2 - \hat{n}_1^2 \sin^2 \varphi_1} \tag{5.10}$$

where φ_i is the complex angle of refraction.

Since the intensity of the evanescent wave and, thus, the intensity of the reflected beam are attenuated at characteristic wavelengths that correspond to the absorption of radiation in the optically rarer medium, the name *attenuated total reflection* is usually employed for indicating such a physical principle. By exploiting this phenomenon, the layer next to the interface can then be probed.

5.2.2 Propagation of IR radiation through stratified media

The propagation of IR radiation through stratified media is of interest for many engineering applications [14]. In the limiting case of an inhomogeneous (in depth) layer, the propagation of an electromagnetic wave could be investigated by slicing the layer into several infinitely thin layers, which could be assumed to be homogeneous [14].

Abelès and Hansen developed a matrix method that allows the calculation of the overall reflectance of any combination of absorbing and non-absorbing isotropic layers at any angle of incidence as well as the computation of the mean-square electric field (*MSEF*) at any location within a certain layer of a system constituted by N isotropic phases (Figs. 5.4) [14, 34, 35].

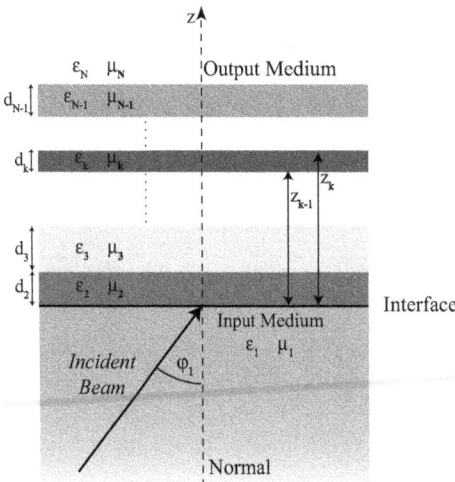

Fig. 5.4: Scheme of a layered system with N phases and $N-2$ interfaces. ε and μ are the permittivity and permeability of the material in the nth layer.

The basic concept of this matrix method involves the definition of a characteristic matrix M_i for each layer ($z_{j-1} < z < z_j$) and for the first and last layers ($z > z_{N-1}$), which are assumed to be semi-infinite and non-absorbing. From the physical point of view, this characteristic matrix M_i for each layer relates the tangential amplitudes of the magnetic and electric field vectors at the input and output film boundaries (for both s-polarized and p-polarized light):

$$M_j^s = \begin{pmatrix} \cos\beta_j & -\frac{i}{p_j}\sin\beta_j \\ -ip_j\sin\beta_j & \cos\beta_j \end{pmatrix}; \quad M_j^p = \begin{pmatrix} \cos\beta_j & -\frac{i}{q_j}\sin\beta_j \\ -iq_j\sin\beta_j & \cos\beta_j \end{pmatrix} \quad (5.11)$$

By defining the generalized complex index of refraction ξ_i as in eq. (5.10), the following relations can be written:

$$p_j = \xi_j; \; q_j = \frac{\xi_j}{\hat{n}_j^2}; \; \beta_j = 2\pi \left(\frac{d_j}{\lambda} \right) \xi_j \tag{5.12}$$

where d_j is the thickness of the jth layer and λ is the wavelength.

The matrix for the entire multilayer system is then computed as the product of the elementary matrices:

$$\hat{M} = M_2 M_3 M_4 \ldots M_{N-1} = \begin{pmatrix} m_{11} & m_{12} \\ m_{21} & m_{22} \end{pmatrix} \tag{5.13}$$

The Fresnel amplitude reflection (r) and transmission (t) coefficients for an N-isotropic-phase system can be calculated using the elements of the characteristic matrix of the overall system:

$$r^s = \frac{(m_{11} + m_{12}p_N)p_1 - (m_{21} + m_{22}p_N)}{(m_{11} + m_{12}p_N)p_1 + (m_{21} + m_{22}p_N)}; \quad r^p = \frac{(m_{11} + m_{12}q_N)q_1 - (m_{21} + m_{22}q_N)}{(m_{11} + m_{12}q_N)q_1 + (m_{21} + m_{22}q_N)} \tag{5.14}$$

$$t^s = \frac{2p_1}{(m_{11} + m_{12}p_N)p_1 + (m_{21} + m_{22}p_N)}; \quad t^p = \frac{2q_1}{(m_{11} + m_{12}q_N)q_1 + (m_{21} + m_{22}q_N)} \tag{5.15}$$

The overall reflectance and transmittance of the stratified medium can be finally derived as

$$R^{s,p} = |r^{s,p}|^2 \tag{5.16}$$

$$T^s = \frac{\mu_1 Re(\hat{n}_N \cos \varphi_N)}{\mu_N n_1 \cos \varphi_1} |t^s|^2; \quad T^p = \frac{\mu_N Re\left(\frac{\hat{n}_N \cos \varphi_N}{\hat{n}_N^2} \right)}{\frac{\mu_N n_1 \cos \varphi_1}{\hat{n}_1^2}} |t^p|^2 \tag{5.17}$$

The matrix formalism developed by Hansen could also be used for calculating the *MSEF* at any location within a certain layer of an N-isotropic-phase system [14, 35]. To compute the *MSEF* at a position z within the kth layer of a system with N isotropic layers, the column vectors $Q_1^{s,p}$ and $Q_{N-1}^{s,p}$ (eqs. (5.18) and (5.19)), which characterize the amplitudes of the tangential components of the electric and magnetic fields at the first and last boundaries, have to be defined first:

$$Q_1^s = \begin{pmatrix} E_{y1}^0 \\ H_{x1}^0 \end{pmatrix} = \begin{pmatrix} 1 + r^s \\ p_1(1 - r^s) \end{pmatrix}; \quad Q_{N-1}^s = \begin{pmatrix} E_{y(N-1)}^0 \\ H_{x(N-1)}^0 \end{pmatrix} = \begin{pmatrix} t^s \\ p_N t^s \end{pmatrix} \tag{5.18}$$

$$Q_1^p = \begin{pmatrix} H_{y1}^0 \\ E_{x1}^0 \end{pmatrix} = \begin{pmatrix} 1 + r^p \\ q_1(1 - r^p) \end{pmatrix} n_1; \quad Q_{N-1}^p = \begin{pmatrix} H_{y(N-1)}^0 \\ E_{x(N-1)}^0 \end{pmatrix} = \begin{pmatrix} t^p \\ q_N t^p \end{pmatrix} n_1 \tag{5.19}$$

where E and H are the electric and magnetic fields, respectively.

For calculating the *MSEF* magnitude at the position z of the kth layer, the matrix $N_j^{s,p}$, which is the reciprocal of the characteristic matrixes of the same layers $M_j^{s,p}$, should be generated. For the kth phase, the matrix $N_k^{s,p}$ is constructed as

$$N_k^s(z) = \begin{pmatrix} \cos[2\pi v \xi_k(z-z_{k-1})] & \frac{i}{p_k}\sin[2\pi v \xi_k(z-z_{k-1})] \\ ip_k \sin[2\pi v \xi_k(z-z_{k-1})] & \cos[2\pi v \xi_k(z-z_{k-1})] \end{pmatrix} \tag{5.20}$$

$$N_k^p(z) = \begin{pmatrix} \cos[2\pi v \xi_k(z-z_{k-1})] & \frac{i}{q_k}\sin[2\pi v \xi_k(z-z_{k-1})] \\ iq_k \sin[2\pi v \xi_k(z-z_{k-1})] & \cos[2\pi v \xi_k(z-z_{k-1})] \end{pmatrix} \tag{5.21}$$

For the final phase ($k = N$), the resultant vectors $Q_N^{s,p}(z)$ are defined as (at $z \geq z_j = \sum_{h=2}^{N-1} d_h$)

$$Q_N^{s,p}(z) = N_N^{s,p}(z)Q_{N-1}^{s,p} \tag{5.22}$$

By defining the resultant vector $Q_k^{s,p}(z)$ as

$$Q_k^{s,p}(z) = \begin{pmatrix} U_k^{s,p}(z) \\ V_k^{s,p}(z) \end{pmatrix} \tag{5.23}$$

and introducing the quantity $W_k(z) = ((n_1 \sin\varphi_1 U_k^p(z))/\hat{\varepsilon}_k)$ ($\hat{\varepsilon}_k$ complex dielectric constant), the *MSEF* magnitude can be calculated as follows:

$$\langle E_k^{s2}(z) \rangle = \langle E_{yk}^2(z) \rangle = |U_k^s(z)|^2 \tag{5.24}$$

$$\langle E_k^{p2}(z) \rangle = \langle E_{xk}^2(z) \rangle + \langle E_{zk}^2(z) \rangle = |V_k^p(z)|^2 + |W_k^p(z)|^2 \tag{5.25}$$

The formalism developed by Abelès and Hansen has also been generalized for anisotropic layers, making it possible to carry out simulations of uniaxial ($\hat{n}_x = \hat{n}_y \neq \hat{n}_z$) and biaxial ($\hat{n}_x \neq \hat{n}_y \neq \hat{n}_z$) layers [36]. The matrix method described above was shown to be a powerful tool for clarifying unknown optical behaviors: by numerically calculating the dependence of the reflectance on several parameters (*e.g.*, angle of incidence, electronic configuration and optical constants, layer thickness), physicochemical changes occurring at the interface between the internal reflection element (IRE) and the outer medium could be explained [21, 22, 30, 35].

5.2.3 Penetration depth and effective thickness

As pointed out in Section 5.2.1, by solving Maxwell's equations while accounting for the continuity of the electric and magnetic fields across the interface, an electromagnetic wave is found to penetrate into the rarer medium and its electric field amplitude falls off exponentially with the distance from the boundary [8, 11, 12, 14]:

$$E = E_0 \exp(-\gamma z) \tag{5.26}$$

where E_0 is the electric field amplitude at the interface ($z = 0$) and γ is the exponential constant (also called *electric field amplitude decay coefficient*), which is defined as follows:

$$\gamma = \frac{2\pi n_1 \sqrt{\sin^2 \varphi - n_{21}^2}}{\lambda} \tag{5.27}$$

where φ is the angle of incidence.

The quantity d_p, usually referred to as *penetration depth*, was introduced by Harrick to assess the penetration of the evanescent wave in the rarer phase [8]. This parameter is defined as the distance along the normal to the boundary between the two media at which the electric field amplitude decays to e^{-1} of its value at the interface (E_0) and is given by

$$d_p = \frac{1}{\gamma} = \frac{\lambda_1}{2\pi n_1 \sqrt{\sin^2 \varphi - n_{21}^2}} \tag{5.28}$$

where λ_1 is the wavelength (in the denser medium), while n_{21} is the ratio between the refractive index of the rarer medium and the refractive index of the denser phase ($n_{21} = n_2/n_1$). Notably, some authors have mathematically expressed the penetration depth replacing the factor 2 in the denominator of eq. (5.28) by a factor 4 [10, 37].

The penetration depth divided by the wavelength λ_1 is plotted as a function of the angle of incidence in Fig. 5.5 for a number of interfaces (i.e., different n_{21} values) [8, 38]. The depth of penetration increases as the angle of incidence decreases and becomes infinitely large as φ approaches the critical angle φ_c. At a fixed angle, d_p is larger for better index matching, i.e., as $n_{21} \to 1$. As a consequence of the dependence of the penetration depth on the radiation wavelength, the bands at longer wavelengths tend to be more intense compared to those at shorter wavelengths. Moreover, a broadening of the absorption bands on their high-wavelength side together with a shift to longer wavelengths is usually observed in comparison with transmission measurements.

According to Harrick, even if the reflection loss due to the interaction of the evanescent wave with the rarer medium could be exactly calculated starting from either Maxwell's or Fresnel's equations, such calculations do not provide any insights into the absorption mechanism or the interaction of the penetrating electromagnetic field with the absorbing phase [8]. Because of this, the concept of the effective thickness (d_e) was introduced to express the strength of the coupling of the exponentially decaying wave with the rarer phase and compare internal reflection spectra to the one acquired in transmission mode.

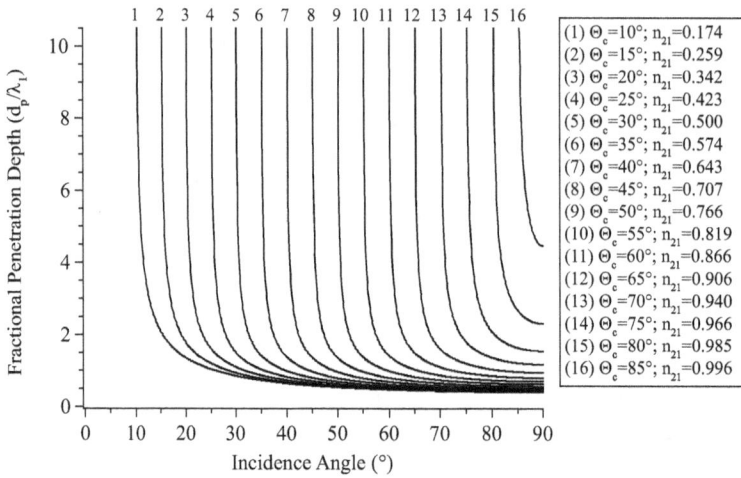

Fig. 5.5: Fractional penetration depth of electromagnetic field in an optically rarer medium in the case of total internal reflection versus angle of incidence for a number of interfaces [8, 38].

In conventional transmission IR spectroscopy, the Beer–Lambert (BL) law is used to relate the transmittance (T) at a given wavelength ($\tilde{\lambda}$) to the sample thickness (d) [1]:

$$T = \frac{I}{I_0} = \exp(-\alpha d) \tag{5.29}$$

where I_0 and I are, respectively, the incident and transmitted intensities at wavelength $\tilde{\lambda}$ and α is the absorption coefficient. Under the assumption of low absorption ($\alpha d < 0.1$), the BL law can be expressed as follows:

$$T = \frac{I}{I_0} \cong 1 - \alpha d \tag{5.30}$$

In a similar way, the reflectivity in internal reflection spectroscopy can be written as

$$R = 1 - \alpha d_e \tag{5.31}$$

where d_e is the effective thickness. For multiple reflections (N number of reflections), the reflected power is given by

$$R^N = (1 - \alpha d_e)^N \tag{5.32}$$

In the case of low absorptions ($\alpha d_e \ll 0.1$):

$$R^N \cong 1 - N\alpha d_e \tag{5.33}$$

i.e., the reflection loss is increased by the number of reflection N.

The comparison of the expressions obtained in case of the low-absorption approximation for transmission (eq. (5.30)) and reflection (eq. (5.33)), where it should be noted that the same absorption coefficient α appears, clearly suggests that the effective thickness (d_e) represents the actual film thickness that would be required in a transmission measurement to measure an absorption value equal to that obtained in a reflection acquisition using a semi-infinite bulk sample [8].

The expressions derived by Harrick, which are different in the case of a semi-infinite bulk material and in the case of a thin film, do not only give an insight into the nature of the interaction of the electromagnetic wave in internal reflection measurements, but are helpful for choosing the refractive index of the IRE and the angle of incidence of the IR beam.

In the case of bulk materials, the sample thickness is much larger than the penetration depth, implying that the electric field exponentially falls to very low values within the rarer medium. The low-absorption approximation for the effective thickness can, then, be computed from the electric fields for zero absorption:

$$ d_e = \frac{n_{21}}{\cos \varphi} \int_0^\infty E^2 dz = \frac{n_{21} E_0^2 d_p}{2 \cos \varphi} \qquad (5.34) $$

where the electric field E in the rarer medium is expressed as in eq. (5.26) and a unit electric field amplitude is assumed for the incident wave in the denser phase.

Due to the dependency of the electric field amplitude in the rarer medium on the polarization of the electromagnetic wave, the effective thickness for parallel and perpendicular polarization turns out to be different. Equation (5.34) also shows that the strength of the coupling of the evanescent wave to the absorbing rarer medium is determined by the penetration depth, which is proportional to the wavelength. Therefore, the effective thickness depends on the wavelength of the radiation as well, which explains the appearance of more intense absorption bands at longer wavelengths. Two peaks with the same intensity in transmission spectra will exhibit unequal intensity in spectra acquired in internal reflection mode. Such a wavelength dependence of the effective thickness also results in greater absorption on the high-wavelength side of a certain band, which, then, appears distorted.

In the case of thin films, the penetration depth is much larger than the layer thickness (d), i.e., $d_p \gg d$. As a consequence of this, the electric field of the radiation can be assumed to be constant over the film thickness and the corresponding effective thickness can be expressed as

$$ d_e - \frac{n_{21} E_0^2 d}{\cos \varphi} \qquad (5.35) $$

By comparing eqs. (5.34) and (5.35), it is evident that the same factors controlling the effective thickness for bulk materials control the effective thickness for thin films, except the penetration depth. Since d_e, in the case of thin films (thickness <1–2 μm, depending on the optical properties of the system under investigation), is not dependent on the penetration depth and, thus, is not proportional to the wavelength, the internal reflection spectra of thin layers exhibit neither stronger absorption bands at longer wavelengths nor peaks distorted on the longer wavelength side. Because of this, internal reflection spectra of thin films resemble those acquired in transmission mode more closely than internal reflection spectra of bulk materials do.

The equations proposed by Harrick for the penetration depth (eq. (5.28)) and for the effective thickness (eqs. (5.34) and (5.35)) were defined in the case of total internal reflection, i.e., under the assumption of non-absorbing media ($k = 0$). However, it has been demonstrated that these simplified expressions can be satisfactorily applied to weakly absorbing materials, such as organic molecules [8, 11]. Besides Harrick, Hansen also derived an expression for the effective thickness on the basis of a Taylor expansion of Fresnel's equations in the absorption coefficient [39]. The resulting equations were found to be a good approximation, which, however, was again valid only for weakly absorbing materials.

In the case of ATR, i.e., in the presence of an absorbing phase, a new parameter (d_p'), which describes the exponential decay of the electromagnetic wave in the rarer medium, could be defined upon inserting the complex index of refraction ($\hat{n} = n + ik$) into the characteristic equation of the evanescent wave (eq. (5.28)) [40–42]:

$$d_p' = \frac{\lambda}{2\pi n_1} \frac{1}{\omega_2^+} \tag{5.36}$$

where

$$\omega_2^\pm = \frac{1}{\sqrt{2}} \sqrt{\sqrt{v^2 + \mu^2} \pm v_2}$$

$$v_2 = \sin^2\varphi - n_{21}^2 \left(1 - \kappa_2^2\right)$$

$$\mu = 2n_{21}^2 \kappa_2 \tag{5.37}$$

and κ is the absorption index ($\kappa = (k/n)$).

The values of d_p and d_p' were found to be roughly the same for angles of incidence well above the critical angle (φ_c). In contrast, while the depth of penetration defined as in eq. (5.28) increases infinitely at the critical angle, d_p' stays finite.

Even if the foregoing equations were defined to establish if a system could be described as consisting of two or three layers, the formalism developed by Müller et al. was based on the assumption that the second and third layers have lower refractive index than the IRE [40–42]. Therefore, internal reflection takes place at the interface between the first/denser (IRE) phase and the second one.

In a similar way to the penetration depth, Müller et al. also suggested different expressions for the effective thickness in the case of absorbing samples [40–42].

5.2.3.1 Penetration depth in ATR/IR spectroscopic analysis of polymers

In the following, the penetration depth of the evanescent wavein a typical ATR/IR experiment will be discussed. A model system, consisting of a germanium ATR crystal (refractive index n_{Ge} equal to 4.0) and a bulk specimen with refractive index ($n_{specimen}$) equal to 1.46 (a typical value for the refractive index of polymers [43]), is considered in the calculations (Fig. 5.6a). The penetration depth (d_p), calculated using the formula proposed by Harrick (eq. (5.28)) [8], is displayed as a function of wavenumber in Fig. 5.6b. For the model system considered in the calculations, d_p increases from 164 nm at 4000 cm^{-1} to 1095 nm at 600 cm^{-1}. This strong wavelength dependence of the penetration depth results in clear differences between ATR and transmission IR spectra. Section 5.2.4 will discuss in detail the differences between the outcomes of transmission and ATR IR experiments. As a consequence of the wavelength dependence of the penetration depth of the evanescence wave, the surface sensitivity in ATR/IR spectroscopy is not constant over the typical mid-IR region, a factor to be considered when interpreting ATR/IR results.

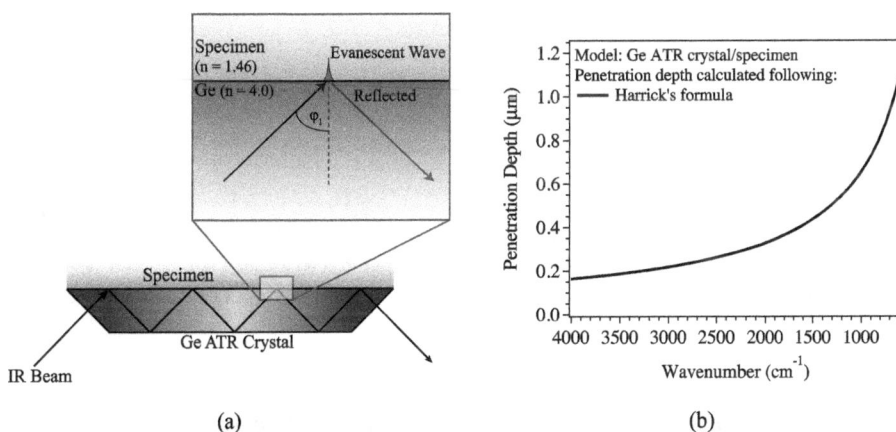

(a)   (b)

Fig. 5.6: (a) Schematic of the two-layered system (Ge ATR crystal/specimen); (b) penetration depth (µm) as a function of wavenumber in the case of a two-layered system (Ge ATR crystal/specimen). In the calculations, a specimen with refractive index equal to 1.46, a typical value for the refractive index of polymers [43], was considered. The d_p values were computed using the formula introduced by Harrick (eq. (5.28)) [8].

As pointed out in Section 5.2.2, the matrix formalism developed by Hansen could also be used for evaluating the *MSEF* at any location within a given layer of a *N*-isotropic-phase system (eqs. (5.11)–(5.25)) [14, 35].

In the case of a two-layered system (Ge ATR crystal/specimen), the *MSEF* magnitude can be numerically calculated at different positions within the rarer medium (i.e., the sample under investigation) and along the normal to the boundary between the two phases (z-axis). By fitting the *MSEF* values using an exponential function ($E = E_0\exp(-kz)$), where the fitting parameters are the decay constant (k) and the electric field amplitude at the interface (E_0), the penetration depth can be determined ($d_p = k^{-1}$). Since in the case of p-polarized light the calculation of the *MSEF* magnitude requires the knowledge of the complex dielectric constant, the *MSEF* magnitude was only evaluated for s-polarized radiation. The penetration depth values, determined using the aforementioned approach for the case of a two-layered system consisting of a Ge ATR crystal and a specimen with a bulk refractive index equal to 1.46, are in good agreement with those calculated by means of the formula proposed by Harrick (eq. (5.28)) (Fig. 5.7).

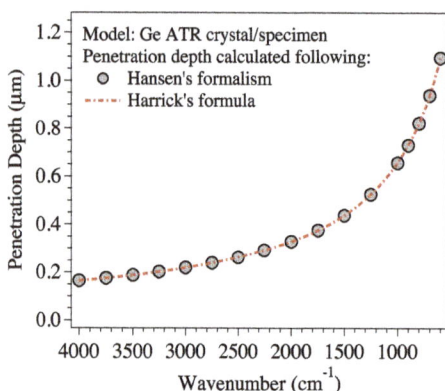

Fig. 5.7: Penetration depth (μm), calculated using the formalism developed by Hansen (eqs. (5.11)–(5.25)) [14, 35], as a function of the wavenumber in the case of a two-layered system (Ge ATR crystal/specimen with refractive index equal to 1.46). The d_p values calculated using the formula introduced by Harrick (eq. (5.28)) [8] are included for comparison.

5.2.4 Transmission IR versus ATR/IR spectroscopy

Since the development of the first IR spectrometers, transmission-sampling techniques have been widely used for acquiring IR spectra, thanks to their simplicity and to the introduction of few artifacts into the spectra [1]. Consequently, most of the published tables reporting characteristic group frequencies and most of the commercially available software libraries are derived from transmission IR experiments.

In the last decades, ATR/IR has become the most common sampling technique in IR spectroscopy, as demonstrated by the profusion of ATR accessories on the market. The IR spectra obtained by the ATR/IR technique turned out not to be identical to those obtained by transmission. Studying the physical principles of ATR/IR spectroscopy helped in understanding the resulting relative shifts in band intensity and absolute shifts in frequency, which characterize the ATR/IR spectra. The interpretation of these spectra on the basis of published frequency tables or software libraries might, then, be misleading and should be performed with care. In order to help users interpret the analytical results, most spectroscopic software packages provide algorithm for correcting the intensity and frequency shifts [11, 44]. In the following, the differences between ATR/IR and transmission IR spectra are outlined and discussed in detail.

As pointed out in Section 5.2.1, Harrick introduced the effective thickness (d_e) in order to express the strength of the coupling of the exponentially decaying wave with the rarer phase in ATR/IR spectroscopy and compare internal reflection spectra to the one acquired in transmission mode [8]. In the case of bulk materials, the effective thickness is proportional to the penetration depth (eq. (5.34)). As the penetration depth depends on the wavelength, the effective thickness increases with the wavelength, resulting in the appearance of more intense absorption bands at longer wavelengths (lower wavenumbers). Two peaks with the same intensity in transmission IR spectra will appear with unequal intensity in internal reflection IR spectra. Such a wavelength dependence of the effective thickness also induces a broadening of the absorption bands on their high-wavelength side together with a shift to longer wavelengths [8]. Conversely, in the case of thin films, no relative distortion or shift of the absorption bands is observed as the penetration depth does not affect the effective thickness (eq. (5.35)). The resulting ATR/IR spectra of thin films closely resemble those acquired by transmission [8].

Dispersion effects (see Section 5.3.1) should also be considered when comparing ATR/IR to transmission IR spectra. In ATR/IR spectroscopy, the variation of the refractive index of a sample in the vicinity of an absorption band results in: (a) the detection of absorption bands whose intensity appears higher than it should normally be; (b) distortions of the shape of the peaks; and (c) small displacements of the band maximum. Such effects become more pronounced for molecular vibrations showing strong absorption bands (i.e., high values of the extinction coefficient) [13] and for incident angles close to the critical one. In order to minimize the distortion of the peaks appearing in ATR/IR spectra, incident angles well above the critical angle should be employed [45] (see Section 5.3.1 for more details).

5.3 Experimental methods in ATR/IR spectroscopy

5.3.1 Internal reflection elements (IREs)

IREs, or ATR crystals, are optical elements transparent in the IR region (or in limited ranges of the IR region of the electromagnetic spectrum) used in ATR/IR spectroscopy for establishing the conditions necessary to obtain ATR/IR spectra of the materials under investigation [8, 11, 12].

The IR radiation, which propagates in the IRE by means of total internal reflection, interacts with the sample, which is in contact with the reflecting surface.

The acquisition of an ATR/IR spectrum and the information provided by its acquisition strongly depends on the characteristics of the IRE, i.e., angle of incidence, number of reflections, aperture, number of passes, surface preparation, and material which the IRE is made of.

The number of reflections is calculated from simple geometrical considerations. The so-called *skip distance*, equal to $t \cdot \tan \varphi$ (t thickness of the IRE; φ angle of incidence) defines the distance the beam advances for each reflection. The total number of reflections for a single-pass IRE is given by

$$N = \frac{l}{t} \cot \varphi \tag{5.38}$$

where l is the IRE length, measured, in the case of a single-pass crystal, from the center of the entrance aperture to the center of the exit aperture.

A large variety of IRE have been developed and reported in the literature. The geometry of the reflection element is crucial, since it determines the ease with which ATR/IR spectra can be obtained. A thorough discussion of the different types of IREs with their uses, advantages, and disadvantages can be found in [8].

Since in the case of thin films and, in particular, of monomolecular layers, single-reflection IREs do not provide spectra with good signal-to-noise ratio, multiple-reflection elements are commonly employed. Among the different geometries, single-pass (i.e., where the incident IR radiation enters the plate, propagates via multiple internal reflections, and leaves through the opposing aperture) trapezoids and parallelograms with an angle of incidence of 45° are the most widely employed IREs.

Besides geometrical issues, the choice of the IRE for a specific analytical study should be carried out considering the material the crystal is made of and the surface preparation. As for the former, transparency, hardness, brittleness, conductivity, and chemical inertness are all factors that play a fundamental role in the accessibility of an ATR crystal, which is suitable for a specific investigation. The material types available include glass-like materials, pressed ones, polycrystalline, and highly pure single crystals. The latter are mainly required for surface studies, since grain boundaries in polycrystalline materials may result in radiation losses. The hardness and brittleness play an important part in the ease of preparation of the IRE as well as in the resistance to surface damage. Conductivity and chemical inertness are two additional factors that have to be taken into consideration in the determination of the species that can be studied.

When choosing the optical material for an IRE, the optical properties of the analyte should be also carefully considered. In fact, it has to be emphasized that the refractive index of a medium strongly varies in the vicinity of an absorption band [8, 13, 33, 45]. This variation is known as *dispersion*. Due to such a variation of the real part of the refractive index of the phase under investigation in correspondence of an absorption band, it may happen that some peaks do not satisfy the total reflection condition (eq. (5.6)). This could result in bands whose intensity appears higher than it should normally be, distortions of the shape of the peaks, and small displacements of the band maximum. Because of the foreseeable effects of dispersion on ATR/IR spectra, ATR crystals with a refractive index considerably higher than that of the sample should be chosen [8, 13, 45].

Since a change in the real part of the refractive index of the sample under investigation induces a variation of the critical angle (eq. (5.6)), dispersion effects in ATR/IR spectra are dependent on the incident angle of the IR radiation. In order to minimize such effects, angles of incidence well above the critical angle should be employed [45]. The rule of thumb is $\varphi \geq \varphi_c + 15°$.

There are several reasons, which can contribute to the poor performance of an ATR crystal, such as poor optical material (defects in the quality of the material), inadequate surface polishing (which causes scattering losses), poor tolerance on lengths and angles, as well as non-parallelism of the reflecting surfaces. A procedure for checking the quality of an IRE is suggested in [38]. In the last few years, the input–output apertures of ATR crystals are coated by antireflection material in order to enhance their optical efficiency [14].

5.3.2 Internal reflection attachments (IRAs)

As in the case of other reflection–absorption IR spectroscopic techniques, the acquisition of ATR/IR spectra requires specific accessories, which are mounted in the sample compartment of an IR spectrometer and hold the IRE. The optical design of an internal reflection attachment (IRA) is primarily determined by the configuration of the IRE. As a consequence of this, the following types of IRA are usually distinguished: (a) variable-angle single internal reflection; (b) fixed-angle multiple internal reflection; and (c) variable-angle multiple internal reflection [38]. All the accessories contain two systems of mirrors: the first one focuses and directs the IR beam into the IRE, while the second one collects the reflected radiation leaving the IRE and directs it toward the IR detector.

The principal requirements for ATR/IR accessories are: (a) high accuracy in adjusting the angle of incidence, and (b) low radiation losses. To simplify routine analyses of bulk samples, ATR units with optimized optical configurations have been developed over the last few years. While these attachments turn out to be the IR analytical tool of choice for investigating bulk samples due to the predetermined incidence angles, they are less effective for studying thin films.

The major issue of ATR/IR spectroscopy lies in achieving an optimal optical contact between the sample and the IRE. To solve this problem [14], several approaches are commonly used, including:

a. the use of an IRE made from the material of the film substrate, which is practical for dielectrics, *e.g.*, sapphire and calcite, and weakly doped semiconductors (germanium and silicon);

b. coating the IRE with a film of the desired material (by means, for example, radio-frequency (RF)/direct current magnetron reactive sputtering) when the IRE cannot be produced out of a particular material;

c. the use of liquid cells, which allow IRE/solution interfaces to be continuously monitored; and

d. pressing the sample against the IRE to improve the optical contact (common for polymer surfaces).

5.3.3 Metal underlayer ATR/IR spectroscopy

In spite of the plethora of analytical techniques available to experimentalists for studying the surface chemistry of materials in vacuum or in ambient air [46], the development of spectroscopic methods that allow for the investigation of the chemical processes at solid–liquid interfaces has been a long-standing challenge for surface scientists. Vibrational spectroscopy, providing information about the metal–adsorbate bonds, local molecular orientations, and intermolecular interactions within the adsorbate layer, has been recognized to be a powerful tool for probing solid–liquid interfaces [14].

Recently, ATR/IR spectroscopy proved to be a suitable analytical method for probing solid–gas and solid–liquid interfaces in situ [15, 20, 21, 47]. Since common technologically relevant materials are not suitable for IREs, thin metal films, deposited by physical vapor deposition onto the IRE, became an important type of model system. The combination of an IRE and such a thin metal film is usually referred to as metal underlayer (Kretschmann) ATR (MUATR) configuration [14] (Fig. 5.8).

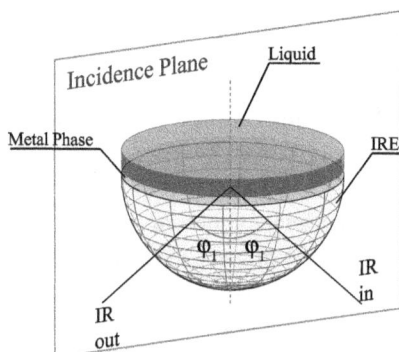

Fig. 5.8: Schematic representation of the metal underlayer (Kretschmann) ATR (MUATR) configuration.

Reproducibility of film preparation and mechanical stability of the resulting metal layer are two fundamental issues for practical applications. Cleaning the IRE before depositing any metal film plays a decisive role in determining reproducibility. In spite of this, metal layers deposited on IREs may be still mechanically unstable and peel off [48, 49]. In order to overcome this problem, a metal oxide support material, such as Al_2O_3 (which has a low absorption coefficient in the mid-IR region), is commonly evaporated first. Such a deposition also results in covering or displacing the contaminants, which might be on the IRE surface. The presence of a 50–100 nm thick Al_2O_3 layer, deposited by electron beam physical vapor deposition, on a Ge IRE was found not to affect the reflectivity of the ATR system and enhance the stability of the overlying platinum film under catalytic reaction conditions [21].

The metal film deposited onto the IRE should not be too thick so that the electric field is not completely attenuated in it, and the electromagnetic wave can still reach the metal–liquid or metal–gas interface [21]. In the case of a Ge/Pt/vacuum system, it has been shown that for increasing platinum-layer thicknesses, the reflectance first decreases, reaches a minimum, and then starts increasing [21]. In the latter regime, the film acts as a mirror and is therefore useless for spectroscopy. It has to be emphasized that, in the case of multiple reflection elements, the total reflectivity is the reflectivity of a single reflection to the power of the number of reflections N ($R_{TOT} = R^N$). On the other hand, the intensity of an absorption band linearly increases with the number of reflections ($A_{TOT} = A \cdot N$). Hence, the experimental setup should be optimized in order to attain a suitable signal-to-noise ratio.

The deposition of thin films onto IREs can also result in the formation of inhomogeneous layers composed of islands smaller than the wavelength of the radiation concerned. The morphology of these layers strongly depends on the substrate (IRE), the metal, and the deposition conditions (temperature, background gases, etc.) [21, 50, 51]. These metal films consisting of islands have been widely employed in IR spectroscopy, since the intensity of the characteristic absorption bands of molecules in contact with the metal can be considerably enhanced compared to the signals in the absence of the metal film. This effect is commonly referred to as *surface-enhanced IR absorption* (SEIRA) [50, 51]. Since the pioneering work of Hartstein et al. in the early 1980s [52], the SEIRA phenomenon has been employed in (bio)chemical IR sensors, the study of electrode–electrolyte interfaces, Langmuir–Blodgett films, and self-assembled monolayers [14, 50, 51]. Over the years, the same effect has been reported with materials different from those used by Hartstein et al. (Au and Ag) [52], such as Cu, In, Li, Sn, Pb, Fe, Pt, Pd, and Pt–Fe alloys [51]. The enhancement factor was, however, found to strongly depend on the morphology of the islands, their size and shape, the spacing between the islands, and the material they are made of. In general, less noble metals show smaller enhancements than noble metals in air. Enhancement factors up to 1000 have been reported for Au films. Research on the mechanism of the enhancement led to a general consensus that two different mechanisms, i.e., electromagnetic and chemical, contribute to the total enhancement. A detailed discussion about these mechanisms is beyond the scope of this chapter and can be found in [50, 51].

It has to be emphasized that, in the case of thin films exhibiting island structure, the ATR system can still be modeled as being layered. However, the optical constants of bulk materials and thickness values must be replaced by effective optical constants and optical thickness values of the island film. According to the Maxwell-Garnett [53], the Bruggeman [54], and other effective medium theories [55], the effective optical constants depend on several parameters, such as packing density of the particles, polarizability of the interacting metal islands, and optical constants of the surrounding material. As a consequence of this, they may significantly differ from the bulk values. However, in the case of thick films, the optical constants approach the bulk values: for silver this is achieved at a thickness of around 10 nm [56], while for iron it was found to be at around 3 nm on MgO (001) surfaces [57].

The presence of a metal layer on the IRE has been found to drastically affect the spectrum of the medium under investigation and of the adsorbate [21, 28]. As already pointed out in Section 5.2.4, ATR/IR spectra show distortions of the shape of the peaks and small displacements of the band maximum due to the variation of the refractive index of a sample in the vicinity of an absorption band. While in the case of IRE/adsorbate/solvent systems, distortions of the peaks are minimized by using incident angles well above the critical angle [45], in the case of IRE/metal/adsorbate/solvent systems strongly distorted bands appear in the ATR/IR spectra even when the incident angle is far from the critical angle [21].

Piras et al. reported that in the case of a Ge/Fe film (10 nm)/zinc dialkyldithiophosphate (ZnDTP) system the ATR/IR spectra show a shift of the characteristic P–O–(C) and C–O absorption bands of ZnDTP to lower wavenumbers together with a distortion of these bands on the lower wavenumber side for an incident angle well above the critical angle [24, 28]. Besides optical arguments, the authors also suggested that the adsorption of ZnDTP on the iron surface might contribute to the observed shift of peaks [28].

The shape of absorption bands measured by ATR/IR for the case of thin films has been experimentally and theoretically investigated by Bürgi [19, 21]. The theoretical results showed that strongly distorted absorption bands appeared at incident angles well above the critical angle in the case of both bulk media and thin adsorbate films: a negative-going peak at high wavenumbers and an increased background at low wavenumbers were always observed. At the same time, the band maximum shifted to lower energy. The theoretical results also revealed that the band shape in MUATR/IR spectroscopy is mainly influenced by the optical properties of the metal film: the distortion of the bands increases upon increasing thickness and refractive index, or upon decreasing the absorption index of the metal layer. Such theoretical results were found to agree well with measured experimental spectra in the case of thick (10 nm) metallic Pt layers.

5.3.3.1 Effect of metal underlayer on penetration depth in ATR/IR spectroscopic analysis of polymers

In the following, the penetration depth of the evanescent wave in a typical MUATR/ IR experiment is discussed. A model system, consisting of a germanium ATR crystal (refractive index n_{Ge} equal to 4.0) coated with a thin iron layer (10 nm) and a bulk specimen with refractive index ($n_{specimen}$) equal to 1.46 (a typical value for the refractive index of polymers [43]), is considered in the calculations.

Since the iron refractive index (n_{Fe}) is higher than that of germanium ($n_{Ge} = 4.0$) over the whole spectral region (Fig. 5.9a), when the IR radiation strikes the Ge/Fe interface, the electromagnetic wave splits into the reflected and transmitted (refracted) components. The angle of refraction (φ_2) is related to the incidence angle (φ_1) by Snell's law (eq. (5.3)), whereas the intensities of reflected and refracted beams are interconnected by the so-called Fresnel amplitude coefficients (eqs. (5.4) and (5.5)). It has to be emphasized that, since the iron refractive index varies with the wavelength (Fig. 5.9a), the angle of refraction (φ_2) is not constant over the mid-IR region.

At the iron/specimen interface, the IR radiation is totally reflected only if the angle of incidence (equal to the angle of refraction (φ_2) of the IR beam as it passes from Ge to Fe) exceeds the critical angle (φ_c) (eq. (5.6)). Due to the frequency dependence of the iron refractive index (Fig. 5.9a), the critical angle also varies with the wavelength.

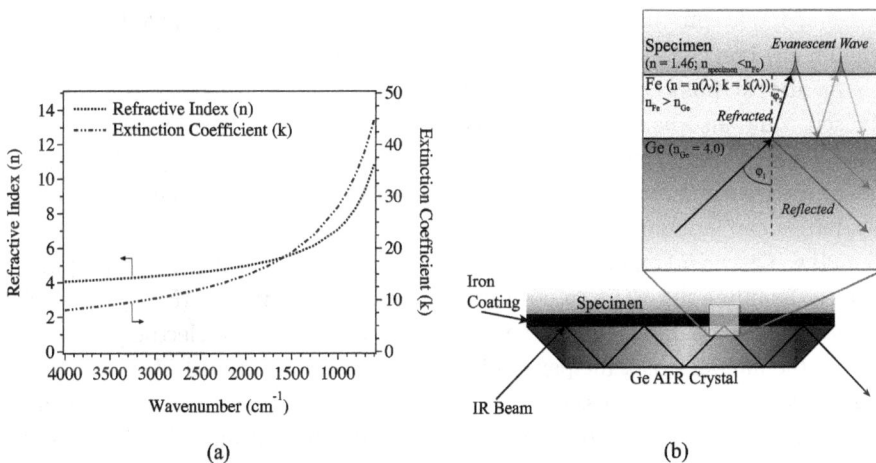

Fig. 5.9: (a) Metallic iron optical constants (refractive index (*n*) and extinction coefficient (*k*)) [58] used for modeling the ATR/IR system employed in the present work; (b) schematic of the three-layered system (Ge ATR crystal/Fe/specimen with refractive index equal to 1.46, a typical value for the refractive index of polymers [43]).

The penetration depth in the case of the three-layered system (Ge ATR crystal/Fe/specimen) can be determined using the formula proposed by Harrick (eq. (5.28)) [8] and considering the frequency dependence of both the iron refractive index and the incidence angle of the IR beam at the Fe/specimen interface (equal to the angle of refraction of the IR beam as it passes from Ge to Fe):

$$d_p = \frac{\lambda}{2\pi n_1 \sqrt{\sin^2\varphi_2 - n_{21}^2}}$$

where

$$n_1 = n_{Fe}(\lambda); \quad n_{21} = \frac{n_2}{n_1} = \frac{n_{specimen}}{n_{Fe}(\lambda)}$$

$$\varphi_2 = \arcsin\left(\sin(\varphi_1)\frac{n_{Ge}}{n_{Fe}(\lambda)}\right); \quad \varphi_1 = 45° \tag{5.39}$$

Due to the increase of the iron refractive index with the wavelength, the angle of refraction (φ_2) significantly decreases as the wavelength increases, but it always exceeds the critical angle (φ_c) (Fig. 5.10). Therefore, the IR beam undergoes total reflection at the iron/specimen interface.

The penetration depth values, determined using the formula proposed by Harrick (eq. (5.28)) [8], in the case of the three-layered system Ge ATR crystal/Fe/specimen are comparable with those calculated in the case of the corresponding ATR/IR system without any iron coating (i.e., two-layered system, Ge ATR crystal/specimen).

As already pointed out in Section 5.2.3.1, the matrix formalism developed by Hansen could also be used for evaluating the MSEF at any location within a certain layer of a system made of N isotropic phases (eq. (5.11)–(5.25)) [14, 35].

In the case of a three-layered system (Ge ATR crystal/MU/specimen), the MSEF magnitude can be numerically calculated at different positions within the rarer medium (i.e., specimen) and along the normal to the boundary between the two phases (z-axis). By fitting the MSEF values using an exponential function ($E = E_0\exp(-kz)$), where the fitting parameters are the decay constant (k) and the electric field amplitude at the interface (E_0), the penetration depth can be determined ($d_p = k^{-1}$). In the present work, since in the case of p-polarized light the calculation of the MSEF magnitude requires the knowledge of the complex dielectric constant, the MSEF magnitude was only evaluated for s-polarized radiation.

The penetration depth values, determined using the aforementioned approach, in the case of the three-layered system Ge ATR crystal/Fe (10 nm)/specimen with refractive index equal to 1.46 are comparable with those calculated using the same formalism, but for the corresponding ATR/FT-IR system without any iron coating, i.e., for the two-layered system Ge ATR crystal/specimen (Fig. 5.11a).

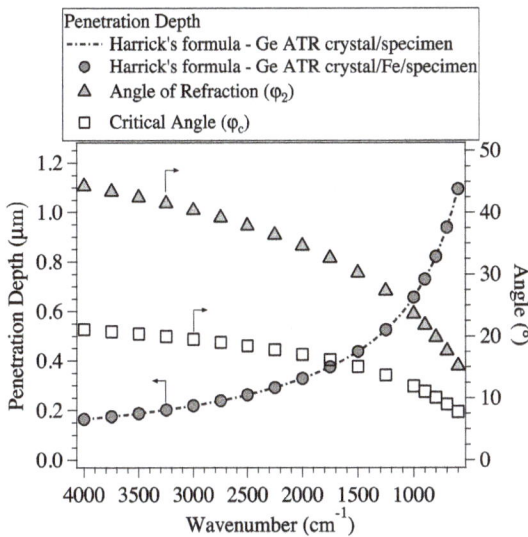

Fig. 5.10: Penetration depth (µm), calculated using the formula introduced by Harrick [8], as a function of the wavenumber in the case of the three-layered system Ge ATR crystal/Fe/specimen with refractive index equal to 1.46, a typical value for the refractive index of polymers [43]. As a consequence of the variation of the iron reflective index with the wavelength (Fig. 5.9), the angle of incidence of the IR beam at the Fe/specimen interface (equal to the angle of refraction of the IR beam as it passes from Ge to Fe) and the critical angle for total reflection at the Fe/specimen interface also change with the wavenumber.

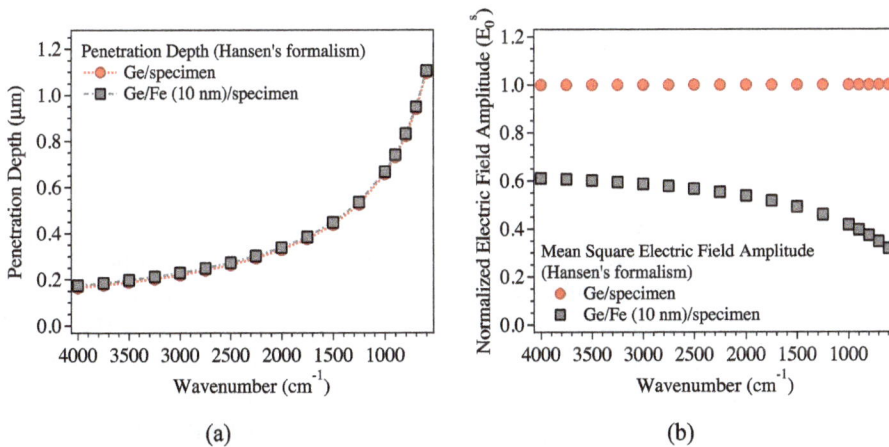

(a)   (b)

Fig. 5.11: (a) Penetration depth (µm), calculated using the formalism developed by Hansen [14, 35], as a function of the wavenumber in the case of a three- and two-layered system (Ge ATR crystal/Fe (10 nm)/specimen and Ge ATR crystal/specimen); (b) normalized electric field amplitude at the Fe/specimen (two-layered system) and iron/specimen (three-layered system) interface (E_0^s) as a function of the wavenumber. In the calculations, a refractive index equal to 1.46, a typical value for the refractive index of polymers [43], was considered for the rarer medium (specimen).

The numerical calculation of the *MSEF* does not only lead to the determination of the penetration depth of the evanescent wavein the rarer medium, but it also allows the evaluation of the electric field amplitude at the interface between the denser and rarer medium (E_0). Compared to the case of a two-layered system (Ge ATR crystal/specimen), a significant reduction of the amplitude of the electric field at the interface between the denser and rarer medium is observed when an iron coating is present on the IRE (Fig. 5.11b).

5.4 Potentials and limitations of ATR/IR spectroscopy

The capability of yielding surface chemical information, information about the kinetics of adsorption and reaction at solid–liquid interfaces, and the possibility of performing ATR/IR analysis using metal-coated IREs have made ATR/IR spectroscopy the analytical technique of choice in different research fields, such as polymer science [16], electrochemistry [17, 59], tribology [24–26, 28–32], the science of bio-membranes and biofilms [18], catalysis [15, 20, 21], and environmental chemistry [23].

In the case of solid–liquid interfaces (*e.g.*, during polymerization studies), since ATR/IR spectroscopy *simultaneously* provides information about the chemistry of both dissolved species, such as reactants and products in the bulk phase, and species adsorbed on the IRE/metal-coated IRE, such as reaction intermediates, it is *unspecific* in the sense that a signal cannot a priori be assigned to dissolved or adsorbed compounds. Moreover, since ATR/IR probes the solid–liquid interface, where surface reactions occur, the concentrations of the species generated at the interface and dissolved in the bulk phase might differ from the one measured by other bulk analytical methods, such as gas chromatography.

Recently, ATR/IR spectroscopy has been successfully applied to the study of supported metal catalysts. Besides providing real-time monitoring capabilities, it also yields information about changes of the state of supported metal particles or metal films. On the basis of reflectivity calculations, Bürgi et al. demonstrated that small changes of the optical constants of the metal (Pd), deriving from a change in concentration of free electrons, result in changes in reflectivity, in agreement with the measured ATR/IR spectra [22]. This variation in the concentration of free electrons could be due to the application of an electrochemical potential, change of the work function due to the formation of an adsorbed layer, hydrogen adsorption, or oxidation of the metal-supported catalyst.

On the basis of this overview, it can be easily deduced that the main challenge of ATR/IR spectroscopy is associated with the complexity of the results. Since all the chemical species present at the metal–liquid interface, either adsorbed on the surface or dissolved in the medium, contribute to the measured ATR/IR spectrum,

the identification of a particular species and of its role in the surface chemistry become a major challenge.

Besides the difficulties in interpreting the outcomes of ATR/IR analysis, practical challenges also have to be faced [28]. The practical limitations of this technique
are mainly due to the IRE and to the IR detector. As for the former, the most widely
used IREs, i.e., Ge, Si, ZnSe, have a limited useful spectral range, which does not
allow for the acquisition of ATR/IR spectra over the whole mid-IR region. Moreover,
the investigation of surface reactions can also require the acquisition of spectra at
high temperatures, precluding, then, the use of semiconductor materials, such as
Ge, whose useful transmission range becomes narrower with increasing temperature as a consequence of the generation of free electrons. On the other hand, ATR/
IR spectroscopy usually requires the use of detectors having high sensitivity, such
as the mercuric-cadmium-telluride detectors, which, however, have wavenumber
cut-offs between 800 and 600 cm^{-1}.

5.5 Applications of ATR/IR spectroscopy

5.5.1 ATR/IR spectroscopy in polymer science

IR spectroscopy is one of the most powerful analytical techniques for the elucidation
of the structure and composition of polymers and polymer surfaces [1, 16, 60–67].
Among the various experimental methods in IR spectroscopy, ATR/IR spectroscopy
has been the analytical technique of choice for the characterization of a wide variety
of elastomeric materials as well as for the in situ investigation of polymerization
processes.

The power of ATR/IR spectroscopy in probing the surface structure and composition of polymers was already highlighted in the 1980s by Dothee et al. [68], who
investigated commercial polyethylene films and suggested that the crystallinity of
polyethylene was higher in the near-surface region than in the bulk of the material.
The influence of uniaxial and biaxial elongations on the surface orientation of polypropylene was studied by Mirabella in a series of pioneering studies [69, 70]. By
using two distinct features in ATR/IR spectra, information about the surface orientation of the helix axis of polypropylene could be obtained. The inspiring work of
Mirabella stimulated further studies concerning the stretching of polymers and the
influence of the applied mechanical strain to the molecular orientations [71, 72]. In
particular, Walls carefully examined the surface composition and orientation of
uniaxially drawn polyethylene terephthalate films by employing polarized ATR/IR
spectroscopy [72].

The versatility of ATR/IR spectroscopy also opened the path for the investigation of surface modifications and reactions of polymers in liquids as well as at

elevated temperatures. The reaction of polyvinylidene fluoride in aqueous sodium hydroxide and in the presence of a phase transfer catalyst (namely, tetrabutylammonium hydrogen sulfate) was shown to lead to surface dehydrofluorination [73, 74], which significantly affected the electrical conductivity of the material [75, 76]. On the other hand, ATR/IR spectroscopy could also be used to shed light on the influence of thermal history on the surface modification of polyvinyl chloride (PVC): significant differences in the ATR/IR spectra of PVC were observed between hot-pressed PVC specimens and solvent-casted PVC samples [16].

A particular field of research, for which ATR/IR spectroscopy has highly been attractive, deals with multicomponent systems and hybrid materials in which interfacial interactions, stability, and mutual compatibility are crucial. For the investigation of these systems, an inherent drawback of ATR/IR measurements is the lack of an internal calibration method, which makes the construction of calibration curves necessary for performing quantitative analysis [16]. As one particular example, Brandrup et al. investigated the migration of one of components (namely polyamide) in polymeric systems (i.e., epoxy), to the film/air or film/substrate interface due the lower surface tension of polyamide compared to its epoxy counterpart [77].

In light of the importance for real technological applications, the interactions of polymer surfaces with solid substrates have widely been studied in the literature. ATR/IR spectroscopy was effectively used by Nguyen et al. for monitoring the diffusion of water through pigmented and alkyd coatings, which explained the observed interfacial degradation [78]. On the other hand, Urban et al. examined the influence of gas-plasma modifications on silicone elastomers [79]. Among the plasma treatments, ammonia plasma turned out to be the only process inducing an increase of the storage modulus, whereas argon or carbon dioxide treatments resulted in a decrease of the same property.

A powerful aspect of ATR/IR spectroscopy relies on the ability of conducting orientation studies of (macro)molecules on solid surfaces. The use of polarized light can, in fact, provide valuable insights into the structure of adsorbates and molecules in the near-surface region, as well as how it is affected by mechanical strain, temperature, and substrate surface chemistry. For example, the orientation of molecules containing hydrophobic and hydrophilic was found to strongly depend on the wettability of the surface the molecule interacts with [16]. The dependence of the orientation of isotactic polypropylene on the applied uniaxial stress was investigated for the first time by Flournoy et al. [80]: the dichroic ratios in the analysis of the polymer by transmission IR and ATR/IR spectroscopy with parallel-polarized light was equal, thus suggesting that the orientation of the polymer molecules is the same near the surface region and in the bulk of the material.

ATR/IR spectroscopy has also been the analytical technique of choice for investigating polymerization processes in situ. Using this approach, information about the kinetics of the polymerization reaction(s), crystallization processes, chain packing, and orientation could be obtained. A pioneering work in this respect was performed by Ishida and Scott [81], who investigated the polymerization reactions on nylon-6. Information about reaction mixtures can also be obtained by ATR/IR over short time intervals: the rate of consumption of free isocyanate groups could be deduced from the spectra together with the rate of the competitive formation of urethane, isocyanurate, and urea linkages [66]. Curing processes could also be followed in situ by ATR/IR spectroscopy: Snyder et al. reported the in situ investigation of the cross-linking of polyester photoresists, which was attached to an ATR crystal [82].

Recently, ATR/IR spectroscopy was exploited to evaluate the interactions of solvent mixtures with grafted-from polymer brushes [83]. As solvent/brush interactions play a critical role in dictating the tribological and mechanical properties of these materials [84, 85], several techniques have already been applied to evaluate the swelling and collapse of brushes, including the surface force apparatus [86], neutron reflectometry [87], atomic force microscopy (AFM) [88], and quartz crystal microbalance [89]. While ATR/IR was used to quantify the water uptake in silicalite-1 films [90] and the interactions of immobilized serum albumin with cholate [91], no previous studies investigated the interactions of solvent mixtures with polymeric brushes. Mathis et al. used, for the first time, ATR/IR spectroscopy to investigate the temperature-dependent interactions of poly(dodecyl methacrylate) (P12MA) brushes with ethanol and toluene solvent molecules as well as their mixtures [83]. The spectroscopic results indicated a temperature-dependent partitioning of the solvents within the polymeric brushes, thus providing clues to strategies for tailoring polymer/solvent combinations for tuning friction and adhesion properties.

5.5.2 Research applications of ATR/IR spectroscopy for studying the surface properties of hydrogels

Synthetic hydrogels are composed of cross-linked hydrophilic polymer chains and a large amount of water. While they have attracted considerable attention in a range of applications, such as flexible electronic devices [92, 93], soft actuators [94–96], controlled drug delivery systems [97], and tissue engineering [98, 99], most published studies focused on rationally designing the chemical composition of hydrogels to tailor their network structures with the final aim of achieving task-specific bulk properties (*e.g.*, stiffness and stretchability) [100, 101]. In the last couple of decades, researchers started focusing on the evaluation, modification, and functionalization of the polymeric structure at hydrogel surfaces [102–112]. The development of effective methods to tune the surface chemistry of hydrogels, which could provide the possibility of controlling specific surface properties of the hydrogel (*e.g.*, lubricity, diffusivity, and biopassivity), hinges on the use of analytical techniques able to characterize the surface chemistry and structure of a phase (polymer in the case of hydrogels) dispersed in medium (water in the case of hydrogels).

Simič et al. performed time-resolved ATR/IR spectroscopic measurements to investigate water exudation with subsequent polymer densification in the near-surface region of polyacrylamide hydrogels with either brushy or cross-linked surface structures [111]. The experimental approach relied on pressing hydrogel pins against a diamond ATR crystal (Fig. 5.12) at a given applied normal load, which was followed by the acquisition of ATR/IR spectra as a function of time.

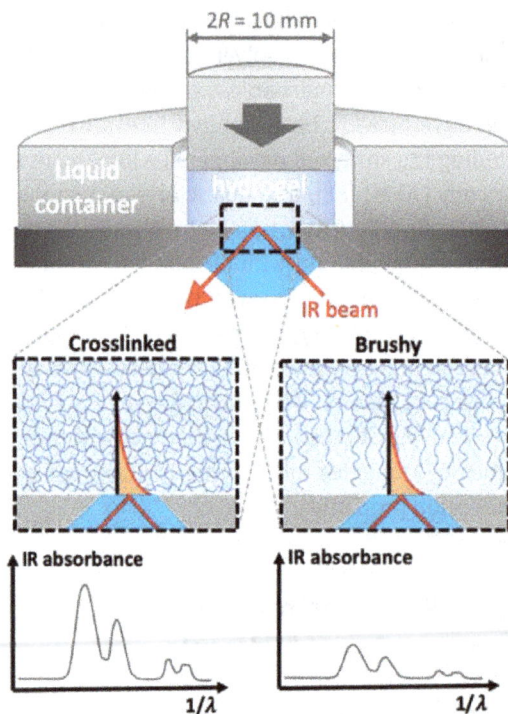

Fig. 5.12: Schematic diagram of the ATR/IR setup used by Simič et al. to evaluate water exudation with subsequent polymer densification in the near-surface region of polyacrylamide (PAAm) hydrogels [111]. A flat pin is pressed against a diamond ATR crystal at various loads and ATR/IR spectra can be acquired as a function of time. The intensity of characteristic ATR/IR absorption features of the hydrogel provides information about the polymer concentration in its near-surface region. (Reprinted with permission from [111]. Copyright (2020) Springer).

Figure 5.13a displays the ATR/IR spectra obtained as a function of time after the application of the normal load onto the pin for both cross-linked and brushy hydrogels, while Fig. 5.13b shows the evolution of the intensity of the amide II peak with time. While the intensity of the amide II absorption feature stabilized within the first minute after the application of the normal load for the case of the cross-linked hydrogel, strong time dependence and normal pressure dependence of the intensity of the same peak were observed in the case of the brushy hydrogel, which indicated a significant increase in the polymer concentration in the near-surface region of this hydrogel. The characteristic densification time extracted from ATR/IR spectroscopic measurements could be correlated with the friction response of the hydrogels under investigation and allowed the authors to draw conclusions about the energy dissipation mechanism that dominates while sliding.

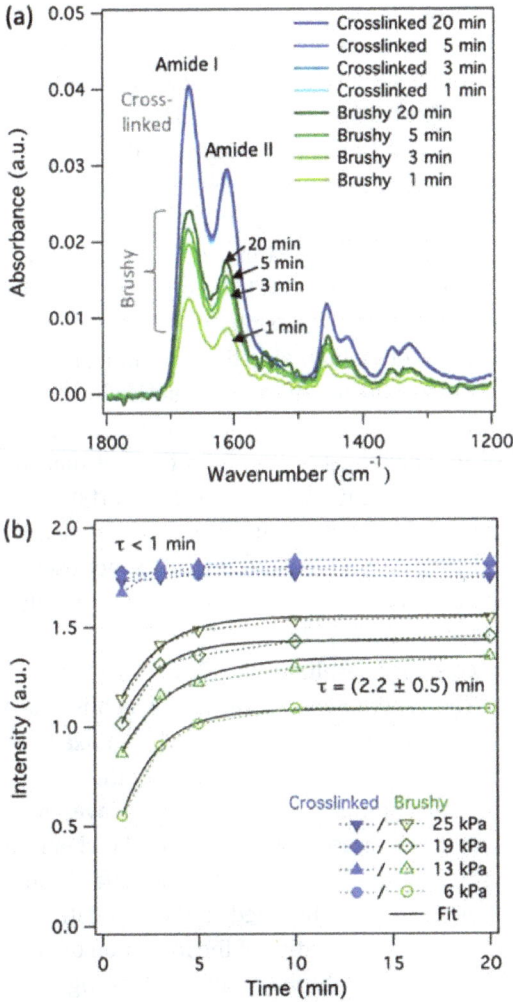

Fig. 5.13: (a) ATR/IR spectra of polyacrylamide (PAAm) hydrogels with either brushy or cross-linked surface structures [111]. The spectra were acquired at different times after pressing the hydrogel pin against the diamond ATR crystal at a normal pressure of 6 kPa; (b) intensity of the amide II absorption feature as a function of time after the hydrogel pin was pressed against the ATR crystal at different pressures. The solid line is the exponential fit used to obtain the characteristic drainage time. (Reprinted with permission from [111]. Copyright (2020) Springer).

5.5.3 In situ ATR/IR spectroscopy of tribochemical phenomena

Lubricating fluids for mechanical components (*e.g.*, gear boxes and internal combustion engines) usually consist of a synthetic or mineral oil and an additive package, which includes several molecules used to impart or improve specific properties. Among them, anti-wear additives are used to reduce material removal from components in relative motion when applied loads are too high and sliding speeds are too low for maintaining a full fluid film at sliding interfaces [113–118]. While ZnDTPs have extensively been employed in engine oils since the 1940s owing to the combination of excellent anti-wear and good anti-oxidation properties [113, 115, 118–120], insights into its mechanism of action could only be gained in the last four decades, thanks to the use of surface-analytical techniques, including X-ray photoelectron spectroscopy (XPS) [121–128], time-of-flight secondary ion mass spectrometry [128, 129], and AFM [130]. The generally accepted view of the anti-wear mechanism of ZnDTPs involves its surface reaction to form patchy, amorphous films [113, 119, 120], which consist of an innermost layer made of short-chain poly(thio)phosphates containing zinc/iron cations and an outermost one made of long-chain zinc poly(thio)phosphates. Despite the remarkable advancements in our understanding of the lubrication mechanism of ZnDTPs, significant open questions still remain about its reaction mechanism(s) at sliding interfaces. A significant challenge derives from the fact that the interface is buried and inaccessible to most conventional analytical tools [131]. This has called for the development, extension, and use of in situ approaches to identify and quantify the underlying chemical processes [131–134]. ATR/IR spectroscopy turned out to be a powerful tool for the in situ analysis of boundary layers generated by lubricant additives between steel and iron surfaces [24–32, 135–139].

Piras et al. developed an ATR/IR tribotester to evaluate chemical changes occurring at oil/metal interfaces [24–26, 28]. The experimental apparatus, which is based on the MUATR/IR configuration, allowed the acquisition of spectra while a fixed cylinder was slid on top of an iron-coated germanium crystal, thus enabling the identification of changes in lubricant chemistry and/or formation of any reaction layer upon sliding. While the experimental results did not indicate any thermo- or tribo-chemical reactions of ZnDTP on iron at temperatures up to 353 K, the ATR/IR spectra demonstrated a rearrangement of the molecule at 423 K, which led to the generation of P–O–P moieties under purely thermal conditions (indicative of the formation of polyphosphates), and the formation of phosphate groups during tribological testing.

As engine exhaust after-treatment systems for diesel and gasoline engines are harmed by phosphorus, sulfur, and zinc, significant research has been performed over the last two decades on identifying chemical compounds that could replace ZnDTP in its multifactorial roles [140]. Rossi et al. combined ATR/IR tribometry, XPS, and temperature-programmed reaction spectroscopy to shed light on the surface reactivity of a metal-free thiophosphate, namely tributyl thiophosphate (TBT), on air-oxidized iron [27]. The spectroscopic results indicated the thermo- and tribo-chemical formation of reaction layers made of iron polyphosphate and sulfate. However, the sulfate concentration was found to be higher in the reaction layers generated upon sliding.

The authors of this chapter further exploited ATR/IR spectroscopy to evaluate in situ the surface reactivity of triphenyl phosphorothionate (TPPT) on air-oxidized iron under pure thermal and tribological conditions [29–32]. The ATR/IR spectra obtained while heating at 423 K a solution of TPPT in synthetic oil (poly-α-olefin, PAO) on iron-coated germanium ATR crystal (Fig. 5.14) indicated the formation of

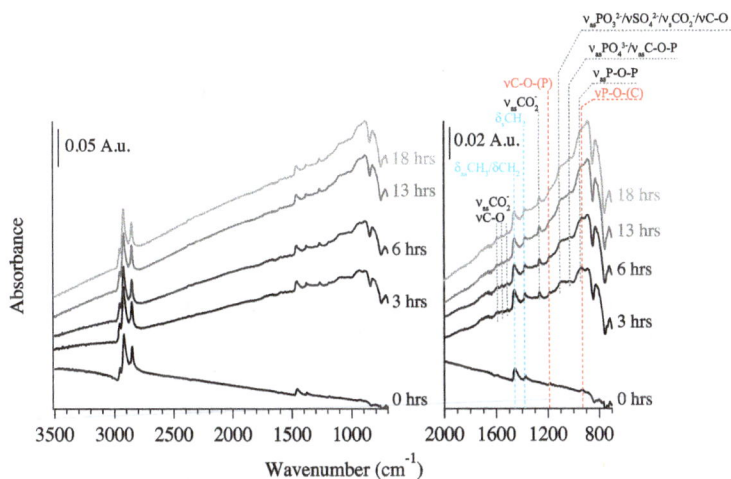

Fig. 5.14: ATR/IR spectra collected during a thermal test performed at 423 K in the presence of a solution of TPPT in PAO (concentration of TPPT: 0.044 mol dm^{-3}) on a germanium ATR crystal coated with iron (10 nm) [30]. Left: whole spectral region. Right: fingerprint region. The ATR/IR spectrum of the unheated solution (0 h) showed the characteristic peaks of TPPT (red vertical lines) and PAO (blue vertical lines) [2, 141–143]. Upon heating at 423 K, additional bands were detected in the fingerprint region: in the 1600–1500 cm^{-1} region, where the characteristic vibration of carboxylate salts ($v_{as}CO_2^-$) as well as the vibrations of monodentate ($v_{as}CO_2^-$) and bidentate ($vC-O$) carbonate complexes occur [5, 143–147], three peaks appeared at 1588, 1551, and 1512 cm^{-1}. An absorption band was detected at 1263 cm^{-1} and assigned to the bidentate $v_{as}CO_2^-$ vibration in carbonate complexes [5, 143–147]. A broad feature appeared at 1087 cm^{-1} and had a shoulder on the lower wavenumber side, which increased in intensity with heating time and became a well-defined peak at 1022 cm^{-1} after 13 h of heating at 423 K. These two bands at 1087 and 1022 cm^{-1} could be assigned to the stretching vibration of PO_3^{2-}/SO_4^{2-} and $PO_4^{3-}/C-O-P$ groups, respectively [143, 148, 149]. A further contribution to the peak at 1087 cm^{-1} might come from the characteristic $v_sCO_2^-$ (bidentate)/$vC-O$ (monodentate) vibrations of carbonate complexes [5, 144–147]. The appearance of a band at 939 cm^{-1}, which could be assigned to the $v_{as}P-O-P$, suggests the generation of pyrophosphate groups at the iron/oil solution interface [143, 148]. The ATR/IR spectra were also characterized by the presence of two peaks at 881 and 822 cm^{-1}, which correspond to the bending vibration of the carbonate anion ($\delta C-O$) [5, 144–147]. While the intensity of the peaks in the fingerprint region increased upon heating at 423 K, a progressive change of the baseline of the spectra was observed. (Reprinted with permission from [30]. Copyright (2011) American Chemical Society).

sulfates, pyrophosphates, and organophosphates together with the generation of carbonates and carboxylates, whose presence derives from the surface adsorption of oxygenated compounds produced as a result of the thermal oxidation of the synthetic oil [30]. The ATR/IR results were corroborated by ex situ angle-resolved XPS analyses, which provided additional information about the in-depth distribution of the reaction products.

To provide a physical basis for the measured baseline changes in the ATR/IR spectra collected during thermal testing, reflectivity calculations were carried out using the matrix formalism developed by Hansen (see Section 5.2.2) [14, 35]. The qualitative comparison between the overall reflectance computed considering a model simulating the growth of a iron phosphate/sulfate layer (Fig. 5.15) with the ATR/IR spectra (Fig. 5.14) suggested that the variations in baseline in the experimental results originate from the generation of an optically different thin film at the iron/oil interface [30].

Even though the development of an ATR/IR tribotester allowed F. Piras et al. to assess the oil/metal interface in situ through a thin iron film (MUATR/IR configuration), which constituted one surface of the tribopair [24–26, 28], no force sensor was included in the experimental apparatus, which did not allow the friction forces to be measured during tribological tests and correlated with any chemical changes occurring at the sliding interface. The authors of this chapter developed a new tribometer that includes a load cell, thus enabling spectroscopic and tribological measurements to be performed concurrently (Fig. 5.16) [29].

The extended ATR/IR tribometer was employed for the in situ study of the tribochemical reactivity of TPPT on air-oxidized iron [31, 32]. While the tribological results indicated that TPPT is not an effective friction modifier, the ATR/IR spectra acquired in situ during sliding experiments performed at 423 K on air-oxidized iron indicated the surface reaction of TPPT to form ortho- and pyro-phosphates, organophosphates, and sulfates (Fig. 5.15) together with the thermo-oxidative aging of the base oil, which led to the generation of carbonate and carboxylate groups. The decrease in absorbance of the characteristic ATR/IR absorption features of the reaction layer upon sliding (Fig. 5.17) provided evidence that the rate at which the surface layer formed was lower than its removal rate.

Altogether, the outcomes of the in situ ATR/IR study demonstrated that the composition of the reaction layers formed by TPPT significantly differs from the composition of the boundary layer generated by ZnDTP, while demonstrating the usefulness of the approach for evaluating tribochemical phenomena and correlating them with changes in friction and/or wear.

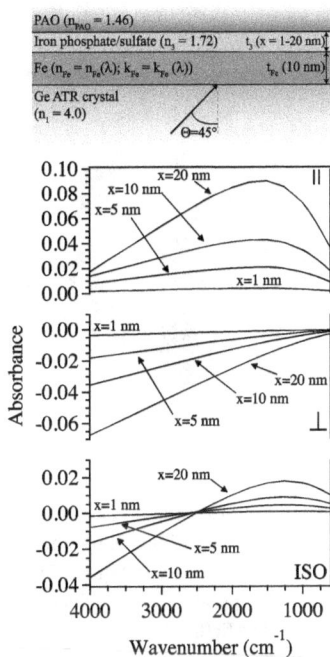

Fig. 5.15: Reflectivity calculation for parallel, perpendicular, and isotropically polarized light considering a model that simulates the growth of a reaction layer made of iron phosphate/sulfate on a germanium ATR crystal coated with iron (10 nm) [30]. In the model, the first (germanium ATR crystal) and last (PAO) layers were assumed to be semi-infinite, transparent (i.e., $k = 0$), and with refractive indexes of 4.0 and 1.46 [150, 151], respectively. The calculations accounted for the dependence of the refractive index and extinction coefficient of iron on the wavelength (Fig. 5.9a) [58]. As for the reaction layer, the average value of the refractive index of the two compounds [150] constituting the phase was calculated. Reflectivity calculations were carried out for both parallel and perpendicular polarized light. The results are displayed in absorbance ($A = -\log(R/R_0)$, where R and R_0 are the reflectivities of the sample and of the reference, respectively. The reflectivity of the reference was computed considering the presence of no reaction layer on top of the iron-coated Ge ATR crystal). After computing the reflectivity of parallel ($R_{//}$) and perpendicular ($R_\perp$) polarized light, the reflectivity of isotropically polarized light was calculated as $R = mR_{//} + (1-m)R_\perp$, where m, which describes the polarization state of the spectrometer and should be in principle frequency-dependent, was assumed to be equal to 0.5 over the whole spectral range. (Reprinted with permission from [30]. Copyright (2011) American Chemical Society).

5.5.4 Recent advances in time-resolved ATR/IR spectroscopy

While traditional FT-IR spectrometers rely on Globar IR sources, the recent development of mid-IR laser sources with a wide spectral coverage and fast tuning speed
has opened the path toward the acquisition of absorption spectra at high rates
[153], thus allowing time-resolved chemical kinetic studies to be performed. Among

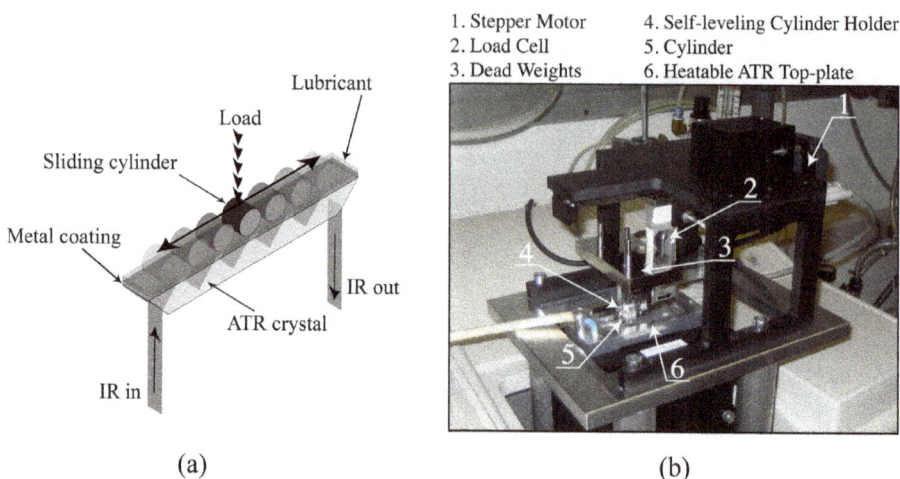

(a)   (b)

Fig. 5.16: (a) Principle of ATR/IR tribometry: a fixed cylinder slides in a reciprocating motion across the metal-coated ATR crystal. The periodic acquisition of ATR/IR spectra during sliding experiments allows chemical changes occurring at the metal/lubricant interface to be identified and quantified as a function of critical tribological parameters, such as applied load, temperature, and sliding speed, and to be correlated with friction results; (b) picture of the in situ ATR/IR tribometer on top of the ATR/IR (Reprinted with permission from [29]. Copyright (2012) Springer).

the several laser sources in the mid-IR region that have already been introduced, such as sources based on nonlinear crystals, lead-salt diode lasers, and quantum cascade lasers (QCLs) [154], QCLs have attracted considerable attention owing to their compactness and robustness. While QCLs have already been employed for time-resolved speciation analyses [155–162], they do have a narrow spectral range, which constrains their use to single species detection. Recently, QCL-based dual-comb spectroscopic (DCS) tools have been developed for mid-IR absorption measurements [163]. These spectrometers are based on the principle of DCS [164], which uses two optical frequency combs having slightly different repetition rates. The two combs interfere on an IR detector and generate an RF comb spectrum with a spacing equal to the difference in repetition rates. Spectroscopic measurements are performed by passing one of the combs through the sample. The absorption of radiation by the sample induces an attenuation of the comb intensity, and thus a change in the RF beat signal. The powerfulness of this approach relies in the fact that the minimum acquisition time reaches sub-microsecond, while the typical values of frequency resolution and comb spacing are $<10^{-4}$ and $\sim 0.3\ \mathrm{cm}^{-1}$.

A table-top QCL-based dual-comb spectrometer has recently been made commercially available and includes two broadband QCL frequency combs that enable a spectral coverage larger than 60 cm^{-1} with a wavenumber center across the mid-IR [165–170]. Even though QCL-based DCS has already been exploited for monitoring the kinetics of formaldehyde formation and oxidation at high temperatures in a

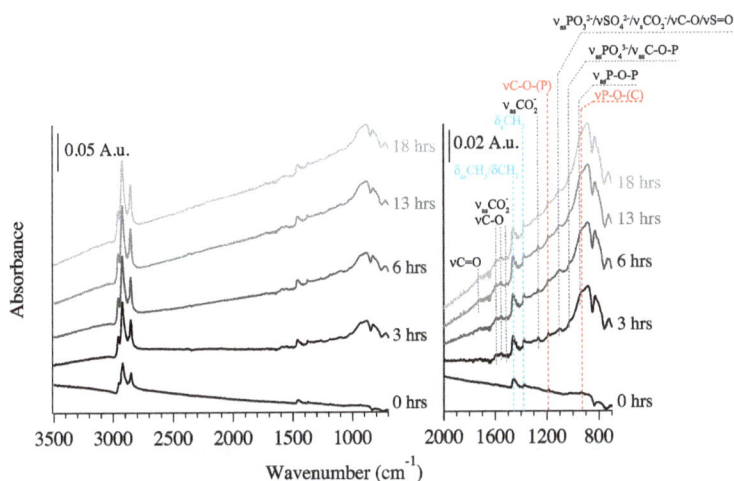

Fig. 5.17: ATR/IR spectra acquired during a tribological experiment (normal load: 6.9 N; sliding speed: 0.5 mm/s; countersurface: 18Cr10Ni steel) at 423 K in the presence of a solution of TPPT in PAO (concentration of TPPT: 0.044 mol dm^{-3}) on a germanium ATR crystal coated with iron (10 nm) [31, 32]. Left: whole spectral region. Right: fingerprint region. The spectrum of the unheated solution (0 h) showed the characteristic peaks of TPPT (red vertical lines) and PAO (blue vertical lines) [2, 141–143]. Upon sliding for 3 h at 423 K, new bands were detected in the fingerprint region. In the 1600–1500 cm^{-1} region, where the characteristic vibration of carboxylate salts ($v_{as}CO_2^-$) and the vibrations of monodentate ($v_{as}CO_2^-$) and bidentate ($vC - O$) carbonate complexes are found [5, 143–147], three peaks appeared at 1588, 1550, and 1511 cm^{-1}. An absorption band at 1265 cm^{-1} was also detected and assigned to the $v_{as}CO_2^-$ (bidentate) vibration in carbonate complexes [5, 143–147]. Additionally, a broad band appeared at 1098 cm^{-1}, which could be assigned to the stretching vibration of PO_3^{2-}/SO_3^{2-} groups [143, 148, 149]. A contribution to this peak at 1098 cm^{-1} might originate from the $v_{as}CO_2^-$ (bidentate)/$vC - O$ (monodentate) vibrations of carbonate complexes [5, 144–147] and from the characteristic stretching vibration of the S = O group in organic sulfoxides [2, 143, 152]. The detection of a band at 938 cm^{-1}, assigned to the $v_{as}P - O - P$ [143, 148], suggests the generation of pyrophosphate groups at the iron/oil interface. The peak at 938 cm^{-1} was found to be asymmetric and to have a shoulder on the high-wavenumber side, which could be assigned to the stretching vibration of PO_4^{3-}/C–O–P groups [143, 148, 149]. The ATR/IR spectra also showed the presence of two strong peaks at 880 and 820 cm^{-1}, which correspond to the bending vibration of the carbonate anion ($\delta C - O$) [5, 144–147]. The intensity of the absorption bands appearing in the fingerprint region of the spectra after 3 h of sliding decreased upon rubbing at 423 K. After 13 h a new weak peak, which became more intense after 18 h, was detected at 1722 cm^{-1} and assigned to the C = O stretching vibration [2, 142, 143]. At the end of the test, the characteristic $vC - O - (P)$ vibration of the TPPT molecule at 1187 cm^{-1} could still be clearly detected. (Reprinted with permission from [29]. Copyright (2012) Springer.).

shock tube [169] as well as carrying out real-time breath analysis as a noninvasive tool in medical diagnostics [171], its application to polymer science is still very limited. A QCL-based dual-comb spectrometer equipped with an ATR accessory was recently employed to carry out noninvasive in situ measurements of the curing process of an UV-activated adhesive [172]. The spectra, which were acquired with a time increment of

25 ms, allowed the kinetics of the curing process to be determined through the quantification of its characteristic lifetime value. While additional experiments are required to identify the reaction pathway, these preliminary results clearly demonstrate the potential of QCL-based dual-comb spectrometers for addressing significant scientific questions of technological relevance across several sectors, including in polymer science and technology.

References

[1]   Stuart BH. Infrared Spectroscopy: Fundamentals and Applications, 1st, New York, John Wiley & Sons, 2004.
[2]   Colthup NB, Daly LH, Wiberley SE. Introduction to Infrared and Raman Spectroscopy, 3rd, London, Academic Press, 1990.
[3]   Griffiths PR, De Haseth JA. Fourier Transform Infrared Spectroscopy, 1st, John Wiley & Sons, New York, 1986.
[4]   Kellner R, Mermet J-M, Otto M, Valcárcel M, Widmer HM, eds Analytical Chemistry: A Modern Approach to Analytical Science, 2nd, New York, John Wiley & Sons, 2004.
[5]   Nakamoto K. Infrared and Raman Spectra of Inorganic and Coordination Compounds, Part A: Theory and Applications in Inorganic Chemistry, 5th, New York, John Wiley & Sons, 1997.
[6]   Smith BC. Fundamentals of Fourier Transform Infrared Spectroscopy, Boca Raton, CRC Press, 1996.
[7]   Harrick NJ. Study of physics and chemistry of surfaces from frustrated total internal reflections. Phys Rev Lett 1960, 4(5), 224.
[8]   Harrick NJ. Internal Reflection Spectroscopy, New York, Interscience Publishers, 1967.
[9]   Fahrenfort J. Attenuated total reflection: A new principle for the production of useful infra-red reflection spectra of organic compounds. Spectrochim Acta 1961, 17(7), 698–709.
[10]   Hansen WN, Delahay P, Tobias CW, eds. Advances in Electrochemistry and Electrochemical Engineering, New York, John Wiley & Sons, 1973.
[11]   Mirabella FM, ed. Internal Reflection Spectroscopy: Theory and Applications, New York, Dekker, 1993.
[12]   Mirabella FM, ed. Modern Techniques in Applied Molecular Spectroscopy, New York, John Wiley & Sons, 1998.
[13]   Suëtaka W, Yates JTJ. Surface Infrared and Raman Spectroscopy: Methods and Applications, New York, Plenum Press, 1995.
[14]   Tolstoy VP, Chernyshova I, Skryshevsky VA. Handbook of Infrared Spectroscopy of Ultrathin Films, 1st, John Wiley & Sons, New York, 2003.
[15]   Fringeli UP, Baurecht D, Siam M, et al. ATR spectroscopy of thin films. In: Nalwa HS, ed. Handbook of Thin Film Materials, San Diego, USA, Academic Press, 2002, 191–229.
[16]   Urban MW. Attenuated Total Reflectance Spectroscopy of Polymers: Theory and Practice, Washington DC, American Chemical Society, 1996.
[17]   Hansen WN. Internal reflection spectroscopy in electrochemistry. In: Muller RH, ed. Optical Techniques in Electrochemistry, New York, John Wiley & Sons, 1973, 1–60.
[18]   Fringeli UP, Günthard H, Baurecht D. Biophysical infrared modulation spectroscopy. In: Gremlich HU, Yan B, eds. Infrared and Raman Spectroscopy of Biological Materials, New York/Basel, Marcel Dekker AG, 2000, 143–192.

[19] Bürgi T. ATR-IR spectroscopy at the metal-liquid interface: Influence of film properties on anomalous band-shape. Phys Chem Chem Phys 2001, 3(11), 2124–2130.
[20] Bürgi T. In situ spectroscopy of catalytic solid–liquid interfaces and chiral surfaces. CHIMIA Int J Chem 2003, 57, 623–627.
[21] Bürgi T, Baiker A. Attenuated total reflection infrared spectroscopy of solid catalysts functioning in the presence of liquid-phase reactants. In: Bruce CG, Helmut K, eds. Advances in Catalysis, Academic Press, 2006, 227–283.
[22] Bürgi T, Wirz R, Baiker A. In situ attenuated total reflection infrared spectroscopy: a sensitive tool for the investigation of reduction-oxidation processes on heterogeneous Pd metal catalysts. J Phys Chem B 2003, 107(28), 6774–6781.
[23] Eggleston CM, Hug S, Stumm W, Sulzberger B, Dos Santos Afonso M. Surface Complexation of Sulfate by Hematite Surfaces: FTIR and STM Observations. Geochim Cosmochim Acta 1998, 62(4), 585–593.
[24] Piras FM, Rossi A, Spencer ND. Growth of tribological films: in situ characterization based on attenuated total reflection infrared spectroscopy. Langmuir 2002, 18(17), 6606–6613.
[25] Piras FM, Rossi A, Spencer ND In situ attenuated total reflection (ATR) spectroscopic analysis of tribological phenomena. In: Dowson D, ed. Boundary and Mixed Lubrication: Science and Applications, Proceedings of the 28th Leeds-Lyon Symposium. Lyon: Elsevier Science B.V.; 2002:199–206.
[26] Piras FM, Rossi A, Spencer ND. Combined *in situ* (ATR FT-IR) and *ex situ* (XPS) Study of the ZnDTP-iron surface interaction. Tribol Lett 2003, 15(3), 181–191.
[27] Rossi A, Piras FM, Kim D, Gellman AJ, Spencer ND. Surface reactivity of tributyl thiophosphate: Effects of temperature and mechanical stress. Tribol Lett 2006, 23(3), 197–208.
[28] Piras FM. In situ attenuated total reflection tribometry. Ph.D. Thesis. Department of Materials. Zurich: ETH Zurich; 2002:248.
[29] Mangolini F, Rossi A, Spencer N. In situ attenuated total reflection (ATR/FT-IR) Tribometry: A powerful tool for investigating tribochemistry at the lubricant-substrate interface. Tribol Lett 2012, 45(1), 207–218.
[30] Mangolini F, Rossi A, Spencer ND. Chemical reactivity of triphenyl phosphorothionate (TPPT) with Iron: An ATR/FT-IR and XPS Investigation. J Phys Chem C 2011, 115(4), 1339–1354.
[31] Mangolini F, Rossi A, Spencer ND. Tribochemistry of triphenyl phosphorothionate (TPPT) by in situ attenuated total reflection (ATR/FT-IR) Tribometry. J Phys Chem C 2012, 116(9), 5614–5627.
[32] Mangolini F. Reactivity of environmentally compatible lubricant additives: An in situ and ex situ investigation. Ph.D. Thesis. Department of Materials. Zurich: Swiss Federal Institute of Technology (ETH); 2011:517.
[33] Hecht E. Optics, San Francisco, Pearson Education Inc., 2002.
[34] Abèles F. Investigation on propagation of sinusoidal electromagnetic wave in stratified media application to thin film. Ann Phys (Paris) 1950, 5, 596–640.
[35] Hansen WN. Electric fields produced by the propagation of plane coherent electromagnetic radiation in a stratified medium. J Opt Soc Am 1968, 58(3), 381–390.
[36] Yamamoto K, Ishida H. Interpretation of reflection and transmission spectra for thin films: reflection. Appl Spectrosc 1994, 48(7), 775–787.
[37] Hirschfeld T. Relationships between the Goos-Hänchen shift and the effective thickness in attenuated total reflection spectroscopy. Appl Spectrosc 1977, 31(3), 243–244.
[38] ASTM International. Standard Practices for Internal Reflection Spectroscopy. ASTM International; 2001.

[39] Hansen WN. Expanded formulas for attenuated total reflection and the derivation of absorption rules for single and multiple ATR spectrometer cells. Spectrochim Acta 1965, 21(4), 815–833.

[40] Müller G, Abraham K, Schaldach M. Quantitative ATR spectroscopy: Some basic considerations. Appl Opt 1981, 20(7), 1182–1190.

[41] Müller GJ, Abraham-Fuchs K. Matrix dependence in single and multilayer IRS spectra. Optik 1991, 88(3), 83–95.

[42] Müller GJ, Abraham-Fuchs K. Matrix dependence in single and multilayer internal reflection spectra. In: Mirabella FM, ed. Internal Reflection Spectroscopy: Theory and Applications, New York, Dekker, 1993.

[43] Callister WDJ. Materials Science and Engineering – an Introduction, 7th, John Wiley & Sons, New York, 2007.

[44] Nunn S, Nishikida K. Advanced ATR Correction Algorithm. Thermo Fisher Scientific Inc. Application Note 50581. 2008.

[45] Belali R, Vigoureux J-M, Morvan J. Dispersion effects on infrared spectra in attenuated total reflection. J Opt Soc Am B 1995, 12(12), 2377–2381.

[46] Bubert H, Jenett H, eds Surface and Thin Film Analysis: Principles, Instrumentation, Applications, 1st, New York, John Wiley & Sons, 2002.

[47] Ryczkowski J. IR spectroscopy in catalysis. Catal Today 2001, 68(4), 263–381.

[48] Ishida KP, Griffiths PR. Theoretical and experimental investigation of internal reflection at thin copper films exposed to aqueous solutions. Anal Chem 1994, 66(4), 522–530.

[49] Zippel E, Kellner R, Breiter MW. Instability of sputtered platinum films on germanium reflection elements in aqueous solutions. J Electroanal Chem 1990, 289(1–2), 297–298.

[50] Osawa M. Surface-enhanced infrared absorption. In: Kawata S, ed. Near-Field Optics and Surface Plasmon Polaritons, Berlin/Heidelberg, Springer, 2001, 163–187.

[51] Osawa M. Surface-Enhanced Infrared Absorption Spectroscopy, John Wiley & Sons, Ltd, 2006.

[52] Hartstein A, Kirtley JR, Tsang JC. Enhancement of the infrared absorption from molecular monolayers with thin metal overlayers. Phys Rev Lett 1980, 45(3), 201.

[53] Garnett JCM. Colours in metal glasses and in metallic films. Philosophical Transactions of the Royal Society of London Series A. Containing Papers Math Phys Character 1904, 203 (359–371), 385–420.

[54] Bruggeman DA. Berechnung verschidener physikalischer konstanten von heterogenen substanzen. Ann Phys, NY 1935, 24, 636–679.

[55] Niklasson GA, Granqvist CG. Optical properties and solar selectivity of coevaporated Co-Al $_2O_3$ composite films. J Appl Phys 1984, 55(9), 3382–3410.

[56] Ishiguro K, Kuwahara G. Determination of the optical constants of Ag films from the measurements of intensity and phase change [I]. J Phys Soc Japan 1951, 6(2), 71, 75.

[57] Fahsold G, Bartel A, Krauth O, Magg N, Pucci A. Infrared optical properties of ultrathin Fe films on MgO(001) beyond the percolation threshold. Phys Rev B 2000, 61(20), 14108.

[58] Ordal MA, Bell R,J, Alexander RW, Newquist LA, Querry MR. Optical properties of Al, Fe, Ti, Ta, W, and Mo at submillimiter wavelengths. Appl Opt 1988, 27(6), 1203–1209.

[59] Bockris JOM, Khan SUM. Surface Electrochemistry: A Molecular Level Approach, New York, Plenum Press, 1993.

[60] Boerio FJ. Measurements of the chemical characteristics of polymers and rubbers by vibrational spectroscopy. In: Chalmers JM, Griffiths PR, eds. Handbook of Vibrational Spectroscopy, Chichester, John Wiley & Sons, 2002, 2419–2536.

[61] Bower DI, Maddams WF. The Vibrational Spectroscopy of Polymers, Cambridge, Cambridge University Press, 1989.

[62] Chalmers JM. Infrared spectroscopy in analysis of polymers and rubbers. In: Meyers RA, ed. Encyclopedia of Analytical Chemistry, Chichester, John Wiley & Sons, 2000, 7702–7759.
[63] Chalmers JM, Everall NJ. Qualitative and quantitative analysis of polymers and rubbers by vibrational spectroscopy. In: Chalmers JM, Griffiths PR, eds. Handbook of Vibrational Spectroscopy, Chichester, John Wiley & Sons, 2002, 2389–2418.
[64] Chalmers JM, Hannah RW, Mayo DW. Spectra–structure correlations: polymer spectra. In: Chalmers JM, Griffiths PR, eds. Handbook of Vibrational Spectroscopy, Chichester, John Wiley & Sons, 2002, 1893–1918.
[65] Koenig JL. Spectroscopy of Polymers, 2nd, Amsterdam, Elsevier, 1999.
[66] Siesler HW, Holland-Moritz K. Infrared and Raman Spectroscopy of Polymers, New York, Marcel Dekker, 1980.
[67] Stuart BH. Polymer Analysis, Chichester, John Wiley & Sons, 2002.
[68] Dothee D, Vigoureux J-M, Camelot M. Application of attenuated total reflection (ATR) infrared spectroscopy to the study of microstructural changes in high density polyethylene during sliding friction. Polym Degrad Stab 1988, 22(2), 161–174.
[69] Mirabella FM. Surface orientation of polypropylene. I. Theoretical considerations for the application of internal reflection spectroscopy. J Polym Sci Polym Phys Ed 1984, 22(7), 1283–1291.
[70] Mirabella FM. Surface orientation of polypropylene. II. Determination for uniaxially and biaxially oriented films using internal reflection spectroscopy. J Polym Sci Polym Phys Ed 1984, 22(7), 1293–1304.
[71] Nishio E, Morimoto M, Nishikida K. Nondestructive quantitative analysis of ethylene-propylene blend polymer by the NIR/ATR/FT-IR Method. Appl Spectrosc 1990, 44(10), 1639–1640.
[72] Walls DJ. Application of ATR-IR to the analysis of surface structure and orientation in uniaxially drawn poly(ethyleneterephthalate). Appl Spectrosc 1991, 45(7), 1193–1198.
[73] Lovinger A, Reed D. Inhomogeneous thermal degradation of poly(vinylidene fluoride) crystallized from the melt. Macromolecules 1980, 13(4), 989–994.
[74] Wentink T, Willwerth LJ, Phaneuf JP. Properties of polyvinylidene fluoride. Part II. Infrared transmission of normal and thermally decomposed polymer. J Polym Sci 1961, 55(162), 551–562.
[75] Kuhn KJ, Hahn B, Percec V, Urban MW. Structural and quantitative analysis of surface modified poly(vinylidene fluoride) films using ATR FT-IR Spectroscopy. Appl Spectrosc 1987, 41(5), 843–847.
[76] Urban MW, Salazar-Rojas EM. Ultrasonic PTC modification of poly(vinylidene fluoride) surfaces and their characterization. Macromolecules 1988, 21(2), 372–378.
[77] Brandrup J, Immergut EH. Polymer Handbook, New York, John Wiley & Sons, 1989.
[78] Nguyen T, Byrd E, Lin C. A spectroscopic technique for in situ measurement of water at the coating/metal interface. J Adhes Sci Technol 1991, 5(9), 697–709.
[79] Urban MW, Stewart MT. DMA and ATR FT-IR studies of gas plasma modified silicone elastomer surfaces. J Appl Polym Sci 1990, 39(2), 265–283.
[80] Flournoy PA. Attenuated total reflection from oriented polypropylene films. Spectrochim Acta 1966, 22(1), 15–20.
[81] Ishida H, Scott C. Fast polymerization and crystallization kinetic studies of nylon 6 by combined use of computerized micro-rim machine and FT-IR. J Polym Eng 1986, 201.
[82] Snyder RW, Fuerniss SJ. ATR/IR spectroscopic method for following photo-polymer curing. Appl Spectrosc 1992, 46(7), 1113–1116.
[83] Mathis CH, Divandari M, Simic R, et al. ATR-IR investigation of solvent interactions with surface-bound polymers. Langmuir 2016, 32(30), 7588–7595.

[84] Nomura A, Okayasu K, Ohno K, Fukuda T, Tsujii Y. Lubrication mechanism of concentrated polymer brushes in solvents: effect of solvent quality and thereby swelling state. Macromolecules 2011, 44(12), 5013.

[85] Chen M, Briscoe WH, Armes SP, Klein J. Lubrication at physiological pressures by polyzwitterionic brushes. Science (Washington, DC, U S) 2009, 323(5922), 1698.

[86] Forster AM, Mays JW, Kilbey SM. Effect of temperature on the frictional forces between polystyrene brushes. J Polym Sci, Part B: Polym Phys 2006, 44(4), 649.

[87] Karim A, Satija SK, Douglas JF, Ankner JF, Fetters LJ. Neutron reflectivity study of the density profile of a model end-grafted polymer brush: influence of solvent quality. Phys Rev Lett 1994, 73(25), 3407.

[88] Sui X, Chen Q, Hempenius MA, Vancso GJ. Probing the collapse dynamics of poly(N-isopropylacrylamide) brushes by AFM: effects of co-nonsolvency and grafting densities. Small 2011, 7(10), 1440.

[89] Nalam PC, Daikhin L, Espinosa-Marzal RM, Clasohm J, Urbakh M, Spencer ND. Two-fluid model for the interpretation of quartz crystal microbalance response: tuning properties of polymer brushes with solvent mixtures. J Phys Chem C 2013, 117, 4533.

[90] Farzaneh A, Zhou M, Potapova E, et al. Adsorption of water and butanol in silicalite-1 film studied with in situ attenuated total reflectance–Fourier transform infrared spectroscopy. Langmuir 2015, 31(17), 4887.

[91] Hassler N, Baurecht D, Reiter G, Fringeli UP. In situ FTIR ATR spectroscopic study of the interaction of immobilized human serum albumin with cholate in aqueous environment. J Phys Chem C 2011, 115(4), 1064.

[92] Lin S, Yuk H, Zhang T, et al. Stretchable hydrogel electronics and devices. Adv Mater 2016, 28, 4497.

[93] Xu Y, Lin Z, Huang X, Liu Y, Huang Y, Duan X. Flexible solid-state supercapacitors based on three-dimensional graphene hydrogel films. ACS Nano 2013, 7, 4042.

[94] Hines L, Petersen K, Lum GZ, Sitti M. Soft actuators for small-scale robotics. Adv Mater 2017, 29, 1603483.

[95] Ionov L. Hydrogel-based actuators: possibilities and limitations. Mater Today 2014, 17, 494.

[96] Liu Z, Calvert P. Multilayer hydrogels as muscle-like actuators. Adv Mater 2000, 12, 288.

[97] Li J, Mooney DJ. Designing hydrogels for controlled drug delivery. Nat Rev Mater 2016, 1, 16071.

[98] Annabi N, Tamayol A, Uquillas JA, et al. 25th Anniversary article: rational design and applications of hydrogels in regenerative medicine. Adv Mater 2014, 26, 85.

[99] Slaughter BV, Khurshid SS, Fisher OZ, Khademhosseini A, Peppas NA. Hydrogels in regenerative medicine. Adv Mater 2009, 21, 3307.

[100] Mitragotri S, Lahann J. Physical Approaches to Biomaterial Design. Nat Mater 2009, 8, 15.

[101] Zhang YS, Khademhosseini A. Advances in Engineering Hydrogels. Science 2017, 356, eaaf3627.

[102] Gong JP, Kurokawa T, Narita T, et al. Synthesis of hydrogels with extremely low surface friction. J Am Chem Soc 2001, 123, 5582.

[103] Kii A, Xu J, Gong JP, Osada Y, Zhang X. Heterogeneous polymerization of hydrogels on hydrophobic substrate. J Phys Chem B 2001, 105, 4565.

[104] Lin P, Zhang R, Wang X, et al. Articular cartilage inspired bilayer tough hydrogel prepared by interfacial modulated polymerization showing excellent combination of high load-bearing and low friction performance. ACS Macro Lett 2016, 5, 1191.

[105] Peng M, Ping Gong J, Osada Y. Substrate effect on the formation of hydrogels with heterogeneous network structure. Chem Rec 2003, 3(1), 40–50.

[106] Peng M, Kurokawa T, Gong JP, Osada Y, Zheng Q. Effect of surface roughness of hydrophobic substrate on heterogeneous polymerization of hydrogels. J Phys Chem B 2002, 106(12), 3073–3081.

[107] Gombert Y, Simič R, Roncoroni F, Dübner M, Geue T, Spencer ND. Structuring hydrogel surfaces for tribology. Adv Mater Interfaces 2019.

[108] Zhang K, Simič R, Yan W, Spencer ND. Creating an interface: rendering a double-network hydrogel lubricious via spontaneous delamination. ACS Appl Mater Interfaces 2019, 11(28), 25427–25435.

[109] Meier YA, Zhang K, Spencer ND, Simič R. Linking friction and surface properties of hydrogels molded against materials of different surface energies. Langmuir 2019, 35(48), 15805–15812, doi: 10.1021/acs.langmuir.9b01636.

[110] Zhang K, Yan W, Simič R, Benetti EM, Spencer ND. Versatile surface modification of hydrogels by surface-initiated, Cu(0)-mediated controlled radical polymerization. ACS Appl Mater Interfaces 2020, 12(5), 6761–6767.

[111] Simič R, Yetkin M, Zhang K, Spencer ND. Importance of hydration and surface structure for friction of acrylamide hydrogels. Tribol Lett 2020, 68, 2.

[112] Zhang K, Simič R, Spencer ND. Imparting ultralow lubricity to double-network hydrogels by surface-initiated controlled radical polymerization under ambient conditions. Biotribology 2021.

[113] Gellman AJ, Spencer ND. Surface chemistry in tribology. J Eng Tribol 2002, 216(6), 443–461.

[114] Hutchings IM. Tribology, Friction and Wear of Engineering Materials, Oxford, Butterworth-Heinemann, 1992.

[115] Mang T, Dresel W, eds Lubricants and Lubrication, 2nd, New York, John Wiley & Sons, 2007.

[116] Mortier RM, Orszulik ST, eds Chemistry and Technology of Lubricants, 2nd, Blakie Academic & Professional, London, 1997.

[117] Pawlak Z. Tribochemistry of Lubricating Oils, Amsterdam, Elsevier, 2003.

[118] Rudnick LR, ed. Lubricant Additives: Chemistry and Applications, New York, Marcel Dekker, Inc., 2003.

[119] Nicholls MA, Do T, Norton PR, Kasrai M, Bancroft GM. Review of the lubrication of metallic surfaces by zinc dialkyl-dithiophosphates. Tribol Int 2005, 38(1), 15–39.

[120] Spikes H. The history and mechanisms of ZDDP. Tribol Lett 2004, V17(3), 469–489.

[121] Crobu M, Rossi A, Mangolini F, Spencer ND. Tribochemistry of bulk zinc metaphosphate glasses. Tribol Lett 2010, 39(2), 121–134.

[122] Eglin M, Rossi A, Spencer ND Additive-surface interaction in boundary lubrication: a combinatorial approach. In: Dowson D, ed. Boundary and Mixed Lubrication: Science and Applications, Proceedings of the 28th Leeds-Lyon Symposium. Lyon: Elsevier Science B.V.; 2002.

[123] Eglin M, Rossi A, Spencer ND. X-ray photoelectron spectroscopy analysis of tribostressed samples in the presence of ZnDTP: a combinatorial approach. Tribol Lett 2003, 15(3), 199–209.

[124] Heuberger R, Rossi A, Spencer N. Pressure dependence of ZnDTP tribochemical film formation: a combinatorial approach. Tribol Lett 2007, 28(2), 209–222.

[125] Heuberger R, Rossi A, Spencer ND. XPS study of the influence of temperature on ZnDTP tribofilm composition. Tribol Lett 2007, 25(3), 185–196.

[126] Martin JM, Grossiord C, Le Mogne T, Bec S, Tonck A. The two-layer structure of ZnDTP tribofilms: Part I: AES, XPS and XANES analyses. Tribol Int 2001, 34(8), 523–530.

[127] Ito K, Martin JM, Minfray C, Kato K. Formation mechanism of a low friction ZDDP tribofilm on iron oxide. Tribol Trans 2007, 50(2), 211–216.

[128] Minfray C, Martin JM, Esnouf C, Le Mogne T, Kersting R, Hagenhoff B. A multi-technique approach of tribofilm characterisation. Thin Solid Films 2004, 447–448, 272–277.

[129] Equey S, Roos S, Mueller U, Hauert R, Spencer ND, Crockett R. Tribofilm formation from ZnDTP on diamond-like carbon. Wear 2008, 264(3–4), 316–321.

[130] Gosvami NN, Bares JA, Mangolini F, Konicek AR, Yablon DG, Carpick RW. Mechanisms of antiwear tribofilm growth revealed in situ by single-asperity sliding contacts. Science 2015, 348(6230), 102–106.

[131] Jacobs TDB, Greiner C, Wahl KJ, Carpick RW. Insights into tribology from in situ nanoscale experiments. MRS Bulletin 2019, 44(06), 478–486.

[132] Spikes HA. In situ methods for tribology research. Tribol Lett 2003, 14(1), 1–1.

[133] Sawyer WG, Wahl KJ. Accessing inaccessible interfaces: in situ approaches to materials tribology MRS. Bulletin 2008, 33, 1145–1150.

[134] Donnet C. Problem-solving methods in tribology with surface-specific techniques. In: Rivière JC, Myhra S, eds. Handbook of Surface and Interface Analysis: Methods and Problem-Solving, New York, Marcel Dekker Inc., 1998, 968.

[135] Cann PM. In-contact molecular spectroscopy of liquid lubricant films. MRS Bulletin 2008, 33, 1151–1158.

[136] Cann PM, Spikes HA. In-contact IR spectroscopy of hydrocarbon lubricants. Tribol Lett 2005, 19(4), 289–297.

[137] Olsen JE, Fischer TE, Gallois B. In situ analysis of the tribochemical reactions of diamond-like carbon by internal reflection spectroscopy. Wear 1996, 200(1–2), 233–237.

[138] Sasaki K, Inayoshi N, Tashiro K. Development of new in situ observation system for dynamic study of lubricant molecules on metal friction surfaces by two-dimensional fast-imaging Fourier-transform infrared-attenuated total reflection spectrometer. Rev Sci Instrum 2008, 79 (12), 123702–123707.

[139] Sasaki K, Inayoshi N, Tashiro K. In situ FTIR-ATR observation of phase transition behavior of n-alkane molecules induced by friction motion on a metal interface. J Phys Chem C 2009, 113(8), 3287–3291.

[140] Spikes H. Low- and zero-sulphated ash, phosphorus and sulphur anti-wear additives for engine oils. Lubr Sci 2008, 20(2), 103–136.

[141] Mangolini F, Rossi A, Spencer ND. Reactivity of triphenyl phosphorothionate in lubricant oil solution. Tribol Lett 2009, 35(1), 31–43.

[142] Roeges NPG. A Guide to the Complete Interpretation of Infrared Spectra of Organic Structures, Chichester, John Wiley & Sons, 1994.

[143] Socrates G. Infrared and Raman Characteristic Group Frequencies, 3rd, Chichester, John Wiley & Sons, 2001.

[144] Bargar JR, Kubicki JD, Reitmeyer R, Davis JA. ATR-FTIR spectroscopic characterization of coexisting carbonate surface complexes on hematite. Geochim Cosmochim Acta 2005, 69(6), 1527–1542.

[145] Rèmazeilles C, Refait P. Fe(II) hydroxycarbonate Fe2(OH)2CO3 (chukanovite) as iron corrosion product: synthesis and study by Fourier transform infrared spectroscopy. Polyhedron 2009, 28(4), 749–756.

[146] Roonasi P, Holmgren A. An ATR-FTIR study of carbonate sorption onto magnetite. Surf Interface Anal 2010, 42(6–7), 1118–1121.

[147] Wijnja H, Schulthess CP. Carbonate adsorption mechanism on goethite studied with ATR-FTIR, DRIFT, and proton coadsorption measurements. Soil Sci Soc Am J 2001, 65(2), 324–330.

[148] Efimov AM. IR fundamental spectra and structure of pyrophosphate glasses along the 2ZnO. P_2O_5-$2Me_2O$.P_2O_5 join (Me being Na and Li). J Non Cryst Solids 1997, 209(3), 209–226.

[149] Ross SD. Inorganic Infrared and Raman Spectra, London, McGraw-Hill, 1972.

[150] Lide DR, ed The CRC Handbook of Chemistry and Physics, 72nd, Cleveland, Ohio, USA, CRC Press, 1991–1992.

[151] Oligomers I. Durasyn(R) Polyalphaolefins, South Shore Boulevard, League City, Texas, INEOS Oligomers.

[152] Silverstein RM, Webster FX, Kiemle DJ. Spectroscopic Identification of Organic Compounds, 5th, New York, John Wiley & Sons, 2005.

[153] Griffiths PR, De Haseth JA. Fourier Transform Infrared Spectrometry, 2nd, Hoboken, Wiley, 2007.

[154] Tittel FK, Richter D, Fried A. Mid-infrared laser applications in spectroscopy. In: Sorokina IT, Vodopyanov KL, eds. Solid-State Mid-Infrared Laser Sources, Berlin, Heidelberg, Springer Berlin Heidelberg, 2003, 458–529.

[155] Ren W, Farooq A, Davidson DF, Hanson RK. CO concentration and temperature sensor for combustion gases using quantum-cascade laser absorption near 4.7 μm. Appl Phys B 2012, 107(3), 849–860.

[156] Sun K, Wang S, Sur R, Chao X, Jeffries JB, Hanson RK. Time-resolved in situ detection of CO in a shock tube using cavity-enhanced absorption spectroscopy with a quantum-cascade laser near 4.6μm. Opt Express 2014, 22(20), 24559–24565.

[157] Sajid MB, Es-sebbar E, Javed T, Fittschen C, Farooq A. Measurement of the rate of hydrogen peroxide thermal decomposition in a shock tube using quantum cascade laser absorption near 7.7 μm. Int J Chem Kinet 2014, 46(5), 275–284.

[158] Sajid MB, Javed T, Farooq A. High-temperature measurements of methane and acetylene using quantum cascade laser absorption near 8μm. J Quant Spectrosc Radiat Transf 2015, 155, 66–74.

[159] Nasir EF, Farooq A. Time-resolved temperature measurements in a rapid compression machine using quantum cascade laser absorption in the intrapulse mode. Proc Combust Inst 2017, 36(3), 4453–4460.

[160] Zhang G, Khabibullin K, Farooq A. An IH-QCL based gas sensor for simultaneous detection of methane and acetylene. Proc Combust Inst 2019, 37(2), 1445–1452.

[161] Welzel S, Gatilova L, Röpcke J, Rousseau A. Time-resolved study of a pulsed dc discharge using quantum cascade laser absorption spectroscopy: NO and gas temperature kinetics. Plasma Sources Sci Technol 2007, 16(4), 822–831.

[162] Spearrin RM, Li S, Davidson DF, Jeffries JB, Hanson RK. High-temperature iso-butene absorption diagnostic for shock tube kinetics using a pulsed quantum cascade laser near 11.3μm. Proc Combust Inst 2015, 35(3), 3645–3651.

[163] Consolino L, Nafa M, De Regis M, et al. Quantum cascade laser based hybrid dual comb spectrometer. Commun Phys 2020, 3, 1.

[164] Coddington I, Newbury N, Swann W. Dual-comb spectroscopy. Optica 2016, 3(4), 414–426.

[165] Klocke JL, Mangold M, Allmendinger P, et al. Single-shot sub-microsecond mid-infrared spectroscopy on protein reactions with quantum cascade laser frequency combs. Anal Chem 2018, 90(17), 10494–10500.

[166] Szczepaniak U, Schneider SH, Horvath R, Kozuch J, Geiser M. Vibrational stark spectroscopy of fluorobenzene using quantum cascade laser dual frequency combs. Appl Spectrosc 2019, 74(3), 347–356.

[167] Gianella M, Nataraj A, Tuzson B, et al. High-resolution and gapless dual comb spectroscopy with current-tuned quantum cascade lasers. Opt Express 2020, 28(5), 6197–6208.

[168] Lins E, Read S, Unni B, Rosendahl SM, Burgess IJ. Microsecond resolved infrared spectroelectrochemistry using dual frequency comb IR lasers. Anal Chem 2020, 92(9), 6241–6244.

[169] Fjodorow P, Allmendinger P, Horvath R, et al. Monitoring formaldehyde in a shock tube with a fast dual-comb spectrometer operating in the spectral range of 1740–1790 cm−1. Appl Phys B 2020, 126, 12.

[170] Zhang G, Horvath R, Liu D, Geiser M, Farooq A. QCL-based dual-comb spectrometer for multi-species measurements at high temperatures and high pressures. Sensors (Basel) 2020, 20(12).

[171] Ghorbani R, Schmidt FM. Real-time breath gas analysis of CO and CO_2 using an EC-QCL. Appl Phys B 2017, 123(5), 144, doi: 10.1007/s00340-017-6715-x.

[172] Irag I. Rapid Reaction Monitoring of Curing Processes with the IRis-F1. 2019.

Dalia Yablon

6 Scanning probe microscopy of polymers

6.1 Introduction

Scanning probe microscopy is a field that was invented in the early 1980s with the invention of the scanning tunneling microscopy by Binnig and Rohrer [1]. This field has spawned a powerful set of characterization microscopies that today have been incorporated into the standard toolkit for nanoscale characterization. Atomic force microscopy, or AFM, was invented in 1986 [2] and is perhaps the most versatile and widely used member of this family.

The power of AFM lies in its high-resolution capability to map a variety of material properties on the nanoscale under a variety of conditions including ambient conditions. A schematic of AFM operation is shown in Fig. 6.1 and described briefly. AFM operation is based on tracking the motion of a very small cantilever/tip assembly as it interacts with a sample. The cantilever is typically very small (length of a few hundred microns, width of a tens of microns, and thickness of a few microns), and a very sharp tip is grown off the end of the cantilever. The motion of the cantilever/tip is tracked through an optical detection system comprised of a laser that is bounced off the back of the lever and then detected toward a position-sensitive detector. The AFM

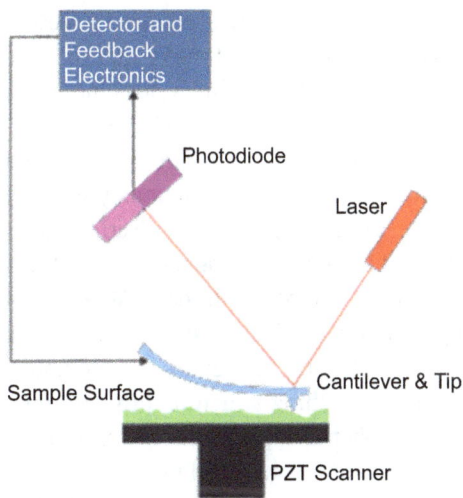

Fig. 6.1: A simplified scheme of an AFM apparatus showing a cantilever/tip assembly in contact with the sample. The cantilever motion is tracked through an optical detection mechanism, whose signal is fed through to the controller allowing for active feedback control on the tip–sample interaction.

https://doi.org/10.1515/9783110701098-006

tip can then sensitively interact with the substrate through a variety of mechanisms to probe a variety of substrate properties.

The series of properties that can be probed by AFM include mechanical, magnetic, electrical, optical/spectroscopic, and thermal; the AFM offers additional capabilities such as lithography and pulling experiments. For polymer applications, the most widely characterized properties by AFM is mechanical such as modulus, adhesion, and viscoelastic moduli. For polymer applications, the AFM now resides alongside optical microscopy and electron microscopy (SEM – scanning electron microscopy and TEM – transmission electron microscopy) as essential tools for characterization. The AFM provides specific advantages over photon and electron-based microscopies to probe materials. Because the AFM provides mechanical interaction between the tip and sample, it often provides contrast in situations where electron or photon-based microscopies struggle or even fail. For example, most polymer materials are composed of primarily carbon and hydrogen so that most chemically sensitive microscopies cannot differentiate between them. However, as long as there is some mechanical difference between the materials (e.g., stiffness or energy dissipation), the AFM will be able to contrast the materials with ease.

There are many AFM imaging modes available for mechanical contrast and specifically to measure polymer materials. The aim of the AFM, like any microscopy, is multifold: mapping, discrimination, and identification. In a multicomponent material, the AFM can discriminate among all the different components that vary in size, morphology, and material properties. Unambiguous identification of the actual mechanical properties of the nanoscale component is sometimes required together with discrimination. The latter relies on the AFM's ability to make quantitative measurements of the parameters of interest (e.g., storage modulus, loss modulus, and loss tangent). There are several reasons why one might want to have a quantitative nanomechanical measurement in polymer studies. The first reason is to aid in the discrimination capability described above to help unambiguously identify components in the AFM image. The second reason is independent of discrimination and is based on the need to measure mechanical properties on the nanoscale. Often, there is no access to a bulk analogue of the material probed with the AFM, and so a thin film or a domain of 10 nm or 1 μm may be all the available material to test. Bulk instruments are not suitable for such measurements, and so a nanoscale method is required. Finally, AFM plays a critical role in establishing structure–property relationships in polymer-based materials. In many polymer applications, the product is designed around a mechanical property requirement – e.g., tensile pulling, impact strength, and tear strength. Often nanoscale and microscale structures play an important role in determining macroscopic properties, and so the ability of the AFM to discriminate and identify the various structural properties of the material on this length scale (e.g., phase segregation, continuity of phases, and dispersion) is important to its ultimate performance.

This chapter will describe the main available AFM capabilities for imaging polymers, illustrated by applications taken from the literature. The chapter begins in Section 6.2 where all AFM measurements begin: sample preparation! Then the most popular method of imaging polymer materials is described in Section 6.3, phase imaging, along with its benefits and shortcomings. Section 6.4 describes advanced, state-of-the-art multifrequency-based imaging. Section 6.5 covers methods measuring mechanical properties of polymers. Section 6.6 is a brief section describing electrical property measurements by AFM. Section 6.7 follows with some "alternative" (not strictly mechanical characterization based) AFM-based methods that are very useful to polymers including nanothermal analysis and spectroscopic-based methods. Finally, Section 6.8 concludes with environmental-based measurements of polymer systems as a function of temperature and deformation, an area in which the AFM provides unique capabilities.

6.2 Sample preparation

AFM imaging requires smooth and flat samples. An AFM can image at most approximately a 100 μm × 100 μm field of view, depending on the instrument and the scanner used. This field of view is actually quite small, especially compared to its electron and photon-based microscopy counterparts. Vertically, the range is confined to features or tilt of less than 15 μm (5 μm in some high-resolution instruments) over the 100 μm × 100 μm region; these limitations are a function of the instrument's hardware.

Due to these limitations, great care has to be taken to prepare smooth and flat samples appropriate to AFM imaging. There are many methods available from the general microscopy community to prepare samples. Preparing smooth thin films through spin coating is a popular method, though not always practical. For materials that are extruded or molded, cryomicrotomy and specifically cryofacing is a preferred method. In cryomicrotomy, the sample is cooled down below the glass transition temperature (T_g) and then cut with a very sharp edge of a glass knife, followed by a sharp edge of a diamond knife. Several commercial vendors make suitable instruments, and the method is described in more detail elsewhere [3]. Cryomicrotomy can yield two kinds of surfaces: a smooth face or a thin section. For AFM imaging, cryofacing – where a pellet or component is simply smoothed out by the knives – is sufficient. Thin sections are essential for TEM imaging but are not required for AFM imaging. Once a sample has been cryomicrotomed, it can be immediately imaged by the AFM without further preparation (e.g. staining, again a common method required for SEM and TEM imaging is not needed for AFM imaging). Samples can age rapidly after cryomicrotoming, especially materials where oxidation and free radical formation upon microtoming can damage the samples. Imaging as soon as possible after cryomicrotoming is generally suggested. The generally minimal sample preparation required by AFM is another one of its many attractive features.

6.3 Phase imaging

6.3.1 Background on phase imaging

Phase imaging is perhaps one of the best known and most widely used modes in AFM. It is a simple method in that it is a channel collected in tapping mode (also called amplitude modulation (AM) mode, intermittent contact mode, and dynamic mode) and requires no post processing. There has been a significant amount of research trying to relate the phase signal to actual material properties. However, phase imaging is rather complicated and interpretation of its meaning is still widely debated and researched [4–7]. Fundamentally, the phase shift measured in a phase image is a measure of the energy dissipated by the cantilever into the material. It is a measure of both conservative and non-conservative tip–sample interaction forces which depends on a number of factors including viscoelasticity, adhesion, capillary forces, and contact area.

Phase shift (φ) is always measured with respect to a reference, and in this case, to the driven cantilever. The cantilever is driven at a resonant frequency and interacts with the sample at a given oscillation amplitude set by the user. The phase shift is then induced by interacting with the sample and is mapped as a channel while the tip raster scans over the surface.

At resonance, the phase lag of a driven harmonic oscillator with damping is 90° with respect to the driving or excitation force; a phase lag of 90° is easily observed at the frequency resonance peak. To understand the concept of phase shift and a 90° shift at resonance, one can try the following experiment at home. Begin by shaking a flexible ruler. Once the ruler is at resonance, watch the motion of the wrist with respect to the peak of the ruler or string. The peak of one's wrist will be 90° out of phase with the peak of the ruler or string's motion at resonance.

Phase imaging is to be used as a qualitative imaging tool to obtain materials-based contrast. As mentioned above, there have been many attempts at extracting mechanical properties from the phase signal such as associating it with the loss tangent [8]. Ultimately, the phase signal is too complex and convoluted to derive any meaningful quantitative mechanical properties [9, 10].

6.3.2 Applications of phase imaging to polymer materials

Phase imaging is most useful to measure the morphology, dispersion, and on occasion, high-resolution structure of multicomponent polymer samples that contain components with widely varying loss tangents and/or moduli [8–10]. The next two figures show examples of polymer blends where phase imaging is ineffective (Fig. 6.2) and effective (Fig. 6.3). A phase image of an 80/20 by weight blend of two plastics, polypropylene (PP) and polystyrene (PS), is shown in Fig. 6.2 where the polystyrene is in the upper half of the image while the PP is in the bottom half of the image; the two domains are separated by a curved border. Note that the phase contrast is not very strongly differentiated for these two materials. This is due to a similarity in storage moduli as well as both materials having very small and similar loss tangents of <0.1.

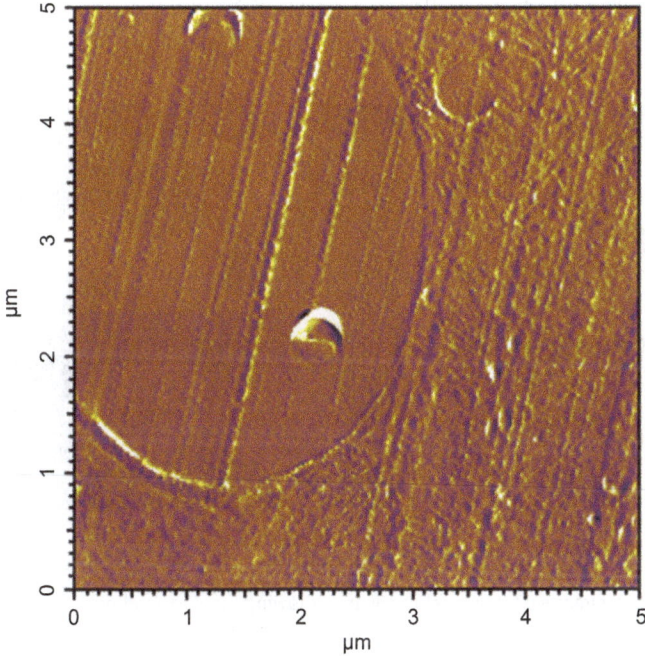

Fig. 6.2: A 5 μm × 5 μm phase image of a blend of polypropylene and polystyrene. The polystyrene is in the left half of the image, differentiated by a smooth topography. The right half of the image contains the polypropylene with a more textured topography. There is a pronounced curved border between the two materials, primarily due to the delamination between the two materials which induces a topographical depression and hence artifact in the phase image.

In contrast to the example in Fig. 6.2, a good example of the utility of phase imaging can be seen in its application to image a category of thermoplastic olefins called impact copolymer (ICP or impact modified polymer) that contains a stiff thermoplastic matrix and elastomeric rubber domains. These materials are toughened plastics with little elasticity but with excellent impact toughness properties.

The example shown in Fig. 6.3 is an AFM phase image of a PP-based ICP with a PP matrix and ethylene–propylene rubber (EPR) domains. The EPR domains are easily identified as the bright, circular regions in the image. The morphology, size, and dispersion of the EPR domains can be quickly ascertained and also quantitatively analyzed, if so desired. These characteristics of the EPR within ICP are essential to determining the material's ultimate mechanical properties such as stiffness, toughness, and strength, which are critical to understanding structure–property relationships as new materials in this class are being developed. High-resolution structures within the ICP are visible in Fig. 6.3b both within the PP and the EPR domains. The lamellar structure within the PP is observed while the EPR shows inclusions indicating a further level of complexity within the EPR. Finally, the interface between the PP and EPR can be explored for mechanical strength, as discussed in Section 6.8.

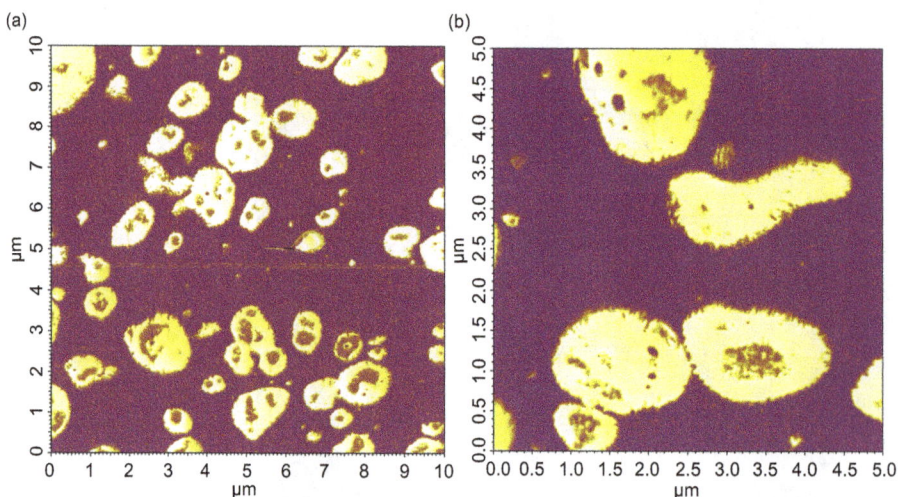

Fig. 6.3: a) 10 μm × 10 μm phase image of an impact copolymer showing the surrounding polypropylene matrix (dark) and the bright patches of ethylene–propylene rubber. Bright inclusions are visible inside the ethylene–propylene rubber. b) High-resolution 5 μm × 5 μm image of the impact copolymer highlighting lamellar structure within the polypropylene and inclusions in the ethylene–propylene rubber.

Another category of polymer materials for which AFM phase imaging has been very popular and useful is block copolymers, where small nanoscale domains of various polymer are assembled together. Beautiful phase images of materials like poly(styrene-b-ethylene-co-butylene-b-styrene) are now routine differentiating the hard and soft blocks [11–14]. Phase imaging was also used to image PolyStyrene-Poly Acrylic Acid (PS-PAA) amphiphilic block copolymer films to visualize the different components [15]. However, the use of phase to discriminate the different components in block copolymers is very extensive and these two areas are just offered as examples.

The limits of resolution continue to be pushed to achieve direct imaging of chains in semicrystalline polymers. One of the first studies of single chains, buried folds, and chain ends was accomplished in the seminal work by Hobbs and coworkers on a polyethylene (PE) film using torsional resonance imaging – where the cantilever's twisting (as opposed to up and down flexural) behavior is measured, providing exquisite sensitivity [16]. Figure 6.4a shows a high-resolution torsional resonance phase image resolving single chains in the (010) surface accompanied by (b) a line profile across these chains showing a spacing of 3.69 Å between chains. The image in Fig. 6.4(c) reveals free loops from the crystalline area into the amorphous region as indicated by the white arrow.

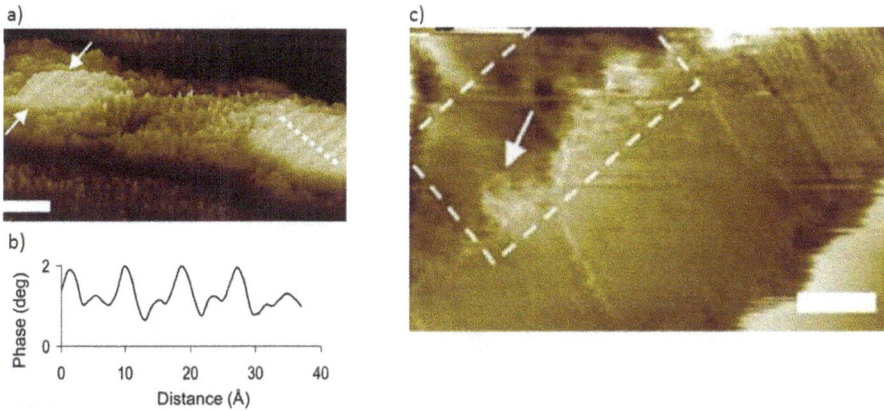

Fig. 6.4: (a) High-resolution phase image of a lamella of polyethylene resolving single chains in the (010) surface accompanied by (b) a line profile across these chains showing a spacing of 3.69 Å between chains. (c) Phase image of polyethylene film showing the interface between the edge of a crystalline lamella and the amorphous region revealing free loops from the crystalline area into the amorphous area. Scale bar in (a) is 100 nm and in (c) is 10 nm Reprinted with permission from Hobbs et al., *Physical Review Letters*, 107, p. 197,801.

More recently, crystallization process of a folded chain crystal in an isotactic polymethyl methacrylate (PMMA) monolayer was visualized [17] and submolecular resolution of thin films and single strands of polythiophene were observed using multifrequency AFM methods outlined in the next section [18].

6.4 Multifrequency imaging

Typically an AFM cantilever is excited at its first or fundamental eigenmode for conventional tapping or intermittent contact mode imaging. The phase imaging discussed in Section 6.3 is based on oscillation of the fundamental mode as well. Recently, with advances in hardware and software, it has become possible to oscillate the AFM cantilever at higher frequencies and even simultaneously multiple frequencies if the instrument has access to multiple lock-in amplifiers. Simultaneous oscillation at multiple frequencies can either occur at higher order harmonics or at higher order eigenmodes (see, e.g., bimodal AC mode by Asylum Research). More information about multifrequency operation and theory can be found elsewhere

[19–21]. Two main multifrequency methods are reviewed below including bimodal dual AC where there is no feedback on the higher order mode and AM–FM (frequency modulation) where there is feedback on the frequency of the higher order mode.

The advantage conferred by bimodal dual AC is still a greatly researched topic; however, there have been noticeable improvements in discrimination capabilities with its operation. Multifrequency imaging has shown improvements in imaging a variety of materials including polymers [22, 23] and biological samples [24]. Recent results reveal bistability in the higher order mode with multiple regimes of operation depending on the selection of operating parameters [25, 26]. One of the challenges with bimodal dual AC is selection of the relative drive amplitudes of the two eigenmodes, but some guidance has been provided in a study of a low-density PE (LDPE)/PS polymer blend [27].

Figure 6.5 shows the ability of multifrequency imaging to improve upon the contrast discrimination in complicated polymer samples.

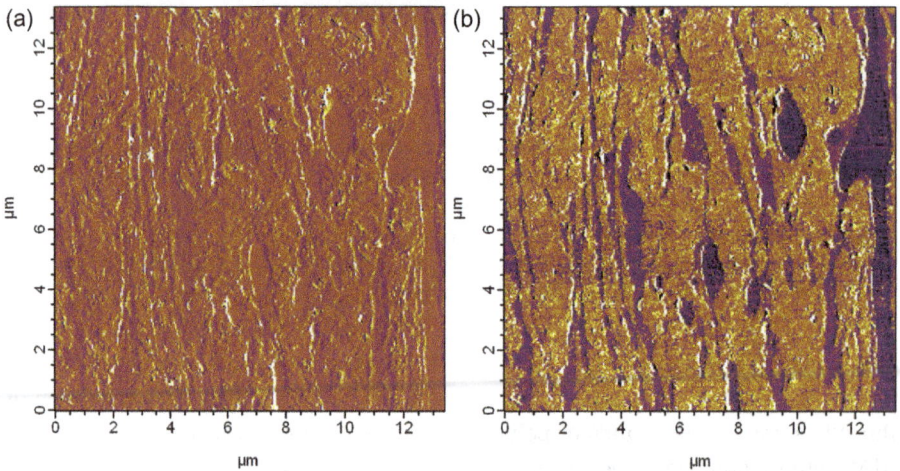

Fig. 6.5: (a) A 13 μm × 13 μm mode 1 phase image of a blend of 60% polyethylene, 20% polypropylene, and 20% polystyrene showing little contrast between the three materials. (b) Mode 2 phase of the same location showing good contrast between the three materials where the polystyrene is darkest and polyethylene (matrix background) is lightest.

Figure 6.5a shows the typical phase image of a three-component 60%/20%/20% (by weight) polymer blend of PP, PS, and PE obtained from the first-order eigenmode AM image. The three different components cannot be differentiated by contrast in the phase image, although the topography does facilitate some differentiation showing the PP as the matrix material. An image of second mode phase, collected in bimodal dual AC (where the cantilever is oscillated simultaneously at two eigenmodes with feedback occurring off the first eigenmode) is shown in Fig. 6.5b. Now the three components of PP, PS, and PE are easily differentiated by contrast where the PE appears bright, PP dark vertical "rivers," and the PS the darkest contrast of patches on the right and in the middle of the image. Clearly the higher order eigenmode clearly differentiates the three materials based on the phase contrast, thereby facilitating interpretation of dispersion and morphology of the various components.

A more advanced multifrequency method involves feedback on the frequency of the second mode in a technique known as amplitude modulation–frequency modulation (AM–FM) microscopy. This technique can be used for nanomechanical mapping where the higher order mode is used to map the Young's modulus of the sample. This has been applied to various polymer materials including a multilayer polymer film [28] and a blend of PS/PMMA [29]. There are certain challenges specific to the application of AM–FM to polymers. First, it measures the mechanical properties at very high frequencies (typically megahertz), which is not a rheologically relevant frequency space. Second, it uses the Hertzian model to calculate the modulus, which does not account for any adhesion between the tip and sample. This limitation is serious for many polymers that do exhibit such adhesion, especially the more rubbery and viscoelastic materials [29]. There are more appropriate methods to measure viscoelastic properties of polymers as described in the next section.

6.5 Mapping of mechanical properties

One of the areas of greatest growth in AFM for polymer measurements has been quantitative measurements of viscoelastic properties. First, quantitative measurements can facilitate unambiguous identification of components in various complex materials. Second, very often there is either not enough material available for a bulk viscoelastic measurement, or material is made in a reactor so that its final form, e.g., micron-sized domains, is the only form that is available. Thus, the ability to quantify properties such as the storage modulus, loss modulus, and loss tangent on the nanoscale and use them to develop structure–property relationships can be very useful.

An important note of caution when making viscoelastic measurements with the AFM with regard to frequency space: AFM measurements are typically made in a frequency space that is not rheologically relevant comprised of resonance frequencies of the AFM cantilevers, which occur at tens of kilohertz to megahertz. These frequencies are substantially higher than corresponding bulk viscoelastic measurements conducted by, for example, dynamic mechanical analysis (DMA), which occur at a few Hertz up to a few hundred Hertz. Time–temperature superposition can be used to convert viscoelastic measurements between time and frequency domains in order to extrapolate appropriately to the high frequency of AFM measurements. The challenges of the frequency space that AFM operates in for measuring viscoelastic materials with tapping mode AFM have been carefully outlined [30].

Another significant challenge of any quantitative viscoelastic measurements with AFM is the accurate measurement of the contact radius between the tip and the sample. This parameter figures prominently into all contact mechanics models that used by AFM to extract elastic and viscoelastic properties. While there are several ways to calibrate the contact radius through both relative and absolute methods such as direct imaging or reverse imaging of the tip, the problem remains that the tip radius is dynamic and inevitably changes over the course of the experiment. On a very hard surface the tip radius might wear down while on a soft or adhesive surface such as a polymer the tip might contaminate. Thus, in order to truly mitigate the contact radius problem, it needs to be "checked" throughout the measurement for any changes.

Dynamic contact methods have emerged as a useful potential tool for quantitative viscoelastic measurements. Specifically, recent progress has been made with contact resonance to quantitatively probe viscoelastic properties of thermoplastics. In contact resonance, a tip oscillates with very small amplitudes while in contact with a sample, resulting in a contact resonance that is at a frequency that is typically substantially higher than the frequency of the freely oscillating frequency [31]. Operation in the contact regime under a constant force set by the user results in a linear force displacement relationship between the tip and sample that is simple to model (as opposed to nonlinear dynamic in tapping mode), resulting in the ability to extract viscoelastic parameters [32, 33]. The contact resonance frequency and line shape are then measured and are a function of the substrate's material properties.

Contact resonance has thus been used to measure the storage modulus, loss modulus [34, 35], and loss tangent [36] of a blend of PP, PE, and PS. The values measured by contact resonance were in good agreement with the values measured for bulk individual components as measured by DMA. An important distinction between the storage/loss modulus measurement and the loss tangent measurement is that the former requires a reference for absolute quantification, whereas the latter loss tangent measurement is reference free. A current limitation of the contact resonance technique is in its application to measuring elastomers. As shown in [35], the adhesion of the elastomer, coupled with a most certainly higher tip contact area due to the tip sinking into the elastomer, results in incorrect measurements on such materials. A recent implementation of contact resonance includes an improved and straightforward method to sample multiple modes for the optimum performance [37].

Force–distance curves or force spectroscopy, which are single-point measurements, are a common tool for mechanical measurements as well. In a force–distance curve, the AFM tip is lowered onto the surface and then retracted while the force vs. z piezo position (that can be ultimately converted to a tip–sample distance) is monitored. Force–distance curves exist in two forms: static and dynamic. Static curves are the force–distance curves described above. In dynamic force curves, the cantilever is oscillated at its resonance frequency as it is lowered onto the surface, and so the amplitude and phase are measured as a function of z piezo position. More details about these single-point measurements can be found elsewhere [38] (see Chapter 3 specifically). Force–distance curves can also be acquired in mapping mode called "force volume imaging." There are several approaches for rapid acquisition of force–distance curves so that it approaches real-time imaging speeds. The most popular rapid force curve acquisition methods are PeakForce QNM (from vendor Bruker) and fast force volume (from vendor Oxford Instruments).

Once a force–distance curve has been measured, it then needs to be analyzed in order to extract viscoelastic properties. Several models exist including Hertz model (no adhesion), DMT (Derjaguin-Muller-Toporov) model (adhesion within the contact), and JKR (Johnson-Kendall-Roberts) model (adhesion outside the contact); a complete treatment of these models is beyond the scope of this work but can be found elsewhere [38] (see Chapter 2 specifically). Extracting viscoelastic materials from force curves is very challenging, but recent progress has been made in the methodology and applied to nylon 6,6 [39].

The power and utility of force curves are shown in Fig. 6.6 with sample force curves from the different components of the impact copolymer described in Fig. 6.3. The force curves plot the force exerted by the cantilever (in nN) vs. the tip–sample indentation (in microns) as the tip approaches the surface from the right (solid arrows), interacts with the sample through the repulsive wall (sloped area), and then withdraws from the surface in the pullout (dashed arrows). The red curve was collected from the rubber domain while the black curve collected from the PP matrix. Note the obvious visual differences between these two curves. First, the repulsive wall on the propylene has a higher slope than the repulsive wall on the rubber, indicating a higher modulus. Second, the curve on the rubber has more separation between the approach and retract, due to viscoelasticity and dissipation. Finally, on the pullout the rubber also exhibits more adhesion than the PP. All of these metrics can be quantified through measurement and application of contact mechanics models described above.

Using force curves, the elastic modulus of a nanocomposite based on natural rubber and multiwalled carbon nanotubes (CNT) was thus measured [40]. These studies used a JKR model in combination with a "two-point method" [41] to reveal a decrease in stiffness on the CNT values and an intermediate modulus region around the CNT. This method was also used to investigate interfacial properties and compatibilization effect in a blend of polyolefin elastomer and polyamide [42]. Moduli have been measured on a variety of polymers including LDPE, acrylonitrile butadiene styrene, PS, and PMMA [43], polyurethanes and PS [44].

A similar method but now with a superimposed oscillation while the cantilever was maintained on the specimen surface was used to map the frequency dependence of the viscoelastic properties of various rubbers that compare well with bulk DMA values [45]. A modified version of this AFM method measured viscoelastic properties of rubbers over a wide frequency and temperature range to probe the interfacial rubber region of silica-filled styrene–butadiene rubber with nanoscale resolution [46].

A new nonresonant AFM mode to measure viscoelastic properties of polymers with nanoscale resolution that measures E', E'', and tan δ at a wide frequency range was recently introduced in the form of AFM-nDMA. This technique also builds on

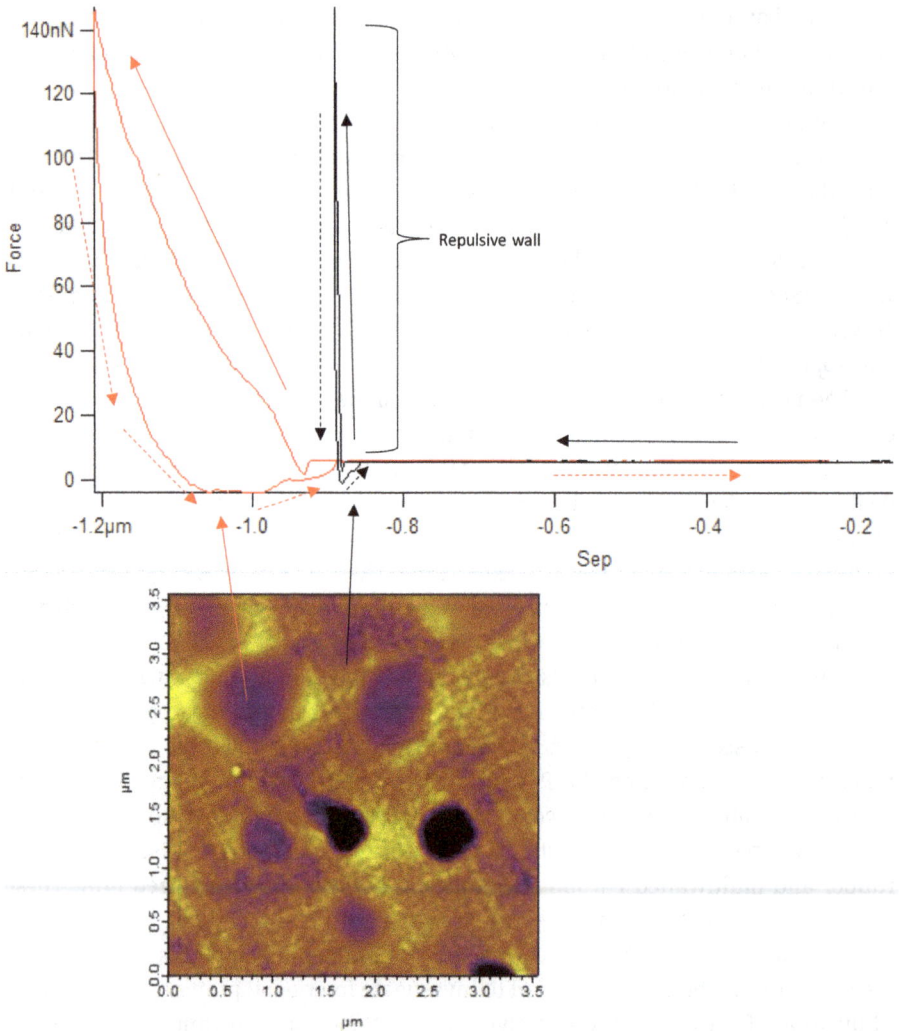

Fig. 6.6: Two force curves showing force (nN) vs. tip–sample separation (µm) collected from two different components of the impact copolymer described in Figure 6.3. The force curve from the rubber domain is shown in red and to the left. The force curve from the polypropylene matrix is to the right and shown in black. The force curve includes both the approach (arrows pointing to the left) and the retract of the cantilever as it approaches the sample and then withdraws. Force curves are collected from two different areas of the impact copolymer.

the force curve platform described above and uses well-defined, calibrated probes to obtain quantitative results at rheologically relevant frequencies with 10 nm spatial resolution [47]. AFM-nDMA also integrates reference measurements throughout the frequency sweep to check the contact radius value and accommodate any changes to it that may occur over the course of the experiment.

AFM-nDMA results collected at 100 Hz of a blend of PP and cyclic olefin copolymer (COC) are shown in Fig. 6.7 at three temperatures: room temperature, 160 °C, and 175 °C.

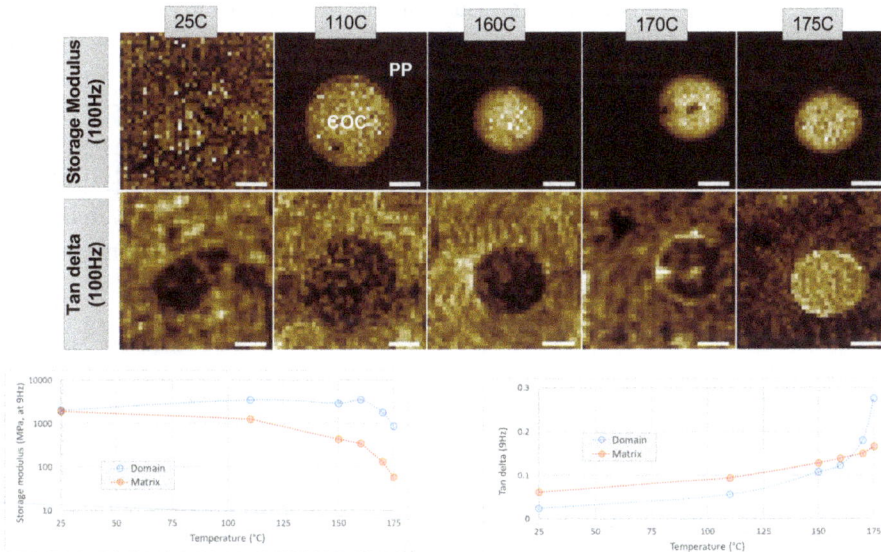

Fig. 6.7: Storage modulus and loss tangent of PP-COC blend as a function of temperature. Top row: maps of storage modulus at 100 Hz with increasing temperature (600 nm scale bar). Middle row: maps of loss tangent with increasing temperature. Bottom row: storage modulus and loss modulus plots at 10 Hz vs. temperature. Blue circles were measured within the COC domain and red triangles were measured on the PP matrix. Reprinted with permission from Bruker.

At room temperature, the loss tangent of the COC is slightly lower than the PP. As the temperature increases, the storage modulus of the two materials diverges as the PP softens more quickly than the COC, ultimately reversing at 170 °C. This technique has also been used to probe the viscoelastic properties of LDPE-modified asphalt binders [48].

6.6 Electrical measurements with AFM

AFM has a whole suite of electrical modes to probe various properties such as conductance, work function, capacitance, and electrostatics. Some of these modes are conductive AFM (c-AFM), electrostatic force microscopy, Kelvin probe force microscopy (KPFM), and scanning capacitance microscopy. For polymer applications, the most common use of such modes is to characterize conductive polymers that can be used in polymer-based solar cells. Poly(3-hexylthiophene) – P3HT – is a popular material studied for these applications.

Measurements are possible both in imaging modes (the focus here), as well as the analogue to force–distance curves in some of the modes. So for example in c-AFM, current–voltage (I–V) curves can be measured at individual points to provide information on electron conductivity and mobility of nanostructures such as in neat P3HT films [49], or P3HT:PCBM blends [50].

The conductivity of thin films of P3HT were explored with c-AFM [51, 52]. Charge transport characteristics of phase-separated blend films of electron donor P3HT and electron acceptor benzothiadiazole films were also observed by c-AFM [53]. KPFM measured a modification of work function of P3HT crystalline nanofibers with the degree of polymer packing order and coupling type [54] as well as the electronic properties of layered P3HT nanowhiskers [55].

6.7 Thermal/spectroscopic measurements

In addition to topography and mechanical-based measurements, AFM methods have been developed to probe thermal properties (scanning thermal microscopy) and even spectroscopic properties of polymers. Measurements to measure localized thermal properties take advantage of the cantilever to detect thermal transitions by several different methods including using an oscillating cantilever to sense temperature changes and using heated probes that can heat very small areas under the AFM tip. In this latter method, the cantilever deflection can then be measured as a function of localized heating so that when the material under the tip softens or melts, the tip will drop or sink into the material. The cantilever sinking results in a large deflection change at the thermal transition point. Localized thermal properties and transitions of a variety of polymers have been measured [56–58].

More recently, hybrid AFM-spectroscopic methods are being developed to measure chemical spectra with very high nanoscale resolution that can be correlated with simultaneously obtained topographic images. Near-field optical techniques to measure Raman and fluorescence on the nanoscale have been around for over almost two decades [59]. For the polymer community, however, infrared (IR) measurements to probe various organic stretches of interest have started to be the focus of much recent research and commercial activity. Typical IR measurements are limited by the diffraction of light resulting in a best-case resolution of several microns (for attenuated total reflection measurements). Early work successfully measured IR spectra on the submicron scale of a thin film of PS and polyethyl acrylate blend [60]; however, these measurements required the use of a very expensive and complicated laser source and detectors. There are now two main current approaches to measure IR spectra on the nanoscale via either a thermomechanical approach or a scattering

a) Topography

b) 3026 cm⁻¹

c) 2920 cm⁻¹

5 µm

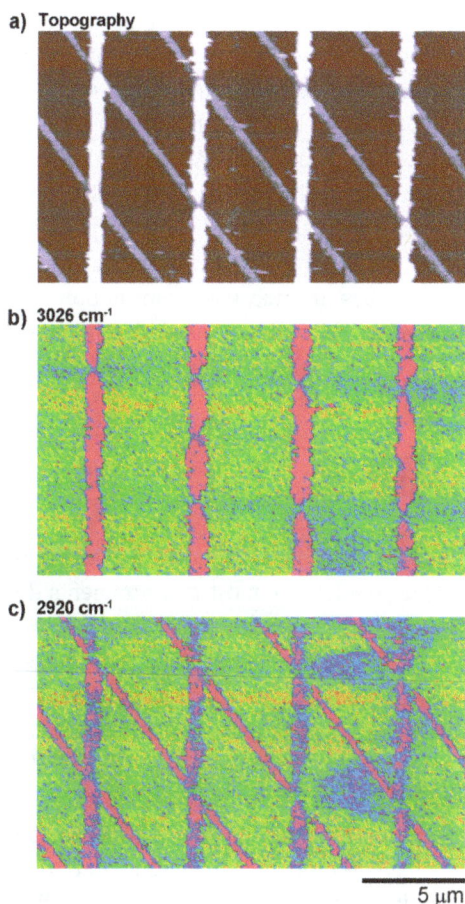

Fig. 6.8: (a) Topography and (b) single wavenumber absorption AFM-IR image at 3026 cm^{-1} image of polystyrene nanowires (vertical features) showing only polystyrene absorption. (c) Single wavenumber absorption image at 2920 cm^{-1} showing absorption of both polystyrene (vertical) and polyethylene (diagonal) wires. Reprinted with permission from Felts et al., *ACS Nano,* vol 6, p. 8015.

near-field scanning optical measurement (NSOM). The thermomechanical approach has been successful at IR measurements of PE nanostructures [61, 62], with the latest results reporting measurements of 15 nm structures, and a multilayer film of nylon and ethylene acrylic acid [63]. This AFM-IR approach was also used to study the degradation of polyester fibers [64].

An example of this capability is show in Fig. 6.8 showing AFM-IR single wavenumber absorption images of PS and PE nanowires where the 3026 cm^{-1} absorption in Fig. 6.8b shows only PS nanowires but absorption image at 2920 cm^{-1} in Fig. 6.8c reveals absorption of both PS and PE wires [62]. When initially developed, the photothermal IR could only operate in AFM's contact mode, which was not particularly conducive to many soft materials. But recently photothermal IR with tapping mode capability has been developed, enabling much broader application toward polymeric systems such as the study of polymeric nanoparticles [65] and fibers [64].

Scattering NSOM measurements are also being developed with polymer-specific applications reporting IR spectroscopic measurements of a PMMA/PS blend resolving several hundred nm domains of PS [66]. This scattering technique has also been used to probe mixed matrix membranes [67] and nanostructured polymer particles [68]. The ability to collect IR spectra at every pixel point in an image is a form of hyperspectral imaging. These kinds of image cubes are rich with information and can be mined to extract spatially resolved chemical information, as shown in an application to polymer blends [69].

6.8 Environmental measurements

The advantage of the scanning probe microscopy is that it can operate in a variety of environments including vacuum, fluid, ambient (which are most measurements described within this chapter), and under a variety of conditions including heating/cooling and stretching. By taking advantage of heating stages, AFM has imaged in situ crystallization of a variety of polymers including PE, polyhydroxybutyrate, PP, and PMMA [70] crystallization of polycaprolactone (PCL) [71], and near-surface crystallization in polyethylene terephthalate and poly(ethylene 2,6-naphthalate) (PEN) [72] – to name only a few of the many examples out there – by following real-time processes such as nucleation, growth, and crystal thickening. AFM can even combine mechanical measurements with a hot stage to follow mechanical properties as a function of temperature. This was done in the AFM-nDMA results described above (Fig. 6.7) as well as mapping isotactic polybutene to compare the surface layer with the bulk [73]. Visualization of polymer crystallization has even been accomplished by combining AFM with fast scanning calorimetry to function as a fast hot stage to view crystallization of polyamide 66, poly(ether ether ketone), polybutylene terephthalate, and PCL [74].

The effect of environmental factors of humidity on temperature on polymer has been explored for polymer based biomaterials of polybutyl methacrylate and polylauryl methacrylate [38] (see specifically Chapter 15 in reference 38). These polymers coated with drugs showed changes in topography and adhesion with temperature, important behavior for performance testing of its application as a drug-eluting stent in the human body. Similar studies were conducted to study the effect of humidity and immersion in water of these coatings.

Finally, AFM imaging of an ICP of PP with an EPR was studied as a function of tensile stress. As the material was stretched, AFM imaged crazing in the PP and equatorial stretching between the PP and rubber domains, as shown in Fig. 6.9a, b, respectively. In situ AFM measurements have been made on other polymer systems, including studying the effect on spherulite morphology in polybutene [75].

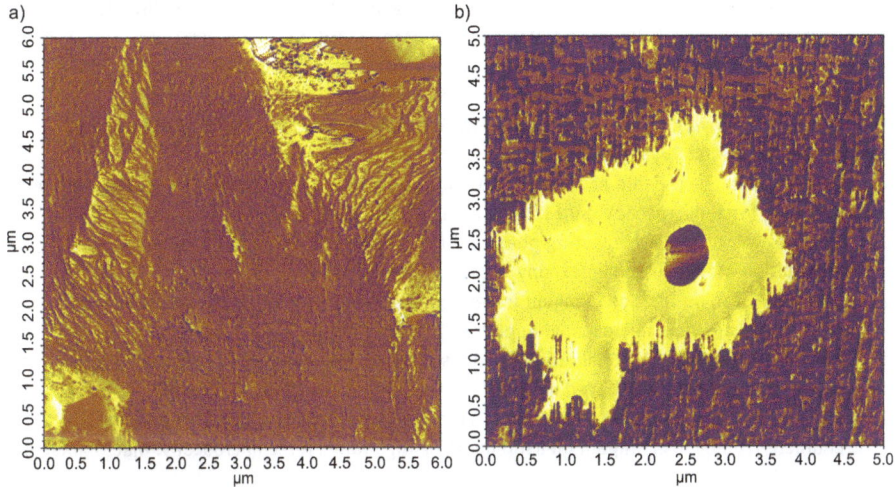

Fig. 6.9: a) A 6 μm × 6 μm phase image of an impact copolymer film under tensile stress showing crazing (circled) in the polypropylene and in a 5 μm × 5 μm image of a different location b) stretch marks between the ethylene–propylene rubber and polypropylene matrix.

6.9 Conclusions

AFM has emerged as a powerful tool for nanoscale characterization of polymers. AFM can probe many material properties, where mechanical properties are the most widely used for polymer applications with the ultimate goal of nanoscale imaging, discrimination, and quantitative mechanical measurements to guide, among other goals, formation of structure–property relationships. Mechanical measurements that can be made with AFM include Young's modulus, adhesion, and viscoelastic moduli critical for polymers: storage modulus, loss modulus, and loss tangent. Several key AFM-based methods to image via mechanical contrast are described here including the popular phase imaging, advanced multifrequency methods, and AFM-nDMA with applications to plastic blends, plastic/elastomer blends, nanocomposites, and thin films, in addition to high-resolution studies of individual chains in semicrystalline polymers. Electrical property measurements and AFM to study thermal and chemical properties of materials using techniques such as AFM-IR are described briefly. The chapter concludes with the ability to operate AFM under environmental conditions such as heating and deformation to capture processes such as crystallization and deformation under high resolution.

References

[1]   Binnig G, Rohrer H. Scanning tunneling microscopy. Surf Sci Rep 1983, 126, 236–244.

[2]   Binnig G, Quate CF, Gerber C. Atomic force microscopy. Phys Rev Lett 1986, 56, 930–933.

[3]   Sawyer LC, Grubb DT, Meyers GF. Polymer Microscopy, 3rd, Springer, 2010.

[4]   Cleveland JP, Anczykowski B, Schmid AE, Elings VB. Energy dissipation in tapping mode atomic force microscopy. Appl Phys Lett 1998, 72, 2613–2615.

[5]   Garcia R, Tamayo J, Calleja M, Garcia F. Phase contrast in tapping-mode scanning force microscopy. Appl Phys A 1998, 66, s309–s312.

[6]   Garcia R, Tamayo J, Paulo AS. Phase contrast and surface energy hysteresis in tapping mode scanning force microscopy. Surf Interface Anal 1999, 27, 312–316.

[7]   Martinez NF, Garcia R. Measuring phase shifts and energy dissipation with amplitude modulation atomic force microscopy. Nanotechnology 2006, 17, S167–172.

[8]   Proksch R, Yablon DG. Loss tangent imaging: Theory and simulations of repulsive mode tapping atomic force microscopy. Appl Phys Lett 2012, 100.

[9]   Yablon DG, Grabowski J, Chakraborty I. Measuring the loss tangent of polymer materials with atomic force microscopy based methods. Meas Sci Technol 2014, 25, 055402.

[10] Nguyen HK, So Fujinami MI, Nakajima K. Viscoelasticity of inhomogeneous polymers characterized by loss tangent measurements using atomic force microscopy. Macromolecules 2014, 47, 7971–7977.

[11] Ganguly A, Sarkar MD, Bhowmick AK. J Polym Sci B Polym Phys 2007, 45, 52–66.

[12] Han X, Hu J, Liu HL, Hu Y. Langmuir: ACS J Surf Colloids 2006, 22, 3428–3433.

[13] Ganguly A, Bhowmick AK. Macromolecules 2008, 41, 6246–6253.

[14] Wang L, Hong S, Hu HQ, Zhao J, Han CC. Langmuir: ACS J Surf Colloids 2007, 23.

[15] Park JH, Sun Y, Goldman YE, Composto RJ. amphiphilic block copolymer films: Phase transition, stabilization, and nanoscale templates. Macromolecules 2009, 42, 1017–1023.

[16] Mullin N, Hobbs JK. Direct imaging of polyethylene films at single-chain resolution with torsional tapping atomic force microscopy. Phys Rev Lett 2011, 107.

[17] Kumaki Y, OAJ. In situ real-time observation of polymer folded-chain crystallization by atomic force microscopy at the molecular level. Macromolecules 2018, 51, 7629–7636.

[18] Korolkov VV, Summerfield A, Murphy A, Amabilino DB, Watanabe K, Taniguchi T, Beton PH. Ultra-high resolution imaging of thin films and single strands of polythiophene using atomic force microscopy. Nat Commun 2019, 10.

[19] Lozano JR, Garcia R. Theory of multifrequency AFM. Phys Rev Lett 2008, 100, 076102.

[20] Rodriguez TR, Garcia R. Compositional mapping of surfaces in atomic force microscopy by excitation of the second normal mode of the microcantilever. Appl Phys Lett 2004, 84, 449.

[21] Proksch R. Multifrequency repulsive mode amplitude modulated atomic force microscopy. Appl Phys Lett 2006, 89, 113121–113121.

[22] Gigler AM, Dietz C, Baumann M, Martinez NF, Garcia R, Stark RW. Repulsive bimodal atomic force microscopy on polymers. Beilstein J Nanotechnol 2012, 3, 456–463.

[23] Dietz C, Zerson M, Riesch C, Gigler AM, Stark R, Rehse N, Magerle R. Nanotomography with enhanced resolution using bimodal atomic force microscopy. Appl Phys Lett 2008, 92, 143107.

[24] Martinez NF, Lozano JR, Herruzo ET, Garcia F, Richter C, Sulzbach T, Garcia R. bimodal AFM imaging of isolated antibodies in air and liquids. Nanotechnology 2008, 19, 384001.

[25] Kiracofe D, Raman A, Yablon DG. Multiple regimes of operation in bimodal AFM: Understanding the energy of cantilever eigenmodes. Beilstein J Nanotechnol 2013, 4, 385–393.

[26] Chakraborty I, Yablon DG. Cantilever energy effects on bimodal AFM: Phase and amplitude contrast of multicomponent samples. Nanotechnology 2013.

[27] Mehrnoosh Damircheli BE. Enhancing phase contrast for bimodal AFM imaging in low quality factor environments. Ultramicroscopy 2019, 204, 18–26.

[28] Marta Kocun AL, Meinhold W, Revenko I, Proksch R. Fast, high resolution, and wide modulus range nanomechanical mapping with bimodal tapping mode. ACS Nano 2017, 11, 10097–10105.

[29] Nguyen1 HK, MI., Nakajima1,2* K. Elastic and viscoelastic characterization of inhomogeneous polymers by bimodal atomic force microscopy. Jpn J Appl Phys 2016, 55.

[30] Lopez-Guerra EA. On the frequency dependence of viscoelastic material characterization with intermittent-contact dynamic atomic force microscopy: Avoiding mischaracterization across large frequency ranges. Beilstein J Nanotechnol 2020, 11, 1409–1418.

[31] Hurley DC. Contact Resonance Force Microscopy Techniques for Nanomechanical measurements. In: Bhushan B, Fuchs H, eds. Applied Scanning Probe Methods Vol. XI, Springer-Verlag, 2009, 97–138.

[32] Yuya PA, Hurley DC, Turner JA. Contact resonance atomic force microscopy for viscoelasticity. J Appl Phys 2008, 104.

[33] Yuya PA, Hurley DC, Turner JA. Relationship between Q-factor and sample damping for contact resonance atomic force microscope measurement of viscoelastic properties. J Appl Phys 2011, 109, 113528.

[34] Killgore JP, Yablon DG, Tsou AH, Gannepalli A, Yuya PA, Turner JA, Proksch R, Hurley DC. Viscoelastic property mapping with contact resonance force microscopy. Langmuir: ACS J Surf Colloids 2011, 27, 13983–13987.

[35] Yablon DG, Grabowski J, Killgore JP, Hurley DC, Proksch R, Tsou AH. Quantitative mapping of viscoelastic properties of polyolefin blends with contact resonance atomic force microscopy. Macromolecules 2012, 45, 4363–4370.

[36] Yablon DG, Grabowski J, Chakraborty I. Measuring the loss tangent of polymer materials with atomic force microscopy based methods. submitted 2013.

[37] Pittenger B, Yablon D. *Quantitative Measurements of Elastic and Viscoelastic Properties with FASTForce Volume CR2017*. Bruker AN 148.

[38] Yablon DG ed. Scanning Probe Microscopy in Industrial Applications: Nanomechanical Characterization, 1st, New York, Wiley, 2013.

[39] Parvini CH. Extracting viscoelastic material parameters using an atomic force microscope and static force spectroscopy. Beilstein J Nanotechnol 2020, 11, 922–937.

[40] Wang D, Fujinami S, Nakajima K, Inukai S, Ueki H, Magario A, Noguchi T, Endo M, Nishi T. Visualization of nanomechanical mapping on polymer nanocomposites by AFM force measurement. Polymer 2010, 51, 2455–2459.

[41] Sun Y, Walker GC. Langmuir: ACS J Surf Colloids 2004, 20, 5837.

[42] Wang D, Fujinami S, Liu H, Nakajima K, Nishi T. Investigation of reactive polymer–polymer interface using nanomechanical mapping. Macromolecules 2010, 43, 5521–5523.

[43] Young TJ, Monclus MA, Burnett TL, Broughton WR, Ogin SL, Smith PA. The use of the PeakForceTM quantitative nanomechanical mapping AFM-based method for high-resolution young's modulus measurement of polymers. Meas Sci Technol 2011, 22, 125703.

[44] Dokukin ME, Sokolov I. Quantitative mapping of the elastic modulus of soft materials with HarmoniX and PeakForce QNM AFM modes. Langmuir: ACS J Surf Colloids 2012, 28, 16060–16071.

[45] Igarashi T, Fujinami S, Nishi T, Asao N, Nakajima AK. Nanorheological mapping of Rubbers by atomic force microscopy. Macromolecules 2013, 46, 1916–1922.

[46] Eijun Ueda XL, Ito M, Nakajima K. Dynamic moduli mapping of silica-filled styrene–butadiene rubber vulcanizate by nanorheological atomic force microscopy. Macromolecules 2019, 52.

[47] Bede Pittenger SO, Yablon D, Mueller T. Nanoscale DMA with the atomic force microscope: A new method for measuring viscoelastic properties of nanostructured polymer materials. JOM 2019, 71, 3390–3398.

[48] Mohammad Fuad Aljarrah EM. Nanoscale viscoelastic characterization of asphalt binders using the AFM-nDMA test. Mater Struct 2020, 53.

[49] Kondo YO, Benten M, Ohkita H, Ito H, S. Electron transport nanostructures of conjugated polymer films visualized by conductive atomic force microscopy. ACS Macro Lett 2015, 4, 879–885.

[50] Dante MP,J, Nguyen TQ. Nanoscale charge transport and internal structure of bulk heterojunction conjugated polymer/fullerene solar cells by scanning probe microscopy. J Phys Chem C 2008, 112, 7241–7249.

[51] Osaka MB,H, Lee L-T, Ohkita H, Ito S. Development of highly conductive nanodomains in poly (3-Hexylthiophene) films studied by conductive atomic force microscopy. Polymer 2013, 54.

[52] Wood DH,I, Jones TS, Wilson NR. Quantitative nanoscale mapping with temperature dependence of the mechanical and electrical properties of Poly(3-Hexylthiophene) by conductive atomic force microscopy. J Phys Chem C 2015, 119, 11459.

[53] Osaka MB, Ohkita H, Ito S. Intermixed Donor/Acceptor region in conjugated polymer blends visualized by conductive atomic force microscopy. Macromolecules 2017, 50, 1618–1625.

[54] Baghgar MB, MD. Work function modification in P3HTH/J aggregate nanostructures revealed by Kelvin probe force microscopy and photoluminescence Imaging. ACS Nano 2015, 9, 7105–7112.

[55] McFarland FMB,B, Guo S. Layered poly(3-hexylthiophene) nanowhiskers studied by atomic force microscopy and Kelvin probe force microscopy. Macromolecules 2015, 48, 3049–3056.

[56] Fryer DS, Nealey PF, Pablo JJD. Thermal probe measurements of the glass transition temperature for ultrathin polymer films as a function of thickness. Macromolecules 2000, 33, 6439–6447.

[57] Ihalainen P, Backfolk K, Sirvio P, Peltonen J. Thermal analysis and topographical characterization of latex films by scanning probe microscopy. J Appl Phys 2007, 101, 043505/ 043501.

[58] Zhou J, Berry B, Douglas JF, Karim A, Snyder CR, Soles C. Nanoscale thermal-mechanical probe determination of softening transitions in thin polymer films. Nanotechnology 2008, 19, 495703/495701.

[59] Lewis A, Lieberman K. Near field optical imaging with a non-evanescently excited high brightness light source of sub-wavelength dimensions. Nature 1991, 354, 214–216.

[60] Michaels CA, Gu X, Chase B, Stranick SJ. Appl Spectrosc 2004, 58, 257.

[61] Felts JR, Cho H, Yu MF, Bergman LA, Vakakis AF, King WP. Atomic force microscope infrared spectroscopy on 15 nm scale polymer nanostructures. Rev Sci Instrum 2013, 84, 023709.

[62] Felts JR, Kjoller K, Lo M, Prater CB, King WP. Nanometer scale infrared spectroscopy of heterogeneous polymer nanostructures fabricated by tip-based nanofabrication. ACS Nano 2012, 6, 8015–8021.

[63] Kjoller K, Felts JR, Cook D, Prater CB, King WP. High sensitivity nanometer scale infrared spectroscopy using a contact mode microcantilever with an internal resonator paddle. nanotechnology 2010, 21, 18705.

[64] Phuong Nguyn-Tri REPH. Nanoscale analysis of the photodegradation of polyester fibers by AFM-IR. J Photochem Photobiol A Chem 2019, 371, 196–204.

[65] Jeremie Mathurin EP, Deniset-Besseau A, Kjoller K, Prater CB, Gref R, Dazzi A. How to unravel the chemical structure and component localization of individual drug-loaded polymeric nanoparticles by using tapping AFM-IR. Analyst 2018.

[66] Taubner T, Hillenbrand R, Keilmann F. Nanoscale polymer recognition by spectral signature in scattering infrared near-field microscopy. Appl Phys Lett 2004, 85, 5064–5066.

[67] Etxeberria-Benavides M, Johnson OD,T, Łozińska MM, Orsi A, Wright PA, Mastel S, Hillenbrand R, Kapteijn F, Gascon J. High performance mixed matrix membranes (MMMs) composed of ZIF-94 filler and 6FDA-DAM polymer. J Membr Sci 2017, 550, 198–207.

[68] Monika Goikoetxea IA, Chimenti S, Paulis M, Leiza JR, Hillenbrand R. Cross-sectional chemical nanoimaging of composite polymer nanoparticles by infrared nanospectroscopy. Macromolecules 2021, 54, 995–1005.

[69] Amenabar I, Goikoetxea SP,M, Nuansing W, Lasch P, Hillenbrand R. Hyperspectral infrared nanoimaging of organic samples based on Fourier transform infrared nanospectroscopy. Nat Commun 2017, 8, 14402.

[70] Hobbs JK, Farrance OE, Kailas L. How atomic force microscopy has contributed to our understanding of polymer crystallization. Polymer 2009, 50, 4281–4292.

[71] Beekmans LGMV. Real-time crystallization study of Poly(ε-Caprolactone) by hot-stage atomic force microscopy. Polymer 2000, 41, 8975–8981.

[72] Assender KSH. In situ AFM study of near-surface crystallization in PET and PEN. J Appl Polym Sci 2016.

[73] Wu X, SS, Yu ZZ, Thomas PR, Wang D. AFM nanomechanical mapping and nanothermal analysis reveal enhanced crystallization at the surface of a semicrystalline polymer. Polymer 2018, 146, 188–195.

[74] Rui Zhang EZ, Androsch R, Schick C. Visualization of polymer crystallization by in situ combination of atomic force microscopy and fast scanning calorimetry. Polymers 2019, 11, 890.

[75] Thomas C, Seguela R, Detrez F, Miri V, Vanmansart C. Plastic deformation of spherulitic semi-crystalline polymers: An in situ AFM study of polybutene under tensile drawing. Polymer 2009, 50, 3714–3723.

Miroslav Šlouf, František Lednický, Petr Wandrol, T. Vacková

7 Polymer surface morphology: characterization by electron microscopies

7.1 Introduction

This chapter deals with the electron microscopy (EM) and its applications on polymer surface characterization. We show that the surface morphology of a polymer system is always closely connected with its composition (types of polymers/additives/fillers and overall composition), with the processing conditions (thermal and mechanical history of the sample), and with the preparation of specimens for microscopic observations (we can observe polymer surfaces either "as received" or, more often, polymer surfaces intentionally prepared by fracturing, cutting, smoothing, etching, etc.). Moreover, we demonstrate how to employ EM in characterization of subsurface and internal structure of polymer materials.

There are two basic types of EM: scanning electron microscopy (SEM) and transmission electron microscopy (TEM). In SEM, we usually work with *bulk* specimens, investigate their surface morphology, and collect signal with detectors placed *above* the sample. In TEM, we almost exclusively work with *ultrathin* specimens, investigate their internal morphology, and collect transmitted electrons using detectors localized *below* the sample. Therefore, SEM is the dominant EM technique for characterization of polymer surfaces. This text focuses on SEM (Section 7.1), briefly mentions TEM (Section 7.2), discusses key step of polymer morphology studies – specimen preparation (Section 7.3), and shows typical applications of EM in polymer science (Section 7.4).

7.2 Scanning electron microscopy

This section summarizes the basic principles of SEM (Section 7.2.1), classical modes of SEM (Section 7.2.2), and new developments in SEM connected with characterization of polymers (Section 7.2.3). Other possibilities of SEM are just briefly mentioned (Section 7.2.4). We explain elements of SEM theory, but more details are to be found in specialized textbooks [1–4].

Acknowledgments: This work was supported by the Technology Agency of the Czech Republic (project TN01000008).

https://doi.org/10.1515/9783110701098-007

7.2.1 SEM: principles

7.2.1.1 Microscope and image formation

Scanning electron microscope consists of an electron column and chamber (Fig. 7.1). At the top of the column, primary electrons (electron beam) are emitted from the electron gun (cathode). The commonly used cathodes include tungsten filament, lanthanum hexaboride (LaB_6), or field-emission gun (FEG). Quality (and price) of the electron guns increases in the following row: tungsten < LaB_6 < FEG. The FEG sources produce the electron beam with the highest brightness, the lowest energy spread and the smallest final probe size, which all leads to the highest resolution of FEG-equipped SEM microscopes (which are occasionally called FEGSEM or FESEM).

In the electron gun, the negatively charged electrons (charge $Q = e$) are emitted. They move to the positively charged anode due to the accelerating voltage (U) and the potential energy of the electron at the anode ($E_1 = eU$) changes to its kinetic energy ($E_2 = 1/2m_e v^2$; m_e = electron mass; v = velocity of the accelerated electron). Supposing that the potential energy is completely transferred to the kinetic energy ($E_1 = E_2$) and neglecting relativistic effects, we calculate the velocity, v, of the accelerated electrons by a straightforward combination of the above relations:

$$v = \sqrt{2eU/m_e} \tag{7.1}$$

Once we know the accelerated electrons velocity, we can calculate their wavelength using de Broglie formula (which employs Planck's constant h):

$$\lambda = h/m_e v \tag{7.2}$$

In the column, the accelerated electrons travel through the set of electromagnetic lenses/coils (from top to bottom: condenser lenses, scanning coils, and objective lens) and are focused to a small *spot* (typical spot size ranges from 1 μm to 1 nm) on the surface of the sample. The scanning coils deflect the electron beam in the x and y directions so that the beam scans the sample surface and the micrograph is collected pixel by pixel. Typically, both the column and the chamber are kept under high vacuum ($p < 10^{-3}$ Pa), because any gas molecules spread and attenuate the electron beam. *Magnification* of an SEM microscope is given by the ratio of the scanned area size to the final micrograph size (typical magnifications range from 10 to 100,000×). *Resolution* of an SEM microscope depends on a number of parameters as discussed in the following sections.

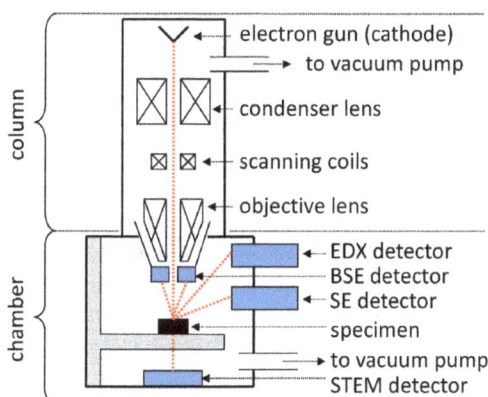

Fig. 7.1: Scheme of the scanning electron microscope. Primary electrons (PE) are emitted from the electron gun and interact with the specimens, generating secondary electrons (SE), backscattered electrons (BSE), characteristic X-rays (collected with EDX detector) and transmitted electrons (collected with STEM detector).

7.2.1.2 Interaction of electron beam with specimen

The primary electrons (PE) in the electron beam travel at very high speed. Typical accelerating voltages in SEM range from 30 to 1 kV. This corresponds, according to eq. (7.1), to velocities of 0.06 c and 0.34 c, respectively, where c is the velocity of light. Despite these extreme speeds, the PE do not penetrate too deep into the sample. The maximum penetration depth of PE, called the maximum electron range R, can be estimated using the Kanaya–Okayama formula [5]:

$$R = 27.6 \frac{M}{\rho Z^{8/9}} U^{5/3} \tag{7.3}$$

where R, M, ρ, Z, and U are the electron range in nanometers, the molecular weight in g/mol, the density in g/cm^3, the sum of atomic numbers of consisting atoms, and the accelerating voltage of PEs in kV, respectively. Calculation of R for polyethylene ($M = 28.0$ g/mol, $\rho = 0.96$ g/cm^3, $Z = 16$) gives the result ~20 μm and ~1 μm for 30 kV and 5 kV, respectively. Analogous calculation for gold ($M = 197.0$ g/mol, $\rho = 19.32$ g/cm^3, $Z = 79$) yields ~1.6 μm and ~0.08 μm. Several important conclusions follow from these simple calculations: (i) certain part of SEM signal may come from sub-surface layers in light-element polymer systems, (ii) lower accelerating voltage reduces penetration depth and thus it enhances the signal from surface, and (iii) very thin polymer films, fibers and particles may appear translucent in SEM, as a part of the electrons manage to penetrate through.

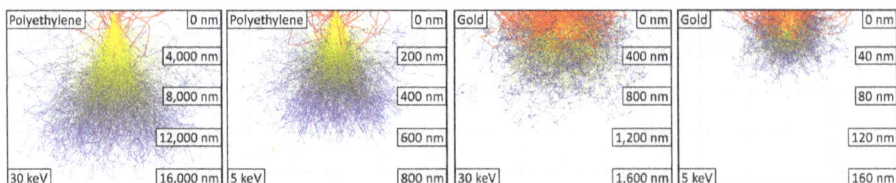

Fig. 7.2: Monte Carlo simulations for 1,000 electron trajectories through polyethylene and gold, using accelerating voltage 30 kV and 5 kV, and probe diameter 10 nm in all cases. From left to right: polyethylene, 30 kV; polyethylene, 5 kV; gold, 30 kV; gold, 5 kV. The thick red trajectories indicate backscattered electrons that escape from the surface of the sample. All other trajectories represent electrons that are absorbed in the sample. Note the scale change in the images.

Trajectories of PEs penetrating into a sample can be also modeled with Monte Carlo simulation programs [6]. The simulations for polyethylene and gold are shown in Fig. 7.2. We note that: (i) the overall size of PE interaction volume correlates with predictions based on eq. (7.3), (ii) the shape of interaction volume changes from pear shape in case of light-element materials to half apple shape for heavy-element materials, and (iii) the interaction volume decreases dramatically with the decreasing accelerating voltage U.

A scheme of electron beam interaction with the specimen is given in Fig. 7.3a. When the electron beam hits the specimen surface, numerous processes take place, the first four of which are of particular importance in polymer science:

- Emission of low-energy (<50 eV), secondary electrons (SE).
- Emission of high-energy (>50 eV), backscattered electrons (BSE).
- Emission of X-radiation (characteristic X-rays).
- Penetration of electrons through an ultrathin sample and their detection in the form of transmitted electrons (TE).
- Emission of Auger electrons (AE).
- Emission of light – cathodoluminiscence (CL).
- Absorption of electrons.

SEM resolution is closely connected with interaction volume and escape depth, which differs for different signals as illustrated in Fig. 7.3b. The highest resolution (~1 nm in top microscopes) is achieved in SE imaging, because the SE are emitted from the very subsurface region. Lower resolution (usually units to tens of nm) is obtained in BSE imaging, as the BSE electrons have higher escape depth, which results in wider lateral spread. The lowest resolution (usually units to tens of μm) is obtained in energy-dispersive analysis of X-rays (EDX), because the characteristic X-rays are able to escape from the highest depths, where the beam is widely spread. STEM resolution is comparable to that of SE, as the samples for STEM must be ultrathin (<100 nm) so that they were electron-transparent and, as a result, the electrons leave the sample before being spread laterally.

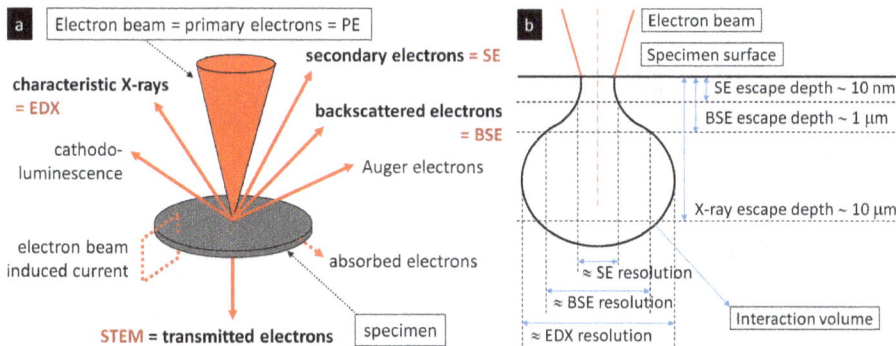

Fig. 7.3: Electron beam interaction with the specimen in SEM: (a) overview of signals generated during the interaction and (b) typical escape depths of the three most important signals – secondary electrons, backscattered electrons, and X-rays. The signals appear simultaneously, but we usually detect just one of them at the time. Real escape depths depend strongly on the accelerating voltage and specimen (eq. (7.3), Fig. 7.2).

7.2.2 SEM: classical modes

The above-listed four main SEM signals (SE, BSE, X-ray, and TE) correspond to four classical modes of SEM microscopy, which are applied for polymer morphology investigations: imaging with SE (SE imaging), imaging with BSE, energy-dispersive analysis of X-rays (EDX; also called as energy-dispersive spectroscopy, EDS, or electron microanalysis) and imaging with transmitted electrons (scanning transmission electron microscopy, STEM). SE micrographs (Fig. 7.4a) exhibit topographic contrast, i.e. the morphology of the surface. BSE micrographs (Fig. 7.4b) display materials contrast, i.e. the difference between atomic numbers of the components. EDX spectra (Fig. 7.4c) yield information about elemental composition of the sample. STEM micrographs (Fig. 7.4d) visualize the internal morphology of samples.

7.2.2.1 SE imaging

SE imaging (Fig. 7.4a) represents the most common mode of SEM. SE micrographs show *topographic contrast*. This is due to the fact that SE have low energies (<50 eV) and so they can escape only from the thin surface layer (~10 nm). The geometry of the surface layer is closely associated with the number of emitted electrons (Fig. 7.5). Surfaces perpendicular to electron beam yield average SE signal (Fig. 7.5a), which increases with the inclination of the surface (Fig. 7.5b), and achieves maximum at surfaces parallel with the electron beam (Fig. 7.5c) – this extreme intensity of SE signal on sharp edges of the specimen is called *edge effect*. Holes in the sample surface

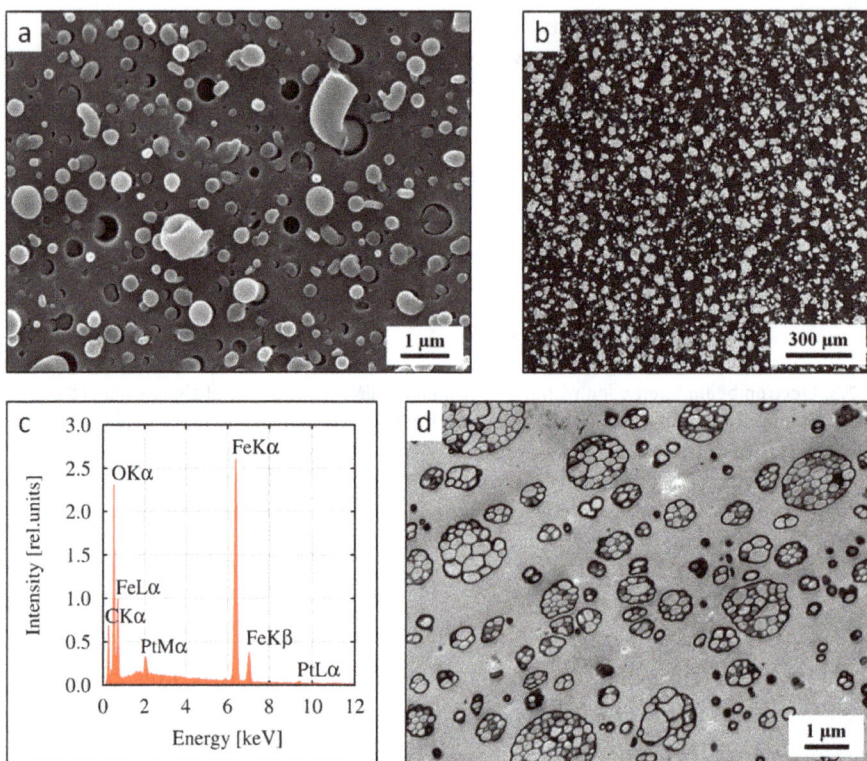

Fig. 7.4: Examples of four main modes of SEM microscopy in polymer science. (a) SE imaging: fracture surface of polypropylene/cycloolefin copolymer blend – topographic contrast. (b) BSE imaging: smooth surface of epoxy composite filled with aluminum oxide particles – material contrast. (c) EDX spectrum: microanalysis of polymer microspheres filled with magnetic nanoparticles – elemental analysis. (d) STEM micrograph: OsO$_4$-stained ultrathin section of high-impact polystyrene – internal morphology.

exhibit decrease of SE intensity due to SE electron absorption (Fig. 7.5d), whereas hills may show intensity increase if their dimension is comparable to probe size (Fig. 7.5e).

SE micrograph in Fig. 7.4a exhibits all types of contrast shown in Fig. 7.5: Flat surfaces have medium intensity (effect in Fig. 7.5a). Inclined areas and edges exhibit increased intensity (effects in Fig. 7.5b,c). The hills above the surface and the holes below the surface display higher and lower intensity, respectively (effects in Fig. 7.5d,e). Three additional pieces of information are worth mentioning here: (i) There are yet another, not-so-frequent types of SE contrast [7], but those shown in Fig. 7.5 suffice for interpretation of most SE micrographs of polymer materials. (ii) Lower accelerating voltage increases surface detail at the expense of resolution. (iii) Edge effect weakens with decreasing accelerating voltage.

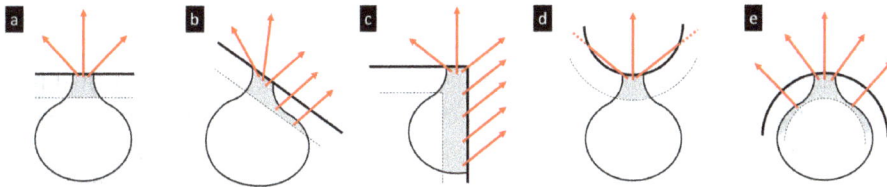

Fig. 7.5: Origin of topographic contrast in SE imaging. Thick black line denotes sample surface, thinner black dashed line marks the SE escape depth and the arrows represent SE emitted from the surface.

7.2.2.2 BSE imaging

BSE imaging (Fig. 7.4b) is applied mostly for multicomponent systems, because number of backscattered electrons depends on the chemical composition – this is called *material contrast*. Let us first consider the difference between the SE and BSE: SE are regarded as the electrons detached from specimen atoms during collisions of PE with the specimen. BSE are supposed to be the original PE, which were so deflected by the collisions with atoms that they were turned back out of the specimen.

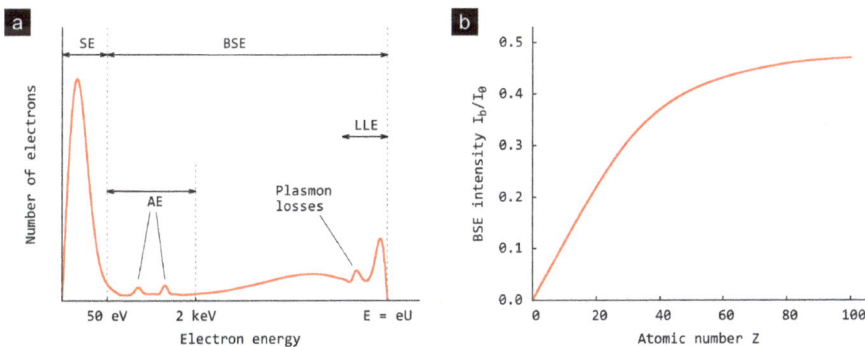

Fig. 7.6: Properties of backscattered electrons: (a) The energy of BSE ranges from 50 eV to energy of primary electrons ($E = eU$, Section 7.2.1.1) and (b) the intensity of BSE signal increases with atomic number Z. The figures were redrawn according to [2].

The SE exhibit low energies because most of PE energy is consumed for the detachment from the atom, while the BSE electrons have higher energies, depending on their collision history. By convention, the limit between SE and BSE is set at 50 eV (Fig. 7.6a). It has been shown experimentally [8] and later also calculated theoretically that the intensity of backscattering increases with increasing atomic number Z

(Fig. 7.6b). The fact that the yield of BSE electrons increases with Z is evident also from Monte Carlo simulations shown above (Fig. 7.2).

Material contrast in BSE imaging is demonstrated in Fig. 7.4b: we clearly differentiate aluminum oxide filler (bright) from the polymer matrix (dark). At low accelerating voltages (~1 kV), the relationship in Fig. 7.6b ceases to hold, and increased backscattering may not indicate higher Z [8]. If BSE detector is composed of two semi-annular segments A and B, it can show either material contrast (signal from A + B) or topographic contrast (signal from A–B) [7]; however, the topographic contrast (and higher resolution; see Fig. 7.3b) is usually obtained with SE imaging.

7.2.2.3 EDX spectra

Energy-dispersive analysis of X-rays (EDX; Fig. 7.4c) is a spectroscopic method, which yields elemental composition in microscopic scale. Typical EDX resolution in polymer materials ranges from 10 to 20 μm in both horizontal and vertical direction (Figs. 7.2 and 7.3). Somewhat lower resolution can be achieved with lower accelerating voltages (Figs. 7.2 and 7.3), but the energy of the electrons (in keV), which equals to the accelerating voltage (in kV), must be ca 3× higher than the rightmost peak in the EDX spectrum in order to get correct intensities for quantitative analysis [9].

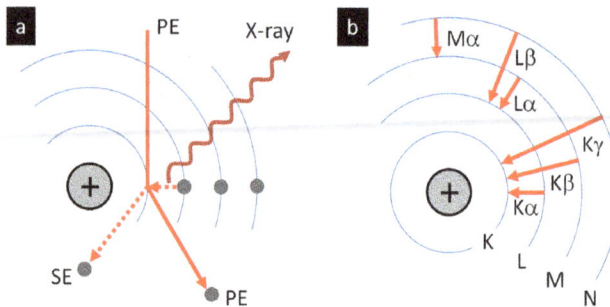

Fig. 7.7: Origin of the EDX signal (a): PE ejects the electron from the inner shell, the electron from outer shell fills the hole and the energy difference is emitted as a characteristic X-ray. Simplified nomenclature of X-ray emission lines (b): the lines are named according to the shell in which the initial vacancy occurs (K, L, M, etc. shells correspond to principal quantum number n = 1, 2, 3, etc.) and the shell from which the electrons fills the hole (α, β, γ, etc. correspond to the 1st, 2nd, 3rd, etc. adjacent shell). The emission/transition shown in (a) is Kα; in real atoms the nomenclature becomes more complex due to the complexity of their electronic structure, but the main emissions/peaks in EDX spectra are Kα, Kβ, Lα, Lβ, and Mα (Fig. 7.4c, [9]).

Origin of characteristic X-rays, which are detected in EDX, is shown schematically in Fig. 7.7a; the image illustrates that the energy of emitted X-rays corresponds to the energy difference between the electron shells. In one-electron atom (rough approximation, but sufficient for showing the principle), the energy of the electron, E, depends only on principal quantum number, n [9]:

$$E(n) = -\mathrm{Ry}\frac{Z^2}{n^2} \tag{7.4}$$

where $\mathrm{Ry} = 13.60$ eV is Rydberg unit of energy and Z is the atomic number. The transitions in one-electron atom are shown in Fig. 7.7b. The energy corresponding to those transitions can be estimated by means of eq. (7.4). For instance, energy of FeKα transition (main peak in Fig. 7.4c) is calculated as $E = E(2)-E(1) = -13.60(26^2/2^2-26^2/1^2) =$ 6.9 keV, which reasonably agrees with experiment ($E = 6.4$ keV; Fig. 7.4c). More precise energies of X-ray emission lines can be calculated using Moseley's laws [10] and precise values are to be found in specialized software, databases, and tables [10].

Modern EDX detectors can detect the lightest elements starting from lithium. *Qualitative* EDX analyses are usually reliable, but experienced user should always check the computer-generated results. *Quantitative* analysis can be done without standards (standardless analysis; results are only semiquantitative) or with standards (more precise, but suitable standards are rarely available for newly developed polymer systems). Commercial EDX systems perform analysis from selected point or user-defined area on previously saved SE/BSE/STEM micrograph. It is also possible to measure line scans or two-dimensional maps (i.e. concentration profiles/maps of selected elements); these are rarely applied to polymer systems because the longer data collection times cause sample damage.

7.2.2.4 Scanning transmission electron microscopy

STEM imaging (Fig. 7.4d) in SEM is regarded as an additional technique of minor interest in most fields of *materials science* [2], but it belongs among useful microscopic methods in *polymer science* [3, 4]. STEM requires electron-transparent, *ultrathin* samples (thickness < 100 nm) with sufficient contrast between the components (contrast of one component is frequently increased by means of *staining*). Preparation of these ultrathin and stained specimens is discussed in Section 7.4.

Thicker, heavier-atom-containing or diffracting areas of the sample appear dark on bright background if we collect transmitted electrons (bright-field imaging; BF) or bright on dark background if we detect scattered electrons (dark-field imaging; DF). Segmented STEM detectors make it possible to record both BF and DF image from the same location. STEM/BF micrographs are more common, but STEM/DF micrographs may exhibit higher contrast at the expense of total intensity (higher data collection times, increased risk of sample damage).

In the field of polymer science, STEM can be applied to: (i) visualization of internal morphology using ultrathin sections from the interior of the material (Fig. 7.8a), (ii) visualization of surface and subsurface morphology using ultrathin cross-sections from the surface of the material (Fig. 7.8b), and (iii) direct visualization surface/internal morphology of thin nanoparticles, nanofibers or nanofilms deposited on electron-transparent carbon films (Fig. 7.8c).

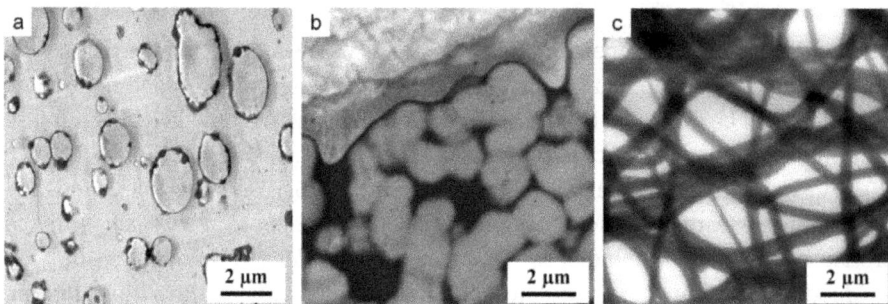

Fig. 7.8: Typical applications of STEM in polymer science. (a) OsO_4-stained ultrathin section from the interior of the sample, (b) RuO_4-stained ultrathin cross-section showing surface and subsurface region of the sample, and (c) polymer nanofibers on an electron transparent carbon film. The first micrograph (a) shows localization of styrene–butadiene block copolymer on the surface of polyethylene microparticles in polystyrene matrix [11]. The second micrograph (b) displays surface and subsurface phase structure of polypropylene/polyethylene copolymer microparticle embedded in epoxy resin [12]. The third micrograph (c) visualizes surface and overall morphology of polystyrene nanofibers [13]. All micrographs are bright-field images.

7.2.3 SEM: Modern trends

7.2.3.1 Variable-pressure SEM

Variable-pressure SEM (VP-SEM) can be used for observation of non-conductive specimens, frozen specimens, and wet-specimens. In VP-SEM, the electron column is kept under high vacuum ($p \leq 10^{-3}$ Pa), whereas the pressure in the specimen chamber is higher ($p > 10$ Pa). The gas in the specimen chamber (usually H_2O or N_2) removes the charge from a non-conductive specimen (typical chamber pressure $p \approx 100$ Pa; method called low-vacuum SEM = LV-SEM) or allows observation of wet/frozen samples (typical pressures $p \approx 1,000$ Pa; method called environmental SEM = ESEM).

In standard, high-vacuum SEM (HV-SEM), the negative charge delivered by the primary electrons (PE) to the surface of the insulator is not conducted off. This effect is called *charging*, influences both PE and signal electrons, and causes artifacts in the image (Fig. 7.9a). The sample can be covered with an ultrathin conductive layer, which usually solves the charging problem (Section 7.4.2.1). If covering of the sample is impossible or impractical, the surface can be visualized by means of LV-SEM (Fig. 7.9b). The negative charge on the specimen surface in LV-SEM is removed thanks to *cascade ionization*: signal electrons ionize gas molecules in the specimen chamber and produce positive ions and further electrons for ionization – *ionization avalanche*. The *positive* ions from the avalanche are attracted towards the negatively charged specimen surface and neutralize the surface charge, while the *negative* electrons are usually used as the imaging signal. Primary electrons in LV-SEM interact with the gas molecules as well, being scattered due to collisions with the gas molecules in the experimental chamber (*skirt effect*). The skirt effect negatively influences resolution and noise, but can be minimized by means of very short sample-objective distance (minimal PE gas path length).

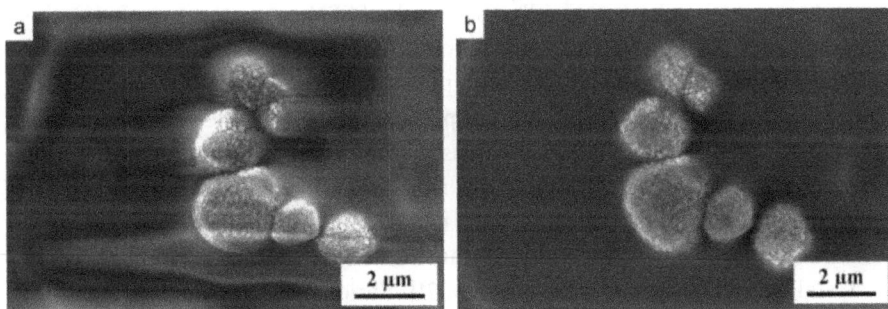

Fig. 7.9: Polymer microcomposite, uncoated, accelerating voltage 5 kV: (a) high-vacuum SEM with over- and under-saturated areas caused by sample charging and (b) the same sample observed in LV-SEM at $p = 90$ Pa.

VP-SEM is also applied on observation of samples containing liquids. The technique is usually called ESEM, as outlined above. Relative humidity around 100% at the sample area must be maintained to keep the sample hydrated. The humidity depends on the temperature (T) of the sample and pressure (p) in the specimen chamber. At room temperature ($T = 22$ °C), we need high pressure ($p = 2{,}700$ Pa) to keep the sample 100% hydrated. Such a pressure prevents high resolution imaging with low noise. Therefore, the sample is usually cooled to lower temperatures ($T \leq 5$ °C; using a cooling stage inside the specimen chamber), at which lower pressures (600–900 Pa) are sufficient to maintain the 100% humidity, and quality of the image improves significantly. Thanks to the pressure and temperature control, ESEM is suitable for both *static* experiments such as visualization of morphology of polymer hydrogels and *dynamic* experiments such as hydration and contact angle measurements [14].

In VP-SEM we may use SE and BSE imaging, as well as EDX analysis. Special SE and BSE detectors exist for both LV-SEM ($p \approx 100$ Pa) and ESEM ($p \approx 1{,}000$ Pa), while EDX detectors work at all pressures (HV-SEM, LV-SEM, and ESEM). Finally, the VP-SEM can be applied on frozen samples (in combination with cooling/freezing stage; Section 7.2.3.2, Fig. 7.10b,c) or on polymer nanoparticles in solution (cryo-SEM [15] or STEM in ESEM [16, 17]). Interestingly, the STEM-in-ESEM method (also known as Wet-STEM) can even visualize *in situ* self-assembling of nanoparticles during the controlled solvent evaporation [18]. Detailed explanation of VP-SEM principles and practice can be found in [14]; a few polymer-related applications are shown in Section 7.5.5.1.

7.2.3.2 Variable-temperature SEM

Sample can be cooled, frozen, or heated inside the SEM chamber. **Sample cooling** to temperatures around 0 °C (typically by means of a simple cooling stage inside the microscope chamber) is usually used in the ESEM to control sample humidity (Section 7.2.3.1, Fig. 7.10a). Fast **sample freezing** at very low temperature (typically by means of liquid N_2, using special equipment both outside and inside the microscope [15]) can be applied to solidify samples containing volatile components, such as polymer hydrogels (Fig. 7.10b) or self-assembled polymer nanoparticles, micelles or vesicles in solution (Fig. 7.10c), which may contain >90% of water [15–17]. More detailed description of the cryogenic SEM can be found in [19]. **Sample heating** enables us to study temperature-induced transformations of sample morphology such as crystallization or phase transitions [20, 21]. Various heating modules capable to heat the sample up to 1,500 °C are commercially available [22, 23]. The heating experiments can be performed in the high vacuum or in a gas atmosphere (H_2O, N_2, air, He) influencing the observed processes, such as corrosion [24] or reduction and oxidation reactions [25]. The monitoring of processes at high temperatures and/or pressures is often called in situ SEM [14].

Fig. 7.10: SEM micrographs of (a, b) poly(2-hydroxyethyl methacrylate) hydrogel and (c) poly(4-hydroxystyrene)-block-poly(ethylene oxide) block copolymer nanoparticles. The first two micrographs show the same sample imaged by ESEM at +2 °C (a) and LV-SEM at −10 °C (b) [26]. The last micrograph is a cryo-HV-SEM image at −135 °C, displaying the spherical nanoparticles in flash-frozen solution/ice [15].

7.2.3.3 Low-voltage SEM

Observation of the sample at accelerating voltages below 5 kV [27] is called *low-voltage SEM* (LV-SEM; note that the same abbreviation is used also for low-vacuum SEM); occasionally, the technique is referenced as *low-energy SEM* (LE-SEM). In principle, it is possible to perform LV-SEM with a standard/thermionic SEM, but the resolution and image quality rapidly decrease with decreasing kV. High-resolution LV-SEM investigations require instruments equipped with an FEG (Section 7.2.1.1) [28] and immersion final lens (ultrahigh-resolution SEM). Further improvements in the electron optics [29] together with deceleration of electrons just above the sample [30] equalize resolution at high and low kV completely. State-of-the-art LV-SEM provides both SE and BSE information at the accelerating voltages down to hundreds of volts [31]. Main benefit of LV-SEM is *detailed information from the sample surface* due to lower penetration depths at low kV, which means that signal comes from the very surface layer (Section 7.2.1.2, Fig. 7.2). The second important advantage of LV-SEM consists in *lower charging and lower sample damage,* because the difference between the number of primary electrons (PE) reaching the sample surface and the number of signal electrons (SE, BSE, etc.) leaving the surface decreases with lower kV. Recently, also low-voltage EDX analysis bringing higher lateral resolution became available for the material research [32]. LV-SEM is of particular importance for non-conductive, electron-beam-sensitive materials such as synthetic polymers, as exemplified in Section 7.5.5.2.

7.2.3.4 Multidimensional SEM

SEM yields two-dimensional (2D) information of sample topography and composition. However, real samples are three-dimensional (3D) and the depth information is often important. Another dimension of the information that SEM can provide is the angle and/or energy of detected electrons.

3D-visualization of surface morphology can be obtained by a method called **stereomicroscopy.** Two images of the sample at two different angles/tilts (*stereo images*) are taken and processed by specialized software [33]. Quantitative depth information can be extracted from the stereo images by sophisticated mathematical algorithms [34]. Another way to receive 3D information of the sample topography is to combine images from two differently located SE detectors [35] or images from SE and BSE detectors [36].

Three-dimensional visualization of subsurface morphology is achieved by means of methods that incorporate either **mechanical sectioning** or **focused ion beam cutting.** Mechanical sectioning (or slicing) is made using ultramicrotomy (Section 7.4.3.4). The ultramicrotome [37] is located either outside the microscope (ultrathin sections in given order are observed using TEM or SEM) or even inside the specimen chamber (such a method is called serial block face SEM and the surfaces are observed with SE or BSE). Focused ion beam SEM (also called FIB-SEM or dual-beam SEM; [38]) uses focused beam of heavy ions (ion beam; the first beam) to etch off a thin layer from a selected area on the specimen surface and the electron beam (the second beam) to observe the subsurface morphology. In the final step of all 3D methods, the set of 2D images is reconstructed into a final 3D model.

Nondestructive 3D-SEM was introduced recently: the area of interest is scanned repetitively by primary electrons of different energies and hence different depth of origin of detected BSE, which is employed in the final 3D reconstruction [39]. *Three-dimensional-TEM microscopy* is used mostly in biology: set of 2D micrographs from one (ultrathin) sample at various tilt angles (typically ±70°, step 1°) is reconstructed into the final 3D model. *Time-resolved SEM/TEM* (where the third dimension is time) can monitor in situ processes in liquid or gas environment in both SEM and TEM [40].

Multidimensional sample information can also be obtained by energy and/or angular filtration of collected electrons. Standard SEM has two detectors inside the specimen chamber (in-chamber SE detector and in-chamber BSE detector). High-resolution field-emission SEMs are equipped with another two or three detectors in the final lens (in-lens detectors). These detectors can simultaneously collect SE and BSE and filter them according to their energy and/or emission angle. This is possible even at low beam energies and currents which makes them suitable for surface sensitive imaging of polymers. Typical in-lens detector setup contains two detectors, one for the SE and second for the BSE. On top of that a third detector can be added to enable angularly selective BSE detection. BSE emitted at smaller angles with respect to the primary beam axis provide stronger material contrast. The material contrast decreases and topographical information becomes more important with increasing angle of BSE emission as illustrated in Fig. 7.11. Another example is energy-selective SE detection providing images from the uppermost surface layer (detector 1 collecting low-energy SE) and topography (detector 2 detecting higher energy SE) [41]. This approach can be further extended towards SE spectroscopy and energy-filtered SE imaging capable to highlight minor material differences in polymer blends [42]. Another emerging technique is based on pixelated detectors [43]. These detectors are made as an array of unit cells (pixels). Signal from each

Fig. 7.11: SEM micrographs of polyimide/SiO$_2$ composite: (a) in-lens SE, (b) mid-angle BSE, (c) small-angle BSE. Source: archive of the first author (MŠ); unpublished data.

cell can be processed independently. This provides new opportunities for SEM imaging, such as diffraction by transmitted [44] or backscattered electrons [45].

7.2.4 SEM: further possibilities

The remaining modes of SEM microscopy, which have not been mentioned in the previous sections, are rarely used in polymer science. We list them for the sake of completeness. The methods can be divided into three groups: **Spectroscopic methods.** EDX (or EDS, Section 7.2.2.3) belongs among classical modes of SEM. WDX (or WDS; wavelength-dispersive analysis of X-rays) is analogous to EDX (the same signal is detected), but WDX adjustment and data collection times are longer (main disadvantage for electron-beam sensitive polymer materials), while WDX results are more precise (better signal-to-noise ratio, narrower peaks). AES (Auger electron spectroscopy) detects emitted electrons with characteristic energies, which are used for microanalysis analogously to EDX and WDX. CL detects electron-beam-generated light (luminescence) and is quite often applied to analysis of luminescent minerals. **Methods connected with crystal structure of samples** are hardly applicable on polymer materials, because synthetic polymers are almost never completely crystalline and, moreover, the polymer microcrystals tend to decay rapidly under electron beam (amorphization). The methods related with crystal structure are: EBSD (electron backscattered diffraction), ECCI (electron channeling contrast imaging), and ECP (electron channeling pattern); in special microscopes it is possible to use LEED (low-energy electron diffraction) and RHEED (reflection high-energy electron diffraction) [1, 2]. **Methods employing sample-specific interactions with electron beam.** In semiconductors, PE can generate electron-hole pairs, which results in measurable electron beam induced current (EBIC mode of SEM). In magnetic specimens, the PE and SE are deflected due to the magnetic fields, which makes even the observation of magnetic fields possible. Other special SEM methods are discussed elsewhere [2].

7.3 Transmission electron microscopy

TEM is applied on polymer characterization analogous to STEM (Section 7.2.2.4). Polymer sample preparation is the same for both methods. If we observe the same sample in STEM and TEM, the STEM micrographs usually exhibit higher contrast (connected with lower accelerating voltages of SEM microscopes in comparison with conventional TEMs) and lower resolution (which is limited by spot size in SEM, but not in conventional TEM [1]). Typical applications of STEM (Fig. 7.8) and TEM (Fig. 7.12) in polymer science are analogous: TEM can be applied to study the detailed internal morphology using ultrathin sections from the interior of samples, such as polymer nanocomposites (Fig. 7.12a), surface and subsurface morphology using ultrathin cross-sections perpendicular to sample surface (Fig. 7.12b), and morphology of polymer nanoparticles or nanofibers which are tiny enough to be directly deposited and visualized on electron-transparent carbon film (Fig. 7.12c).

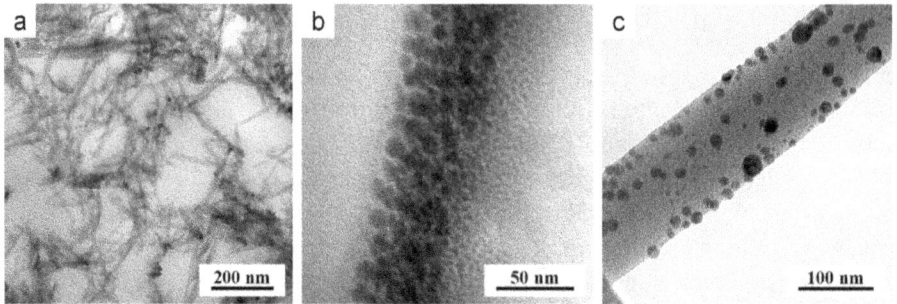

Fig. 7.12: Typical applications of TEM in polymer science: (a) Ultrathin section from the interior of a sample, (b) ultrathin cross-section showing surface and subsurface regions, and (c) polymer nanoparticles on an electron transparent carbon film. The first micrograph (a) shows dispersion of titanate nanotubes [46] in polypropylene matrix. The second micrograph (b) displays cross-section of poly(ethylene terephthalate) embedded in epoxy resin; the polymer surface contains implanted nickel nanoparticles [47]. The third micrograph (c) visualizes polyurethane nanofiber covered with Ag nanoparticles [48]. All micrographs are bright-field images.

As for *polymer morphology*, bright-field imaging (shown in Fig. 7.12) is by far the most important mode of TEM. EDX in TEM is based on the same principles as in SEM (Section 7.2.2.3), but higher resolution can be achieved [1]. Other modes of TEM (electron diffraction, dark-field imaging, scanning transmission mode of TEM in combination with electron spectroscopy, high-resolution TEM etc.) are used mostly for analysis of inorganic crystalline materials [49]. In the field of polymer science, electron diffraction can characterize inorganic nanoparticles employed in functional polymer systems [50–52].

As for *polymer surface morphology*, SEM is the dominant technique and TEM plays just a supplementary role. Nevertheless, for polymer surfaces decorated with nanoparticles it is usually advantageous to combine SEM and TEM results [53]: SEM micrographs show the nanoparticles on the polymer surface (Fig. 7.13a, b), while TEM/BF micrographs of ultrathin sections perpendicular to the surface reveal possible penetration of nanoparticles inside the polymer matrix (Fig. 7.13c). The polymer films were immobilized in an embedding resin before cutting ultrathin sections from TEM (Section 7.4.3.4, [50, 53]). Details concerning TEM microscopy (such as scheme of the microscope, principle of imaging, electron diffraction etc.), which are of minor interest for polymer surface characterization, are to be found elsewhere [1, 3, 49, 54, 55].

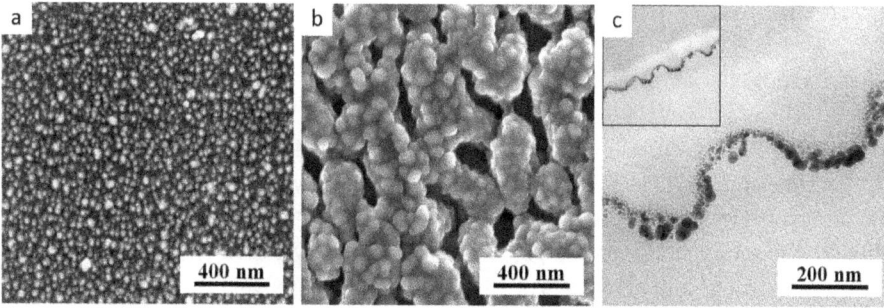

Fig. 7.13: Micrographs showing poly(ethylene terephthalate) with Ag nanoparticles immobilized on its surface by means of laser-induced optomechanical processing [53]: (a, b) FEGSEM micrograph (SE imaging) showing the PET surface with Ag nanoparticles immobilized at laser fluences 14 and 22 mJ/cm^2, respectively. (c) TEM micrograph (BF imaging) of the cross-section perpendicular to the surface of the second sample, displaying penetration of Ag nanoparticles into PET film.

7.4 Sample preparation

7.4.1 Overview of polymer materials

From the point of view of morphology and sample preparation, polymer materials can be divided into several groups (Fig. 7.14). Single-phase systems (amorphous polymers, miscible polymer blends) are morphologically homogeneous and so the microscopic studies are limited to phenomena like orientation, deformation and fracture. Multiphase polymer materials containing only polymer components (semi-crystalline polymers, block copolymers and immiscible blends) usually require specific sample preparations due to low contrast between the phases. Multiphase systems with non-polymer components (filled systems and composites) must be prepared by different techniques, taking into account different mechanical properties of the polymer matrix and the filler.

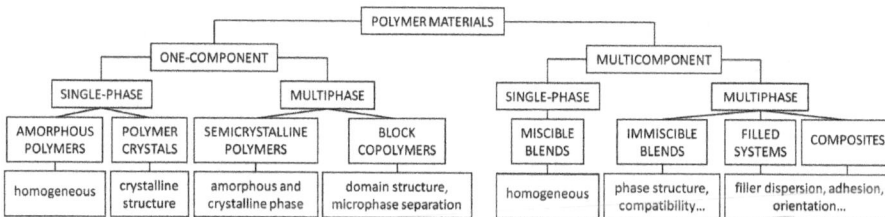

Fig. 7.14: Types of polymer materials according to their initial composition and resulting morphology.

7.4.2 Specific features of polymer materials

7.4.2.1 Charging and electron-beam damage

In a typical SEM experiment (nonconductive material such as polymer, bulk sample, accelerating voltage >5 kV) the amount of primary electrons penetrating into the sample is higher than the amount of SE and BSE emitted from the surface. Consequently, the sample tends to *charge*, which results in lower quality of micrographs or in image artifacts (subsections 7.2.3.1 and 7.2.3.3). Moreover, the non-conductive polymer specimens are more sensitive to electron-beam induced damage (surface deformation, cracking and/or contamination). We list the most frequent ways how to avoid charging and electron-beam damage: **Coating of the sample** with a thin layer of a (semi)conductive material by means of commercially available sputter-coating or thermal evaporation devices [56]. In SEM, we usually coat the samples with heavy elements such as Au (average grain size ~10 nm), Pt (~4 nm) or Cr (~2 nm), which also increase the signal, i.e. the number of emitted SE and BSE electrons due to higher Z (Fig. 7.2). In both SEM and TEM we use a carbon (semiconductive light element, amorphous layer), which does not hide the observed structures and increases sample stability under electron beam. **Variable pressure SEM** may be used for elimination of charging (Section 7.2.3.1). **LV-SEM**, which is more-and-more common in modern SEM microscopes, usually means lower sample damage and increase in surface detail at the expense of resolution (Section 7.2.3.3.). In TEM, lower acceleration voltage (below ~5 kV) usually means higher contrast, but also higher sample damage because low energy electrons are more absorbed in the sample.

7.4.2.2 Skin-core effect

Great majority of bulk synthetic polymer products is made from the melt using processes such as mixing, molding, compression, extrusion, etc. During these processes, the polymer melts are subjected to flow (different flow rates at the surface and inside) and thermal treatment (different cooling rates at the surface and inside). Consequently, the morphology close to the surface (skin) frequently differs from the morphology of the interior (core). The morphological changes include size, concentration and/or orientation of lamellae in semicrystalline polymers (Fig. 7.15a), particles in polymer blends (Fig. 7.15b) or fillers in polymer composites (Fig. 7.15c). The skin-core effects are observed also in solution-prepared polymer materials (Fig. 7.15d) and natural polymers.

Two important conclusions concerning sample preparation follow from the skin-core effects: (i) If we study the external surface, we should be aware of the fact that the internal morphology may differ significantly. (ii) If we analyze cross-sections (such as fracture surfaces, cut surfaces or sections), we should take micrographs

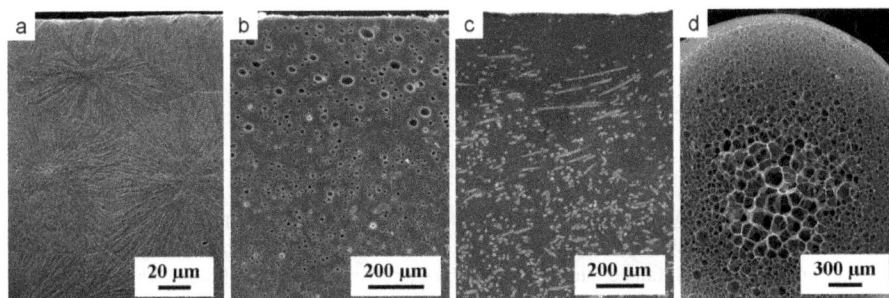

Fig. 7.15: Typical skin-core effects in polymers: (a) Crystalline lamellae in polypropylene crystallized on highly oriented pyrolytic graphite, (b) smooth surface of polystyrene with styrene maleic anhydride, (c) short glass fibers embedded in polyamide polymer matrix, and (d) cross-section of polystyrene foam bead.

from at least two regions – close to the original surface (skin morphology) and from the middle of the specimen (core morphology).

7.4.2.3 Low contrast between components

From the point of view of elemental composition, synthetic polymers are frequently quite similar. Consequently, the yield of SE, BSE, TE and characteristic X-rays from the individual components of multiphase polymer systems is almost the same. Therefore, the polymer systems are rarely observed "as received". Instead, their morphology has to be made visible by suitable preparation techniques (fracturing, etching, staining etc.) as described below.

7.4.3 Preparation techniques for polymer materials

The polymer materials can be observed "as received" (Fig. 7.16a) or their morphology is revealed by means of preparation techniques, the most important of which are fracturing, etching, or cutting (Fig. 7.16b–d). The direct observation of polymer surfaces (Fig. 7.16a) is usually limited to the observation of overall shape of polymer micro/nanoparticles (Section 7.4.3.1.). More often we "open" the structure, preparing *fracture surfaces* (Fig. 7.16b), *etched surfaces* (Fig. 7.16c), and *cut surfaces or sections* (Fig. 7.16d). With those surfaces, we usually achieve a higher topographic contrast in SE (fracture surfaces, Section 7.4.3.2) or higher material contrast in BSE (cut and stained surfaces, Sections 7.4.3.3–7.4.3.4).

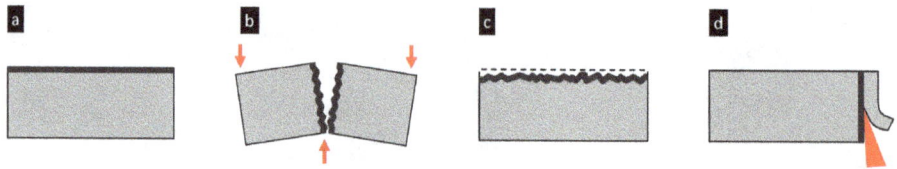

Fig. 7.16: Main types of polymer surfaces from the point of view of sample preparation: (a) original surface, observed directly; (b–d) surfaces prepared by fracturing (b), etching (c) or cutting (d), which often show more details of surface, subsurface and/or interior morphology. The observed surfaces are denoted with thick line.

7.4.3.1 Direct observation of polymer surface

Morphology and surface of polymer microparticles (Fig. 7.17a), nanoparticles (Fig. 7.17b) or fibers (Fig. 7.17c) is usually observed without complicated sample preparation.

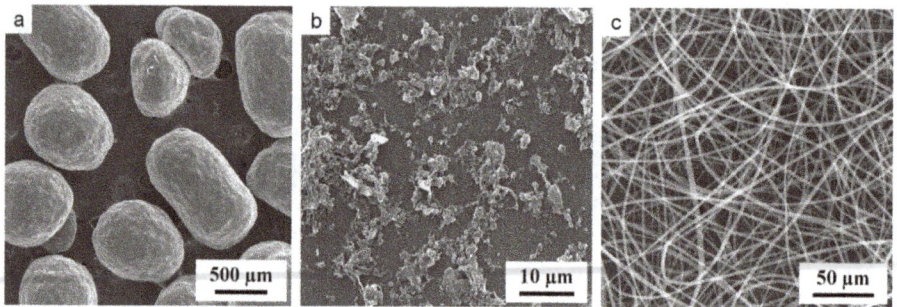

Fig. 7.17: Typical examples of the direct observation of surface morphology: (a) polymer microparticles for drug delivery, (b) isolated ultrahigh molecular weight polyethylene wear micro- and nanoparticles from periprosthetic tissues, and (c) poly-gamma-benzyl-L-glutamate nanofibers.

The sample is just placed onto a flat surface, fixed with a paste or glue or double adhesive tape, surface-coated and observed in SE mode of SEM. Nanoparticles do not have to be glued, the surface-coating usually fixes them sufficiently. If we want to observe sample without surface coating, conductive support (such as mica pre-sputtered with platinum), conductive glue (such as commercially available silver paste), lower accelerating voltage (lower penetration depth and charging) and/or low-vacuum microscopy have to be applied. Direct observation of bulk polymer surfaces is less frequent. The surfaces of bulk polymers only reflect the processing-associated features (such as roughness of the molds or scratches due to subsequent machining or polishing). If these features are of interest, the sample preparation is analogous to that of polymer micro/nanoparticles.

7.4.3.2 Fracturing

Fracture of any material is controlled by its mechanical properties and applied forces. In polymer sample preparation we differentiate between brittle fractures (sharp fracture lines, no deformation textures – Fig. 7.18a) and ductile fractures (fracture surface with visible plastic deformations/textures, no sharp fracture lines – Fig. 7.18c). If the fractured material contains heterogeneities, these frequently become centers of secondary fractures (Fig. 7.18b) and typical conic fracture lines are formed on the brittle fracture surface [57].

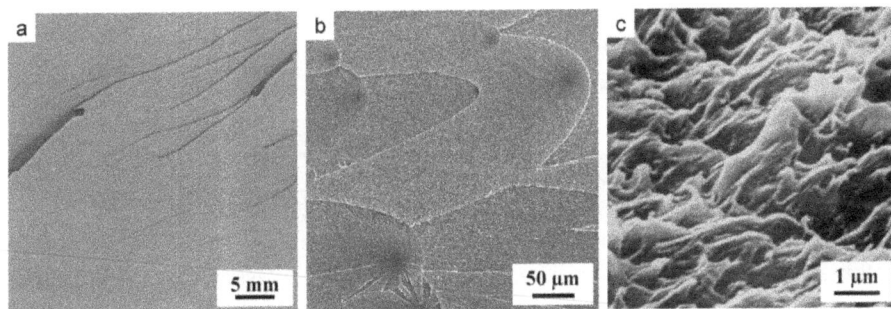

Fig. 7.18: Fracture surfaces: (a) brittle fracture in epoxy resin, (b) brittle fracture in epoxy resin/rubber polymer with centers of secondary fracture, (c) ductile fracture with microplastic deformations in acrylonitrile-butadiene-styrene copolymer [58].

As the (micro)plastic deformations hide original morphology of the sample, the brittle fracture surfaces are usually preferred. Fracturing at low temperature (mostly in liquid nitrogen) is used with those polymer materials which are not sufficiently brittle at room temperature. Care should be taken to cool the sample thoroughly (samples completely submerged in liquid nitrogen bath, cooling time ≥5 min, tools for fracturing cooled in the same way as the sample). After fracturing, the polymer sample is fixed on a conductive support, sputter-coated and observed in SEM.

The fracturing is quite universal and widely used technique in polymer microscopy. It reveals morphology of polymer blends (Fig. 7.4a, [59]) and composites (Fig. 7.9 and Fig. 7.19, [60]). Fracture surfaces of polymer systems after impact testing help to distinguish various types of fracture mechanism (such as linear elastic and elastic-plastic with unstable or stable crack propagation) [61–63]. Fracturing is also a unique method how to find minute amount of heterogeneities or defects present in material, as the defects are easily localized on fracture surface due to the fact that they are centers of secondary fracture (Fig. 7.19a, [57]). Special fracturing conditions can be employed when classical preparations fail: for instance, fracturing with elastomeric matrix (at elevated temperatures) is advantageous for amorphous polymers with low amount of inorganic filler (Fig. 7.19b) [64].

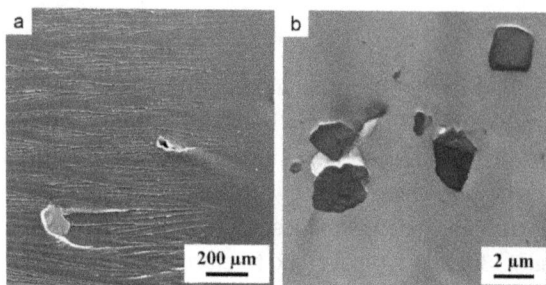

Fig. 7.19: (a) Brittle fracture in an epoxy resin, with inorganic inhomogeneities acting as sources of secondary fracture. (b) Soft matrix fracture surface of ethylene-propylene-diene-monomer rubber with a low content of $CaCO_3$ particles.

7.4.3.3 Etching

Etching consists in the selective removal of component(s) from a multiphase polymer system (Fig. 7.14) by chemical or physical treatment. The surface for etching is obtained by any of the ways listed above (Fig. 7.16 – "as received", fractured, etched or cut surfaces). The chemical etching means treating the polymer surface with a sample-specific chemical, called *etching agent*. The physical etching of the surface is carried out in special devices, where the surface is exposed to high energy beam, and the processes are called *ion beam etching* or *plasma etching*.

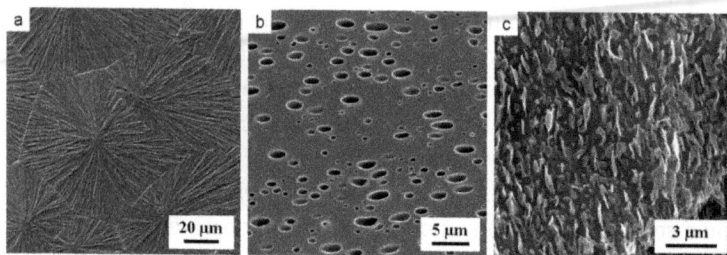

Fig. 7.20: SEM micrographs (SE imaging) of (a) PP spherulites etched with permanganic mixture (b) PP/EPR polymer blend etched with n-hexane, and (c) PP/MMT composite fracture surface etched with plasma.

Figure 7.20 shows three typical examples of etching, applied on multiphase polymer systems: semicrystalline polymer (Fig. 7.20a), immiscible polymer blend (Fig. 7.20b) and polymer composite (Fig. 7.20c). The polypropylene in Fig. 7.20a was crystallized on mica decorated with gold nanoparticles, the mica was peeled off and the unmodified surface was etched with permanganic mixture [65]. The polypropylene/ethylene propylene rubber (PP/EPR) polymer blend in Fig. 7.20b was cut with a microtome,

the cut surface was smoothed under liquid nitrogen [66] and etched with permanganic mixture [67]. The polypropylene/montmorillonite (PP/MMT) composite in
Fig. 7.20c was fractured and etched by plasma [68].

Permanganic mixture is a quite universal etching agent, which etches almost
all polymers and the different rate of etching visualizes their morphology. In semicrystalline polymers (Fig. 7.20a), the amorphous phase is etched faster and the crystalline lamellae emerge [69]. In immiscible polymer blends (Fig. 7.20b), the different
etching rate of the components reveals phase structure [70]. Other etching agents applied on specific polymers include toluene [71], n-hexane [72], tetrahydrofuran [73],
hydrochloric acid [74] etc. Plasma etching gives the best result if one component is
much more resistant to ion bombardment – typical example is a polymer composite
with hard inorganic filler (Fig. 7.20c; [68]).

7.4.3.4 Cutting and staining

Cutting is a frequently used technique that yields *cut surfaces* for SEM and *ultrathin
sections* for (S)TEM. In most cases, cutting is combined with *etching* (described in
previous section) or *staining*, in order to achieve sufficient contrast. Two most important staining agents for polymer materials are OsO_4 and RuO_4. During staining,
the heavy atoms (Os or Ru) selectively binds to one phase, which then yields higher
signal in SEM/BSE (Fig. 7.21a) or (S)TEM (Fig. 7.21b) due to higher atomic number Z.

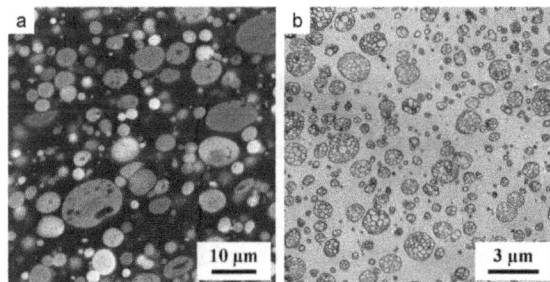

Fig. 7.21: SEM micrographs of high-impact polystyrene prepared by ultramicrotomy followed by
OsO_4 staining: (a) cut surface, BSE imaging; (b) ultrathin section, TE micrograph, bright field. The
stained particles appear darker than unstained matrix in TE micrograph and brighter in BSE
micrograph due to higher atomic number of the staining agent.

OsO_4 staining is more specific because osmium attacks selectively the components
containing double bonds [75]. RuO_4 staining is more universal as ruthenium stains
almost all common polymers (exception is polymethylmethacrylate and its derivatives) [59]. Other staining agents containing heavy elements (uranyl acetate, phos-
phorotungstenic acids) are used less often [76, 77].

Cut surfaces and/or thin sections can be prepared by three different techniques: (a) microtomy, (b) ultramicrotomy, and (iii) smoothing in liquid nitrogen. Other methods are unsuitable for polymer systems because they produce rough and/or deformed surfaces. *Microtomy* yields cut surfaces and *thin sections* (thickness ~10 μm) that are thin enough for light microscopy. *Ultramicrotomy* gives even smoother cut surfaces and *ultrathin* sections (thickness ~50 nm), which are transparent in STEM and TEM; polymer thin films, fibers and micro- or nanoparticles can be immobilized in an embeding resin in order to make ultramicrotomy feasible [50, 53]. *Smoothing* consists in polishing the sample submerged in liquid nitrogen using a freshly broken glass. This method yields only surfaces, but it is simpler and faster than (ultra)microtomy, and yields high-quality surfaces for morphology studies of polymer blends (Figs. 7.20b and 7.21a) [66, 67]. It is worth noting that smoothing and etching works well also for popular biopolymer systems, such as TPS/PCL or PCL/PLA blends (where TPS is thermoplastic starch, PCL is poly(ε-carprolactone) and PLA is poly(lactic acid)) [62, 73, 74].

7.4.3.5 Special techniques

In addition to the classical sample preparation techniques mentioned above, any other physical or chemical procedures can be applied in order to reveal the surface or subsurface morphology. The techniques can be divided into several groups: (i) combined techniques, (ii) less frequently used techniques, and (iii) specific techniques for wet specimens. Typical examples follow.

Combinations of various preparation techniques may be advantageous, if not necessary. The combinations are not limited to the typical examples mentioned above (cutting/etching, Fig. 7.20b; fracturing/etching, Fig. 7.20c; and cutting/staining, Fig. 7.21). Other examples of combined techniques are cutting/annealing [78] or heating/fracturing [64]. Possible combinations and other procedures depend solely on the investigated systems, their structure, and properties.

Less frequently used techniques may improve characterization of specific samples. Interfacial adhesion between polymer matrix and filler is usually examined on fracture surfaces (Fig. 7.30), but it can be also monitored by scratching the surface and SE imaging of the scratch morphology [79, 80]. Phase structure of multicomponent systems is revealed by fracturing, cutting, staining, etching and their combinations, as described above. However, the surface relief on the cut surface can also be induced by selective swelling [81]. Heterogeneous phase structure of polymer composites can be revealed by fracturing at the temperature, at which the polymer matrix is soft (above its glass transition temperature, T_g) while the dispersed phase is hard (below its T_g) [64]. In the cases when neither cutting nor fracturing is applicable (hard and resistant polymer microcomposites), sample polishing in combination with BSE imaging can solve the problem (Fig. 7.15c). Abrasive papers are to be used

with successively finer graininess similarly to the metallographic sample preparation procedures [82, 83]. As for STEM and TEM, two less common but useful techniques are worth mentioning: (i) negative staining facilitates visualization of polymer
micro/nanoparticles [76] and (ii) surface replication makes it possible to visualize
surface at high resolution using TE imaging [84].

Specific preparation techniques for wet specimens have already been discussed
above, because they are closely connected with variable-pressure and variable-tem-
perature SEM (Sections 7.2.3.1–7.2.3.2). To summarize, water-containing specimens
in EM can be observed in dried, frozen, or hydrated form. The *drying* of water-containing specimens must be carried out in such a way that their morphology did not
collapse [85]. The *freezing* must be performed quickly (liquid N_2, liquid propane) so
that the ice crystals were negligible and did not destroy the internal structure [15,
86]. The direct observation of *hydrated/wet samples* requires a specialized SEM microscope, which allows controlling pressure and temperature in the specimen
chamber (Section 7.2.3.1, [87]). Summary and detailed comparison of the preparation and visualization techniques for polymer hydrogels with high content of water
can be found in recent review [88].

7.5 Applications

7.5.1 Homopolymers

Homopolymers contain just one type of repeating unit in the polymer chain. Very
definite differences in morphology and properties occur between *amorphous homopolymers* (which cannot crystallize) and *semicrystalline homopolymers* (which crystallize to significant extent). *Fully crystalline homopolymers* (or polymer single
crystals) are exceptional and their morphology is discussed elsewhere [89]. The
ability of a polymer to form a crystal depends on the regularity, symmetry and chi-
rality of the monomer units forming the homopolymer chain [90].

Amorphous polymers (such as polystyrene) are morphologically homogeneous (Fig. 7.14). They do not exhibit any ordered structures in WAXS (wide-angle
X-ray scattering) and do not have a first-order melting transition in DSC (differential
scanning calorimetry). Several structure models for amorphous polymers have been
suggested [91], but numerous measurements indicated that the amorphous state is
homogeneous with no anisotropic structures [92]. SANS (small-angle neutron scattering) measurements demonstrated that amorphous polymers contain randomly
oriented molecules in the form of Gaussian coils [93]. On the other hand, *aging*
(change of macroscopic properties as a function of time) of amorphous polymers
indicates that some structural changes and chain rearrangements take place [94,
95]. Consequently, the main morphological features of amorphous polymers are

density fluctuations and entanglements between the polymer chains which exhibit very low contrast and cannot be observed by SEM or TEM [96, 97]. Therefore, the morphology of amorphous polymers limits itself to studies of deformation and fracture (type of fracture – Fig. 7.18; deformation zones, crazes, deformation, and shear bands [98]).

Semicrystalline polymers (such as polyethylene) contain amorphous and crystalline regions/phases (Fig. 7.14). The polymer chains are either randomly oriented (Fig. 7.22a) or packed in lamellae (Fig. 7.22b–c). Real conformations of the polymer chains are usually a mixture of all three cases (Fig. 7.22d) and depend on processing history (temperatures, pressures, nucleating agents/fillers, etc.). Typical bulk semicrystalline polymer consists of amorphous phase (Fig. 7.22a) and crystalline lamellae (Fig. 7.22b) in the form of flat crystallites organized into polymer spherulites (scheme in Fig. 7.23). Example of polymer spherulites, observed in SEM microscope, was given in Fig. 7.15a. Other common morphologies of semicrystalline polymers include randomly oriented lamellae in amorphous matrix, oriented lamellae and/or fibrous structures, epitaxial lamellar overgrowth on microfibrils, hedrites or axialities, dendrites, and single lamellae [99].

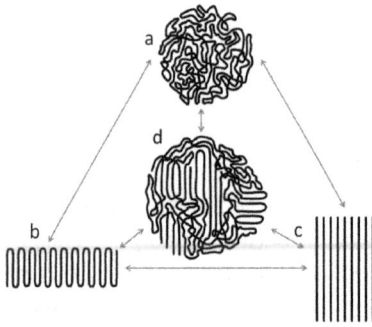

Fig. 7.22: Chain conformations in semicrystalline polymers: (a) random coils, (b) folded, and (c) extended chain lamella, and (d) mixture of (a + b + c).

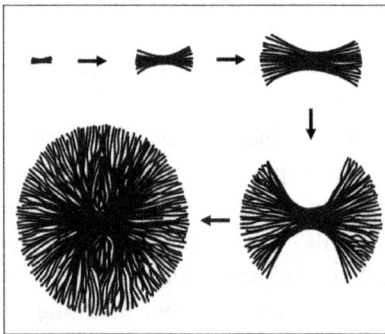

Fig. 7.23: Flat crystalline lamella (that corresponds to Fig. 7.22b) and its growth with branching, which results in formation of polymer spherulite [100].

Ultrahigh-molecular-weight polyethylene (UHMWPE) is a semicrystalline polymer used as bearing material in total joint replacements [101]. It contains crystalline lamellae randomly oriented in the amorphous matrix (Fig. 7.24). Weight fraction and thickness of the lamellae depend on the processing and influence mechanical properties of the polymer [102, 103]. The lamellae can be visualized by EM: SEM micrographs (Fig. 7.24a, c) show cut surfaces (prepared by microtomy) etched by permanganic mixture, while TEM micrographs (Fig. 7.24b, d) display ultrathin sections (prepared by ultramicrotomy) stained with oleum (details on sample preparation are given in [102]). The difference between the virgin polymer (Fig. 7.24a, b) and remelted polymer (Fig. 7.24c, d) is clearly visible.

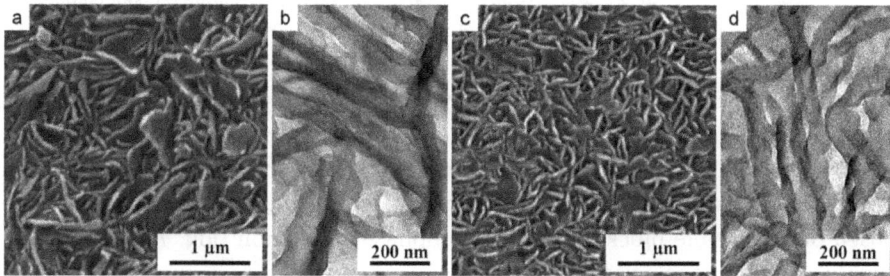

Fig. 7.24: SEM (a, c) and TEM micrographs (b, d) micrographs of UHMWPE crystalline lamellae in amorphous matrix: (a, b) virgin UHMWPE and (c, d) remelted UHMWPE (sample thermally treated at 150 °C).

7.5.2 Copolymers

Copolymers contain two or more repeating units in the polymer chain. *Statistical/ random copolymers* have statistically/randomly linked repeating units in the linear chain, while in *alternating copolymers* the repeating units alternate regularly. *Block copolymers* contain two (or more) long blocks of the same repeating units. Other geometries include branched and star copolymers [104].

Fig. 7.25: Microphase separation in block copolymers composed of two immiscible blocks A and B. The blocks A and B tend to minimize mutual contacts and separate from each other, forming various structures shown in Fig. 7.26 [104].

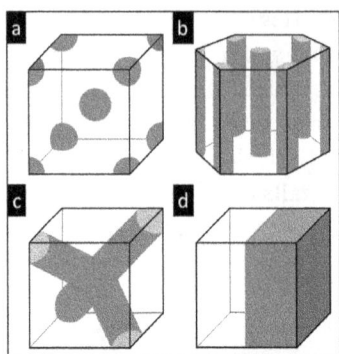

Fig. 7.26: Morphology of block copolymers A-*block*-B, composed of two immiscible blocks A and B. As the concentration of block A increases, the structure changes from cubic (a), to hexagonal (b), gyroid (c), and lamellar (d) [104].

From the point of view of morphology, statistical, random, branched, and star copolymers are usually amorphous, as their irregular molecular structure prevents crystallization. Alternating copolymers can be both amorphous and semicrystalline, depending on the symmetry and regularity of their chains. Block copolymers, composed of immiscible blocks, exhibit microphase separation (Fig. 7.25) [104] and wide range of morphologies (Fig. 7.26) [104].

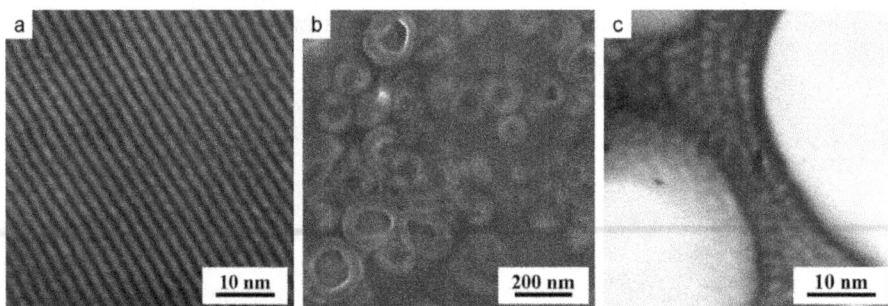

Fig. 7.27: Samples showing the variability of copolymer structures: (a) TEM micrograph of lamellar pentablock polystyrene (PS) and polybutadiene (PB) copolymer with lamellar morphology (Fig. 7.26d); (b) SEM micrograph of dried polymer vesicles, which are formed by copolymer of polystyrene and poly(acrylic acid), PS-b-PAA; (c) TEM micrograph of PB particles with PS-block-PB compatibilizer in PS/PB polymer blend; the PS-b-PB compatibilizer between the particles exhibits hexagonal morphology (Fig. 7.26b).

Figure 7.27 shows three typical examples and applications of block copolymers. The pentablock copolymer of polystyrene and polybutadiene (PS-*b*-PB; Fig. 7.27a) belongs to an important class of polymers called *thermoplastic elastomers*, i.e. materials exhibiting both thermoplastic and elastomeric properties [96]. The bicontinuous lamellar structure of block copolymer in Fig. 7.27a) was obtained from 50% of PS and 50% of PB; the sample was prepared by ultramicrotomy, stained with OsO_4 and observed in TEM using bright field imaging.

The copolymer of polystyrene and poly(acrylic acid) (PS-*b*-PAA; Fig. 7.27b) exhibits self-assembling in solution and forms polymer vesicles [16]. Block copolymer self-assemblies, such as micelles and vesicles, are studied due to their possible applications in drug delivery. Self-assembled PS-*b*-PAA vesicles can be observed in dried state using standard SEM microscopy (as illustrated in Fig. 7.27b) or in hydrated state using wet-STEM microscopy (as shown in our study [16]).

Diblock copolymer of polystyrene and polybutadiene (PS-*b*-PB; Fig. 7.27c) was used as a *compatibilizer* in PS/PB polymer blend. The block copolymer *compatibilizers* are used to increase interphase adhesion between the components of polymer blends, stabilize the morphology and improve final mechanical properties [105]. Figure 7.27c shows a TEM micrograph of PS-*b*-PS copolymer with hexagonal morphology (Fig. 7.26b) between two polystyrene particles. The bulk PS/PB blend sample was prepared by melt-mixing, the specimen for TEM microscopy was prepared by ultramicrotomy in combination with OsO$_4$ staining (Section 7.4.3.4).

7.5.3 Polymer blends

Polymer blends consist of two or more different polymers. Great majority of polymers are mutually incompatible [105] and so the polymer blends are multiphase systems (Fig. 7.14). Phase morphology of the immiscible polymer blends (particulate, fibrous, co-continuous) is closely connected with their properties (modulus, toughness, gas permeability) [105]. Therefore, blends morphology control (by optimizing composition, processing conditions and/or compatibilization) and visualization by EM have been attracting the attention of numerous researchers [11, 63, 97, 106].

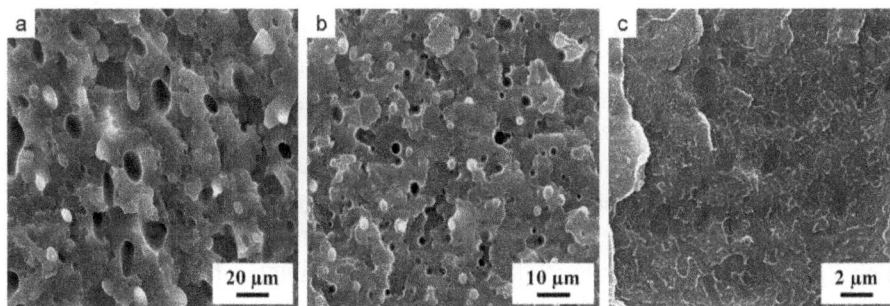

Fig. 7.28: SEM micrographs (SE imaging) of fracture surfaces of PS/LDPE (80/20) blends: non-compatibilized (a) and compatibilized with 5% (b) or 10% (c) of polystyrene-*block*-polybutadiene copolymer [11].

Compatibilization of polymer blends increases adhesion between the blend components (interphase adhesion) and improves mechanical performance. Block copolymers

(Section 7.5.2) with blocks that are identical, miscible or adhering to the related components of a blend, are frequently used as *compatibilizers* [11]. SEM micrographs in Fig. 7.28 illustrate typical changes of blend morphology as a function of compatibilizer concentration: non-compatibilized polystyrene/low-density polyethylene blend (PS/LDPE) exhibits strong phase separation with formation of large particles and fracture running along the interface (Fig. 7.28a). Addition of 5% of polystyrene-block-polybutadiene compatibilizer (PS-*b*-PB) decreases average particle size (Fig. 7.28b), but the fracture still propagates along the interface. Further increase in compatibilizer concentration to 10% leads to even smaller average particle size and improvement of interphase adhesion, which is evidenced by the fact that the fracture runs directly through the LDPE particles and not along the interface (Fig. 7.28c).

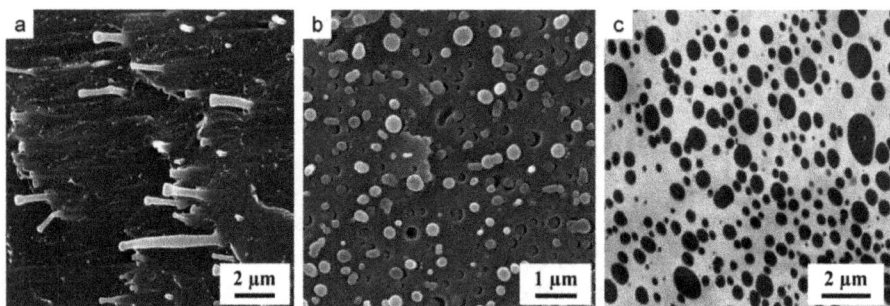

Fig. 7.29: SEM micrographs of PP/COC (70/30) polymer blends: (a) SE imaging of fracture surface parallel to the injection molding direction, (b) SE imaging of fracture surface perpendicular to the injection direction, and (c) STEM micrograph of RuO₄-stained ultrathin section, perpendicular to the injection molding direction.

Phase morphology is another important parameter influencing final properties of polymer blends [97, 105]. SEM microscopy of fracture surfaces, cut surfaces, and/or ultrathin sections has become a key tool for investigation of the polymer blend morphology [107, 108]. Figure 7.29 shows selected results from morphological study of polypropylene/cycloolefin copolymer blend (PP/COC) [59]. Morphology of the blend was optimized by careful selection of processing conditions so that the minority component (COC) could form reinforcing fibers, which imparted to the PP matrix higher modulus and yield strength [109]. Fibrous morphology of the PP/COC blends was clearly evidenced in SEM micrographs of fracture surfaces parallel with injection molding direction (Fig. 7.29a). SEM micrographs of fracture surfaces perpendicular to injection direction confirmed uniaxial orientation of the fibers (Fig. 7.29b). Finally, STEM micrographs of RuO₄-stained ultrathin sections perpendicular to injection direction yielded detailed information about average diameters of COC fibers in the PP matrix. Compatibilization of PP/COC blends (and analogous PE/COC blends, where PE is high-density polyethylene [71, 107]) was not necessary as the SEM micrographs, tensile testing results and comparison with theoretical predictions confirmed very high interphase adhesion [71].

7.5.4 Polymer composites

Polymer composites represent a popular group of polymer materials, in which isometric particles, fibers or plates dispersed in the polymer matrix act as reinforcing elements. In a typical polymer composite, stiff filler (glass fibers, clay platelets, etc.) is embedded in tough polymer matrix (either thermoplastic or thermosetting). The final polymer composite properties can be adjusted to final application by suitable combination of the components and processing [96, 97].

Polymer microcomposites contain filler particles with average size >1 µm. Homogeneous microfiller dispersion and good matrix-filler interphase adhesion are required in order to achieve good mechanical performance. Both dispersion and interphase adhesion can be visualized by SE imaging of liquid-nitrogen fracture surfaces of the composite as illustrated in Fig. 7.30, which shows recycled poly(ethylene terephthalate) filled with basalt fibers (Fig. 7.30a, b; sample 1) and glass fibers with coupling agent (Fig. 7.30c, d; sample 2). In both samples 1 and 2, the fibers are dispersed homogeneously in the polymer matrix. In sample 1, the fracture is brittle (sharp fracture lines in Fig. 7.30a) and the interphase adhesion is low (distinct interface between fibers and matrix in Fig. 7.30b). In sample 2, the fracture is semi-brittle (microplastic deformations in Fig. 7.30c) and the interphase adhesion is high (matrix adhering to the fiber in Fig. 7.30d). The improved interphase adhesion in sample 2 resulted in higher modulus, tensile and impact strength [60].

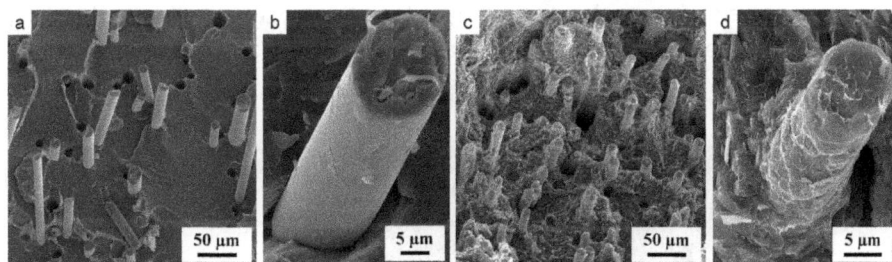

Fig. 7.30: SEM micrographs of fracture surfaces of recycled PET polymer composites with embedded: (a, b) basalt fibers [110] and (c, d) glass fibers with coupling agent [60].

Polymer nanocomposites contain particles with at least one dimension <1 µm. If the adhesion between embedded nanofiller and polymer matrix is strong, further improvement of mechanical properties can be achieved because nanoparticles have higher specific surface than microparticles. Polymer nanocomposites are studied mostly by TEM microscopy of ultrathin sections from the interior of the sample, as briefly mentioned in Section 7.3. This can be attributed to the fact that nanofillers are best observed at higher resolutions in TEM microscopes.

It is worth noting that some researchers [111] differentiate between **polymer composites** and **filled systems** (see also Fig. 7.14). The basic difference between polymer composites (or reinforced systems) and filled systems consists in that in composites/reinforced systems the fillers are used to improve properties, whereas in filled systems the fillers are used to reduce cost. However, the distinction between reinforced and filled systems is sometimes unclear [112].

7.5.5 Special applications

Synthetic polymers have several specific features (see also Section 7.4.2), which are addressed in the following three sections. LV-SEM helps to overcome problems with non-conductivity and beam sensitivity of the polymers and enhances surface structures (Section 7.5.5.1). Variable-pressure SEM makes it possible to visualize wet, hydrated, or frozen polymer specimens close to their natural state (Section 7.5.5.2). Other special applications include microscopy at elevated temperatures, under deformation, and morphology changes associated with processing (Section 7.5.5.3).

7.5.5.1 Low-voltage SEM in polymer science

Most of the samples in the polymer science are insulators which are difficult to observe in the SEM due to their charging and electron beam-induced damaging. Typical way of observation is covering their surface by a thin conductive layer. However, such a layer can cover details on the surface and therefore coating should be minimized if details on the surface are to be investigated.

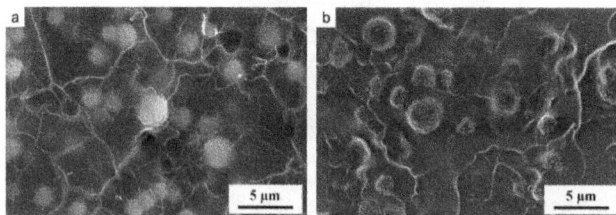

Fig. 7.31: Influence of the penetration depth on the visualization of a polyurethane sample filled with silica microparticles observed at accelerating voltage of 20 kV (a) and at 1 kV (b). The sample was uncoated for the 1 kV and carbon coated for the 20 kV imaging.

The LV-SEM enhances details on the surfaces of uncoated or mildly coated insulators as explained in Section 7.2.3.3. The LV-SEM surface sensitivity is demonstrated in the images of silica particles embedded in a polyurethane matrix captured at the accelerating voltage of 20 kV (Fig. 7.31a) and 1 kV (Fig. 7.31b). Morphology of the silica particles is less clear and particles embedded below the surface are recognizable due to high penetration depth of 20 kV electrons (see also eq. (7.3) in Section 7.2.1.2), while at lower accelerating voltage of 1 kV the surface detail is increased.

Sensitivity of the polymers to electron beam is also associated with sample contamination and/or electron beam induced sample damage. Both effects are best visualized if we observe the specimen at higher magnification and then (without moving the stage) at lower magnification. If the sample is contaminated or damaged, the lower magnification SEM micrograph shows dark rectangle or an imprint in the center of the image. The sample damage can be minimized by using lower accelerating voltages in combination with low doses (small spot size and/or probe current) [113, 114].

Some polymer specimens may charge so extensively that low accelerating voltage and electron dose do not help even if the standard way of image acquisition is used. The SEM image is usually acquired by *sequential scanning*: the frame is scanned pixel by pixel with the electron beam staying at each pixel long enough in order to obtain low noise image; each pixel of the image is irradiated just once, but using relatively high dose. Another image acquisition strategy is *integration method*: the beam dwells at each pixel a few hundred nanoseconds at most, a series of quickly scanned frames is collected, and in the last step they are integrated together. Image quality improvement due to integration method is demonstrated in Fig. 7.32. Although the total electron dose necessary to acquire images in Fig. 7.32 is the same, the integration method (Fig. 7.32b) distributes the total dose over the whole acquisition time and hence it decreases the charging. Similar, but even more advanced scanning strategies such as scan interlacing and drift correction have recently been described [115].

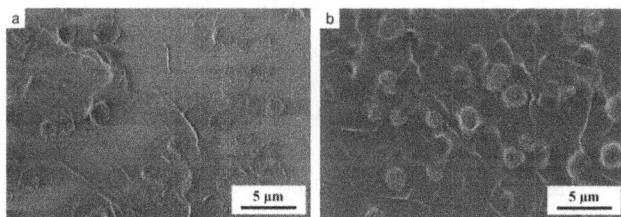

Fig. 7.32: Image of a polymer composite (polyurethane/silica microparticles; the same sample as in Fig. 7.31) acquired by sequential scanning (a) and by integration of 64 frames (b). Total acquisition time, accelerating voltage, and electron dose were identical.

All improvements described above must be used to image highly beam sensitive polymers, such as biodegradable thermoplastic aliphatic polyester–polylactic acid (PLA) treated by plasma shown in Fig. 7.33. While beam energy of 1 keV and dose of

8.8 pC/µm² mitigate sample charging, the beam energy must be decreased down to 300 eV to reduce the radiation damage. Boundary between the irradiated and virgin area is clearly visible and detailed surface structure is melted away in the 1 keV image (bottom right part of Fig. 7.33a). These artifacts do not occur at beam energy of 300 eV where the difference between the virgin and irradiated area is not noticeable (Fig. 7.33b).

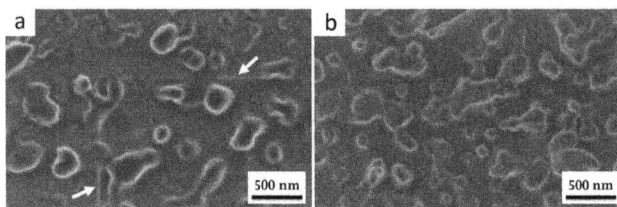

Fig. 7.33: Comparison of the level of the radiation damage of the plasma treated polylactic acid (PLA) imaged at beam energy of 1 keV (a) and 300 eV (b).

7.5.5.2 Wet specimens in polymer science

Wet polymer specimens, such as polymer hydrogels (for tissue engineering), polymer nanoparticles (micelles and vesicles for drug delivery), or polymer nanofibers (scaffolds for antimicrobial photodynamic therapy), represent important and intensively studied materials. These samples frequently contain up to 90 wt% of water and so they could not be observed in their natural state in standard (high-vacuum) SEM (HV-SEM). Several examples have already been shown above (Section 7.2.3). Here we summarize how the morphology of the samples can be visualized with EM. In the first step, we should verify if the samples can be dried in such a way that their structure does not break down (air drying, freeze drying, critical point drying [85, 116]) and observed using standard HV-SEM microscopy. In the second step, if the drying destroys the structure, the samples can be flash-frozen and observed at low temperatures ($T < 0$ °C; cryo-microscopy) using either HV-SEM or low-vacuum SEM (LV-SEM). HV-SEM can be applied if the samples are frozen at very low temperatures, where the rate of sublimation of ice is negligible ($T < -100$ °C; [15]), while LV-SEM can be applied at samples frozen just a few degrees below zero ($T \sim -10$ °C; [86]). If even the fast freezing of the samples results in artifacts, the samples can be observed in the hydrated/wet state at close-to-room temperature ($T > 0$ °C) using either SE/BSE imaging (ESEM; [26, 88]) or TE imaging (STEM-in-ESEM or wet-STEM; [16, 17]). Difficulty of the sample preparation and observation techniques usually increases in the row: HV-SEM < LV-SEM < ESEM < wet-STEM. Technical details about the methods can be found above (Sections 7.2.3.1 and 7.2.3.2); sample micrographs of wet polymer specimens are shown in Fig. 7.10.

7.5.5.3 Further applications

The above text was focused on typical applications of EM on polymer surface and/ or subsurface morphology characterization. A few other, more specialized applications are listed in this section for the sake of completeness. Polymer morphologies associated with processing, deformation and fracture are frequently studied, being focused on craze development, shear band formation, and fracture mechanics [98]. Three-dimensional microscopy of subsurface structures has become quite popular recently, but has not been discussed here as this contribution concentrates on polymer surfaces (basic information was given in Section 7.2.3.4). Polymer liquid crystals are usually studied by polarized light microscopy in combination with DSC and X-ray scattering, but EM methods can be applied as well [117].

References

[1]   Brandon D, Kaplan WD. Microstructural Characterization of Material, Chippenham, Wiltshire, UK, John Wiley & Sons Ltd, 2008.
[2]   Reimer L. Scanning Electron Microscopy, Berlin, Germany, Springer, 1998.
[3]   Watt IM. The Principles and Practice of Electron Microscopy, 2nd edn, Cambridge, UK, Cambridge University Press, 1997.
[4]   Sawyer LC, Grubb DT. Polymer Microscopy, 2nd edn, London, UK, Chapman & Hall, 1996.
[5]   Kanaya K, Okayama S. Penetration and energy-loss theory of electrons in solid targets. J Phys D: Appl Phys 1972, 5, 43–48.
[6]   Drouin D, Couture AR, Gauvin R, et al. Program Casino – Monte Carlo simulation of electron trajectory in solids (version 2.48, 2011; http://www.gel.usherbrooke.ca/casino/).
[7]   Reimer L. Image contrast and signal processing. In: Scanning Electron Microscopy, Berlin, Germany, Springer, 1998, 207–252.
[8]   Watt IM. Electron-specimen interactions: Processes and detectors. In: The Principles and Practice of Electron Microscopy, Cambridge, UK, Cambridge University Press, 1997, 30–57.
[9]   Brandon D, Kaplan WD. Microanalysis in electron microscopy. In: Microstructural Characterization of Material, 2nd edn, Chippenham, Wiltshire, UK, John Wiley & Sons Ltd, 2008, 333–386.
[10] Fultz B, Howe JM. Characteristic radiation. In: Transmission Electron Microscopy and Diffractometry of Materials, Berlin, Germany, Springer, 2001, 16–19.
[11] Fortelný I, Šlouf M, Sikora A, et al. The effect of the architecture and concentration of styrene-butadiene compatibilizers on the morphology of polystyrene/low-density polyethylene blends. J Appl Polym Sci 2006, 100, 2803–2816.
[12] Lednický F, Hromádková J, Šlouf M, Dybal J. Uložení E/P Kaučuku V Houževnatých E/P Kopolymerech. Research Report for Company Polymer Institute Brno, Czech Republic, Edition, 2008, Macro T-727. (in Czech).
[13] Henke P, Lang K, Kubát P, Sýkora J, Šlouf M, Mosinger J. Polystyrene nanofiber materials modified with an externally bound porphyrin photosensitizer. ACS Appl Mater Interfaces 2013, 5, 3776–3783.

[14] Stokes DJ. Principles and Practice of Variable Pressure/environmental Scanning Electron Microscopy (VP-ESEM), Padstow, Cornwall, UK, John Wiley & Sons Ltd, 2008.

[15] Štěpánek M, Hajduová J, Procházka K. Association of poly(4-hydroxystyrene)-block-poly (ethylene oxide) in aqueous solutions: Block copolymer nanoparticles with intermixed blocks. Langmuir 2012, 28, 307–313.

[16] Šlouf M, Lapčíková M, Štěpánek M. Imaging of block copolymer vesicles in solvated state by wet scanning transmission electron microscopy. Eur Polym J 2011, 47, 1273–1278.

[17] Hajduova J, Prochazka K, Slouf M, et al. Polyelectrolyte–surfactant complexes of poly[3,5- bis (dimethylaminomethyl)-4-hydroxystyrene]-block-poly(ethylene oxide) and sodium dodecyl sulfate: Anomalous self-assembly behavior. Langmuir 2013, 29, 5443–5449.

[18] Novotny F, Wandrol P, Proska J, Slouf M. In situ wet STEM observation of gold nanorod self-assembly dynamics in a drying colloidal droplet. Microsci Microanal 2014, 20, 385–393.

[19] Goldstein J, Newbury D, Joy D, et al. Low-temperature specimen preparation. In: Scanning Electron Microscopy and X-ray Microanalysis, 3rd edn, NY, USA, Kluwer Academic/Plenum Publishers, 2003, 621–646.

[20] Seward GGE, Celotto S, Prior DJ, Wheeler J, Pond RC. In situ SEM-EBSD observations of the hcp to bcc phase transformation in commercially pure titanium. Acta Mater 2004, 52, 821–832.

[21] Srinivasan NS. Dynamic study of changes in structure and morphology during the heating and sintering of iron powder. Powder Technol 2002, 124, 40–44.

[22] Slouf M, Krejcikova S, Vackova T, Kratochvil J, Novak L. In situ observation of nucleated polymer crystallization in polyoxymethylene sandwich composites. Front Mater 2015, 2, 23.

[23] Novak L, Wu M, Wandrol P, Kolibal M, Vystavel T. New approaches to in-situ heating in FIB/ SEM systems. Microsci Microanal 2017, 23(Suppl 1), 928–929.

[24] Reichmann A. High-temperature corrosion of steel in an ESEM with subsequent scale characterisation by Raman microscopy. Oxid Met 2008, 70, 257–266.

[25] Appel CC, Rasmussen A-M, Ullmann S, Hansen PL. Experiments with a modified heating stage on an environmental scanning electron microscope. In: Engelbrecht J, editor. International Congress on Electron Microscopy. Proceedings of the 15th International Congress on Electron Microscopy; 2002, Sep 1–6, Durban, South Africa, 2002, 227–228.

[26] Karpushkin E, Dušková-Smrčková M, Šlouf M, Dušek K. Rheology and porosity control of poly (2-hydroxyethyl methacrylate) hydrogels. Polymer 2013, 54, 661–672.

[27] Reimer L. Image Formation in Low-Voltage Scanning Electron Microscopy, Bellingham, Washington, USA, The international society for optical engineering, 1993.

[28] Goldstein J, Newbury D, Joy D, et al. Low-voltage operation. In: Scanning Electron Microscopy and X-ray Microanalysis, 3rd edn, NY, USA, Kluwer Academic / Plenum Publishers, 2003, 55–56.

[29] Young R, Templeton T, Roussel L, et al. Extreme high resolution SEM: A paradigm shift. Microsc Today 2008, 16, 24–28.

[30] Müllerová I, Frank L. Scanning low-energy electron microscopy. Adv Imag Elect Phys 2003, 128, 309–443.

[31] Frank L, Zadražil M, Müllerová I. Low energy imaging of nonconductive surfaces in SEM. Mikrochim Acta 1996, 13, 289–298.

[32] Xiaobing L, Bhadare S, Statham P, Burgess S, Holland J, Rowlands N. Improving low energy sensitivity of EDS detectors – towards lithium detection. In: Prince RL, editor. Instrumental Symposium. Proceedings of Microscopy and Microanalysis; 2012, Jul 29-Aug 2, Arizona, USA, Phoenix, 2012.

[33] Scandium software, module Stereo; electronic manual available at www.soft-imaging.net.

[34] Danzl R, Schroettner H, Helmli F, Scherer S. Coordinate measurement with nano-metric resolution from multiple SEM images. In: The European Microscopy Society Proceedings of the 15th European Microscopy Congress; 2012, Sep 16–21, Manchester UK, 2012.

[35] Suganuma T. Measurement of surface topography using SEM with two secondary electron detectors. J Electron Microsc 1985, 34, 328–337.

[36] Kodama T, Li X, Nakahira K, Ito D. Evolutionary computation applied to the reconstruction of 3-D surface topography in the SEM. J Electron Microsc 2005, 54, 429–435.

[37] Denk W, Horstmann H. Serial block-face scanning electron microscopy to reconstruct three-dimensional tissue nanostructure. PLoS Biol 2004, 2, 1900–1909.

[38] Yao N, Imanishi N, Kang HHC, et al. Focused Ion Beam Systems: Basics and Applications, Cambridge, Cambridge University Press, 2007.

[39] Boughorbel F, Zhuge X, Potocek P, Lich B. SEM 3D reconstruction of stained bulk samples using landing energy variation and deconvolution. In: Prince RL, editor. Instrumental Symposium. Proceedings of Microscopy and Microanalysis; 2012, Jul 29-Aug 2, Phoenix, Arizona, USA, 2012, 560–561.

[40] Wandrol P, Unčovský M, Vystavěl T. In-situ observation of solutions by wet STEM. In: Prince RL, editor. Instrumental Symposium. Proceedings of Microscopy and Microanalysis; 2012, Jul 29-Aug 2, Arizona, USA, Phoenix, 2012, 1090–1091.

[41] Wandrol P. Trinity Detection System for SEM and FIB/SEM. Microsci Microanal 2019, 25(Suppl 2), 458–459.

[42] Masters RC, et al. Sub-nanometre resolution imaging of polymer: Fullerene photovoltaic blends using energy-filtered scanning electron microscopy. Nat Commun 2015, 6, 6928.

[43] Holm JD, Caplins BW. STEM-in-SEM – introduction of scanning transmission electron microscopy for microelectronics failure analysis. ASM International, Ohio, USA, 2020, 20–33.

[44] Vystavel T, et. al Expanding capabilities of low-kV STEM imaging and transmission electron diffraction in FIB/SEM systems. Microsci Microanal 2026, 23(Suppl1), 554–555.

[45] Vystavel T, Stejskal P, Uncovsky M, Stephens C. Tilt-free EBSD. Microsci Microanal 2018, 24(Suppl 1), 1126–1127.

[46] Kralova D, Slouf M, Klementova M, Kuzel R, Kelnar I. Preparation of gram quantities of high-quality titanate nanotubes and their composites with polyamide 6. Mater Chem Phys 2010, 124, 652–657.

[47] Malinsky P, Mackova A, Hnatowicz V. Properties of polyimide, polyether ether ketone and polyethylene terephthalate implanted by Ni ions to high fluences. Nucl Instrum Meth B 2012, 272, 396–399.

[48] Dolina J, Lederer T. Silver particles incorporation to nanofibre structure for surface membrane modification. In: TANGER Ltd, editor. NANOCON 2011. Proceedings of 3rd International Conference on NANOCON; 2011, Sep 21–23, Czech Republic, Brno, 2011, 331–340.

[49] Fultz B, Howe JM. Transmission Electron Microscopy and Diffractometry of Materials, Berlin, Germany, Springer, 2001.

[50] Horak D, Hlidkova H, Trachtova S, Slouf M, Rittich B, Spanova A. Evaluation of poly(ethylene glycol)-coated monodispersed magnetic poly(2-hydroxyethyl methacrylate) and poly(glycidyl methacrylate) microspheres by PCR. Eur Polym J 2015, 68, 687–696.

[51] Kostiv U, Patsula V, Slouf M, Pongrac IM, Skokic S, Dobrivojevic Radmilovic M, Pavicic I, Vinkovic Vrcek I, Gajovic S, Horak D. Physico-chemical characteristics, biocompatibility, and MRI applicability of novel monodisperse PEG-modified magnetic $Fe_3O_4\&SiO_2$ core-shell nanoparticles. RSC Adv 2017, 7, 8786–8797.

[52] Kostiv U, Engstova H, Krajnik B, Slouf M, Proks V, Podhorodecki A, Jezek P, Horak D. Monodisperse Core-Shell $NaYF_4$:Yb^{3+}/Er^{3+}@$NaYF_4$:Nd^{3+}-PEGGGGRGDSGGGY-NH_2

Nanoparticles Excitable at 808 and 980 nm: Design, Surface Engineering, and Application in Life Sciences. Front Chem 2020, 8, 497.

[53] Siegel J, Kaimlova M, Vyhnalkova B, Trelin A, Lyutakov O, Slepicka P, Svorcik V, Vesely M, Vokata B, Malinsky P, Slouf M, Hasal P, Hubacek T. Optomechanical processing of silver colloids: New generation of nanoparticle-polymer composites with bactericidal effect. Int J Mol Sci 2021, 22, 312.

[54] Rossiter BW, Hamilton JF, eds Physical Methods of Chemistry. Volume IV: Microscopy, 2nd edn, New York, USA, John Willey & Sons, 1991.

[55] Reimer L, Kohl H. Transmission Electron Microscopy, New York, USA, Springer, 2008.

[56] Sawyer LC, Grubb DT, Meyers GF. Conductive coatings. In: Polymer Microscopy, 2nd edn, London, UK, Chapman & Hall, 1996, 201–211.

[57] Lednický F. Morphology of coalescence processes: Secondary fracture, cracks and spherulites. Polym Bull 1984, 11, 579–584.

[58] Lednický F, Pelzbauer Z. Specific use of electron microscopic techniques in the characterization of ABS polymers. Polym Test 1987, 7, 91–107.

[59] Šlouf M, Kolařík J, Fambri L. Phase morphology of PP/COC blends. J Appl Polym Sci 2004, 91, 253–259.

[60] Kráčalík M, Pospíšil L, Šlouf M, et al. Effect of glass fibres on rheology, thermal and mechanical properties of recycled PET. Polym Compos 2008, 29, 915–921.

[61] Grellmann W, Seidler S. Deformation and Fracture Behavior of Polymers, Berlin Germany, Springer, 2001.

[62] Ostafinska A, Fortelny I, Hodan J, Krejcikova S, Nevoralova M, Kredatusova J, Krulis Z, Kotek J, Slouf M. Strong synergistic effects in PLA/PCL blends: Impact of PLA matrix viscosity. J Mech Behav Biomed Mater 2017, 69, 229–241.

[63] Fortelny I, Ujcic A, Fambri L, Slouf M. Phase structure, compatibility, and toughness of PLA/PCL blends: A review. Front Mater 2019, 6, 206.

[64] Lednický F, Michler GH. Soft matrix fracture surface as a means to reveal the morphology of multi-phase polymer systems. J Mater Sci 1990, 25, 4549–4554.

[65] Jordan ND, Bassett DC, Olley RH, Smith NG. In-depth morphological changes and embrittlement near the wear surface of UHMWPE inserts from uncemented hip systems. J Biomed Mater Res 2001, 55, 158–163.

[66] Fortelný I, Jůza J, Dimzoski B. Coalescence in quiescent polymer blends with a high content of the dispersed phase. Eur Polym J 2012, 48, 1230–1240.

[67] Šlouf M, Kolařík J, Kotek J. Rubber-toughened polypropylene/acrylonitrile-co-butadiene-co-styrene blends: Morphology and mechanical properties. Polym Eng Sci 2007, 47, 582–592.

[68] Merinska D, Kovarova L, Kalendova A, et al. Polypropylene nanocomposites based on the montmorillonite modified by octadecylamine and stearic acid co-intercalation. J Polym Eng 2003, 23, 241–257.

[69] Galeski A, Bartczak Z, Kazmierczak T, Slouf M. Morphology of undeformed and deformed polyethylene lamellar crystals. Polymer 2010, 51, 5780–5787.

[70] Fortelný I, Jůza J, Vacková T, Šlouf M. The effect of anisometry of dispersed droplets on their coalescence during annealing of polymer blends. Colloid Polym Sci 2011, 289, 1895–1903.

[71] Vacková T, Slouf M, Nevoralová M, Kaprálková L. HDPE/COC blends with fibrous morphology and their properties. Eur Polym J 2012, 48, 2031–2039.

[72] Dimzoski B, Fortelný I, Šlouf M, Nevoralová M, Michálková D, Mikešová J. Coalescence during annealing of quiescent immiscible polymer blends. e-Polymers 2011, 11, 115–126.

[73] Ostafinska A, Fortelny I, Nevoralova M, Hodan J, Kredatusova J, Slouf M. Synergistic effects in mechanical properties of PLA/PCL blends with optimized composition, processing, and morphology. RSC Adv 2015, 5, 98971–98982.

[74] Nevoralova M, Koutny M, Ujic A, Stary Z, Sera J, Vlkova H, Slouf M, Fortelny I, Krulis Z. Structure characterization and biodegradation rate of poly(ε-caprolactone)/starch blends. Front Mater 2020, 7, 141/1–141/14.

[75] Lednický F, Hromádková J, Pientka Z. Ultrathin sectioning of polymeric materials for low-voltage electron microscopy. Polymer 2001, 42, 4329–4338.

[76] Harris JR, Roos C, Djalali R, Rheingans O, Maskos M, Schmidt M. Application of the negative staining technique to both aqueous and organic solvent solutions of polymer particles. Micron 1999, 30, 289–298.

[77] Stara H, Slouf M, Lednicky F, Pavlova E, Baldrian J, Stary Z. New and simple staining method for visualizing UHMWPE lamellar structure in TEM. J Macromol Sci Phys 2008, 47, 1148–1160.

[78] Lednický F, Hromádková J, Kolařík J. Revealing the phase structure of polymer blends via volume relaxation. Polym Test 1992, 11, 205–213.

[79] Kučera J, Lednický F. An investigation of properties of filled polypropylene-elastomer blends (in Czech). Plasty a Kaučuk 1989, 26, 97–102.

[80] Lednický F, Kučera J. Possibilities of visualization of the structures of composites (in Czech). Plasty a Kaučuk 1989, 26, 289–292.

[81] Hromádková J, Lednický F, Kolařík J. Visualization of the phase structure of selected polycarbonate blends using various preparation techniques. Polym Test 1994, 13, 461–478.

[82] Haworth B, Hindle CS, Sandilands GJ, White JR. Assessment of internal stresses in injection-molded thermoplastics. Plast Rubber Proc Appl 1982, 2, 59–71.

[83] Enrique-Jimenez P, Quiles-Díaz S, Salavagione HJ, Fernández-Blázquez JP, Monclús MA, Guzman De Villoria R, Gómez-Fatou MA, Ania F, Flores A. Nanoindentation mapping of multiscale composites of graphene-reinforced polypropylene and carbon fibres. Compos Sci Technol 2019, 169, 151–157.

[84] Clay CS, Peace GW. Ion beam sputtering: An improved method of metal coating SEM samples and shadowing CTEM samples. J Microsc 1981, 123, 25–34.

[85] Meredith P, Donald AM, Payne RS. Freeze-drying: In situ observations using cryoenvironmental scanning electron microscopy and differential scanning calorimetry. J Pharmaceut Sci 1996, 85, 631–637.

[86] Přádný M, Lesný P, Smetana KJ. Macroporous hydrogels based on 2-hydroxyethyl methacrylate Part II Copolymers with positive and negative charges, polyelectrolyte complexes. J Mater Sci: Mater Med 2005, 16, 767–773.

[87] Nedela V. Methods for additive hydration allowing observation of fully hydrated state of wet samples in environmental SEM. Microsc Res Techn 2007, 70, 95–100.

[88] Kaberova Z, Karpushkin E, Nevoralova M, Vetrik M, Slouf M, Duskova M. Microscopic structure of swollen hydrogels by scanning electron and light microscopies: Artifacts and reality. Polymers 2020, 12, 578/1–578/18.

[89] Wittmann JC, Lotz B. Polymer decoration: The orientation of polymer folds as revealed by the crystallization of polymer vapors. J Polym Sci Phys Ed 1985, 23, 205–226.

[90] Woodward AE. Crystallizable polymers. In: Understanding Polymer Morphology, Munich, Germany, Carl Hanser Verlag, 1995, 19–29.

[91] Sperling LH. The amorphous state. In: Introduction to Physical Polymer Science, 4th edn, Hoboken, New Jersey, USA, John Wiley & Sons Ltd, 2008, 197–238.

[92] Fischer EW, Wendorff JH, Dettenmaier M, Lieser G, Voigt-Martin I. Chain conformation and structure in amorphous polymers as revealed by X-ray, neutron, light and electron diffraction. J Macromol Sci B 1976, 12, 41–59.

[93] Bates FS. Small-angle neutron scattering from amorphous polymers. J Appl Cryst 1988, 21, 681–691.

[94] Lednický F. Density measurements of annealed amorphous polymers. Collect Czech Chem Commun 1995, 60, 1935–1940.

[95] Balta-Calleja FJ, Fakirov S. Microhardness of Glassy Polymers, Cambridge, Cambridge University Press, 2000, 46–79.

[96] Sperling LH. Introduction to Physical Polymer Science, Hoboken, New Jersey, USA, John Wiley, John Wiley & Sons Ltd, 2008.

[97] Michler GH. Electron Microscopy of Polymers, Heidelberg (Germany), Springer-Verlag, 2008.

[98] Woodward AE. Morphologies associated with deformation and fracture. In: Understanding Polymer Morphology, Munich, Germany, Carl Hanser Verlag, 1995, 101–112.

[99] Woodward AE. Morphologies of crystallized polymers. In: Understanding Polymer Morphology, Munich, Germany, Carl Hanser Verlag, 1995, 31–56.

[100] Bassett DC. Spherulites. In: Principles of Polymer Morphology, Cambridge, UK, Cambridge University Press, 1981, 16–36.

[101] Kurtz SM, ed UHMWPE Biomaterials Handbook, 2nd edn, Netherlands, Academic Press Elsevier, Amsterdam, 2009.

[102] Slouf M, Kotek J, Baldrian J, et al. Comparison of one-step and sequentially irradiated ultrahigh-molecular-weight polyethylene for total joint replacement. J Biomed Mater Res Part B 2013, 101B, 414–422.

[103] Slouf M, Vackova T, Nevoralova M, Pokorny D. Micromechanical properties of one-step and sequentially crosslinked UHMWPEs for total joint replacements. Polym Test 2015, 41, 191–197.

[104] Woodward AE. Block copolymers. In: Atlas of Polymer Morphology, Munich, Germany, Carl Hanser Verlag, Hanser Publishers, 1989, 189–197.

[105] Horák Z, Fortelný I, Kolařík J, Hlavatá D, Sikora A. Polymer Blends. In: Kroschwitz J, ed. Encyclopedia of Polymer Science and Technology, Indianapolis, John Wiley & Sons, Inc., 2005, 1–59.

[106] Fortelny I, Ostafinska A, Michalkova D, Juza J, Mikesova J, Slouf M. Phase structure evolution during mixing and processing of poly(lactic acid)/polycaprolactone (PLA/PCL) blends. Polym Bull 2015, 72, 2931–2947.

[107] Kolařík J, Kruliš Z, Šlouf M. High-density polyethylene/cycloolefin copolymer blends. Part 1: Phase structure, dynamic mechanical, tensile and impact properties. Polym Engi Sci 2005, 45, 817–826.

[108] Ostafinska A, Vackova T, Slouf M. Strong synergistic improvement of mechanical properties in HDPE/COC blends with fibrillar morphology. Polym Eng Sci 2018, 58, 1955–1964.

[109] Vacková T, Šlouf M, Nevoralová M, Kaprálková L. Processing-improved properties and morphology of PP/COC blends. J Appl Polym Sci 2011, 122, 1168–1175.

[110] Kráčalík M, Pospíšil L, Šlouf M, et al. Recycled poly(ethylene terephthalate) reinforced with basalt fibres: Rheology, structure, and utility properties. Polym Compos 2008, 29, 437–442.

[111] Balta-Calleja FJ, Fakirov S. Microhardness of polymer blends, copolymers and composites. In: Microhardness of Polymers, Cambridge, Cambridge University Press, 2000, 127–175.

[112] Karger-Kocsis J. Composites. In: Salamone JC, ed. Polymeric Materials Encyclopedia, New York, USA, CRC Press, Inc., 1996, 1378–1383.

[113] Stokes DJ. A lab in a chamber – Is situ methods VP-ESEM and other applications. In: Principles and Practice of Variable Pressure/environmental Scanning Electron Microscopy (VP-ESEM), Padstow, Cornwall, UK, John Wiley & Sons Ltd, 2008, 169–214.

[114] Wandrol P, Slouf M. Polymer imaging in SEM – Charge, damage and coating free. Microsci Microanal 2017, 23(Suppl 1), 1816–1817.

[115] Wall D, Bosch E, Sluyterman S, Wandrol P. Advanced technologies for charge mitigation on non-conductive samples. In: Prince RL, editor. Instrumental Symposium. Proceedings of Microscopy and Microanalysis; 2012, Jul 29-Aug 2, Arizona, USA, Phoenix, 2012, 1284–1285.

[116] Watt IM. Preparation of moist specimens. In: The Principles and Practice of Electron Microscopy, 2nd edn, Cambridge, Cambridge University Press, 1997, 169–174.

[117] Woodward AE. Liquid crystalline morphologies. In: Understanding Polymer Morphology, Munich, Germany, Carl Hanser Verlag, 1995, 57–70.

Eva Bittrich, Klaus-Jochen Eichhorn

8 Application of spectroscopic ellipsometry in the analysis of thin polymer films/ polymer interfaces

8.1 Introduction

Spectroscopic ellipsometry (SE), as a noninvasive technique, is well placed for the investigation of organic films and interfaces containing polymers. In an ellipsometric measurement, the change in the polarization state of light upon interaction with a material is detected. Spectroscopic investigations from the ultraviolet (UV) to the infrared (IR) and further to the terahertz (THz) range are possible [1–3], providing a profound set of data for the analysis of polymer film dielectric function and thickness.

Polymer films are prepared in a huge variety of special film structures like polymer brushes [4], multilayers [5, 6], cross-linked polymers [7], polymer blends [8], polymer–nanoparticle (NP) composites (PNC) [9, 10], or hybrid materials [11]. Next to the determination of the dielectric function ε (or the refractive index N) and the film thickness d of dry polymer films, advanced problems in the characterization of such films are addressed: the surface roughness [12, 13], the glass transition and crystallization behavior [14, 15], swelling in aqueous and nonaqueous solvents [16, 17], and adsorption processes at polymer surfaces [18, 19].

Different ellipsometric setups, such as standard polarizer-compensator-sample-analyzer (PCSA) configuration [20], total internal reflection ellipsometry (TIRE) [21], in situ cells [19, 22], imaging setups [23, 24], as well as in-line monitoring [25] are applied to address the various research topics of polymer films by ellipsometry.

After a short introduction on standard SE in reflection mode, we will discuss selected aspects of the research on optical, structural, and thermodynamical properties of polymer thin films by SE focusing on the visible (VIS) and IR spectral range.

8.1.1 Basics of ellipsometry

Light, in its physical nature, is described by the electromagnetic theory for transverse plane waves. Since the induced magnetization can be neglected in most ellipsometric experiments on polymers, the electric field $\vec{E}$ is the most important quantity for the description of the light wave. We refer to the literature for a comprehensive mathematical description [26, 27]. One aspect important for ellipsometry is the polarization state of light. The light wave can be completely unpolarized (e.g., sunlight), partially polarized, or totally polarized, characterized by the degree of polarization. In an unpolarized wave, the orientation of the electric field

https://doi.org/10.1515/9783110701098-008

vector $\vec{E}$ changes randomly with time, where in a polarized wave the orientation of $\vec{E}$ is determined for every time step. The wave can have linear, circular, or, in general, elliptical polarization.

When the light is reflected at a planar surface, the wave vector $\vec{k}$ of the incident beam and the surface normal define the plane of incidence (see Fig. 8.1).

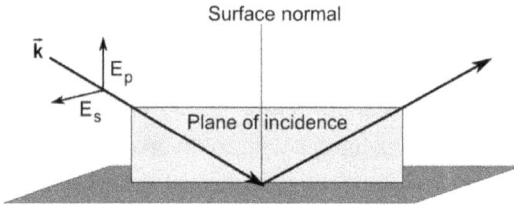

Fig. 8.1: Sketch of the incident and reflected beam and the surface normal spanning the plane of incidence.

With respect to this plane, $\vec{E}$ can be expressed by two components: E_p parallel and E_s perpendicular to the plane of incidence. Both components have an amplitude $A_{s,p}$ and a phase $\delta_{s,p}$. The polarization state of the wave is described by the amplitude ratio A_p/A_s and the phase difference $\delta_p - \delta_s$. Amplitude ratio and phase difference change in the standard ellipsometric reflection experiment. Thus, two ellipsometric angles Δ and ψ are defined as

$$\Delta = \left(\delta_p^r - \delta_s^r \right) - \left(\delta_p^i - \delta_s^i \right) \tag{8.1}$$

$$\tan \psi = \frac{A_p^r/A_s^r}{A_p^i/A_s^i}. \tag{8.2}$$

with the index "r" indicating the reflected and the index "i" the incident beam. In a spectroscopic experiment, they are dependent on the wavelength λ.

Upon interaction with the material of the surface, the wave changes velocity and can be absorbed by the material. This is expressed by the complex index of refraction N or the complex dielectric function ε:

$$N(\lambda) = n(\lambda) + ik(\lambda), \tag{8.3}$$

$$\varepsilon(E) = \varepsilon_1(E) + i\varepsilon_2(E) = (n + ik)^2, \tag{8.4}$$

where n is called the (real) index of refraction, k the extinction coefficient, and ε_1 and ε_2 are real and imaginary parts of the dielectric function, respectively. Usually the material scientist uses the dielectric function ε dependent on the photon energy E, while the polymer chemist or physicist is often interested in the complex refractive index N versus the wavelength λ. Conversion between these quantities is as follows:

$$\varepsilon_1 = n^2 - k^2, \varepsilon_2 = 2nk \qquad (8.5)$$

$$E = \frac{hc}{\lambda}, \qquad (8.6)$$

where h is the Planck constant and c is the velocity of light [28]. Furthermore, the extinction coefficient k can be expressed in terms of the absorption coefficient α (well known in optical spectroscopy) as follows:

$$k = \frac{\lambda}{4\pi}\alpha. \qquad (8.7)$$

The reflection of the light wave at a surface is described by the total complex reflection coefficients R_s for the s- and R_p for the p-component of the electrical field, which depend on the complex index of refraction of the media (ambient, film, and substrate), on incident and refractive angles, and on the wavelength λ of the light (see [29] for details). The total reflection coefficients are connected to the ellipsometric angles by the "fundamental equation of ellipsometry"

$$\tan\psi\, e^{i\Delta} = \rho = \frac{R^p}{R^s}. \qquad (8.8)$$

Please note that both expressions $\tan\psi\, e^{i\Delta}$ and ρ are complex numbers. In practice, an optical model for a substrate-layer system is created, and theoretical values for $\tan\psi$ and Δ are calculated. These values are compared to the measured quantities and the model is iteratively refined by minimizing the mean square error of the fit. Due to the spectroscopic measurement at various wavelengths a comprehensive set of data is available.

In ellipsometry in reflection mode on transparent surfaces ($k = 0$), the incident angle or the range of incident angles in a variable angle (VASE) experiment is often chosen near the Brewster angle of the substrate. At this angle of incidence, the reflectance $\Re_p = |R_p|^2 = 0$ and only the s-polarized wave is reflected from the surface. Additionally, the angle between reflected and transmitted beam is 90°.

Since changes in the polarization state of the light wave are utilized for the characterization of a surface or a film in ellipsometry, some basic components are in principle sufficient for an ellipsometer: a light source, a polarizing element (e.g., polarizer P and compensator C) to determine the polarization state of the light before the sample, and a device to measure changes in the polarization after reflection on the sample (e.g., analyzer A and detector). A typical PCSA ellipsometer configuration is sketched in Fig. 8.2

The compensator enables the modification of the state of polarization before the sample to every possible elliptical polarization, including circular polarization. Modern ellipsometric configurations utilize rotating elements (analyzer, compensator, two rotating compensators), or phase modulation elements [30, 31], The most advanced setups

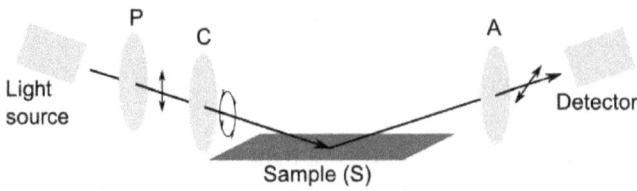

Fig. 8.2: Sketch of a typical ellipsometer setup with light source, polarizer (P), compensator (C), sample (S), analyzer (A), and detector.

allow for the complete analysis of the polarization state of light after interaction with the sample by Mueller matrix ellipsometry [32]. This is often necessary to describe the optical dispersion of biaxial anisotropic layers or materials. Often polymer films show isotropic or uniaxial anisotropic optical properties [33]. However, biaxial optical properties can be found for mechanically stretched polymer samples due to strain-induced crystallization [34]. For spectroscopic ellipsometry in the IR-range (IR SE), an ellipsometer setup is combined with a Fourier transform spectrometer (see [35] for details).

8.2 Specific ellipsometric methods, techniques, and aspects relevant for polymer surface characterization

8.2.1 The optical dispersion

A set of standard optical dispersion relations for (n, k) or $(\varepsilon_1, \varepsilon_2)$ used for the description of polymer films is listed in Tab. 8.1.

Tab. 8.1: Standard dispersion relations for polymer films.

Model	Description	Example
Cauchy	– $n(\lambda)$ for transparent films – $k(\lambda) = 0$	– Fluorohydrocarbon polymer film [36]
Cauchy–Urbach	– $n(\lambda)$ described by Cauchy model – $k(\lambda) \neq 0$: model of absorption tail for energies below the band gap energy of the material	– Poly(tetrafluoroethylene) thin film [37] (Fig. 8.3)
Sellmeier	– Lorentz oscillator model with broadening $\Gamma = 0$, and therefore $k(\lambda) = 0$	– Ultra-thin free-standing films of polyvinyl formal [38]

Tab. 8.1 (continued)

Model	Description	Example
Oscillator	– $k(\lambda)\neq0$, e.g., semiconducting polymers absorbing in the VIS range; description of molecular vibrations in IR range – Defined oscillator line shape in $\varepsilon_2(\lambda)$ (Gaussian, Lorentz, Tauc–Lorentz, etc.) and Kramers–Kronig consistent calculation of $\varepsilon_1(\lambda)$	– Poly(arylenephenylene) film [39] (Fig. 8.4)
B-spline	– $\varepsilon_1(\lambda)$ and $\varepsilon_2(\lambda)$ modeled by mathematical spline curves – Parameters: number of node points/node spacing – Can be calculated Kramers–Kronig consistent	– Any polymer film

The Cauchy dispersion

$$n(\lambda) = A + \frac{B}{\lambda^2} + \frac{C}{\lambda^4} + \cdots, \quad k = 0 \tag{8.9}$$

is commonly used for transparent polymer films [36] and represents an approximation of the Sellmeier model. For polymers with an absorption band in the UV, an absorption tail in the near-UV range is often observed. In the absorption tail, the extinction coefficient $k(\lambda)$ decreases exponentially with the wavelength. In this case, as simplest model, the Cauchy–Urbach dispersion can be used [29, 37] (Fig. 8.3).

The optical dispersion of polymer thin films with $k\neq0$ in the near-UV or VIS range can also be modeled very effectively by a B-spline model [40, 41]. This is a purely mathematical description of the optical dispersion via spline curves (piecewise-defined smooth polynomial functions). Kramers–Kronig (KK) consistency is important to ensure physical validity of the model, where the KK relations connect the imaginary part and the real part of the optical function [26]. The B-spline model can be made KK consistent. However, for Cauchy and Cauchy–Urbach models, KK consistency is not considered. Alternatively, to the Cauchy model, the KK-consistent Sellmeyer model can be used [27].

Different theoretical models exist to model absorption bands of polymers in the UV–VIS (and IR) spectral range [27, 42, 43], among them harmonic oscillator models, the standard critical point (SCP) model, or the Huang–Rhys description. In contrast to the mathematical description with a B-spline, an analytical model of the

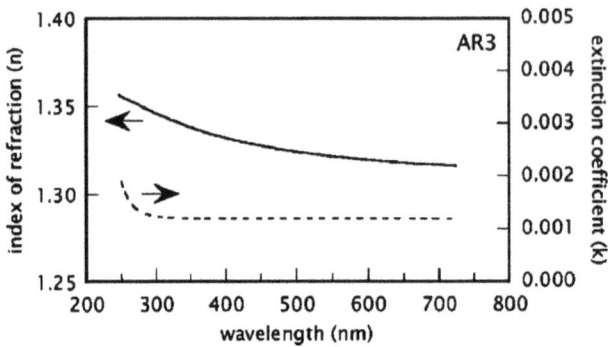

Fig. 8.3: Wavelength-dependent refractive index and extinction coefficient of a poly (tetrafluoroethylene) (PTFE) thin film prepared by hot filament chemical vapor deposition (HFCVD) using hexafluoropropylene oxide (HFPO) as precursor gas. The optical dispersion is described by a Cauchy–Urbach model accounting for an absorption tail ($k > 0$) in the near UV. Reprinted with permission from Lau KKS, Caufield JA, Gleason KK, *J Vac Sci Technol A*, 18, 2404–11, 2000. Copyright 2000 American Vacuum Society [37].

dielectric function provides further information on the electronic/chemical properties of the material, for example, the band gap energy E_g, the energies of electronic excitations, or information on aggregation and order. As an example, Fig. 8.4 shows the optical dispersion of a semiconducting polymer film with an electronic transition in the VIS range [39]. The film is uniaxial isotropic with an absorption band above the fundamental absorption edge due to a π–π* electronic transition. The fundamental band gap energy is the energy necessary to excite an electron from the highest occupied molecular orbital to the lowest unoccupied molecular orbital of the polymer material. In principle, the extinction coefficient k is zero below E_g and has positive values above E_g. However, for thin films, often a nonzero absorption below the fundamental band gap edge is observed, for example, due to localized defect states [39]. In a systematic comparison of oscillator modeling and the SCP model, the latter was found to fit the optical dispersion of uniaxially aligned (poly(9,9-dioctylfluorene)) films best with the minimum set of fit parameters [44].

To describe the effective dielectric functions of heterogeneous media, effective medium approaches (EMA) were developed [45]. The individual dielectric functions of the components a and b contribute to an effective ε according to the underlying model assumptions. Besides a linear combination, more advanced approaches were developed leading to the Maxwell–Garnett, the Bruggeman and the Lorentz–Lorenz expression (Tab. 8.2).

The EMA according to Bruggeman is described by

$$0 = f_a \frac{n_a^2 - n^2}{n_a^2 + 2n^2} + f_b \frac{n_b^2 - n^2}{n_b^2 + 2n^2} \tag{8.10}$$

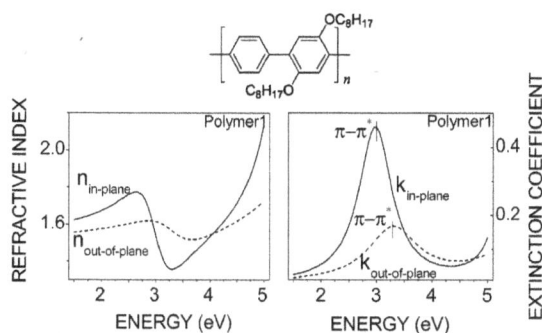

Fig. 8.4: Structure (top) of poly(arylenephenylene) and uniaxial anisotropic refractive index (left) and extinction coefficient (right) of a spin-coated thin film. SE spectra were fitted to two sets of a double Lorentzian oscillator. Adapted by permission from Losurdo M, Giangregorio MM, Capezzuto P, et al., *Macromolecules*, 36, 4492–7, 2003. Copyright 2003, American Chemical Society [39].

Tab. 8.2: Two-component EMA with model assumptions: the Lorentz–Lorenz approach is based on point-like phases a and b, while the other approaches consider microstructures with dimensions d much smaller than the wavelength λ but with own dielectric identity.

Approach	Assumptions
Lorentz–Lorenz	Mixing of a and b on an atomic scale in vacuum as the host medium
Maxwell–Garnett	Microstructure of spherical a ($d_a \ll \lambda$) in the host medium b and $f_b \gg f_a$
Bruggeman	Random-mixture microstructure ($d_a, d_b \ll \lambda$) with $f_b \approx f_a$

where f_a and f_a are the volume fractions of two components a (e.g., solvent) and b (e.g., polymer) and $n_{a,b}$ are the corresponding refractive indices, with n the total effective refractive index of the heterogeneous medium. Typically, f_a and f_b are the fitting parameters and $n(\lambda)_{a,b}$ is known for the components.

Films of semicrystalline polymers or molecules with local ordering can have uniaxial or biaxial optical anisotropy. Hereby uniaxial anisotropy is often found in films of semiconducting polymers for organic electronics [43], while biaxial films are reported for the special case of liquid crystalline polymers, for example, suitable for liquid-crystal displays [46–48]. A uniaxial anisotropic polymer film has its optical axis parallel to the surface normal and has different optical constants in the plane of the film (ordinary: n_o, k_o) and perpendicular to it (extraordinary: n_e, k_e). Uniaxial anisotropy can be modeled from a standard SE experiment, where $\Delta n = |n_o - n_e|$ describes the birefringence of the film. To characterize biaxial polymer films, with different optical constants along all three axes of space, "generalized ellipsometry" (GE) has to be performed [48, 49]. For GE either the Jones matrix or Mueller matrix formalism is used. The first one provides the mathematical description for the reflection at a layered surface in the case of non-depolarizing samples, where the latter is needed if depolarization occurs. The dielectric function is replaced by the full nine-element anisotropic dielectric function tensor (see [49] for details).

8.2.2 In situ setups for experiments in solution

In the context of ellipsometry the term in situ is used for measurements performed during dynamic changes of the properties of a sample surface, like film formation, swelling, or adsorption processes. For polymer thin films the swelling in respective solvents and the adsorption of proteins onto or into the film in an aqueous environment are main topics addressed by in situ SE [19]. For these studies, liquid-cell designs for direct studies in reflection mode, special in situ cells for the IR range, cells allowing for simultaneous quartz crystal microbalance studies (QCMD), and cells for internal reflection mode exist.

8.2.2.1 Liquid cells for measurements in solution in the VIS and the IR range

Liquid cells with a fixed angle of incidence for in situ SE studies in solution are used in the form of batch and flow cells [19, 22], while the angle of incidence is typically set perpendicular to the angle of the cell windows. Batch cells are suitable to study swelling of polymer films but also adsorption of proteins to these films (Fig. 8.5a) [36, 50]. Flow cells are in principle used for the same investigations. Common problems of liquid cells are concentration gradients in the batch geometry, which can have a great effect on kinetics and equilibrium value of adsorption processes, and turbulent flow as well as shearing forces in flow cells, depending on the flow rate. To overcome the problem of concentration gradients in the batch geometry a gentle stirring of the solution in the cell has proven to be efficient [22]. For eliminating the effect of turbulent flow in flow cells a special designed cell using stagnation point flow conditions was reported with well-defined solution and interface conditions [51]. Thus, these stagnation point flow cells are best suitable to measure swelling and adsorption kinetics. The drawback of flow cells is a relatively high need for material (solution and adsorbent). Other important aspects of the in situ cell design are the sample mounting, the means for exchanging liquid, and window effects [22]. It is important to consider imperfections in the cell windows, since they may cause systematic errors in the ellipsometric data. To correct for these window effects, an invariant reference substrate has to be measured in the same solution like the sample. Shifts in the ellipsometric angle Δ that occur due to window imperfections, like stress-induced birefringence, are analyzed and included in the modeling of the in situ data of the sample.

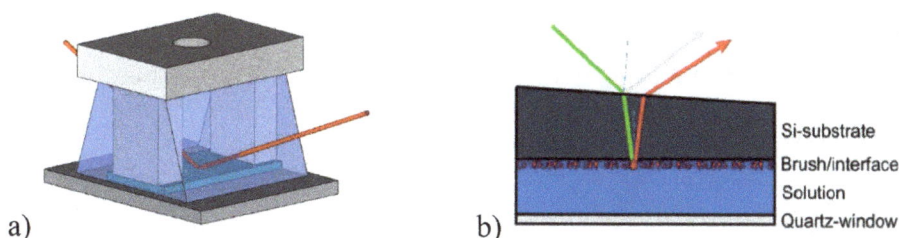

a) b)

Si-substrate
Brush/interface
Solution
Quartz-window

Fig. 8.5: (a) Sketch of an in situ batch cell for the VIS range and (b) sketch of an IR in situ cell. Reprinted with permission from Bittrich E, Furchner, A, König, M, *et al.*, In: Hinrichs K., Eichhorn KJ. (eds) Ellipsometry of Functional Organic Surfaces and Films, Springer Series in Surface Sciences, vol 52, Springer Nature, 2018 [19] and Aulich D, Hoy O, Luzinov I, *et al.*, *Langmuir*, 26, 12926–32, 2010. Copyright 2010, American Chemical Society [52].

For in situ measurements in the IR range, a different cell design than in the VIS-range is needed [52]. An example of such a cell is sketched in Fig. 8.5b. Since the penetration depth of IR radiation in water is very limited (μm range), the investigation of the solid/liquid interface is done via the IR-transparent silicon substrate. Commonly a silicon wedge is used to separate front- and backside reflection. While this cell design needs a solution volume of the order of mL a new optofluidic platform for IR-SE of samples in a microfluidic environment was recently developed [53].

8.2.2.2 Coupling SE with quartz crystal microbalance

The correlation of in situ ellipsometric data with other techniques provides a deeper and often complimentary insight into polymer film properties. Simultaneous application of SE was reported together with quartz crystal microbalance (QCMD) [54, 55], contact angle measurements [56, 57], or electrochemical methods [58, 59].

Quartz crystal microbalance with dissipation monitoring (QCMD) is an acoustic measurement technique [60–62]. The substrate consists of a piezoelectric quartz crystal, where a mechanical oscillation of the crystal lattice is induced by an applied alternating current. Adding or removing a surface layer on this crystal results in changes in the overtone frequencies Δf of the oscillating mass, and in case of a viscous layer dissipation ΔD occurs. These parameters can be measured for a set of overtones and modeled to obtain absorbed mass and viscoelastic data of the layer. Special considerations in the modeling of viscoelastic, hydrated films are necessary. A common approach for such films (e.g., swollen polymer layers) is a continuum viscoelastic model, based on theoretical work of Voinova [55, 60, 63].

a)   b)

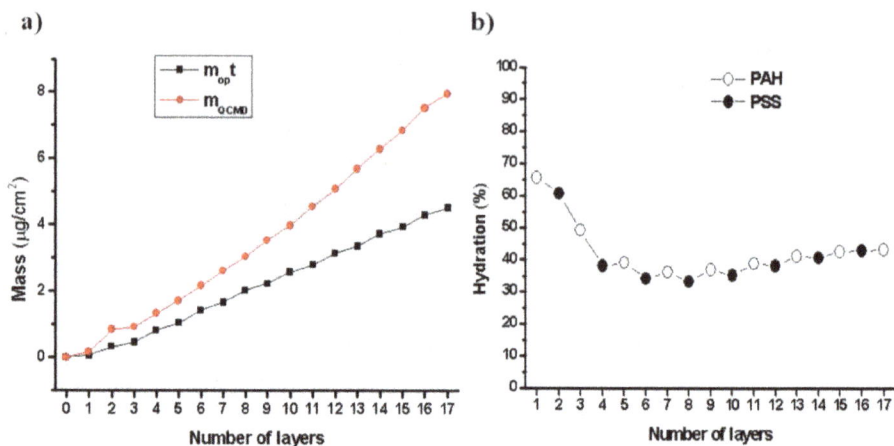

Fig. 8.6: (a) Comparison of m_{opt} (squares) and m_{QCMD} (circles) for poly(allylamine hydrochloride)/poly(sodium 4-styrenesulfonate) (PAH/PSS) multilayers. The polymer surface was washed in water between deposition steps. (b) Hydration in % of the multilayer for each step. Reprinted with permission from Iturri Ramos JJ, Stahl S, Richter RP, Moya SE, *Macromolecules*, 43, 9063–70, 2010. Copyright 2010, American Chemical Society [64].

The combination of SE and QCMD provides an extended set of optical and mechanical data of the film. The amount of adsorbates at a surface, defined as surface mass m or adsorbed amount Γ, can be calculated individually from both techniques and correlated [54, 70]. This was applied, for example, to retrieve stepwise changes in the amount of hydration upon the built up of polyelectrolyte multilayers [64]. The comparison of ellipsometric optical mass m_{opt}, which resembles the mass of the polymer deposited, and the mass m_{QCMD} calculated from QCMD measurements is displayed in Fig. 8.6a. In m_{QCMD} solvent in the film is included. From the comparison of these quantities, the percentage of hydration can be obtained (Fig. 8.6b).

For thick polymer layers with a known bulk dispersion, the solvent content in the film can be retrieved from both the ellipsometric thickness and the viscoelastic thickness, respectively. However systematic differences may occur, for example, if the density of the film is not homogeneous but varies along the surface normal, or if the film/solution interface is diffuse [62]. In principle, also a correlation of the optical (n, k) to the mechanical properties (viscosity η, shear modulus μ) of a polymer film is possible.

8.2.2.3 Total internal reflection ellipsometry (TIRE)

Because of its enhanced sensitivity to changes in the ellipsometric parameters Δ and $\tan \psi$ TIRE is often used to investigate the adsorption of proteins to various surfaces [65], and to utilize surfaces patterned with biomolecules as biosensors [66].

A common setup of a TIRE experiment is sketched in Fig. 8.7. In addition to the ellipsometry setup, a prism and a specially designed flow cell are needed (parts 6 and 8 in Fig. 8.7). In some applications of the TIRE setup, adsorption of biomolecules was performed directly on the glass prism/solution interface [65]. However, in most cases, a thin metal (e.g., gold) layer is coated to the glass prism or to a glass slide (part 7) to utilize the surface plasmon resonance (SPR) technique for enhanced sensitivity [67, 68]. Thus, TIRE commonly refers to ellipsometry in internal reflection mode with SPR. An additional metal-coated glass slide, instead of the metal layer directly on the prism, is favorable because adsorption of molecules is kept away from the prism and single use of the glass slide is possible. Cleaning of the prism or the metal layer on the prism is not trivial. To reduce interface reflections the refractive indices of prism and slide should be the same, and the glass slide is mounted with a matching index oil. Various prism forms are used for the internal reflection including triangular, trapezoidal, and semicylindrical shapes [68, 69].

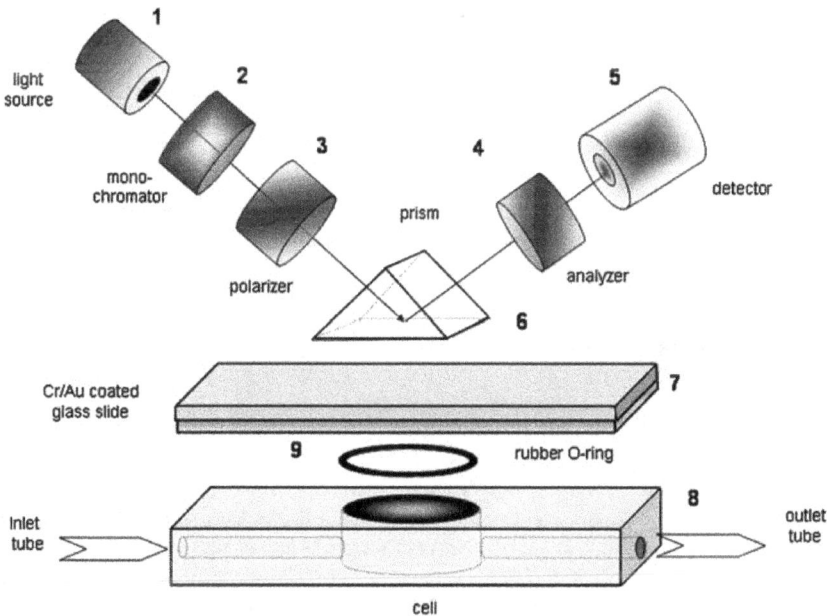

Fig. 8.7: Setup for a TIRE experiment with triangular prism. Reprinted with permission from Basova T, Tsargorodskaya A, Nabok A, et al., *Mater Sci Eng C*, 29, 814–8, 2009. Copyright 2009 Elsevier [67].

For a defined set of the angle of incidence, the metal layer thickness, and the wavelength, the *p*-polarized component of the incident electrical field couples to a surface plasmon wave, a collective oscillation of the conducting electrons, in the metal layer. This excitation can be described in terms of a quasi-particle, the surface plasmon polariton (SPP) [68]. The field of the SPP is maximal at the metal/liquid interface and decreases into both media. This leads to a minimum in the reflectance curve, often called the SPR dip. In principle, for each wavelength, a different set of system parameters (e.g., angle of incidence and metal layer thickness) can be found to optimize the resonance situation.

The high sensitivity of the SPR mode is utilized to monitor adsorption processes to the metal surface. Adsorbing a thin protein film to the metal layer leads to a shift in the wavelength of the SPR dip, as displayed in Fig. 8.8 for the example of ferritin adsorption to a gold layer [70]. Changes in Δ of more than 90° are observed compared to changes of a few degrees in standard external SE experiments [71]. The example in Fig. 8.8 also illustrates the high significance of the correct angle of incidence to increase the sensitivity of the method by SPR. Due to its potential for biosensing, it is suitable to combine imaging ellipsometry with TIRE [66].

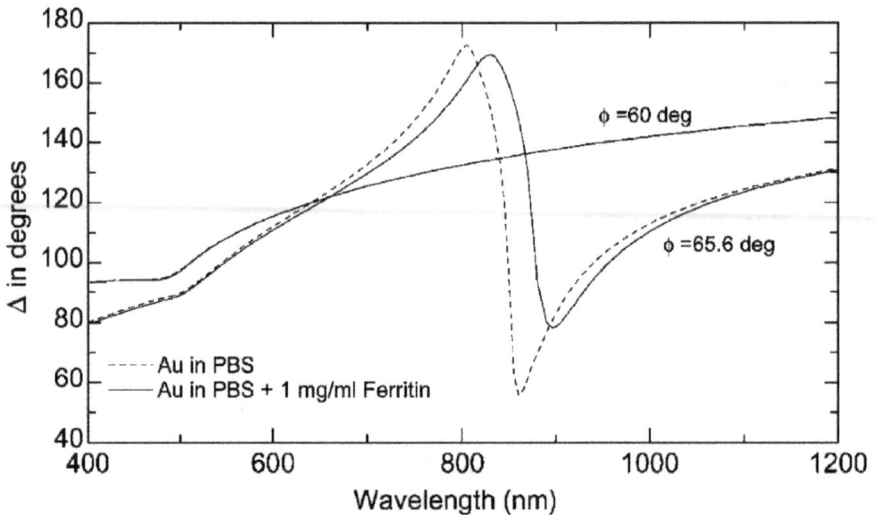

Fig. 8.8: Comparison of Δ for an Au layer with 43.4 nm thickness in PBS buffer (dashed line) and after ferritin adsorption (solid line), at angles of incidence of 65.5° and 60°. SPR occurs for a wavelength of about 830 nm at 65.5° for the given Au thickness. The shift in the SPR mode at this angle of incidence due to the ferritin adsorption is evident, while no significant change in Δ is observed for 60°. Reprinted with permission from Poksinski M, Arwin H, *Thin Solid Films*, 455, 716–21, 2004. Copyright 2004 Elsevier [70].

8.2.3 In-line monitoring of polymer thin films

In-line monitoring with SE is promising due to the noninvasiveness of the technique and the fast recording of measurement data. Here in-line stands for a real-time in situ measurement integrated in a manufacturing process. Real-time in situ SE was extensively applied in the investigation of inorganic film formation for solar cell applications [72, 73]. In the context of polymer thin films, such in situ SE is on the one hand also reported for the monitoring of reaction processes in a vacuum chamber, for example, plasma treatment of polycarbonate substrates for surface functionalization [74]. On the other hand, SE was adapted to the monitoring of roll-to-roll (r2r) fabrication processes done on flexible material, for example, PET or Kapton foils [25, 75, 76].

8.2.3.1 Monitoring processes in a vacuum chamber

Challenges for using an ellipsometer together with a vacuum chamber lie in the minimization of vibrations of the optical elements, in the quality of optical windows from the ellipsometer to the vacuum chamber, and, in general, in the attachment of the ellipsometric parts to the vacuum setup in the correct optical geometry. Suitable optical windows consist of BaF_2 [74], and firm attachment of the optical elements of the ellipsometer to the vacuum chamber by suitable optical ports ensures minimal vibration.

8.2.3.2 Roll-to-roll (r2r) fabrication processes

R2r processes are used in the fabrication of organic electronic devices (organic photovoltaics – OPVs, organic light-emitting diodes – OLEDs) on flexible foils. Here in-line ellipsometry contributes to the quality control of the fabrication process. A simplified fabrication setup with integrated ellipsometer is displayed in Fig. 8.9.

The foil represents a moving sample; therefore, very fast acquisition times are necessary. Logothetidis et al. used an array detector with 32 fiber optics and achieved 100 ms integration time for one multi-wavelength spectrum [75]. To ensure stable measurement conditions, the ellipsometer is located between two directional rolls, where the foil is measured on a vertical path. With this setup, the substrate is ensured to be flat and without wrinkles. The latter would lead to a defocusing of the reflected light beam. Another challenge is represented by vibrations due to the moving rolls of the in-line fabrication setup. Vibration damping for the setup displayed in Fig. 8.9 was done by putting the whole r2r system on an antivibration stage.

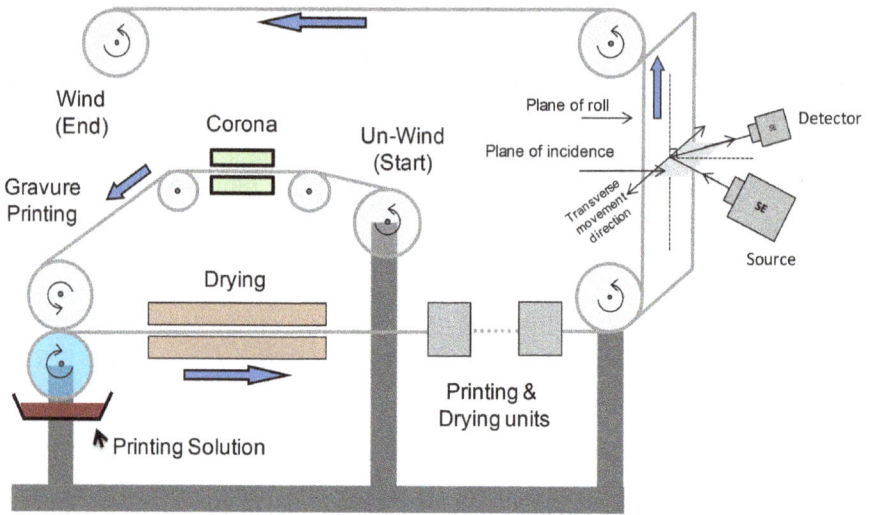

Fig. 8.9: Scheme of a lab-scale r2r fabrication system with in-line ellipsometry unit. Reprinted with permission from Logothetidis S, Georgiou D, Laskarakis A, *et al.*, *Sol Energy Mater Sol Cells* 112, 144–56, 2013. Copyright 2013 Elsevier [75].

8.2.4 Micropatterned films – imaging ellipsometry

Micropatterned polymer/biopolymer films are, for example, appealing for localized cell adhesion [77, 78], tissue engineering [79], or in general for biosensor applications [80]. Here, the ellipsometric characterization of the heterogeneity of the film structure (thickness and optical constants) in the VIS range [77, 79, 81], as well as chemical heterogeneity (IR range) [82] are discussed. Heterogeneity of micropatterned films can be accessed by imaging/mapping ellipsometry.

8.2.4.1 VIS imaging ellipsometry

A typical setup of a VIS imaging ellipsometer includes objective and a spatially resolving CCD detector as displayed in Fig. 8.10 [83, 84].

With the objective, the irradiated area of the sample is projected onto the detector with spatially resolved optical information. Thickness and refractive index maps are recorded with a lateral resolution of 1–2 µm [84]. Imaging is often combined with null ellipsometry, where selected areas of the sample map fulfill nulling conditions of polarizer, compensator, and analyzer. Thus, the light signal is extinguished at the detector (dark image areas). For the example displayed in Fig. 8.10, a laser source in single wavelength mode was used. However, spectroscopic measurements are the state of the art [85]. By combination of polarimetric elements with

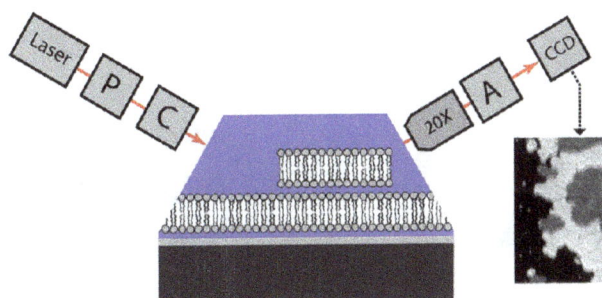

Fig. 8.10: Configuration of a single-wavelength imaging null ellipsometer with polarizer (P), compensator (C), objective (20×), analyzer (A), and CCD camera. Additionally, an example map of a lipid bilayer is displayed. Reprinted with permission from Bruun Nielsen MM, Cohen Simonsen A, *Langmuir*, 29, 1525–32, 2013. Copyright 2013, American Chemical Society [83].

microscopic setups, Mueller Matrix microscopy is utilized to image biological samples [86], for example, for cancer detection, or crystalline samples [87], and birefringent thin films [88].

Also, the combination of imaging ellipsometry and grazing incidence small-angle X-ray scattering at the synchrotron beamline BW4 (Hamburg, Germany) was reported for the investigation of microstructured conducting polymer films [89, 90].

8.2.4.2 Microfocus-mapping IR ellipsometry

For the mapping of structured surfaces in the IR range, the combination of SE with light sources based on synchrotron radiation was developed [91]. These light sources have up to 3 times higher brilliance than IR light sources in the lab and enable spatial resolutions down to 10 times of the wavelength (e.g., down to 100 μm at BESSYII, Berlin, Germany) [92]. Thus, experiments with both high sensitivity and high lateral resolution can be done, not possible with lab instruments. The brilliance of a light source is defined by the number of photons impinging on a unit sample area per time unit.

In Fig. 8.11, an example of the application of microfocus-mapping IR-SE for the study of lateral homogeneity of an organic surface layer is displayed [53]. In another example, the chemical "fingerprint" (characteristic vibrational bands) of a linker and a peptide molecule were used to resolve the lateral position of the surface components, giving insight into the photoelectrochemical reaction process of the linker molecule with the surface [82].

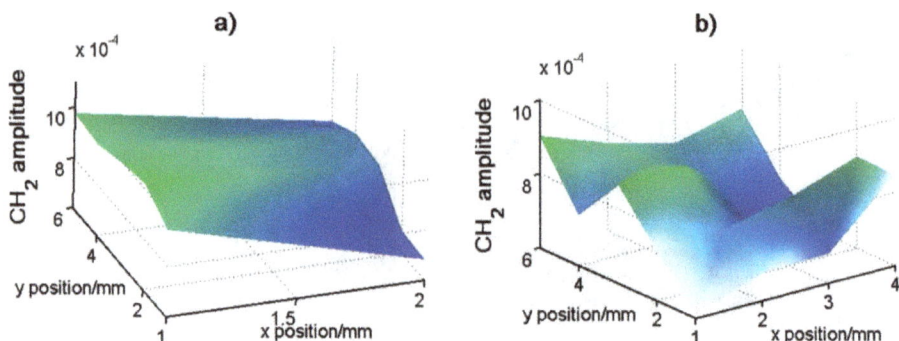

Fig. 8.11: Two-dimensional maps of hexadecanethiol (HDT) on Au (a) and GaAs (b). Displayed is the CH_2 amplitude at 2922 cm^{-1} from the tan ψ spectra. Lateral step width is 1 mm. Reprinted with permission from Rosu DM, Jones JC, Hsu JWP, *et al.*, *Langmuir*, 25, 919–23, 2009. Copyright 2009, American Chemical Society [53].

8.3 Selected architectures of polymer films, blends, and composites

8.3.1 Polymer blends and cross-linked polymer films

Commonly, a combination of VIS-SE with Fourier transform infrared (FTIR) spectroscopy measurements is used to derive the thickness and optical constants of the polymer film (VIS-SE), and to monitor chemical processes via characteristic vibration bands in the cross-linking procedure (FTIR) [93–95]. The composition of polymer blends can, for example, be analyzed by a combination of the methods VIS-SE (EMA approach) [96] and atomic force microscopy coupled with IR (AFM-IR) [97].

Also, IR-SE provides access to structural and chemical properties of polymer materials and is well placed to characterize polymer blends and the cross-linking process in polymer films.

For polymer blends, the miscibility of the single components dependent on temperature was investigated by monitoring the formation of a mixing layer in a bilayer system of miscible polymers [98, 99]. Due to interdiffusion at the polymer/polymer interface, the blend was formed. Hinrichs et al. studied the poly(*n*-butyl methacrylate) (PnBMA)/poly(vinyl chloride) (PVC) blend after annealing of a PnBMA/PVC double layer [98]. In Fig. 8.12, the measured and simulated tan ψ and Δ spectra with the successful three-layer model are displayed. The optical constants n and k for the single phases were determined beforehand by combining IR and VIS-SE. For the mixed phase, n and k were simulated by weighing the individual optical constants with the thickness of the components PnBMA (100 nm) and PVC (106.6 nm) in the original bilayer.

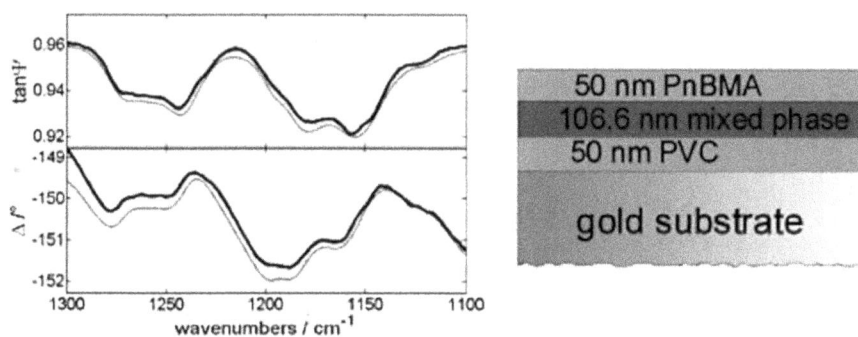

Fig. 8.12: Measured tan ψ and Δ spectra (black) and simulated spectra (gray) of a three-layer model of PnBMA and PVC in a double-layer conformation with a blend layer created by interdiffusion in an annealing process at 110 °C for 10 min. Reprinted with permission from Hinrichs K, Gensch M, Nikonenko N, *et al.*, *Macromol Symp*, 230, 26–32, 2005. Copyright 2005 Wiley-VCH Verlag GmbH & Co. [98].

Cross-linking in polymer films was investigated by IR-SE for catalytical processes [7], and possible cross-linking by plasma treatment [74, 100, 101]. Simpson et al. investigated the catalytical cross-linking of vinyl-terminated poly(dimethyl siloxane) with a trimethylsilyl-terminated poly(hydrogen methylsiloxane) cross-linker at the presence of platinum as catalyst. A cos Δ spectrum of a 14.6 µm thick film at the beginning of cross-linking at 80 °C is displayed in Fig. 8.13a. Vibration bands are: the SiH stretching mode at 2150 cm^{-1} (1), the asymmetric (2) and symmetric (3) CH$_3$ deformations, the Si–O–Si stretching mode (4), the CH$_3$ rocking mode and the Si–C stretching mode (5), and the asymmetric CH$_3$ stretching mode (6). To analyze the completion of the cross-linking reaction they evaluated the decrease in the concentration of unreacted SiH groups in the cross-linking agent by monitoring the decrease of the characteristic SiH stretching vibration band at 2150 cm^{-1} (Fig. 8.13b). Coatings between 1 and 27 µm thickness were investigated, and Simpson et al. found the reaction rate to be dependent on the thickness of the coating [7].

Plasma treatment of polymer surfaces is commonly done to improve the adhesion properties on polymers. With IR-SE chain scissions in an overlayer were detected that could lead to cross-linking [74, 101]. The overlayer thickness depends on the treatment time and on applied voltage for the plasma treatment and can increase to several 100 nm [101]. Logothetidis et al. combined IR-SE with VIS–far UV SE and correlated changes in the chemical bonding in the overlayer of polymer membranes with modifications in characteristic features of electronic transitions [74].

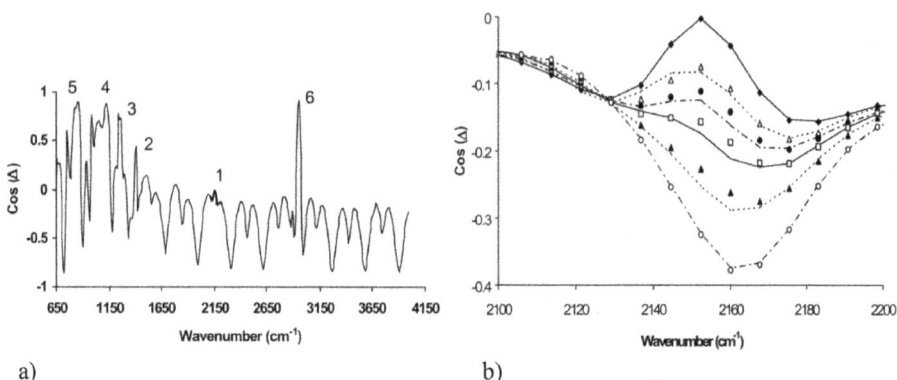

Fig. 8.13: $\cos \Delta$ spectrum of a 14.6 µm thick PDMS coating at the beginning of cross-linking at 80 °C (a), and the decrease of the SiH stretching vibration band at 2150 cm^{-1} during cross-linking (b). Measurements were taken at 0 min (dark squares), 1.2 min (light triangles), 3.7 min (dark circles), 4.9 min (light squares), 13.7 min (dark triangles), and 182 min (light circles). The lines are the best fit to a Bruggeman EMA, where the PDMS is considered to be a blend of a fully cured component (without SiH groups) and an unreacted component (with SiH groups). Adapted with permission from Simpson TRE, Parbhoo B, Keddie JL, *Polymer*, 44, 4829–38, 2003. Copyright 2003 Elsevier [7].

8.3.2 T_g in thin polymer films of different architectures

Thin polymer films can have different properties than in the bulk due to confinement effects. Hereby, thermo-optical properties of phase transitions, like the glass transition temperature T_g or crystallization processes, are of high technological relevance. An overview of temperature-dependent ellipsometry and its application to study thermo-optical properties can be found in [15]. Among other methods, T_g can be determined by analyzing the maximum of the second derivative either of the ellipsometric angles $\Delta(T)$ and $\tan \psi(T)$, of the film thickness $d(T)$ or of the refractive index $n(T)$. Several empirical fit functions for these parameters were proposed [102–104]. While the approach of J. A. Forrest and K. Dalnoki-Veress requires stable films with constant thermal expansion in the glassy and the rubbery state [103], we developed a fit function suitable to analyze T_g for systems with nonstable, but linear, thermal expansion in glassy and rubbery state, for example, due to evaporation of low-molecular-weight compounds [102].

The glass transition of a polymer is not a real phase transition but rather a solidification process. With decreasing temperature, the relaxation time of the polymer molecules increases until it is of the order of the experimental time and the polymer material appears solid. For thin polymer films the surface to volume ratio is significantly higher as compared to bulk material. Thus, polymers can be geometrically confined in the thin film and the importance of interface interactions on the polymer relaxation time increases for the overall behavior of the film [105].

A model system, extensively investigated for its glass transition behavior, is thin films and ultra-thin films (with thicknesses < 10 nm) of polystyrene (PS). Hereby, controversial results exist about the influence of confinement effects on the T_g, and whether T_g changes with decreasing film thickness. Three key reasons were identified for changes in the T_g [106]: (a) free surfaces lead to a reduction of T_g, and free-standing films show significantly higher decrease in T_g than supported films; (b) strong interactions between the polymers and the substrate increase the T_g, and (c) the measured change of T_g often depends on the experiment used. The latter is an effect of how the heterogeneous gradient in T_g within the thin film is averaged by the experimental method, but also stems from the type of measurement principle itself [106]. Additional effects that influence the determination of T_g are preparation and treatment artifacts like chemical degradation [107], water sorption [104], plasticizing due to remaining solvent [108], thermal history of the sample [109], and agglomeration of solvent at the interface between polymer and substrate [14]. Great care has to be taken, and a deep understanding of the chosen measurement technique is necessary to investigate the glass transition of thin films, especially with thicknesses below 10 nm.

To illustrate an example for the T_g determination by VIS-SE, temperature-dependent thicknesses $d(T)$ for PS films for two different molecular weights (a – 27 500 g/mol, b – 58 900 g/mol) and five film thicknesses from 12 to 234 nm are displayed in Fig. 8.14 [104]. The T_g was defined as the maxima of the second derivative of the fit function of $d(T)$, as displayed in the insets of the graphs. Tress et al. found no significant influence of molecular weight and film thickness on the value of T_g down to a thickness of 12 nm. Small deviations from the bulk T_g were within error ranges. They compared the determination of T_g by SE measurements with broadband dielectric spectroscopy results and found coincidence in the absence of significant alterations in T_g for decreasing film thickness.

Erber et al. also investigated the influence of the polymer–substrate interaction on the T_g of polyester thin films [14, 110]. They found significant shifts in the T_g values, when changing the interaction strength between polymer and substrate due to different functional groups for aromatic-aliphatic hyperbranched polyesters (Fig. 8.15). Introducing attractive interactions by benzoyl (-OBz) groups led to an increase of the T_g already for relatively thick films of 200 nm with maximal changes in T_g up to 14 K. Repulsive interactions were created by tert-butyldimethylsilyl (-Si) functionalization and led to a reduction of T_g for film thicknesses below 50 nm.

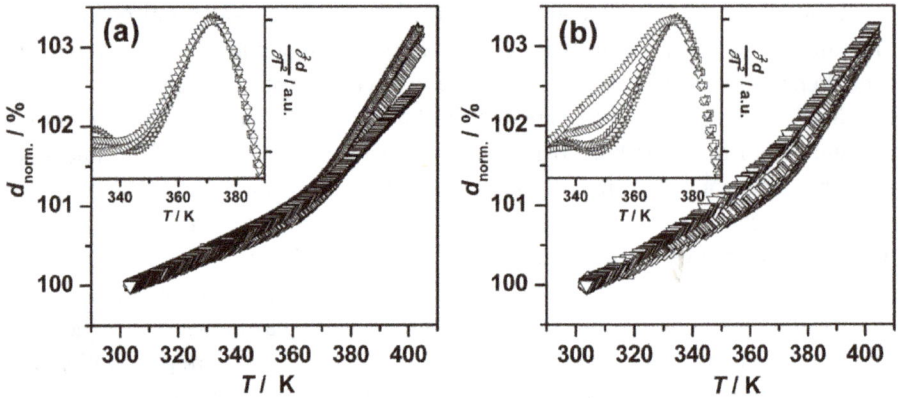

Fig. 8.14: Temperature dependence of the thickness of PS films determined by SE for a molecular weight of (a) 27 500 g/mol (thicknesses: 234 nm (squares), 121 nm (up triangles), 86 nm (stars), 54 nm (diamonds), 26 nm (down triangles)) and (b) 58 900 g/mol (thicknesses: 360 nm (squares), 138 nm (up triangles), 59 nm (stars), 26 nm (diamonds), 12 nm (down triangles)). Thicknesses are normalized at 300 K. Insets show the second derivative of the fitted thickness data (9th-order polynomial fitting). Reprinted with permission from Tress M, Erber M, Mapesa EU, *et al.*, *Macromolecules*, 43, 9937–44, 2010. Copyright 2010, American Chemical Society [104].

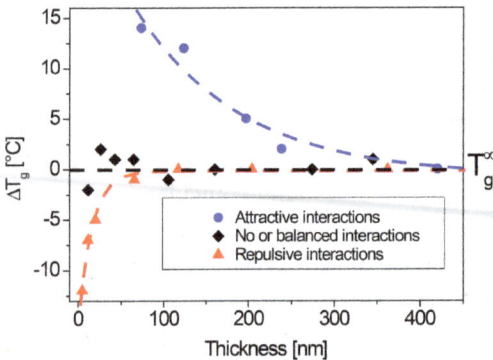

Fig. 8.15: Effect of interfacial interactions between functionalized aromatic-aliphatic hyperbranched polyesters and the substrate (Si/SiO$_x$) on T_g. Shifts of ΔT_g dependent on the film thickness are displayed. Functional groups were: -OH (squares), -OBz (circles), and –OSi (stars). Adapted with permission from Erber M, Tress M, Bittrich E, *et al.*, In: Hinrichs K., Eichhorn KJ. (eds) Ellipsometry of Functional Organic Surfaces and Films, Springer Series in Surface Sciences, vol 52, Springer Nature, 2018 [14].

8.3.3 Polymer–nanoparticle composites (PNC)

NPs are added to a polymer material to change its mechanical, optical, or electrical properties. For example, self-healing materials could be created by controlling particle size and particle coating [111], and noble metal NPs are utilized in plasmonics using the localized SPR effect [112]. Routes to the formation of PNC are, among others, the use of a polymer matrix as a template for NP growth [113] or the introduction of NPs into polymer brushes [114, 115].

From standard VIS-SE data, the volume fraction/fill factor of NPs in the polymer matrix is commonly derived by the Maxwell–Garnett EMA, with the restrictions that particle sizes should be well below the wavelength of light (for the VIS range <20 nm) and that no interactions between particles in the host material occur (low fill factors) [113, 115–118]. Under these prerequisites, the bulk dielectric function of both materials (host matrix and NPs) are used for the EMA [115, 117]. For a nonabsorbing polymer matrix, a two-component Cauchy dispersion can be used, determined for the corresponding thick polymer film without NPs. The determination of the dielectric function of NPs is more difficult than for the polymer, since nanoscale properties, especially of metal NPs, can be considerably different from the bulk. The use of averaged literature values for different forms of iron oxide was reported to model the optical dispersion of iron oxide NPs [115], while for noble metal particles a modified Drude model was used [117].

Oates and Christalle utilized PS and poly(vinyl alcohol) (PVA) matrixes as growing templates for silver NPs [113, 117]. For the growth of Ag NPs in PVA films, they monitored the time-dependent NP formation and observed an increase of the plasmon absorption peak in ε_2, which reflects an increase in the particle size (Fig. 8.16a) [117]. After NP growth, the polymer matrix was evaporated to achieve an array of freestanding particles on the substrate. A considerable shift in the plasmon resonance can be observed upon removing the polymer (Fig. 8.16b).

Free-standing arrays of noble NPs are highly interesting for biosensorics and are used in surface-enhanced Raman scattering [119]. Due to the enhanced local electrical field at the particle surfaces, adsorption of Raman-active biomolecules to the NPs can be detected very sensitively.

8.3.4 Polymers in nanostructured surfaces

Hybrid organic–inorganic thin films with unique optical properties are created by incorporation of polymers into structured surfaces prepared by glancing angle deposition (GLAD) [11, 120]. The GLAD technique employs a physical vapor deposition process, where a particle flux is inserted on a substrate at oblique angles [121]. Due to the self-shadowing effect, slanted columns grow from the surface leading to a "porous" nanostructured slanted columnar thin film (SCTF). By rotation of the

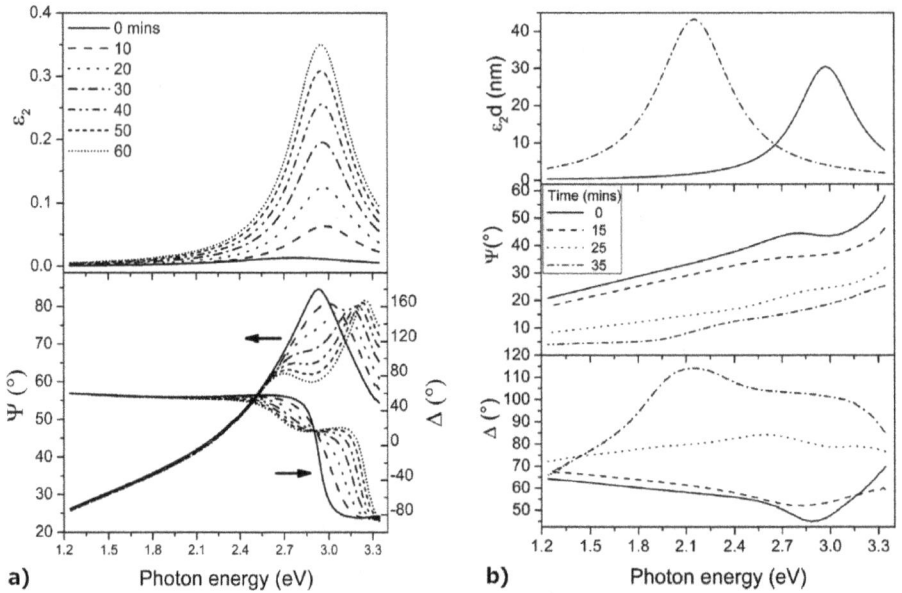

Fig. 8.16: (a) Imaginary part ε_2 of the dielectric function (top) of Ag particles in poly(vinyl alcohol) (PVA) films upon nanoparticle growth with time at 120 °C. Corresponding ellipsometric data ψ and Δ (bottom). (b) Shift of the plasmonic response upon evaporation of the polymer matrix. Displayed are ε_2 (top image) scaled with the film thickness d for 0 min (solid line) and 35 min (dashed line) of the evaporation process. Below ψ and Δ are displayed. Reprinted with permission from Oates TWH, Christalle E, *J Phys Chem C*, 111, 182–7, 2007. Copyright 2007 Elsevier [117].

substrate, more sophisticated structures like chevrons and staircases for rotation in discrete steps, or helices upon continuous rotation can be sculptured. Figure 8.17 shows cross-sectional scanning electron microscopy images of a titanium chevron structure as prepared (a) and after coating with the semiconducting poly (3-dodecylthiophene-2,5-diyl) (P3DDT) (b).

SCTF thin films have high birefringence, tunable by structural parameters such as the distance between columns, the slanting angle, the topology, and the intercolumnar dielectric medium [122]. Also, bimodal SCTF with metal nodes were created, allowing for sensing applications of these surfaces by utilizing the SPR effect [123].

For the optical characterization of these biaxial anisotropic thin films (Fig. 8.17c), GE is used [122]. Schmidt et al. recorded the full Mueller matrix of the Ti chevron SCTF using VIS-SE with dual rotating compensators in the wavelength range of 400–1700 nm at multiple angles of incidence [11]. The film was modeled with two anisotropic layers on silicon substrate. Parameters of the model are the thickness of each biaxial layer and the three major elements ε_a, ε_b, ε_c of the dielectric tensor in the sample coordinate system. The orientation of the nanostructures is described by the three Euler angles that are the elements of the rotation matrix from Cartesian

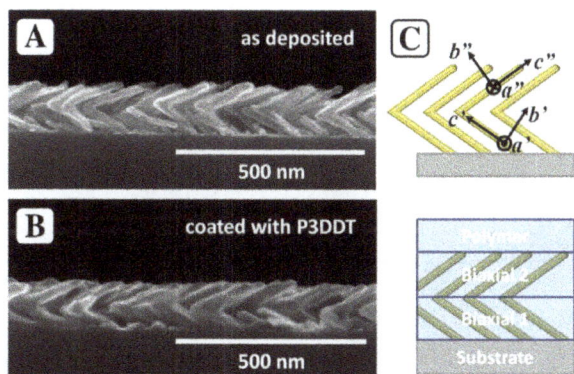

Fig. 8.17: Scanning electron microscopy images of cross-sections of Ti chevron structures as prepared (a), and after infiltration with poly(3-dodecylthiophene) (P3DDT) (b). (c) Scheme of the chevron structure with sample coordinate system along major polarizability axes (top), and ellipsometric layer model (bottom). Reprinted with permission from Schmidt D, Müller C, Hofmann T, *et al.*, *Thin Solid Films*, 519, 2645–9, 2011. Copyright 2011 Elsevier [11].

coordinates to the sample coordinate system of the individual layers. After infiltration of the SCTF structures with polymer an additional polymer top layer was introduced in the model. The geometry of the chevron structures is not affected by the polymer infiltration. A complete set of optical constants (n_i, k_i) of the Ti chevron structures before and after hybridization with polymer could be obtained for all axis a, b, c in the sample coordinate system. SEM investigations are in very good agreement with the structural information derived from the ellipsometric modeling.

In situ GE experiments were performed by Kasputis et al. [120]. They characterized the swelling of poly acrylic acid brushes grafted inside and on top of Si-SCTF. By varying the pH of the ambient buffer solution between pH 7.3 (swollen PAA brush) and pH 3.7 (deswollen brush), they monitored changes in the Mueller matrix elements. After a profound characterization of the optical properties of the dry sample, in situ GE was performed for wavelengths between 400 and 900 nm at a fixed angle of incidence of 65°. With an anisotropic Bruggeman effective medium approximation (AB-EMA), the volume fraction of the brush inside the slanted columns was evaluated (Fig. 8.18, black squares). A top layer above the AB-EMA layer is used in the model to consider polymer chains grafted on top that swell out of the slanted columns (Fig. 8.18, white squares). As they coupled their in situ GE with QCM-D measurements, periodical changes in the surface mass were obtained simultaneously, and confirmed the swelling/deswelling of the PAA brushes.

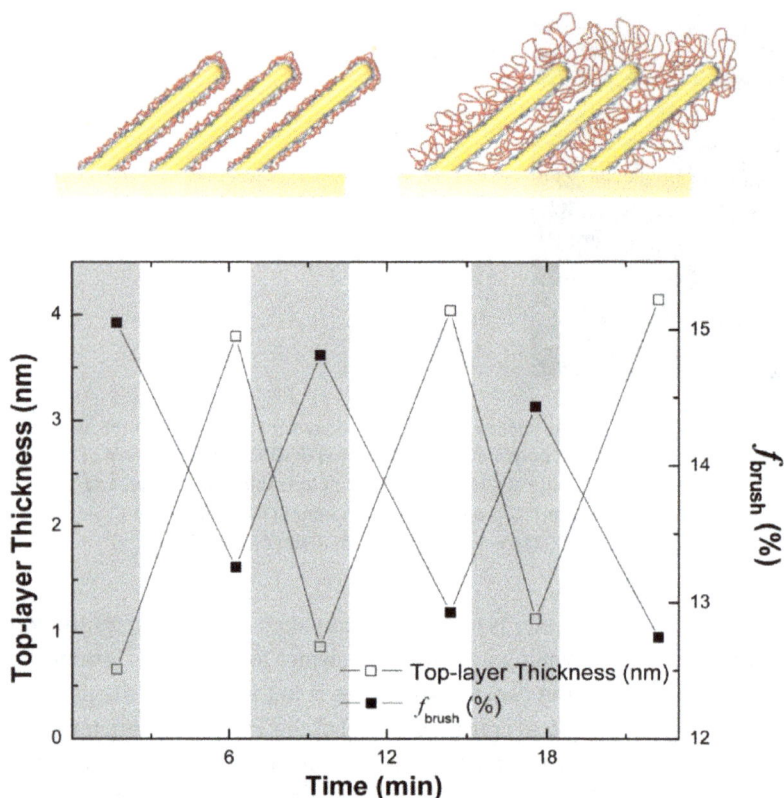

Fig. 8.18: Scheme of the slanted columnar thin films modified by swellable poly(acrylic acid) (PAA) brushes (top), and top-layer thickness as well as the vol% of the brushes for swelling/deswelling cycles (bottom). Parameters of the swollen brush are indicated with light background, of the deswollen brush with gray background. Reprinted with permission from Kasputis T, Koenig M, Schmidt D, et al., *J Phys Chem C*, 117, 13971–80, 2013. Copyright 2013, American Chemical Society [120].

8.4 Polymer films absorbing in the visible spectral range

8.4.1 Chemical modification of polymer films with dye molecules

Including dye molecules into polymer films that absorb in the VIS range is one approach to create functionalized multilayer films for optoelectronics [124], but can also be used to increase the sensitivity of in situ VIS-SE to changes in adsorption or swelling processes [125, 126].

Rauch et al. modified temperature-sensitive poly(*N*-isopropylacrylamide) (PNIPAAm) brushes with the fluorescent dye rhodamine B, and studied temperature-dependent changes in the film thickness, the effective optical properties (n, k) and in the water content upon deswelling of the PNIPAAm brushes by in situ VIS-SE [126]. A tuning of the effective extinction coefficient of the rhodamine-PNIPAAm system dependent on the water content of the swollen polymer brush was observed (Fig. 8.19b). This effect was concluded to be solely due to the change in water content upon the temperature-dependent deswelling of the PNIPAAm brush. This is an example how n and k obtained by SE depend on the film conditions, and how changes in the environment/constitution of the film affect its optical properties, although the dye concentration is constant. The ellipsometric parameters Δ and ψ of the PNIPAAm-rhodamine B film were modeled by a box layer using two Gaussian oscillators.

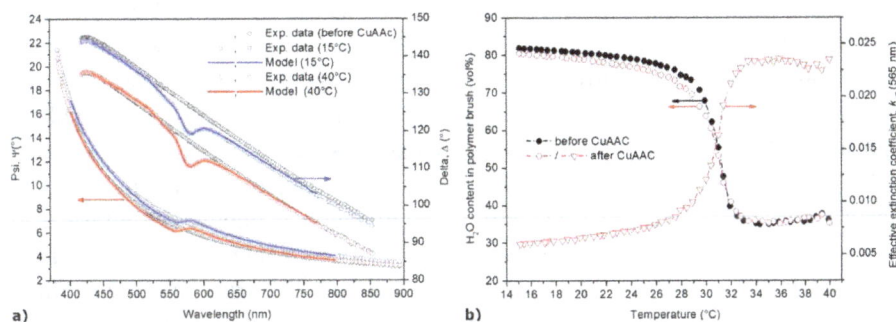

Fig. 8.19: (a) In situ SE parameters Δ and ψ dependent on the wavelength for the rhodamine B-PNIPAAm system swollen at 15 °C, 40 °C, and before modification with dye. (b) Water content of the swollen PNIPAAm film before and after functionalization with rhodamine B (by copper-catalyzed azide–alkyne cycloaddition: CuAAC) and effective extinction coefficient of the rhodamine B-PNIPAAm system dependent on temperature. From Rauch S, Eichhorn K-J, Oertel U, *et al.*, *Soft Matter*, 8, 10260–70, 2012. Reproduced with permission from the Royal Society of Chemistry [126].

An increase of the sensitivity in Δ and ψ was found in the absorbing region of the dye for the modified PNIPAAm brushes (Fig. 8.19a). A similar effect was used to increase the sensitivity of SE in the study of protein adsorption [125]. Garcia and Nejadnik reacted adsorbed protein (bovine serum albumin) with Coomassie brilliant blue G, which enhanced the sensitivity of subsequent measured Δ and ψ values about 2.5 in the absorbing region of the dye compared to the transparent protein film. They used a Lorentz oscillator model for their analysis.

8.4.2 Semiconducting polymers and blends for OPV and OLED

OPV and OLED are two prominent applications for semiconducting polymers [42, 127]. The use of polymers for the preparation of these typically multilayered structures is highly desirable due to the low cost, easy processing, high optical transparency, and tunability of mechanical properties of polymers. Intense research has been dedicated to this field, and a multitude of publications on SE characterization of semiconducting polymers exists. Important examples of the field are SE measurements on polymer–fullerene blends suitable as photoactive nanolayer in OPV [76, 128–132], on polymers for transparent electrodes such as poly(3,4-ethylenedioxythiophene): poly(styrenesulfonate) (PEDOT:PSS) [133–136], and on polyfluorene gain media for OLED [137, 138].

Out of the huge number of published SE studies on semiconducting polymers, we want to focus in this section on polymer:fullerene blends for OPV and want to highlight the principle method of modeling for such polymers, as well as fundamental advances in the investigation of the morphology of the blend films.

The photoactive nanolayer is the core of the OPV, and a donor–acceptor blend is used to absorb photons and to generate excitons, bound electron–hole pairs, most successfully [139]. The blend consists of an electron donor like poly(3-hexylthiophene) (P3HT) and an electron acceptor such as phenyl-C_{61}-butyric acid methyl ester (PCBM) [76]. The optical and electronical properties of the blend depend on its morphology [128]. The regioregular P3HT forms microcrystalline structures by self-organization, and the formation of a phase-separated morphology with PCBM domains in the crystalline P3HT is desired [140]. For optimal charge separation the interfacial area between PCBM and P3HT should be maximized. More fullerene (n-type semiconductor) near the cathode as well as more polymer (p-type semiconductor) near the anode is best for optimal charge collection by the electrodes [127]. Thus, detailed characterization of the morphology is needed for the optimization of the blend structure.

Since semiconducting polymers typically have electronic transitions in the VIS range, VIS is well suited to study the correlation between optical and electronic properties. Optical dispersions for a P3HT:PCBM blend and the corresponding pristine layers are displayed in Fig. 8.20 [76]. Madsen et al. calculated the refractive index and extinction coefficient of the blend layer from the pure phase dispersions using a Bruggeman EMA. To apply the EMA, the dielectric functions of the pristine films have to be known. SE data of pure P3HT and PCBM films were modeled with a generalized oscillator model using up to three Tauc–Lorentz oscillators. Karagiannidis et al. modeled the effective dielectric function of the blend layer directly using five Tauc–Lorentz oscillators [141]. Engmann et al. modeled the optical anisotropy of the P3HT phase and used optical dispersions of anisotropic P3HT (high-ordered and low-ordered phases) and of isotropic PCBM in their EMA approach [141]. Besides the generalized oscillator approach, Campoy-Quiles et al. suggested the SCP

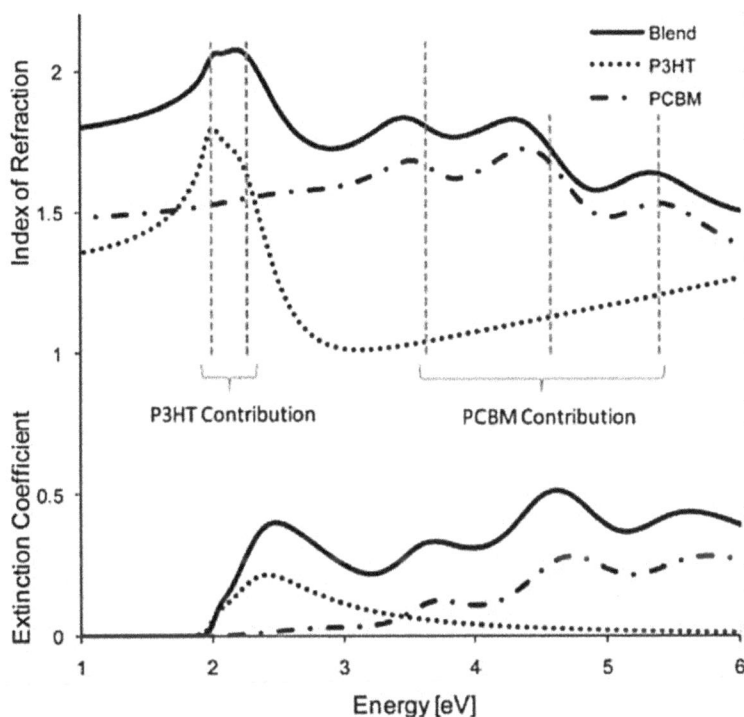

Fig. 8.20: Optical dispersions of pristine P3HT (dotted line), pristine PCBM (dashed/dotted line), and the blend layer (solid line). The optical dispersion for the blend was calculated by an EMA model from the pure phase dispersions. Reprinted with permission from Madsen MV, Sylvester-Hvid KO, Dastmalchi B, et al., *J Phys Chem C*, 115, 10817–22, 2011. Copyright 2011, American Chemical Society [76].

model to be best suited for the description of the optical dispersion of conjugated polymers [43, 138].

Next to the modeling of the dielectric function of the blend, SE provides the means for noninvasive analysis of the blend morphology. A gradient in the vertical composition with higher P3HT concentration at the air–polymer interface and higher PCBM concentration at the blend–electrode interface was found and modeled by using two Bruggeman EMA layers [127, 128]. Aging of the blend morphology, which highly affects the efficiency of the photogeneration of charges, was observed upon annealing of P3HT:PCBM blends. An increase in the degree of crystallinity of P3HT as well as a diffusion of PCBM crystals to the surface of the blend layer were reported.

8.5 Swelling and adsorption processes: proteins and stimuli-responsive polymers

8.5.1 Swelling of stimuli-responsive polymer layers

Stimuli-responsive polymers like polyelectrolytes or polymers with a lower critical solution temperature (LCST) can be used to prepare smart functionalized surfaces [142]. Polyelectrolytes are characterized by ionizable groups at each monomer unit [143], while polymers with LCST deswell above a critical temperature [144]. The unique swelling behavior of these layers can be utilized for various applications such as smart optical systems, drug delivery, tissue engineering, and sensorics [145]. Stimuli-responsive polymer films are prepared in the form of hydrogels [145], polyelectrolyte multilayers [6], or polymer brushes [50, 144, 146].

By the analysis of in situ VIS-SE measurements swollen layer thickness, refractive index, and solvent content can be derived [50, 147, 148]. For the modeling of swollen polymer brushes with in situ thicknesses below ca. 100 nm, the use of a simple box model with sharp interfaces [50], but also a more advanced graded EMA description [148], and a density profile approach [147] were reported. For thicker brushes, a complementary error function $\varphi(z)$ defined as

$$\varphi(z) = \frac{1}{2}\left(1 - erf\left(\frac{z-d}{w}\right)\right)$$

(8.11)

was used to model the brush segment density profile (Fig. 8.21) [149]. φ is the polymer volume fraction, z the coordinate normal to the surface, and d the point of inflection. w is a measure of the smoothness of the brush edge. $\varphi(z)$ approaches the box model for small w.

Kooij et al. reported on the temperature-sensitive deswelling of PNIPAAm brushes with dry thicknesses between 8 and 250 nm [147]. They used a box model with Cauchy dispersion for the description of n of the collapsed brush above the LCST. Below the LCST, a two-layer model with an outer gradient profile in n (exponential decrease in the Cauchy parameter A) was necessary to adequately fit the ellipsometric data (Fig. 8.22a). Thus, a decreasing segment density of the brush with higher distances from the substrate was assumed for the optical model. Figure 8.22b shows the different spectral sensitivity of the ellipsometric parameters with temperature, and in Fig. 8.22c, the temperature dependence of the dense brush layer near the substrate (filled symbols) and the more diluted layer near the brush–solution interface (open symbols) are displayed.

For the modeling of ellipsometric data of swollen hydrogel layers, optical gradient profiles for μm-thick PNIPAAm hydrogels were used as well as a three-parameter error function to model the interface roughness of a 1.2 μm thick hydrogel in water

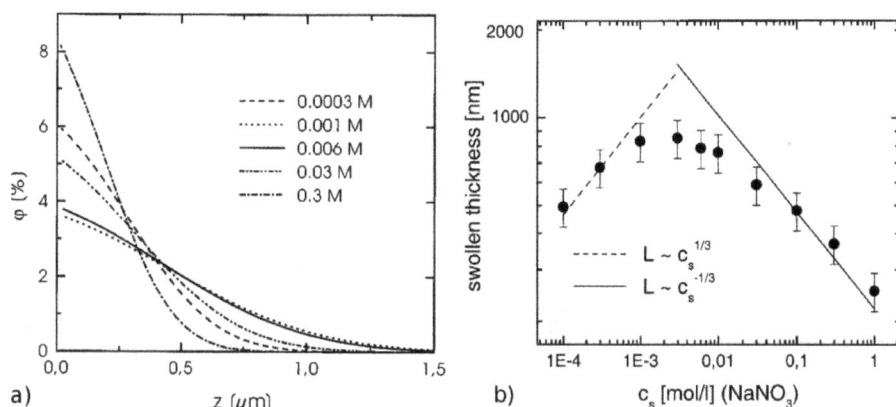

Fig. 8.21: Segment density profiles of a weak polyacid brush (PMAA) swollen at pH 4 in aqueous solution dependent on salt concentration (a). The profiles were obtained from model calculations on ellipsometric spectra using the complementary error function. The swollen thickness at individual salt concentration is determined as twice the first momentum of the segment density profiles (b). Weak polyacid brushes have a maximum swollen thickness at intermediate salt concentrations. The dashed and solid lines in (b) represent the expected scaling behavior. Reprinted with permission from Biesalski M, Johannsmann D, Rühe J, *J Chem Phys*, 117, 4988–15, 2002. Copyright 2002 AIP Publishing LLC [149].

[150, 151]. For very thin hydrogels in the nm range, successful modeling was achieved with a box model [152].

With in situ IR-SE changes in vibration bands due to swelling or dissociation processes can be correlated to thickness and solvent content [52, 153, 154]. Furchner et al. investigated the swelling of PNIPAAm brushes below and above the LCST in the IR range [153]. The effective dielectric function of the PNIPAAm brush layer was modeled by a Bruggeman EMA, using the components polymer and water, in a three-step fitting procedure. First starting values for the swollen brush thickness and the water content were roughly estimated to match the characteristic vibrational bands (amide I and amide II bands of PNIPAAm and the water stretching band $v(H_2O)$ of measured and simulated tan ψ spectra). Then in the dielectric function of PNIPAAm, the oscillator positions of the amide bands had to be shifted, reflecting the interaction of the polymer molecules with water. In the last step of modeling, the thickness of the swollen brush and the water fraction were refined by minimizing the amplitude difference of the characteristic vibrational bands between measured and simulated spectra. Varying the water content in the model proved the high sensitivity of the $v(H_2O)$ band on the amount of water in the brush layer. Distinct changes in the band amplitude can be observed for 2% deviation of the water content (Fig. 8.23).

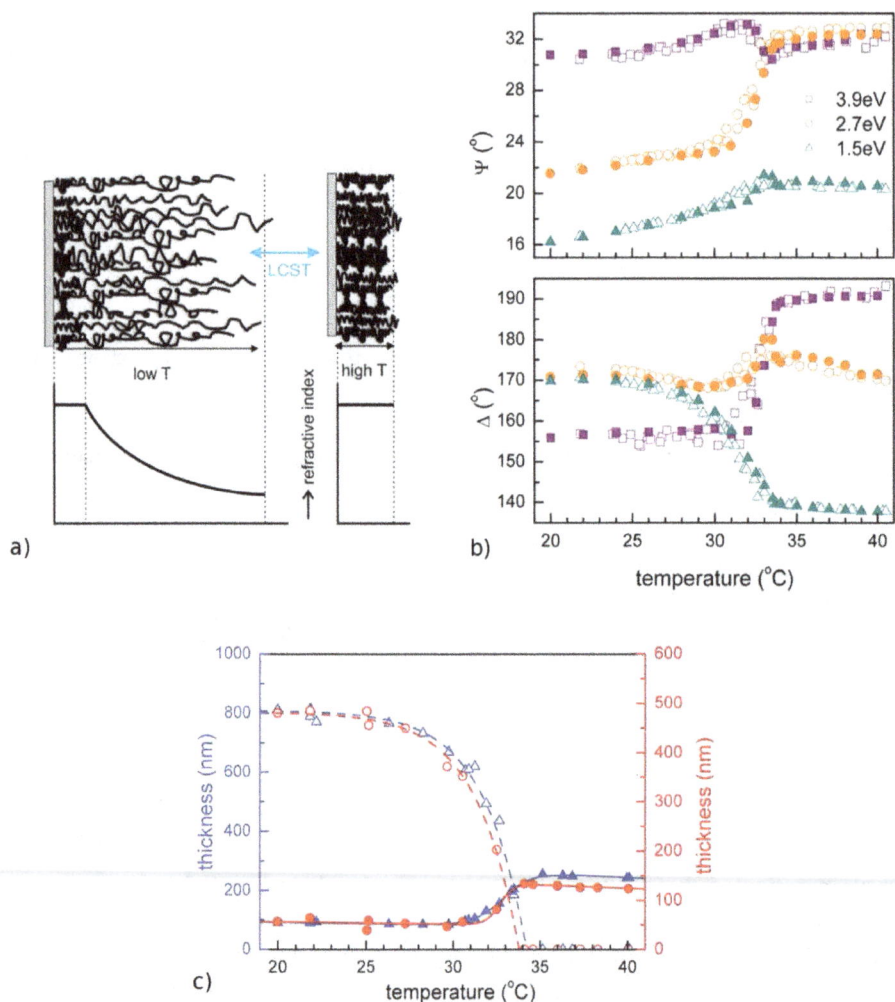

Fig. 8.22: (a) Scheme of the deswelling of PNIPAAm with the LCST and assumed model profile of the refractive index. (b) Measured (open symbols) and simulated (filled symbols) Δ and ψ dependent on temperature for three different photon energies for a brush with a grafting density of approximately 0.27 chains/nm^2. (c) Temperature-dependent swollen brush thickness as modeled from the ellipsometric spectra. Results for brushes with two different grafting densities are displayed: triangles/left/blue: 0.69 chains/nm^2, and circles/right/red: 0.27 chains/nm^2. A two layer model is applied with a dense layer (filled symbols) close to the substrate interface and a diluted layer (open symbols, segment density profile (a) left side) at the brush - solution interface. Adapted with permission from Kooij ES, Sui X, Hempenius MA, *et al.*, *J Phys Chem B*, 116, 9261–8, 2012. Copyright 2012, American Chemical Society [147].

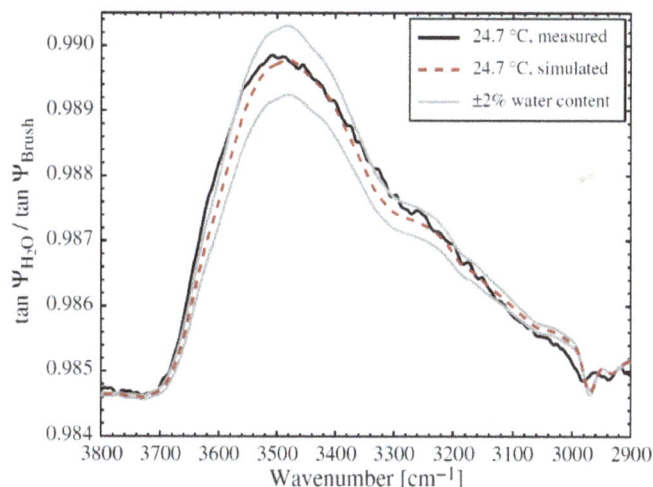

Fig. 8.23: Measured and simulated tan ψ spectrum of the water OH-stretching band $\nu(H_2O)$ referenced to tan ψ_{brush} at 24.7 °C below the LCST of PNIPAAm. Gray lines reflect the sensitivity of $\nu(H_2O)$ towards changes of the water content in the swollen PNIPAAm brush about 2%. Reprinted with permission from Furchner A, Bittrich E, Uhlmann P, et al., *Thin Solid Films*, 541, 41–5, 2013. Copyright 2013 Elsevier [153].

8.5.2 Protein adsorption at soft polymer surfaces

Proteins are biopolymers consisting of polypeptide chains with a defined primary, secondary, and ternary structure. Their physical properties comprise molecular weight, shape, colloidal and amphoteric nature, ion binding capacity, optical activity, and their solubility. The investigation of protein adsorption to solid surfaces by ellipsometry is extensively reviewed in the literature [22, 155–157]. IR-SE was used to investigate the characteristic amide I and amide II IR bands of a protein layer on a silicon substrate as well as to study denaturation due to structural changes in the protein molecules upon heating above 100 °C [158]. Because of its enhanced sensitivity to adsorption processes, TIRE is extensively used to characterize protein adsorption to solid surfaces but also to create biosensors based on the detection of adsorbed protein molecules [65, 66] (see Section 8.2.2).

Investigation of protein adsorption at soft polymer surfaces by in situ VIS-SE provides new challenges due to the swollen state of the polymer layer. For the modeling of the adsorbed amount Γ of protein at soft polymer surfaces, several approaches were reported [19, 159, 160]. The calculations of Γ, presented here, are all based on an adaption of the model of de Feijter. The original equation was formulated for the adsorption of organic molecules at an air–water interface [161]. It was shown that this model can be adopted to adsorption processes of proteins at solid surfaces [156, 162].

At soft surfaces different types of protein adsorption (primary, secondary, and ternary) can occur, and the de Feijter approach has to be modified (Tab. 8.3).

Tab. 8.3: List of models based on the de Feijter approach used to calculate the adsorbed amount Γ for different protein adsorption processes on soft polymer surfaces.

Adsorption process	Formula	Prerequisites
Secondary adsorption Black: substrate Light gray: polymer film Dark gray: protein molecules	$\Gamma_{II} = d\dfrac{n - n_{amb}}{dn/dc}$ – d: layer thickness of protein – n: refractive index of protein layer – n_{amb}: refractive index of liquid ambient – dn/dc: refractive index increment of protein	– dn/dc constant for a wide range of protein concentration – Low swelling of polymer film – Optical dispersion $n(\lambda)$ for protein layer as best-fit result or fixed at constant value from literature, e.g., $n = 1.375$ [157], or $n = 1.575$ [163]
Primary adsorption at substrate or ternary adsorption in film	$\Gamma_{I,III} = d_{pol}\dfrac{n_{ads} - n_{pol}}{dn/dc}$ – d_{pol}: d of swollen polymer layer before adsorption – n_{ads}: n after adsorption – n_{pol}: n of swollen polymer layer	– Thickness of swollen polymer layer in adsorption process constant – One-layer model with fixed thickness: $d_{pol} = $ const – Optical dispersion for $n_{ads}(\lambda)$ and $n_{pol}(\lambda)$
Secondary, primary, and/or ternary adsorption	$\Gamma_{comb} = \Gamma_{I,III} + (d_{ads} - d_{pol})\dfrac{n_{ads} - n_{amb}}{dn/dc}$ – d_{ads}: thickness of composite polymer–protein layer after adsorption	– One-layer model with changing thickness upon adsorption – Optical dispersion for $n_{ads}(\lambda)$ and $n_{amb}(\lambda)$

Formulas based on the de Feijter equation consider the increase of molecule concentration at the adsorbing surface expressed by an increase in the optical density. These equations can be used for adsorption processes in aqueous solution. However, a prerequisite is that dn/dc of the adsorbing biomacromolecule stays constant over a wide range of concentrations c, which was proven for selected proteins [161].

Since the analysis of the adsorbed amount of protein by in situ VIS-SE is model based, reference measurements with alternative methods are necessary to validate the adapted de Feijter approaches for soft polymer surfaces. Good agreement was found for $\Gamma_{I,III}$ between SE and results of radio assays [160] as well as for Γ_{comb} between SE and HPLC analysis after dissolution of the combined protein–polymer layer [164]. For the calculation of $\Gamma_{I,III}$ from SE data the overall layer thickness is assumed to stay constant in the adsorption experiment and only the refractive index of the film is changing due to penetration of molecules into the film. The calculation of Γ_{comb} is more general and considers a change in the film thickness upon adsorption. For both modified de Feijter approaches a combined polymer–protein composite layer is modeled from the ellipsometric data.

Next to protein adsorption at polymer films, protein resistance of surfaces is investigated, and the detection limit of in situ SE discussed. Temperature-sensitive PNIPAAm brushes were found to be highly resistant against the adsorption of the protein human serum albumin for certain molecular weight and grafting densities of PNIPAAm [165], while the adsorption of lysozyme was observed [160]. Changes in the ellipsometric parameters at $\lambda = 546$ nm due to lysozyme adsorption are presented in Fig. 8.24. At this wavelength, the sensitivity of Δ and ψ toward protein adsorption was highest. Xue et al. used the expression for $\Gamma_{I,III}$ to simulate changes in Δ and ψ for given adsorbed amounts of lysozyme between 0 and 3.5 mg/m^2 at two types of surfaces: (1) a PNIPAAm brush and (2) a self-assembled monolayer (SAM) of oligo(ethylene glycol), which was used as an anchoring layer for the brush preparation [160]. Included in Fig. 8.24 are the simulated shifts for 1 mg/m^2 protein adsorption at the brush and for 0.1 mg/m^2 protein adsorption at the SAM underlayer. They concluded that protein adsorption at the PNIPAAm brushes was below 1 mg/m^2 at all conditions. Simulated Δ and ψ shifts of 0.5 mg/m^2 protein adsorption at the brushes were barely distinguishable from the data simulated for 1 mg/m^2. Thus, the sensitivity of the SE measurements was limited to protein amounts higher than the detection limit. For the SAM underlayer, the detection limit was as good as 0.1 mg/m^2,

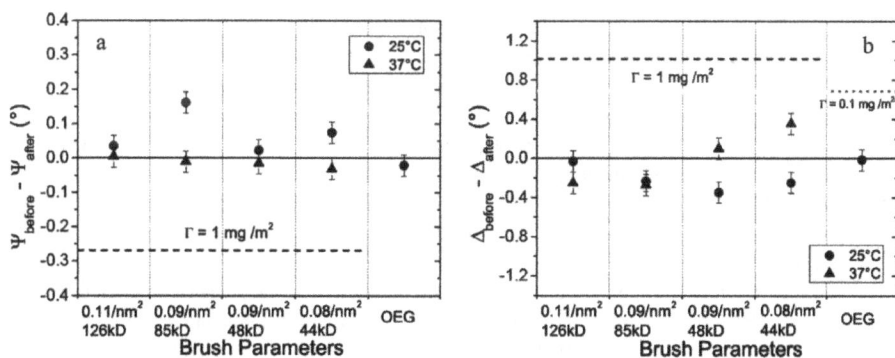

Fig. 8.24: Changes in the ellipsometric parameters ψ (a) and Δ (b) at $\lambda = 546$ nm upon adsorption of lysozyme to PNIPAAm brushes with different molecular weights and grafting densities as well as to the underlying SAM layer of oligo(ethylene glycol) (OEG) for 25 °C (below the LCST of PNIPAAm) and for 37 °C (above the LCST of PNIPAAm). Simulated shift of Δ and ψ for 1 mg/m^2 adsorbed protein at the brush (dashed line) and for 0.1 mg/m^2 at the SAM layer (dotted line) are included. Reprinted with permission from Xue C, Yonet-Tanyeri N, Brouette N, et al., *Langmuir*, 27, 8810–8, 2011. Copyright 2011, American Chemical Society [160].

although this simulated adsorbed amount produced no shift in ψ, but a significant shift in Δ. To overcome this sensitivity limit for the investigation of protein resistant soft polymer films an alternative approach based on the change of the optical thickness contrast $d(n - n_{\mathrm{amb}})$ was proposed [166]. d and n are the in situ layer thickness and the in situ refractive index of the combined polymer–protein layer, respectively. Nevertheless, the highest sensitivity of SE towards protein adsorption can be achieved with TIRE using the SPR effect (see Section 8.2.2.3).

References

[1] Toudert J. Spectroscopic ellipsometry for active nano- and meta-materials. Nanotechnol Rev 2014, 3(3), 223–245.
[2] Woollam JA, Hilfiker JN, Bungay CL, Synowicki RA, Tiwald TE, Thompson DW. Spectroscopic ellipsometry from the vacuum ultraviolet to the far infrared. AIP Conf Proc 2001, 550(1), 511–518.
[3] Neshat M, Armitage NP. Developments in THz range ellipsometry. J Infrared, Millimeter, Terahertz Waves 2013, 34(11), 682–708.
[4] Chen W-L, Cordero R, Tran H, Ober CK. 50th anniversary perspective: Polymer brushes: Novel surfaces for future materials. Macromolecules 2017, 50(11), 4089–4113.
[5] Schaaf P, Voegel J-C, Jierry L, Boulmedais F. Spray-assisted polyelectrolyte multilayer buildup: From step-by-step to single-step polyelectrolyte film constructions. Adv Mater 2012, 24(8), 1001–1016.
[6] Boudou T, Crouzier T, Ren K, Blin G, Picart C. Multiple functionalities of polyelectrolyte multilayer films: New biomedical applications. Adv Mater 2010, 22(4), 441–467.
[7] Simpson TRE, Parbhoo B, Keddie JL. The dependence of the rate of crosslinking in poly (dimethyl siloxane) on the thickness of coatings. Polymer 2003, 44(17), 4829–4838.
[8] Parameswaranpillai J, Thomas S, Grohens Y. Polymer blends: State of the art, new challenges, and opportunities. Charact Polym Blends 2014, 1–6.
[9] Hanemann T, Szabó DV. Polymer-nanoparticle composites: from synthesis to modern applications. Materials 2010, 3(6).
[10] Horechyy A, Nandan B, Zafeiropoulos NE, et al. A step-wise approach for dual nanoparticle patterning via block copolymer self-assembly. Adv Funct Mater 2013, 23(4), 483–490.
[11] Schmidt D, Müller C, Hofmann T, et al. Optical properties of hybrid titanium chevron sculptured thin films coated with a semiconducting polymer. Thin Solid Films 2011, 519(9), 2645–2649.
[12] Koh J, Lu Y, Wronski CR, et al. Correlation of real time spectroellipsometry and atomic force microscopy measurements of surface roughness on amorphous semiconductor thin films. Appl Phys Lett 1996, 69(9), 1297–1299.
[13] Keddie JL, Meredith P, Jones RAL, Donald AM. Kinetics of film formation in acrylic lattices studied with multiple-angle-of-incidence ellipsometry and environmental SEM. Macromolecules 1995, 28(8), 2673–2682.
[14] Erber M, Tress M, Bittrich E, Bittrich L, Eichhorn K-J. Glass transition of polymers with different architectures in the confinement of nanoscopic films -. In: Hinrichs K, Eichhorn K-J, editors. Ellipsometry of Functional Organic Surfaces and Films, Cham, Springer International Publishing, 2018, 97–114.

[15] Hajduk B, Bednarski H, Trzebicka B. Temperature-dependent spectroscopic ellipsometry of thin polymer films. J Phys Chem B 2020, 124(16), 3229–3251.

[16] Habicht J, Schmidt M, Rühe J, Johannsmann D. Swelling of thick polymer brushes investigated with ellipsometry. Langmuir 1999, 15(7), 2460–2465.

[17] Yong H, Bittrich E, Uhlmann P, Fery A, Sommer J-U. Co-nonsolvency transition of poly (N-isopropylacrylamide) brushes in a series of binary mixtures. Macromolecules 2019, 52(16), 6285–6293.

[18] Mora MF, Wehmeyer JL, Synowicki R, Garcia CD. Investigating Protein Adsorption via Spectroscopic Ellipsometry BT – Biological Interactions on Materials Surfaces: Understanding and Controlling Protein, Cell, and Tissue Responses. In: Puleo DA, Bizios R, eds. New York, NY, Springer US, 2009, 19–41.

[19] Bittrich E, Furchner A, Koenig M, et al. Polymer brushes, hydrogels, polyelectrolyte multilayers: Stimuli-responsivity and control of protein adsorption. In: Hinrichs K, Eichhorn K-J, editors. Ellipsometry of Functional Organic Surfaces and Films, Cham, Springer International Publishing, 2018, 115–143.

[20] Tompkins HG. 4 – optical components and the simple PCSA (polarizer, compensator, sample, analyzer) ellipsometer. In: Tompkins HG, Irene EA, editors. Handbook of Ellipsometry, Norwich, NY, William Andrew Publishing, 2005, 299–328.

[21] Arwin H, Poksinski M, Johansen K. Total internal reflection ellipsometry: Principles and applications. Appl Opt 2004, 43(15), 3028–3036.

[22] Arwin H. 12 – Ellipsometry in life sciences. In: Tompkins HG, Irene EA, editors. Handbook of Ellipsometry, Norwich, NY, William Andrew Publishing, 2005, 799–855.

[23] Spandana KU, Mahato KK, Mazumder N. Polarization-resolved Stokes-Mueller imaging: A review of technology and applications. Lasers Med Sci 2019.

[24] Asinovski L, Beaglehole D, Clarkson MT. Imaging ellipsometry: Quantitative analysis. Phys Status Solidi 2008, 205(4), 764–771.

[25] Laskarakis A, Logothetidis S. In-line quality control of organic thin film fabrication on rigid and flexible substrates BT – ellipsometry of functional organic surfaces and films. In: Hinrichs K, Eichhorn K-J, eds. Cham, Springer International Publishing, 2018, 437–458.

[26] Humlíček J. 1 – Polarized light and ellipsometry. In: Tompkins HG, Irene EA, editors. Handbook of Ellipsometry, Norwich, NY, William Andrew Publishing, 2005, 3–91.

[27] Collins RW, Ferlauto AS. 2 – Optical physics of materials. In: Tompkins HG, Irene EA, editors. Handbook of Ellipsometry, Norwich, NY, William Andrew Publishing, 2005, 93–235.

[28] Kurt A, Demirelli K. A study on the optical properties of three-armed polystyrene and poly (styrene-b-isobutyl methacrylate). Polym Eng Sci 2010, 50(2), 268–277.

[29] Tompkins HG, McGahan WA. Spectroscopic Ellipsometry and Reflectometry: A User's Guide, New York, NY, John Wiley & Sons, Ltd, 1999.

[30] Garcia-Caurel E, De Martino A, Gaston J-P, Yan L. Application of spectroscopic ellipsometry and Mueller ellipsometry to optical characterization. Appl Spectrosc 2013, 67(1), 1–21.

[31] Mendoza-Galván A, Muñoz-Pineda E, Ribeiro SJL, Santos MV, Järrendahl K, Arwin H. Mueller matrix spectroscopic ellipsometry study of chiral nanocrystalline cellulose films. J Opt 2018, 20(2), 24001.

[32] Garcia-Caurel E, Ossikovski R, Foldyna M, Pierangelo A, Drévillon B, De Martino A. Advanced Mueller Ellipsometry Instrumentation and Data Analysis BT – Ellipsometry at the Nanoscale. In: Losurdo M, Hingerl K, eds. Berlin, Heidelberg, Springer Berlin Heidelberg, 2013, 31–143.

[33] Koziara BT, Nijmeijer K, Benes NE. Optical anisotropy, molecular orientations, and internal stresses in thin sulfonated poly(ether ether ketone) films. J Mater Sci 2015, 50(8), 3031–3040.

[34] De Vries AJ, Bonnebat C, Beautemps J. Uni- and biaxial orientation of polymer films and sheets. J Polym Sci Polym Symp 1977, 58(1), 109–156.

[35] Röseler A. 11 – Spectroscopic infrared ellipsometry. In: Tompkins HG, Irene EA, editors. Handbook of Ellipsometry, Norwich, NY, William Andrew Publishing, 2005, 763–798.

[36] Werner C, Eichhorn K-J, Grundke K, Simon F, Grählert W, Jacobasch H-J. Insights on structural variations of protein adsorption layers on hydrophobic fluorohydrocarbon polymers gained by spectroscopic ellipsometry (part I). Colloids Surf A Physicochem Eng Asp 1999, 156(1), 3–17.

[37] Lau KKS, Caulfield JA, Gleason KK. Variable angle spectroscopic ellipsometry of fluorocarbon films from hot filament chemical vapor deposition. J Vac Sci Technol A 2000, 18(5), 2404–2411.

[38] Hilfiker JN, Stadermann M, Sun J, et al. Determining thickness and refractive index from free-standing ultra-thin polymer films with spectroscopic ellipsometry. Appl Surf Sci 2017, 421, 508–512.

[39] Losurdo M, Giangregorio MM, Capezzuto P, et al. Study of anisotropic optical properties of poly(arylenephenylene) thin films: Dependence on polymer backbone. Macromolecules 2003, 36(12), 4492–4497.

[40] Johs B, Hale JS. Dielectric function representation by B-splines. Phys Status Solidi 2008, 205 (4), 715–719.

[41] Wang T, Pearson AJ, Lidzey DG, Jones RAL. Evolution of structure, optoelectronic properties, and device performance of polythiophene: fullerene solar cells during thermal annealing. Adv Funct Mater 2011, 21(8), 1383–1390.

[42] Alonso MI, Campoy-Quiles M. Conjugated polymers: Relationship between morphology and optical properties BT – ellipsometry of functional organic surfaces and films. In: Hinrichs K, Eichhorn K-J, eds. Cham, Springer International Publishing, 2018, 335–353.

[43] Campoy-Quiles M, Alonso MI, Bradley DDC, Richter LJ. Advanced ellipsometric characterization of conjugated polymer films. Adv Funct Mater 2014, 24(15), 2116–2134.

[44] Campoy-Quiles M, Etchegoin PG, Bradley DDC. On the optical anisotropy of conjugated polymer thin films. Phys Rev B 2005, 72(4), 45209.

[45] Aspnes DE. Optical properties of thin films. Thin Solid Films 1982, 89(3), 249–262.

[46] Schubert M, Rheinländer B, Cramer C, et al. Generalized transmission ellipsometry for twisted biaxial dielectric media: Application to chiral liquid crystals. J Opt Soc Am A 1996, 13(9), 1930–1940.

[47] Tkachenko V, Marino A, Vita F, et al. Spectroscopic ellipsometry study of liquid crystal and polymeric thin films in visible and near infrared. Eur Phys J E 2004, 14(2), 185–192.

[48] Benecke C, Seiberle H, Schadt M. Determination of director distributions in liquid crystal polymer-films by means of generalized anisotropic ellipsometry. Jpn J Appl Phys 2000, 39(Part 1, No. 2A), 525–531.

[49] Schubert M. 9 – Theory and application of generalized ellipsometry. In: Tompkins HG, Irene EA, editors. Handbook of Ellipsometry, Norwich, NY, William Andrew Publishing, 2005, 637–717.

[50] Bittrich E, Burkert S, Müller M, Eichhorn K-J, Stamm M, Uhlmann P. Temperature-sensitive swelling of poly(N-isopropylacrylamide) brushes with low molecular weight and grafting density. Langmuir 2012, 28(7), 3439–3448.

[51] Mora MF, Reza Nejadnik M, Baylon-Cardiel JL, Giacomelli CE, Garcia CD. Determination of a setup correction function to obtain adsorption kinetic data at stagnation point flow conditions. J Colloid Interface Sci 2010, 346(1), 208–215.

[52] Aulich D, Hoy O, Luzinov I, et al. In situ studies on the switching behavior of ultrathin poly (acrylic acid) polyelectrolyte brushes in different aqueous environments. Langmuir 2010, 26(15), 12926–12932.

[53]   Rosu DM, Jones JC, Hsu JWP, et al. Molecular orientation in octanedithiol and
       hexadecanethiol monolayers on GaAs and Au measured by infrared spectroscopic
       ellipsometry. Langmuir 2009, 25(2), 919–923.
[54]   Bittrich E, Rodenhausen KB, Eichhorn K-J, et al. Protein adsorption on and swelling of
       polyelectrolyte brushes: A simultaneous ellipsometry-quartz crystal microbalance study.
       Biointerphases 2010, 5(4), 159–167.
[55]   Richter RP, Rodenhausen KB, Eisele NB, Schubert M. Coupling Spectroscopic Ellipsometry
       and Quartz Crystal Microbalance to Study Organic Films at the Solid–Liquid Interface
       BT – Ellipsometry of Functional Organic Surfaces and Films. In: Hinrichs K, Eichhorn K-J, eds.
       Cham, Springer International Publishing, 2018, 391–417.
[56]   Hennig A, Eichhorn K-J, Staudinger U, et al. Contact angle hysteresis: study by dynamic
       cycling contact angle measurements and variable angle spectroscopic ellipsometry on
       polyimide. Langmuir 2004, 20(16), 6685–6691.
[57]   Noordmans J, Wormeester H, Busscher HJ. Simultaneous monitoring of protein adsorption at
       the solid–liquid interface from sessile solution droplets by ellipsometry and axisymmetric
       drop shape analysis by profile. Colloids Surf B Biointerfaces 1999, 15(3), 227–233.
[58]   Haberska K, Ruzgas T. Polymer multilayer film formation studied by in situ ellipsometry and
       electrochemistry. Bioelectrochemistry 2009, 76(1), 153–161.
[59]   Ogieglo W, Wormeester H, Eichhorn K-J, Wessling M, Benes NE. In situ ellipsometry studies
       on swelling of thin polymer films: A review. Prog Polym Sci 2015, 42, 42–78.
[60]   Voinova MV, Jonson M, Kasemo B. Dynamics of viscous amphiphilic films supported by
       elastic solid substrates. J Phys Condens Matter 1997, 9(37), 7799–7808.
[61]   Höök F, Kasemo B, Nylander T, Fant C, Sott K, Elwing H. Variations in coupled water,
       viscoelastic properties, and film thickness of a Mefp-1 protein film during adsorption and
       cross-linking: A quartz crystal microbalance with dissipation monitoring, ellipsometry, and
       surface plasmon resonance study. Anal Chem 2001, 73(24), 5796–5804.
[62]   Domack A, Prucker O, Rühe J, Johannsmann D. Swelling of a polymer brush probed with a
       quartz crystal resonator. Phys Rev E 1997, 56(1), 680–689.
[63]   Voinova MV, Jonson M, Kasemo B. 'Missing mass' effect in biosensor's QCM applications.
       Biosens Bioelectron 2002, 17(10), 835–841.
[64]   Iturri Ramos JJ, Stahl S, Richter RP, Moya SE. Water content and buildup of poly
       (diallyldimethylammonium chloride)/poly(sodium 4-styrenesulfonate) and poly(allylamine
       hydrochloride)/poly(sodium 4-styrenesulfonate) polyelectrolyte multilayers studied by an
       in situ combination of a quartz crystal. Microb Macromol 2010, 43(21), 9063–9070.
[65]   Poksinski M, Arwin H. Total internal reflection ellipsometry: Ultrahigh sensitivity for protein
       adsorption on metal surfaces. Opt Lett 2007, 32(10), 1308–1310.
[66]   Liu L, Chen Y, Meng Y, Chen S, Jin G. Improvement for sensitivity of biosensor with total
       internal reflection imaging ellipsometry (TIRIE). Thin Solid Films 2011, 519(9), 2758–2762.
[67]   Basova T, Tsargorodskaya A, Nabok A, et al. Investigation of gas-sensing properties of
       copper phthalocyanine films. Mater Sci Eng C 2009, 29(3), 814–818.
[68]   Arwin H. TIRE and SPR-Enhanced SE for adsorption processes BT – ellipsometry of functional
       organic surfaces and films. In: Hinrichs K, Eichhorn K-J, eds. Cham, Springer International
       Publishing, 2018, 419–435.
[69]   Kalas B, Nador J, Agocs E, et al. Protein adsorption monitored by plasmon-enhanced
       semi-cylindrical Kretschmann ellipsometry. Appl Surf Sci 2017, 421, 585–592.
[70]   Poksinski M, Arwin H. Protein monolayers monitored by internal reflection ellipsometry. Thin
       Solid Films 2004, 455–456, 716–721.
[71]   Mårtensson J, Arwin H, Nygren H, Lundström I. Adsorption and optical properties of ferritin layers
       on gold studied with spectroscopic ellipsometry. J Colloid Interface Sci 1995, 174(1), 79–85.

[72] Saenger MF, Sun J, Schädel M, Hilfiker J, Schubert M, Woollam JA. Spectroscopic ellipsometry characterization of SiNx antireflection films on textured multicrystalline and monocrystalline silicon solar cells. Thin Solid Films 2010, 518(7), 1830–1834.

[73] Fried M. On-line monitoring of solar cell module production by ellipsometry technique. Thin Solid Films 2014, 571, 345–355.

[74] Laskarakis A, Kassavetis S, Gravalidis C, Logothetidis S. In situ and real-time optical investigation of nitrogen plasma treatment of polycarbonate. Nucl Instruments Methods Phys Res Sect B Beam Interact Mater Atoms 2010, 268(5), 460–465.

[75] Logothetidis S, Georgiou D, Laskarakis A, Koidis C, Kalfagiannis N. In-line spectroscopic ellipsometry for the monitoring of the optical properties and quality of roll-to-roll printed nanolayers for organic photovoltaics. Sol Energy Mater Sol Cells 2013, 112, 144–156.

[76] Madsen MV, Sylvester-Hvid KO, Dastmalchi B, et al. Ellipsometry as a nondestructive depth profiling tool for roll-to-roll manufactured flexible solar cells. J Phys Chem C 2011, 115(21), 10817–10822.

[77] Telford AM, Meagher L, Glattauer V, Gengenbach TR, Easton CD, Neto C. Micropatterning of polymer brushes: Grafting from dewetting polymer films for biological applications. Biomacromolecules 2012, 13(9), 2989–2996.

[78] Chien H-W, Chang T-Y, Tsai W-B. Spatial control of cellular adhesion using photo-crosslinked micropatterned polyelectrolyte multilayer films. Biomaterials 2009, 30(12), 2209–2218.

[79] Huang N, Thakar R, Wong M, Kim D, Lee R, Li S. Tissue engineering of muscle on micropatterned polymer films. Conf Proc IEEE Med Biol Soc 2004, 7, 4966–4969.

[80] Niu Y, Jin G. Surface modification methods to improve behavior of biosensor based on imaging ellipsometry. Appl Surf Sci 2013, 281, 84–88.

[81] Schmaljohann D, Nitschke M, Schulze R, Eing A, Werner C, Eichhorn K-J. In situ study of the thermoresponsive behavior of micropatterned hydrogel films by imaging ellipsometry. Langmuir 2005, 21(6), 2317–2322.

[82] Hinrichs K, Gensch M, Esser N, et al. Analysis of biosensors by chemically specific optical techniques. Chemiluminescence-imaging and infrared spectroscopic mapping ellipsometry. Anal Bioanal Chem 2007, 387(5), 1823–1829.

[83] Nielsen MMB, Simonsen AC. Imaging ellipsometry of spin-coated membranes: Mapping of multilamellar films, hydrated membranes, and fluid domains. Langmuir 2013, 29(5), 1525–1532.

[84] Chang M-J, Pang C-R, Liu J, et al. High spatial resolution label-free detection of antigen–antibody binding on patterned surface by imaging ellipsometry. J Colloid Interface Sci 2011, 360(2), 826–833.

[85] Matković A, Beltaos A, Milićević M, et al. Spectroscopic imaging ellipsometry and Fano resonance modeling of graphene. J Appl Phys 2012, 112(12), 123523.

[86] Rivet S, Dubreuil M, Bradu A, Le Grand Y. Fast spectrally encoded Mueller optical scanning microscopy. Sci Rep 2019, 9(1), 3972.

[87] Arteaga O, Baldrís M, Antó J, Canillas A, Pascual E, Bertran E. Mueller matrix microscope with a dual continuous rotating compensator setup and digital demodulation. Appl Opt 2014, 53(10), 2236–2245.

[88] Peev D, others. Anisotropic contrast optical microscope. Rev Sci Instrum 2016, 87(11), 113701.

[89] Meier R, Chiang H-Y, Ruderer MA, et al. In situ film characterization of thermally treated microstructured conducting polymer films. J Polym Sci Part B Polym Phys 2012, 50(9), 631–641.

[90] Körstgens V, Meier R, Ruderer MA, et al. Note: Grazing incidence small and wide angle x-ray scattering combined with imaging ellipsometry. Rev Sci Instrum 2012, 83(7), 76107.

[91] Gensch M. Brilliant infrared light sources for micro-ellipsometric studies of organic thin films BT – ellipsometry of functional organic surfaces and films. In: Hinrichs K, Eichhorn K-J, eds. Cham, Springer International Publishing, 2018, 505–518.

[92] Gensch M, Korte EH, Esser N, Schade U, Hinrichs K. Microfocus-infrared synchrotron ellipsometer for mapping of ultra thin films. Infrared Phys Technol 2006, 49(1), 74–77.

[93] Zhou H, Toney MF, Bent SF. Cross-linked ultrathin polyurea films via molecular layer deposition. Macromolecules 2013, 46(14), 5638–5643.

[94] Lilge I, Schönherr H. Covalently cross-linked poly(acrylamide) brushes on gold with tunable mechanical properties via surface-initiated atom transfer radical polymerization. Eur Polym J 2013, 49(8), 1943–1951.

[95] Griesser T, Wolfberger A, Daschiel U, et al. Cross-linking of ROMP derived polymers using the two-photon induced thiol–ene reaction: Towards the fabrication of 3D-polymer microstructures. Polym Chem 2013, 4(5), 1708–1714.

[96] Kiss É. Characterization of polymer blends: Ellipsometry. Charact Polym Blends 2014, 299–326.

[97] Nguyen-Tri P, Ghassemi P, Carriere P, Nanda S, Assadi AA, Nguyen DD. Recent applications of advanced atomic force microscopy in polymer science: A review. Polymer 2020, 12(5).

[98] Hinrichs K, Gensch M, Nikonenko N, Pionteck J, Eichhorn K-J. Spectroscopic ellipsometry for characterization of thin films of polymer blends. Macromol Symp 2005, 230(1), 26–32.

[99] Duckworth P, Richardson H, Carelli C, Keddie JL. Infrared ellipsometry of interdiffusion in thin films of miscible polymers. Surf Interface Anal 2005, 37(1), 33–41.

[100] Vallon S, Drévillon B, Poncin-Epaillard F. In situ spectroellipsometry study of the crosslinking of polypropylene by an argon plasma. Appl Surf Sci 1997, 108(1), 177–185.

[101] Laskarakis A, Gravalidis C, Logothetidis S. FTIR and Vis–FUV real time spectroscopic ellipsometry studies of polymer surface modifications during ion beam bombardment. Nucl Instruments Methods Phys Res Sect B Beam Interact Mater Atoms 2004, 216, 131–136.

[102] Bittrich E, Windrich F, Martens D, Bittrich L, Häussler L, Eichhorn K-J. Determination of the glass transition temperature in thin polymeric films used for microelectronic packaging by temperature-dependent spectroscopic ellipsometry. Polym Test 2017, 64, 48–54.

[103] Forrest JA, Dalnoki-Veress K. The glass transition in thin polymer films. Adv Colloid Interface Sci 2001, 94(1), 167–195.

[104] Tress M, Erber M, Mapesa EU, et al. Glassy dynamics and glass transition in nanometric thin layers of polystyrene. Macromolecules 2010, 43(23), 9937–9944.

[105] El Ouakili A, Vignaud G, Balnois E, Bardeau J-F, Grohens Y. Multiple glass transition temperatures of polymer thin films as probed by multi-wavelength ellipsometry. Thin Solid Films 2011, 519(6), 2031–2036.

[106] Vogt BD. Mechanical and viscoelastic properties of confined amorphous polymers. J Polym Sci Part B Polym Phys 2018, 56(1), 9–30.

[107] Serghei A, Kremer F. Unexpected Preparative Effects on the Properties of Thin Polymer Films BT – Characterization of Polymer Surfaces and Thin Films. In: Grundke K, Stamm M, Adler H-J, eds. Berlin, Heidelberg, Springer Berlin Heidelberg, 2006, 33–40.

[108] Perlich J, Metwalli E, Schulz L, Georgii R, Müller-Buschbaum P. Solvent content in thin spin-coated polystyrene homopolymer films. Macromolecules 2009, 42(1), 337–344.

[109] Serghei A, Kremer F. Metastable states of glassy dynamics, possibly mimicking confinement-effects in thin polymer films. Macromol Chem Phys 2008, 209(8), 810–817.

[110] Erber M, Khalyavina A, Eichhorn K-J, Voit BI. Variations in the glass transition temperature of polyester with special architectures confined in thin films. Polymer (Guildf) 2010, 51(1), 129–135.

[111] Balazs AC, Emrick T, Russell TP. Nanoparticle polymer composites: where two small worlds meet. Science (80-) 2006, 314(5802), 1107LP –1110.

[112] Zeng S, Baillargeat D, Ho H-P, Yong K-T. Nanomaterials enhanced surface plasmon resonance for biological and chemical sensing applications. Chem Soc Rev 2014, 43(10), 3426–3452.

[113] Oates TWH. Real time spectroscopic ellipsometry of nanoparticle growth. Appl Phys Lett 2006, 88(21), 213115.

[114] Oren R, Liang Z, Barnard JS, Warren SC, Wiesner U, Huck WTS. Organization of nanoparticles in polymer brushes. J Am Chem Soc 2009, 131(5), 1670–1671.

[115] Rauch S, Eichhorn K-J, Stamm M, Uhlmann P. Spectroscopic ellipsometry of superparamagnetic nanoparticles in thin films of poly(N-isopropylacrylamide). J Vac Sci Technol A 2012, 30(4), 41514.

[116] Oates TWH, Wormeester H, Arwin H. Characterization of plasmonic effects in thin films and metamaterials using spectroscopic ellipsometry. Prog Surf Sci 2011, 86(11), 328–376.

[117] Oates TWH, Christalle E. Real-time spectroscopic ellipsometry of silver nanoparticle formation in poly(vinyl alcohol) thin films. J Phys Chem C 2007, 111(1), 182–187.

[118] Yashchenok AM, Gorin DA, Badylevich M, et al. Impact of magnetite nanoparticle incorporation on optical and electrical properties of nanocomposite LbL assemblies. Phys Chem Chem Phys 2010, 12(35), 10469–10475.

[119] Qian X-M, Nie SM. Single-molecule and single-nanoparticle SERS: From fundamental mechanisms to biomedical applications. Chem Soc Rev 2008, 37(5), 912–920.

[120] Kasputis T, Koenig M, Schmidt D, et al. Slanted columnar thin films prepared by glancing angle deposition functionalized with polyacrylic acid polymer brushes. J Phys Chem C 2013, 117(27), 13971–13980.

[121] Hawkeye MM, Brett MJ. Glancing angle deposition: Fabrication, properties, and applications of micro- and nanostructured thin films. J Vac Sci Technol A Vacuum, Surfaces, Film 2007, 25(5), 1317.

[122] Schmidt D, Schubert E, Schubert M. Generalized Ellipsometry Characterization of Sculptured Thin Films Made by Glancing Angle Deposition BT – Ellipsometry at the Nanoscale. In: Losurdo M, Hingerl K, eds. Berlin, Heidelberg, Springer Berlin Heidelberg, 2013, 341–410.

[123] Kılıç U, Mock A, Feder R, et al. Tunable plasmonic resonances in Si-Au slanted columnar heterostructure thin films. Sci Rep 2019, 9(1), 71.

[124] Dragan ES, Schwarz S, Eichhorn K-J. Specific effects of the counterion type and concentration on the construction and morphology of polycation/azo dye multilayers. Colloids Surf A Physicochem Eng Asp 2010, 372(1), 210–216.

[125] Nejadnik MR, Garcia CD. Staining proteins: A simple method to increase the sensitivity of ellipsometric measurements in adsorption studies. Colloids Surf B Biointerfaces 2011, 82(1), 253–257.

[126] Rauch S, Eichhorn K-J, Oertel U, Stamm M, Kuckling D, Uhlmann P. Temperature responsive polymer brushes with clicked rhodamine B: Synthesis, characterization and swelling dynamics studied by spectroscopic ellipsometry. Soft Matter 2012, 8(40), 10260–10270.

[127] Logothetidis S. Polymer blends and composites BT – Ellipsometry of functional organic surfaces and films. In: Hinrichs K, Eichhorn K-J, eds. Cham, Springer International Publishing, 2018, 271–294.

[128] Campoy-Quiles M, Ferenczi T, Agostinelli T, et al. Morphology evolution via self-organization and lateral and vertical diffusion in polymer: Fullerenesolar cell blends. Nat Mater 2008, 7(2), 158–164.

[129] Bednarski H, Hajduk B, Domański M, et al. Unveiling of polymer/fullerene blend films morphology by ellipsometrically determined optical order within polymer and fullerene phases. J Polym Sci Part B Polym Phys 2018, 56(15), 1094–1100.

[130] Engmann S, Singh CR, Turkovic V, Hoppe H, Gobsch G. Direct correlation of the organic solar cell device performance to the in-depth distribution of highly ordered polymer domains in polymer/fullerene films. Adv Energy Mater 2013, 3(11), 1463–1472.

[131] Karagiannidis PG, Kalfagiannis N, Georgiou D, et al. Effects of buffer layer properties and annealing process on bulk heterojunction morphology and organic solar cell performance. J Mater Chem 2012, 22(29), 14624–14632.

[132] Müller C, Bergqvist J, Vandewal K, et al. Phase behaviour of liquid-crystalline polymer/ fullerene organic photovoltaic blends: Thermal stability and miscibility. J Mater Chem 2011, 21(29), 10676–10684.

[133] Herrmann F, Engmann S, Presselt M, Hoppe H, Shokhovets S, Gobsch G. Correlation between near infrared-visible absorption, intrinsic local and global sheet resistance of poly(3,4-ethylenedioxy-thiophene) poly(styrene sulfonate) thin films. Appl Phys Lett 2012, 100(15), 153301.

[134] Schubert M, Bundesmann C, Jakopic G, et al. Infrared ellipsometry characterization of conducting thin organic films. Thin Solid Films 2004, 455–456, 295–300.

[135] Mauger SA, Chang L, Rochester CW, Moulé AJ. Directional dependence of electron blocking in PEDOT:PSS. Org Electron 2012, 13(11), 2747–2756.

[136] Ino T, Hiate T, Fukuda T, Ueno K, Shirai H. Real-time ellipsometric characterization of the initial growth stage of poly(3,4-ethylenedioxythiophene): Poly(styrenesulfonate) films by electrospray deposition using N,N-dimethylformamide solvent solution. J Non Cryst Solids 2012, 358(17), 2520–2524.

[137] Yap BK, Xia R, Campoy-Quiles M, Stavrinou PN, Bradley DDC. Simultaneous optimization of charge-carrier mobility and optical gain in semiconducting polymer films. Nat Mater 2008, 7(5), 376–380.

[138] Campoy-Quiles M, Etchegoin PG, Bradley DDC. Exploring the potential of ellipsometry for the characterisation of electronic, optical, morphologic and thermodynamic properties of polyfluorene thin films. Synth Met 2005, 155(2), 279–282.

[139] Pearson AJ, Wang T, Lidzey DG. The role of dynamic measurements in correlating structure with optoelectronic properties in polymer : Fullerene bulk-heterojunction solar cells. Rep Prog Phys 2013, 76(2), 22501.

[140] Yang X, Loos J, Veenstra SC, et al. Nanoscale morphology of high-performance polymer solar cells. Nano Lett 2005, 5(4), 579–583.

[141] Karagiannidis PG, Georgiou D, Pitsalidis C, Laskarakis A, Logothetidis S. Evolution of vertical phase separation in P3HT:PCBM thin films induced by thermal annealing. Mater Chem Phys 2011, 129(3), 1207–1213.

[142] Russell TP. Surface-responsive materials. Science (80-) 2002, 297(5583), 964LP –967.

[143] Dong R, Lindau M, Ober CK. Dissociation behavior of weak polyelectrolyte brushes on a planar surface. Langmuir 2009, 25(8), 4774–4779.

[144] Toomey R, Tirrell M. Functional polymer brushes in aqueous media from self-assembled and surface-initiated polymers. Annu Rev Phys Chem 2008, 59(1), 493–517.

[145] Stuart MAC, Huck WTS, Genzer J, et al. Emerging applications of stimuli-responsive polymer materials. Nat Mater 2010, 9(2), 101–113.

[146] Brittain WJ, Minko S. A structural definition of polymer brushes. J Polym Sci Part A Polym Chem 2007, 45(16), 3505–3512.

[147] Kooij ES, Sui X, Hempenius MA, Zandvliet HJW, Vancso GJ. Probing the thermal collapse of poly(N-isopropylacrylamide) grafts by quantitative in situ ellipsometry. J Phys Chem B 2012, 116(30), 9261–9268.

[148] Wu T, Gong P, Szleifer I, Vlček P, Šubr V, Genzer J. Behavior of surface-anchored poly(acrylic acid) brushes with grafting density gradients on solid substrates: 1. Experiment. Macromolecules 2007, 40(24), 8756–8764.

[149] Biesalski M, Johannsmann D, Rühe J. Synthesis and swelling behavior of a weak polyacid brush. J Chem Phys 2002, 117(10), 4988–4994.

[150] Toomey R, Freidank D, Rühe J. Swelling behavior of thin, surface-attached polymer networks. Macromolecules 2004, 37(3), 882–887.

[151] Junk MJN, Anac I, Menges B, Jonas U. Analysis of optical gradient profiles during temperature- and salt-dependent swelling of thin responsive hydrogel films. Langmuir 2010, 26(14), 12253–12259.

[152] Gramm S, Teichmann J, Nitschke M, Gohs U, Eichhorn K-J, Werner C. Electron beam immobilization of functionalized poly(vinyl methyl ether) thin films on polymer surfaces – Towards stimuli responsive coatings for biomedical purposes. eExpress Poly Lett 2011, 5, 970–976.

[153] Furchner A, Bittrich E, Uhlmann P, Eichhorn K-J, Hinrichs K. In-situ characterization of the temperature-sensitive swelling behavior of poly(N-isopropylacrylamide) brushes by infrared and visible ellipsometry. Thin Solid Films 2013, 541, 41–45.

[154] Hinrichs K, Aulich D, Ionov L, et al. Chemical and structural changes in a ph-responsive mixed polyelectrolyte brush studied by infrared ellipsometry. Langmuir 2009, 25(18), 10987–10991.

[155] Berlind T, Tengvall P, Hultman L, Arwin H. Protein adsorption on thin films of carbon and carbon nitride monitored with in situ ellipsometry. Acta Biomater 2011, 7(3), 1369–1378.

[156] Arwin H. Application of ellipsometry techniques to biological materials. Thin Solid Films 2011, 519(9), 2589–2592.

[157] Malmsten M. Ellipsometry studies of protein layers adsorbed at hydrophobic surfaces. J Colloid Interface Sci 1994, 166(2), 333–342.

[158] Arwin H, Askendahl A, Tengvall P, Thompson DW, Woollam JA. Infrared ellipsometry studies of thermal stability of protein monolayers and multilayers. Phys Status Solidi C 2008, 5(5), 1438–1441.

[159] Reichelt S, Eichhorn K-J, Aulich D, et al. Functionalization of solid surfaces with hyperbranched polyesters to control protein adsorption. Colloids Surf B Biointerfaces 2009, 69(2), 169–177.

[160] Xue C, Yonet-Tanyeri N, Brouette N, Sferrazza M, Braun PV, Leckband DE. Protein adsorption on poly(N-isopropylacrylamide) brushes: dependence on grafting density and chain collapse. Langmuir 2011, 27(14), 8810–8818.

[161] De Feijter JA, Benjamins J, Veer FA. Ellipsometry as a tool to study the adsorption behavior of synthetic and biopolymers at the air–water interface. Biopolymers 1978, 17(7), 1759–1772.

[162] Benesch J, Askendal A, Tengvall P. Quantification of adsorbed human serum albumin at solid interfaces: A comparison between radioimmunoassay (RIA) and simple null ellipsometry. Colloids Surf B Biointerfaces 2000, 18(2), 71–81.

[163] Arwin H. Optical properties of thin layers of bovine serum albumin, γ-globulin, and hemoglobin. Appl Spectrosc 1986, 40(3), 313–318.

[164] König U, Psarra E, Guskova O, et al. Bioinspired thermoresponsive nanoscaled coatings: Tailor-made polymer brushes with bioconjugated arginine-glycine-aspartic acid-peptides. Biointerphases 2018, 13(2), 21002.

[165] Burkert S, Bittrich E, Kuntzsch M, et al. Protein resistance of PNIPAAm Brushes: Application to switchable protein adsorption. Langmuir 2010, 26(3), 1786–1795.

[166] Bittrich E, Mele F, Janke A, et al. Interactions of bioactive molecules with thin dendritic glycopolymer layers. Biointerphases 2018, 13(6), 06D405.

Rosa Di Mundo, Fabio Palumbo

9 Wettability of polymer surfaces: significance and measurements. Recent developments and applications

9.1 Introduction

Many technological issues as well as natural phenomena are based on the interaction between solids and liquids. *Wetting*, the phenomenon of bringing a liquid in contact with a solid surface, considered both when two like substances are involved, and when two different ones are forced to interact, has been investigated in the last 200 years by many researchers, often in a controversial way. A Good contact between a solid and a liquid is required in case of gluing, painting, printing paper, dyeing, or washing [1, 2]. Conversely, plants leaves resisting water uptake or even surviving underwater, waterproof clothes/plastics, water strider running over a pond surface are examples of anti-adhesive surfaces characterized by a very poor interaction with liquids [3–6]. Similarly, phenomena like metal/polymer adhesion or fluid lubrication between engine parts depend at least in part from the wettability characteristic of the solid surface involved [7, 8]. In this chapter, some highlights will be given on the meaning of wettability and its link with materials surface energy and liquid surface tension, the main properties driving the surface contact-related phenomena. Then, the importance of contact angle and its measurement for the evaluation of the wettability will be analyzed taking into due consideration the definition of hydrophilic and hydrophobic materials.

9.2 Contact angle and surface energy

A wetting event leads to the formation of an interface between a liquid and a solid normally both being in contact with a fluid environment. In fact, the most typical experiment is that of placing a drop of liquid on a solid in presence of air, but not uncommon is the case where the solid is in contact with oil in presence of water, just thinking to the petroleum extraction field [9]. In the first case the fluid environment is air, while in the latter is water. Hence, since an interface is formed between the liquid and the solid, an adhesion issue is implicated as well as surface tension of the various phases interested.

When two semi-infinite bodies made of different materials, *a* and *b*, separated at such a distance that the mutual interaction energy is negligible are brought into contact, the *a*–*b* interface is formed and the interfaces between each material and the fluid, supposing for simplicity being air, which the bodies are immersed in,

https://doi.org/10.1515/9783110701098-009

disappear as shown in Fig. 9.1a. The variation of free energy per unit area of the system, ΔG_{ab}, due to the interfaces energy variation is

$$\Delta G_{ab} = \gamma_{ab} - \gamma_a - \gamma_b \qquad (9.1)$$

where γ_{ab} is the interfacial energy between a and b, while γ_a and γ_b are the surface energies of each material. As a consequence the energy per unit area necessary to achieve the separation of the bodies, called *work of adhesion*, W_{ab}, is equal to the opposite of the free energy variation, reported in (1) and in turn

$$W_{ab} = \gamma_a + \gamma_b - \gamma_{ab} \qquad (9.2)$$

It is interesting to observe that if the two materials are the same, we do not deal with *adhesion* force, that is defined as the force necessary to detach two different materials, but with *cohesion*, and the energy per unit area necessary to overcome the intermolecular (cohesive) forces between molecules of the same species is

$$W_c = 2\gamma_a \qquad (9.3)$$

Here it is necessary to consider that, as depicted in Fig. 9.1b, the surface energy can be considered as the work per unit area, dW, due to the force F necessary to stretch a surface A of the quantity dA, and the units are typically mN/m, or mJ/cm². When dealing with liquids, normally the term *surface tension* is used.

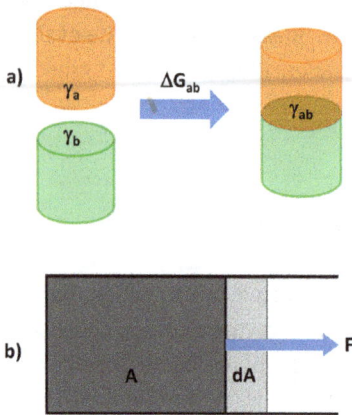

Fig. 9.1: (a) Gibbs energy variation due to the work of adhesion between bodies a and b, and (b) sketch of surface area increase, from A to A + dA, due to a force F.

The shape of a liquid drop wetting a solid surface depends on the balance of the surface energies of the materials in contact with the fluid, as depicted in Fig. 9.2. More than 200 years ago Thomas Young [10], in an essay on the cohesion of fluids, derived the following fundamental equation nowadays used by scientist to describe the phenomenon of wetting between a solid s and a liquid l immersed in a fluid f:

$$\frac{\gamma_{sf} - \gamma_{sl}}{\gamma_{lf}} = \cos\theta \qquad (9.4)$$

where, in the case the fluid corresponds to air, γ_{sf} and γ_{lf} can be replaced with γ_s and γ_l, respectively, indicating the surface energy of the solid and the liquid, eq. (9.4) being transformed in the more familiar form

$$\frac{\gamma_s - \gamma_{sl}}{\gamma_l} = \cos\theta \qquad (9.5)$$

The angle θ, called *contact angle (CA)*, is defined as the angle between the tangent at the liquid–fluid interface and tangent at the solid surface at the three phase contact point. The Young equation has been derived for an ideal surface at equilibrium that is smooth, rigid, chemically homogeneous and not soluble in the liquid phase, where no external forces are applied.

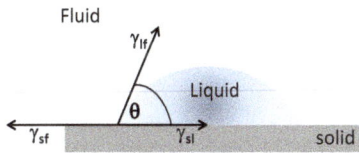

Fig. 9.2: Contact angle, θ, between a liquid l and a solid s, immersed in a fluid f.

To better understand the meaning of the Young equation some qualitative comments can be drawn. The first observation is that for a given solid material a lower contact angle, hence higher spreading of the liquid, indicates a better wetting of the surface, and looking at eq. (9.5) and Tabs. 9.1 and 9.2, this is obtained with apolar liquids, having a lower surface tension. *e.g.* Teflon, a typical hydrophobic material, can be wetted by a hydrocarbon solvent such as *n*-decane.

As well known, water is the most polar and common liquid solvent with a surface tension as high as 72.8 mN/m, thus influencing a number of natural phenomena based on capillarity or water repellency [6, 11, 12]. On the other hand if one considers water as the wetting liquid, as illustrated in Tab. 9.2, surfaces with a higher surface energy will show a lower contact angle, thus being more wettable. Most of the organic polymeric surfaces rarely display water contact angle lower than 70°, and this is the reason for the interest in developing methods for increasing the wettability of polymers. In fact, scarce wettability means poor adhesion of dyes or inks, or of high surface energy metal, e.g. in metallization process for packaging [7].

The correlation of contact angle and adhesion of liquids appears more evident when combining eqs. (9.1) and (9.5), obtaining:

$$W_{sl} = \gamma_l(1 + \cos\theta) \tag{9.6}$$

This relation, known as Young–Dupre's equation, highlights that a liquid drop sticks better on a solid surface where a low contact angle is formed, furthermore polar liquid having higher surface tension will show higher work of adhesion.

Tab. 9.1: Surface tension of some common liquids.

Liquid	γ (mN/m)
n-Hexane	18.4
Methanol	22.5
n-Decane	23.4
Tetrahydrofuran	27.4
Benzene	28.9
Dimethyl sulfoxide	44.0
Diiodomethane	50.8
Formamide	58.0
Glycerol	64.0
Water	72.8
Mercury	484.0

Tab. 9.2: Surface energy of some common polymers, and corresponding water contact angle. Value with * are advancing, the others have been determined in static mode.

Polymer	Surface energy (mN/m)	Water contact angle (°)
PC (polycarbonate)	29	93* [10]
PDMS (polydimethylsiloxane)	20	114 [11]
PE (polyethylene)	27	93 [12]
PMMA (polyethylmethacrylate)	37	65 [13]
PP (polypropylene)	28	95 [14]
PS (polystyrene)	29	90 [15]
PTFE (polytetrafluoroethylene)	16	117 [16]

Tab. 9.2 (continued)

Polymer	Surface energy (mN/m)	Water contact angle (°)
PVA (polyvinylalcohol)	39	51 [17]
PVC (polyvinylchloride)	27	103 [18]
PVDF (polyvinylidenefluoride)	32	85 [17]

9.2.1 Surface energy evaluation

Coming back to the Young equation, it is possible to observe that, in principle, it can be an instrument for the evaluation of the surface energy of polymers from contact angle measurements. However, serious concerns have been faced by investigators in the following two centuries: in fact, supposing that the liquid contact angle, θ, can be measured (Section 9.4) and several methods exist to get the surface energy of a liquid, y_l, two unknowns are present in eq. (9.5), the surface energy of the solid, y_s, and y_{sl}, the liquid–solid interfacial energy, and none can be directly measured, since solids cannot be deformed like liquids [13]. Hence, the Young equation, even if extremely appealing in describing wetting, requires a further equation to describe y_{sl} or y_s, in order to leave just an unknown in eq. (9.5). From an historical point of view the interpretation of the Young equation to get solid surface energy data started at the end of the 1950s with the work of Zisman and coworkers that identified a critical surface energy y_c, as it will be described below [14]. Then in the 1960s and 1970s two main approaches were developed to get solid surface energy from liquid contact angle measurement [15–19]:

a) Multicomponent approach: the total surface energy consists of a sum of different components, each of which arises due to a specific type of intermolecular forces, acid–base, dispersive, hydrogen bonding, and so on

b) Equation of state approach: based on the thermodynamic definition of the work of adhesion and the London theory of dispersion forces, investigators derived a modified Young's equation relating the contact angle to the surface energy of the liquid and of the solid

In the following the Zisman method and the main models of the multicomponent approach will be briefly discussed, leaving the equation of state to appropriate references [16, 20].

Zisman and coworkers observed that for a given solid surface the contact angle, θ, of different liquids changes with $\cos\theta$ decreasing quite linearly with the surface tension of the liquids and they considered that a solid would be completely wetted by a liquid having the same surface tension of the solid. This is clear in a graph like the one in

Fig. 9.3, where the cosine of contact angle of various liquids is plotted vs the liquid surface tension for a polyethylene-like solid. Furthermore, they defined the critical surface energy y_c of a solid as the surface tension of the liquid for which cos θ is 1, a completely wetted surface. The critical surface energy, hence, is obtained from similar plots built from contact angle measurements with different liquids, the obtained line is extrapolated at cos θ equal to 1, and the corresponding y_l is y_c. Obviously, y_c is close, but not the surface energy of the solid, and it has not a physical meaning. However, the Zisman plot methods have been very useful for an applicative comparison of the wetting behavior of solids and still important for lubricants, inks, lacquers, varnishes applications: a liquid having a surface tension below the critical surface energy of the material will wet that surface completely.

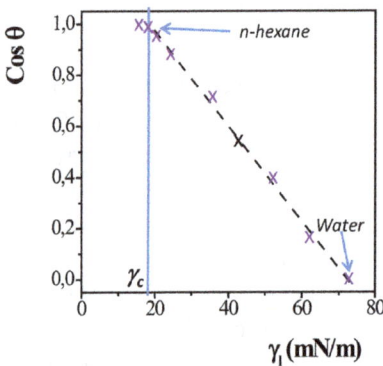

Fig. 9.3: Cosine of water contact angle vs surface tension of different liquids for a generic polyethylene-like polymer. y_c is the critical surface tension of the polymer.

Apart the empirical Zisman method, the main approach to get solid surface energy is the multicomponent one. We can distinguish two broad categories: i) the so-called Fowkes' method [17, 18, 21] and ii) the acid–base theory [22].

According to the Fowkes' method, the surface energy is composed of a *dispersive* and a *polar* component. The former results in matter from the interaction between adjacent molecules arising from the electron dipole fluctuations, namely London forces. The remaining kind of interactions, called polar, are those like van der Waals ones, induction (the Keesom and Debye ones), and acid–base. The main assumption of the method is that such forces operate independently and in turn their contribution to the surface tension is additive and a geometric mean relationship can be used to describe solid-liquid interfacial interactions.

With this assumption one can write the surface energy in the following way:

$$\gamma = \gamma^d + \gamma^p \tag{9.7}$$

where

$$\gamma^p = \gamma^{vdW} + \gamma^i + \gamma^h + \gamma^{AB} + \cdots \tag{9.8}$$

with *d*, *p*, *vdW*, *i*, *h*, *AB* labels indicating, respectively, dispersive, polar, van der Waals, induction, hydrogen bonding, and acid–base interaction. Combining such additive hypothesis with the geometric mean description of the interfacial interaction energy applied to the work of adhesion for a force *f*, $W^f_{ab} = 2(\gamma^f_a \gamma^f_b)^{1/2}$, eq. (9.2) can be written as follows:

$$\gamma_{sl} = \gamma_s + \gamma_l - 2\left[\left(\gamma^d_s \gamma^d_l\right)^{1/2} + \left(\gamma^p_s \gamma^p_l\right)^{1/2}\right] \tag{9.9}$$

Where the apex *p* groups all the non-dispersive interactions. Equation (9.9) is a generalized form coming from the pioneering contribution of Fowkes and then further revised by Owens and Wendt [17, 18], and it represents the desired relation between γ_{sl} and γ_s necessary to solve the problem of the excess number of unknown in the Young equation.

The simpler case for taking advantage of eq. (9.9) is a liquid that can interact with a polymer surface only through dispersive forces, such as diiodomethane and polytetrafluoroethylene. In this case

$$\gamma^p_s = \gamma^p_l = 0$$

And eq. (9.9) can be simplified taking into account eqs. (9.2) and (9.6), leading to

$$\gamma_s = \gamma_l(1 + \cos\theta)^2/4 \tag{9.10}$$

From which it is evident that simply measuring the contact angle of a non-polar liquid of well-known surface tension, the surface energy of a purely dispersive polymer can be obtained.

In a more general case, the polymer surface will express one dispersive and *n* polar components of the surface energy. It can be demonstrated that if n + 1 liquids can be used, each with known corresponding surface tension components, the measurement of each contact angle will allow for the complete assessment of the polymer surface energy [23].

The acid–base theory, developed mainly by Van Oss and coworkers [22, 24], foresees that all the interactions arising from dipole–dipole and dipole-induced dipole interactions, like van der Waals and Keesom/Debye ones and London forces cannot be separated and be considered additive. Hence, a term γ^{LW} should be better considered to embrace such forces effects, the apex LW denoting Lifschitz-van der Waals electromagnetic interactions [25]. Besides γ^{LW}, acid–base interactions, in a general term like the Lewis one, contribute to surface energy. Remember that according to Lewis acid base theory, acid is an electrons acceptor substance, and a base is an electrons donor one. In this frame also an hydrogen bond is an acid–base interaction, even a non conventional acid substance like chloroform, $CHCl_3$, should be considered an acid, or an ether, R-O-R, a base. For sake of clarity of the following considerations, it is possible to distinguish between *monopolar* species, those having only an acid or a base character, *bipolar*, like water, exhibiting both acid and base character (even not of the same entity), and *apolar* having no base or acid character.

Hence, supposing additive the two main kind of interactions, LW and acid–base, the surface energy for the species i, can be written as

$$\gamma_i = \gamma_i^{LW} + \gamma_i^{AB} \tag{9.11a}$$

where y_i^{AB} is the acid–base contribution to the surface energy of the material i.

It is possible to define $y^{(+)}$ and $y^{(-)}$, as the acid and base contribution to the surface energy, respectively, and to derive the following relation:

$$\gamma_i^{AB} = 2\left(\gamma_i^{(+)}\gamma_i^{(-)}\right)^{1/2} \tag{9.11b}$$

It should be considered that:
- the acid component of the surface interact only with the base counterpart of the liquid and vice versa,
- if the solid and the liquid are monopolar in the same direction, no interaction is possible,
- y_i^{AB} is 0 if the acid or base contribution is null (see eq. (9.11b)).

Taking into account these elements an equation analogous to the (9.9) one can be composed to express the interfacial surface tension:

$$\gamma_{sl} = \gamma_s + \gamma_l - 2\left(\gamma_s^{LW}\gamma_l^{LW}\right)^{1/2} - 2\left[\left(\gamma_s^{(+)}\gamma_l^{(-)}\right)^{1/2} + \left(\gamma_s^{(-)}\gamma_l^{(+)}\right)^{1/2}\right] \tag{9.12}$$

With a rationale analogous to what done for the Fowkes theory, one can couple the Young–Dupre equation of interfacial work of adhesion with eq. (9.12) obtaining

$$2\left(\gamma_s^{LW}\gamma_l^{LW}\right)^{1/2} + 2\left(\gamma_s^{+}\gamma_l^{-}\right)^{1/2} + 2\left(\gamma_s^{-}\gamma_l^{+}\right)^{1/2} = \gamma_l(1 + \cos\theta) \tag{9.13}$$

The use of eq. (9.13) allows for the determination of the three components of a solid surface energy, y_s^{LW}, $y_s^{(+)}$, and $y_s^{(-)}$, solving simultaneously the three equations obtained measuring the contact angle with three different liquids whose components of the surface tension are known. Best is to use at least one apolar liquid, for which no acid and base components are present ($\gamma_l^{+} = \gamma_l^{-} = 0$), thus $y_l = y_l^{LW}$, and the Lifschitz–van der Waals component of y_s, measured the contact angle with that liquid, from eq. (9.13) is straightforward.

It is important to highlight that, the set of $y_l^{(+)}$ and $y_l^{(-)}$ for suitable probing liquids are determined by arbitrary attributing to water equal values for the acid and base components. Considering that at 20 °C, the surface tension of water is 72.8 mN/m and the apolar component is 21.8 mN/m, from eqs. (9.11a) and (9.11b) it is possible to obtain that for water $y_{H2O}^{(+)} = y_{H2O}^{(-)} = 25.5$ mN/m. In Tab. 9.3, a selection of probing liquids is reported with the corresponding components of the surface tension. It is useful to observe that the basic component to the surface tension for methanol and dimethylsulfoxide is quite high (77 and 32 mN/m, respectively), but the total γ_l^{AB} is low, being the product of the acid and base ones. However, a strong acid base interaction could develop with a polymer functionalized with acid groups, deeply affecting the wettability.

Tab. 9.3: Surface tension parameters in mN/m of polar and apolar liquid.

Liquid	Total	γ^{LW}	γ^{AB}	$\gamma^{(+)}$	$\gamma^{(-)}$
1-Bromonaphthalene	44.4	44.4	0.0	0.0	0.0
Diiodomethane	50.8	50.8	0.0	0.0	0.0
Dimethyl sulfoxide	44	36	8	0.5	32
Ethanol	21.4	18.8	2.6	0.02	68
Ethylene glycol	48	29	19	3.0	30.1
Formamide	58	39	19	2.28	39.6
Glycerol	64	34	30	3.92	57.4
Methanol	22.5	18.2	4.3	0.06	77
Tetrahydrofuran	27.4	27.4	0.0	0.0	15
Water	72.8	21.8	51.0	25.5	25.5

Even if surface energy evaluation is a quite well-known subject and results interpretation lies on quite well-understood classic thermodynamic principles, debates and different opinions are still present in the scientific community. As an example scientists are divided between those that follows the multicomponent approach and those searching for an equation of state describing the interfacial solid liquid tension, but others are still looking for a more unifying theory [25, 26]. Nevertheless, automatic contact angle apparatus (as described in the following paragraphs) are often equipped with software tools that allows to approach the surface energy determination with all the discussed methods, leaving, thus, the user the decision on the more feasible interpretation.

It should be considered that surface energy evaluation is highly dependent on the following conditions, often very hard to be respected by the real-lab situation:
- no dissolution or swelling of the solid material during the contact angle measurement;
- smoothness of the sample surface, allowing to measure the equilibrium contact angle;
- surface cleanliness;

– surface homogeneity such that sessile drops of test liquids have a hemispherical
   shape.

On the other hand, if these parameters are not compliant with the solid under investigation, the contact angle measurement alone, even when not leading to the surface energy characterization, will give indication on the wettability in the field of
interest of that specific material.

9.3 Contact angle hysteresis

In Section 9.2, the liquid–solid work of adhesion has been derived as a function of
the contact angle θ, eq. (9.6). Such an equation, however, though formally correct,
does not consider that often on real surfaces different values for θ can be detected,
generally retrieved under dynamic conditions, i.e. upon motion of the liquid with
respect to the solid or *viceversa*. Evaluation of dynamics of wetting in addition or
alternative to statics, though studied and reported since the 60s, has been often neglected in experimental research for decades and sometimes still skipped, notwith-
standing the dynamic situations in wetting are those effectively linked to applications.
In particular, a so-called contact angle *hysteresis* exists, defined as the difference between the *advancing* and the *receding* contact angle, observed under dynamic conditions. The former is the angle when the liquid front is advancing over an un-wetted
portion of the surface, and it is the maximum angle measurable on a surface; the latter
is the one observed while the liquid front is retracting from the previously wetted surface, the minimum observable angle. As it will be better explained in the next section
the measurement of hysteresis can be done either forcing the liquid to increase/reduce
the probing volume on a level surface, or evaluating the front and rear angles of a
drop at rest on a tilted plane.

Furmidge [27] first observed that such a difference between the front and the
rear wetting state with respect to liquid (drop) motion generates a force opposing
the weight of the drop and able, if the drop is small enough, to balance it. Thus, he
modified the Young–Dupre equation, in order to take into account the difference
between the advancing, θ_a, and the receding angle, θ_r, as follows:

$$W_{sl} = \gamma_l(\cos\theta_r - \cos\theta_a)$$ (9.14)

Thus a drop will remain stuck on the surface as long as

$$\gamma_l(\cos\theta_r - \cos\theta_a) \geq mg\sin\alpha$$ (9.15)

where α is the angle of inclination, m the drop mass, g the gravity acceleration.
From eq. (9.15) it is evident that when tilting a plane, a drop placed on the top will

start to slide when the angle α will be such that the gravity force pass the adhesion one. In this view the parameter directly correlated to the adhesive or sticky character is the contact angle hysteresis and not the mere single static angle.

But, what about the origin of such an hysteresis? As first observed by De Gennes [28], chemical, topographical, and morphological heterogeneities are at the basis of the phenomenon. An ideal surface, perfectly smooth and made of only one type of atoms/moieties, possibly with no or poor interaction with the liquid, would not have any reason to show a difference between the advancing and receding angle. This, for instance, is almost the case of a (freshly washed) flat polystyrene slab [12]. Instead, since heterogeneities exist on most real surfaces, the liquid–solid interfacial energy is characterized by several metastable states each leading to different apparent contact angle in a range defined by the advancing and receding contact angle.

Another way to look at hysteresis is the following: commonly a surface is composed of a variety of functional groups (the distribution can be more or less broad depending on the surface), though homogeneously distributed in the space. Just considering the simple system where the liquid is water and air is the immersion fluid, as mentioned above, θ_a probe a dry surface while θ_r a surface that has reached a stable state with the liquid. Hence the advancing angle is more sensitive to lower energy, apolar groups, while the receding angle is affected by high-energy, more polar, ones: θ_a and θ_r must be different. For example, the receding angle on a plasma-treated polymer (when the treatment is addressed to hydrophilicity) is the only way to investigate high-energy-grafted functionalities. It should be considered that this is even more important when dealing with treated polymers that have been aged: as illustrated in Fig. 9.4, while the advancing angle can reveal hydrophobic recovery (water contact angle increases with time, as better explained in Section 9.5.2), the receding angle generally remains much lower, since retains the memory of the modification, suggesting that high energy groups have been buried within the reorganized surface [29].

On the other hand, there are also surfaces with little or no expectation of chemical heterogeneity that can nevertheless exhibit strong hysteresis; this might possibly be due to peculiar chain mobility upon wetting, leading to reversible interactions with the liquid drop. Typical is the case of commercial PTFE which, though being the conventional polymer with the highest advancing water contact angle, it also displays a quite high hysteresis (115°/90°) [30, 31]. Taking into consideration eq. (9.14), it is evident that the work of adhesion can be important for PTFE, and such a particular behavior has lead famous wetting scholars to state "Teflon is hydrophilic": among a distinction between shear and tensile hydrophobicity they find that teflon is "shear" hydrophilic (but tensile hydrophobic) since a drop does not slide easily if the surface is tilted [40]. This is also found for the case of thin, smooth Teflon -like films deposited by conventional PECVD processes, which show higher hysteresis than homologous hydrocarbon surfaces [32].

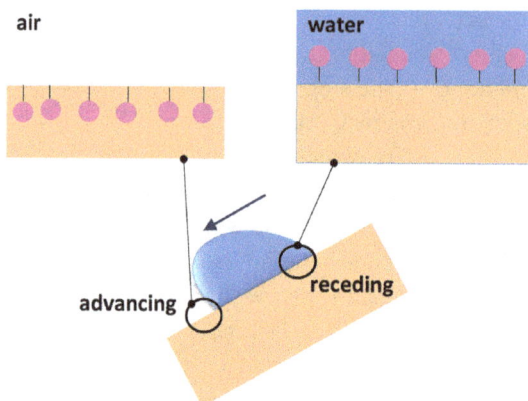

Fig. 9.4: Sketch of the different orientation of polar surface groups (pink circles) grafted on the surface of a un-polar polymer (e.g. by plasma treatment) in air, after a certain time from the treatment: advancing and receding angles are affected by the different configurations that can be achieved by the surface.

Roughness has a strong influence on the static angle and even more it affects sensitively contact angle hysteresis, as it will be better described in Section 9.5. This had been already shown by Johnson and Dettre [33], who reported, as shown in Fig. 9.5 a steep increase of both advancing and receding angle, as a function of roughness, with the latter showing a not monotonous trend. The increase of hysteresis in the first range of that diagram is a consequence of a full wet contact between liquid and solid, so the presence of protrusions hinders motion of the liquid (described as "pinning" of the drop). The following reduction of hysteresis is due to the establishment of non-wet states, where the drop is suspended on the top of the surface protrusions and sliding/rolling is favored. This behavior has been observed by several authors dealing with rough surfaces, in particular when able to tune a topographical parameter (rms roughness, height, aspect ratio, density of the structures) in such a way that the non-wet regime can be established [34, 35].

From the above discussion it is easily understood that the use of a simple static angle (often very close to the advancing angle) is often misleading, and it may hide an adhesive surface (indicated by a low receding angle), a situation that is, for instance, unfavorable for hydro-repellent applications. That is why, advancing and receding angle should be always measured.

The way contact angle hysteresis is represented and reported is particularly relevant when dealing with the concept of adhesion. As a consequence of eq. (9.14), a mere difference between advancing and receding angles is not meaningful: low work of adhesion is achieved when θ_a and θ_r are very close and their absolute values are high. Therefore, when numbers have to be compared it is suggested either to calculate the difference between cosines $(\cos_r - \cos_a)$ or simply to report both absolute values, as done by most scholars [5, 32, 36–40].

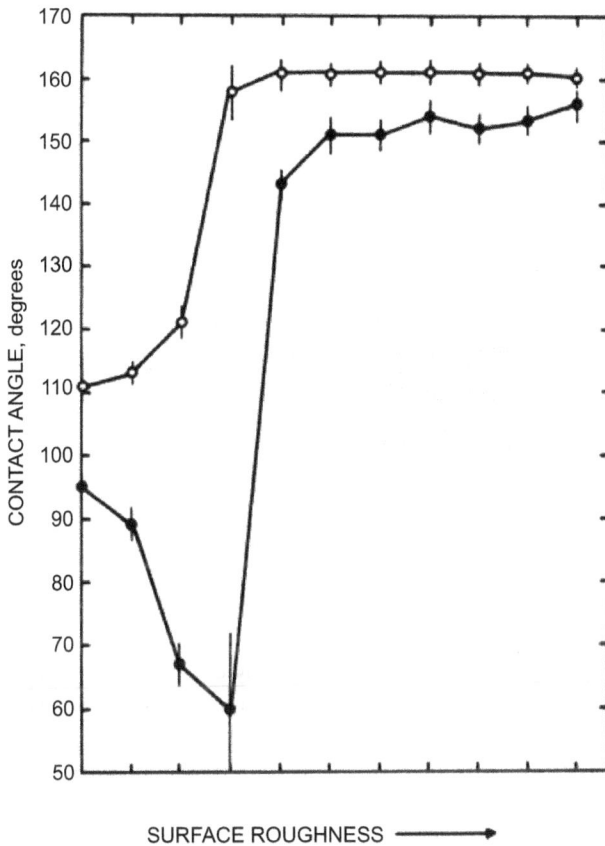

Fig. 9.5: Contact angle of water on wax substrates, as a function of the substrate roughness. Both the advancing angle (∘) and the receding one (•) are reported (adapted with permission from [33], Copyright 1964 American Chemical Society).

When considering adhesion of a material to ice [41] the work by Meuler *et al.* [42] clarified the best way to correlate this quantity to wetting parameters. They compared θ_a (advancing angle), θ_e (equilibrium angle), $\cos\theta_r - \cos\theta_a$ (contact angle hysteresis), $1 + \cos\theta_e$ (dimensionless equilibrium work of adhesion, Young–Dupre's equation, eq. (9.6), Section 9.2), $1 + \cos\theta_r$ (dimensionless practical work of adhesion). The last term, contrary to the equilibrium work of adhesion, has been defined "practical" since it is the actual work required to separate the liquid from the surface, given the explained meaning of the receding process. Results by Meuler *et al.* have confirmed that $1 + \cos\theta_r$ is the parameter best correlated to ice adhesion strength, with the experimental data of ice adhesion strength of various polymers and other materials (flat surface) fitted perfectly by a straight line passing through the origin.

9.4 Measurement methods for contact angle

Different methods have been reported for the measurement of contact angle, and for non-porous solids two main approaches can be identified: i) direct geometric determination through an optical goniometer and ii) force tensiometry. In goniometry, typically the sessile drop of a test liquid on a solid substrate is observed and the tangent in the triple point (e.g. air-liquid-solid contact point) determined. Also, the observation of a captive bubble on a solid, both immersed in a liquid, can be included in this methodology. Tensiometry, instead, is based on the measurement of the force of interaction between a solid and a liquid when in contact. In case the solid is porous, such as powders and alike, the wetting liquid can be absorbed by the material under investigation and another approach is used, based on capillarity and known as the Washburn method.

9.4.1 Direct measurement by optical goniometry

The most common technique of contact angle determination is the direct measurement of the tangent angle at the three-phase contact point on a sessile drop placed on a solid and requires a quite simple apparatus. It consists basically of three components: a horizontal sample stage, where to place the solid, a light source for the illumination of the drop sample region, and on the opposite side an image capture system. In the old systems, the latter consisted of a simple optical observation through a goniometric graduated eyepiece, with a suitable magnification (see Fig. 9.6). Afterwards, to improve reliability and accuracy of the analysis, the equipment has been implemented with a camera that allows acquisition either of photographs or movies, useful in case the contact angle has a dynamic (CA hysteresis) or kinetic behavior (absorbing materials, such as textiles). Furthermore, the goniometer can be equipped with a manual or motor-driven graduated microsyringe for liquid drop dispensing, a suitable system for tilting the sample stage, and a containment chamber to keep the environment around the sample clean and for optional conditioning of the ambient to measure the CA at different temperature or pressure.

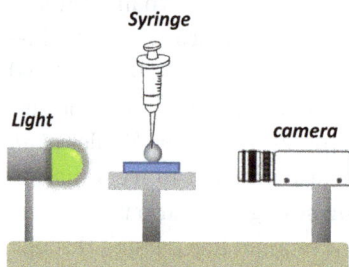

Fig. 9.6: Sketch of a contact angle goniometer.

From the operative point of view, it should be considered that drops might be not symmetrical, so it is advisable to measure contact angle on both sides of the liquid drop profile. It is also recommended to measure the contact angle of drops in more points on the investigated material, for a more accurate characterization of the wetting properties. It is quite recognized that this method cannot lead to accuracy better than ±2°.

In the static mode a drop is placed with a fixed volume (typically 2–5 µL) and then the measurement is done supposing that the drop is not changing during the analysis. When the material is characterized by hysteresis, the advancing and receding contact angle evaluation, can be obtained measuring the CA by increasing and decreasing the volume of the drop during the analysis (keeping the needle of the syringe inside the liquid) or, alternatively, by placing a drop on an inclined plate with variable tilting angle, as illustrated in Fig. 9.7a–c. In the first case, the volume of the liquid drop placed on the sample is slowly increased; for small added volume, the measured contact angle increases but the triple-phase contact line remains stuck. At a certain drop size, the line front starts to move, but the contact angle does not increase more: that is the advancing contact angle (Fig. 9.7a). For the receding contact angle, the situation is reversed (Fig. 9.7b): the liquid is slowly withdrawn with consequent reduction of the volume and of the contact angle. When the contact line starts to recede the acquired contact angle is the receding one. With a motor driven syringe and a fast speed camera this acquisition is straightforward. In the "tilting plate method" (Fig. 9.7c) the liquid drop is placed on the plate, and the top and bottom contact angle are measured during the slow tilting until the drop just begins to slide. At this step, the contact angles measured at the lowest and highest point are the advancing and receding ones, respectively [43]. It should be mentioned here that some authors demonstrated that, in some cases, mostly on hydrophobic surfaces, the tilting plate method could lead to artifacts and the volume addition and withdrawal method should be preferred [44].

For a more accurate determination of the contact angle an image of the drop on the solid surface is acquired, hence, a software for the elaboration of the drop shape image (normally provided with the goniometer) and for the contact line determination is important. Different methods have been developed to fit the contour of the whole drop or part of it, and basically they can be divided in those fitting the curve with a geometric function, and approaches using the Young–Laplace equation, relating the pressure difference (across the liquid/air interface) to the shape of the surface drop [45].

Instead of placing a liquid sessile drop above the solid sample, an air bubble can be formed underneath the surface sample, which is immersed in the testing liquid (Fig. 9.7d): then the contact angle formed by the air bubble in liquid can be directly measured. This method is referred to as the "captive bubble method." The addition of more air to the bubble through the needle leads the surrounding liquid front to recedes, and θ_r is obtained; on the other hand, air withdrawal from the

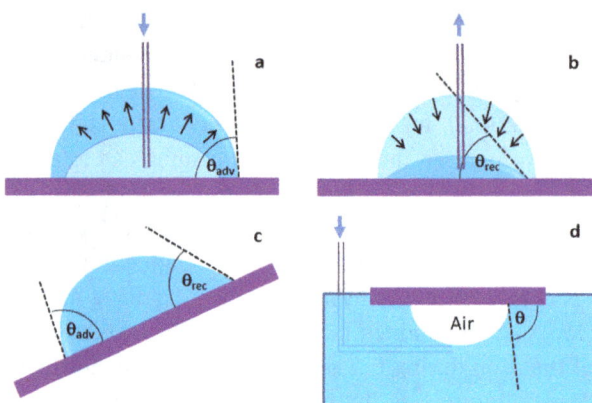

Fig. 9.7: Scheme of the determination of the advancing and receding angle by means of the sessile drop (a and b), and of the tilting method (c), and captive bubble method (d).

bubble causes the liquid front to advance, and θ_a can be determined. The captive bubble method is similar to the sessile drop one and it has the advantage of ensuring that the surface is in contact with a saturated atmosphere. Furthermore, it is much easier to monitor the temperature of the liquid in the captive bubble method than with sessile drops, which makes possible to study the temperature dependence of contact angles. However, the captive bubble method requires far more liquid than the sessile drop method.

Very recently, an important innovation in this field is represented by the use of a smartphone, if provided with a suitable application for the angle calculation. Smartphones give the advantage of their intrinsic portability and, now, of extremely advanced digital optics, almost reaching professional cameras standards, as well as easily accessible accessories such as supports and desk trivets. Chen et al. [46] developed a smartphone-based contact angle measurement instrument, featured with an automatic contact point detection algorithm to allow the instrument to correctly detect the drop angle, reaching accuracy of 0.01% with both the Young–Laplace and polynomial fitting methods. It is agreed that this is a significant advancement in the field since the appearance of the first instruments in 1980s. However, a deep knowledge of the fundamentals about wetting and contact angle is recommended to correctly adjust the setup and catch the significance of the acquired data.

9.4.2 Force tensiometry

Tensiometry indirectly measures contact angles evaluating the forces involved when a sample of solid is brought into contact with a test liquid. A flat and thin sample is typically hung to a Wilhelmy balance and the test liquid is raised to contact the solid. Upon the contact and during the immersion the force measured by the balance changes and it is recorded as a function of the depth. The detected force, F, depends on buoyancy and on wetting (once zeroed for the weight), and it can be defined as

$$F = \gamma_l p \cos\theta - gV\Delta\rho \qquad (9.16)$$

where γ_l is the surface tension of the test liquid, p is the perimeter of the contact line (supposed to be the same of the sample section), θ the contact angle, V the displaced liquid volume, $\Delta\rho$ the difference in density between the liquid and air (or a second liquid), and g the gravity acceleration. Considering Fig. 9.8, during an immersion-emersion cycle in a Wilhelmy balance measurement the following events follow: (A) The sample approaches the liquid, and the force is zero; (B) the sample contacts the liquid surface, the liquid rises up along the surface if $\theta < 90°$, with a positive wetting force (the force is opposite for $\theta > 90°$); (C) the sample is further dipped, and the increase of buoyancy causes a decrease in the force detected on the balance; the force is measured for the advancing angle. (D) The sample is pulled out of the liquid after having reached the desired depth; the force is measured for the receding angle. When a hysteresis is present, the intercept of the advancing and receding curves, obtained with a regression fit, at depth = 0 give the forces corresponding to the advancing and the receding angle. If there is not a hysteresis behavior, the advancing and receding curves should overlap and the force extrapolated at depth 0 directly gives the contact angle.

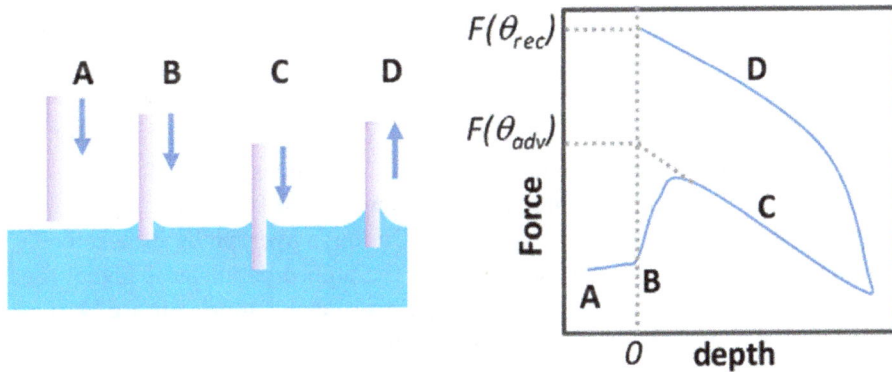

Fig. 9.8: Operating principle of Wilhelmy balance, left, and corresponding force/depth diagram, right. $F(\theta_{adv})$ and $F(\theta_{rec})$ force extrapolated at depth 0 giving the advancing and receding angle, respectively, when hysteresis is present.

9.4.3 Methods comparison

Optical goniometry is usually quite fast (few minutes for the direct acquisition of advancing and receding angle on one sample) and it can be used in many situations where force tensiometry cannot, provided that sample has a relatively flat portion for testing. For samples with complex morphology, as fabrics, the latter can be an issue. Testing can be done with very small amount of liquid, hence small samples are feasible, and suitable chamber allows for measurements in severe conditions.

On the other hand, some limited drawbacks can be highlighted. Identification of the tangent line (especially when an eye determination is operated) can affect the reproducibility of contact angle measurements, mainly at extreme contact angle. The conditions which produce advancing and receding angles are sometimes difficult to reproduce. It is challenging to measure contact angles on single fibers, even if picoliter liquid dispenser are now available [47].

Considering force tensiometry, the main advantage is the direct achievement of an average value both in terms of force and of surface on a sample wider than the ones usually tested with goniometry: This is particularly advantageous in industrial lab, where typical size of the sample is several centimeters. Since a whole force/immersion curve is recorded, values can be averaged all over the surface of the sample or on a portion of it, depending on the region of the graph obtained. A further feature is the possibility of generating multiple wetting/dewetting cycles gaining information on liquid–solid interaction (such as absorption or surface reorientation) [48]. Finally, analysis of fibers can be handled quite easily by force tensiometer.

There are two main drawbacks (more relevant than those of optical goniometry) in contact angle determination by tensiometry. First, enough liquid is necessary to immerse the solid sample (for common solvents this is quite straightforward). Secondly, the sample must satisfy certain constraints: regular shape of well-defined perimeter (see eq. (9.16)) and size compatible with the tensiometer. Here it should be stressed that often the surface morphology can be complex and evaluation of the true perimeter is not trivial: porous, rough samples and in the worst case membranes or textiles have not regular geometry. On rough hydrophilic surface the liquid can permeate through the asperities and the probed perimeter is higher than the geometric one. On the other hand, a superhydrophobic rough surface has a reduced real contact with the liquid. Thus, in this case, sessile drop method should be preferred.

9.4.4 Particular case: granular materials

In the case of granular materials, the measurement of the contact angle with a goniometer is severe because the probing liquid more or less penetrates in the porous structure, and since the surface is not flat and smooth the contact angle is geometrically not accessible. Commonly for granular materials the Washburn method can be used where the powder to be measured is placed into a glass tube closed to the bottom end with a filter. The tube is then hanged at a balance and when the vessel has contacted the liquid, the liquid rise rate through the bulk powder is measured.

The equation describing such phenomenon is as follows:

$$\cos\theta = A1A2(m^2/t) \tag{9.17}$$

where m is the mass of the material, t is the probing time, $A1 = \eta/\rho^2\gamma_l$ and $A2 = (2/\pi^2 r^5 n^2)$. Thus, the parameter $A1$ contains information on the liquid, viscosity, η, density and surface tension, while $A2$ is related to material properties supposed to be composed of n pores of radius r.

The parameter $A2$ is determined for the specific material by a measurement with a liquid that completely spread on that surface ($\cos\theta = 1$). Then the experiment is carried out with the liquid of interest and the parameter $A1$ is calculated using the specific parameters (density, viscosity, and surface tension) for that liquid. It should be considered that no measurement can be obtained with liquid exceeding 90° of contact angle with the granular material under investigation, since in that case the liquid will not rise in the vessel due to the Laplace-Washburn equation defining the capillary rise of a liquid in pores or capillary channels, which will be discussed in Section 9.5.4.

9.5 Application of wettability characterization

In Fig. 9.9, a representation of the contact angle scale in the full 0–180° range is reported. Though based on conventional boundaries, it is useful to use a uniform language of the involved community. When no hysteresis is observed, surfaces with water contact angle (WCA) much lower than 90° (somewhere lower than 60°) are considered hydrophilic. These are defined super-hydrophilic when WCA is lower than 5°, thus when a full spreading of the drop (formation of a liquid film) is observed. Above 90° a surface is hydrophobic; it is considered super-hydrophobic when the angle exceeds 150°, thus when an almost spherical drop is formed. As it will be better discussed in the next section, such a behavior can only arise by a combination of a low surface energy chemistry and a nano- or micro-texture, since the angle limit for a flat surface is 120°.

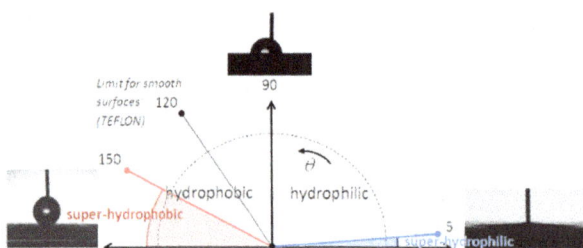

Fig. 9.9: Representation of the contact angle (Θ) scale in the full range 0–180°. The hydrophilic (0–70/90°,), hydrophobic (70/90–180°), superhydrophilic (0–5°), and superhydrophobic (150–180°) ranges are indicated with images of drops with Θ = 2°, 90°, 170°.

9.5.1 From hydrophobic to water- and oil-repellent materials

In Tab. 9.2 the water contact angle for some common polymers has been reported. It is evident that except PVA and at some extent PMMA, the main part of them has a significant hydrophobic behavior. In fact, such polymers are mainly composed of scarcely polar or polarizable groups as CH_x and above all CF_x or $Si-CH_x$ units, leading to low surface energy and this account for a WCA value higher than 110° in the case of polytetrafluoroethylene (PTFE) and polydimethylsiloxane (PDMS) [1]. Nevertheless, quite often we encounter synthetic and natural hydrophilic polymers such as hydrogel: highly cross-linked polymers rich in polar groups, able to entrap a high amount of water forming a gel.

However, for polymers in use in commodities, a hydrophobic behavior is most common, and a similar behavior is found also in polymer films deposited in less conventional methods, such as low temperature plasma processing. Plasma processes, addressed to provide surfaces with such functionalities, either with grafting or, more often with thin film deposition, are extensively applied to a wide range of materials. In the frame of wettability control the plasma enhanced chemical vapor deposition of Teflon-like coatings from fluorocarbon monomers (often with addition of H_2) has been widely investigated in the 1980s–1990s. In any case, it was clear that plasma parameters able to tune the F-to-C ratio of the coatings are in turn active in tuning the surface hydrophobic character (i.e. the value of WCA, in the range 70 to 120°) as shown in Fig. 9.10 [49].

More recently, specifically in the last 10 years, the interest in developing superhydrophobic surfaces, with WCA higher than 150°, increased a lot, as demonstrated by the huge number of published papers on this subject [1, 3, 5, 6, 8, 32, 34, 36, 39, 40]. However, since playing with surface chemistry cannot lead to WCA higher than 120° a trick should be found to fabricate superhydrophobic surfaces. Copying nature, "Biomimetics," is often useful to surface engineers [6, 50]: lotus leaf (microscopy images reported in Fig. 9.11 together with those of alike cases), but many other examples can be

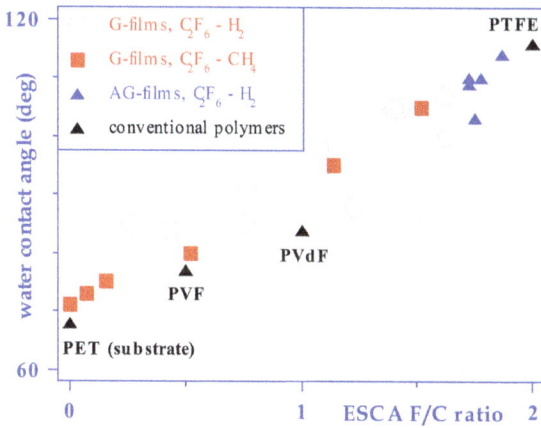

Fig. 9.10: Variation of the static WCA as a function of the fluorine over carbon surface atomic ratio, for coatings plasma deposited in different conditions. G-films and AG-films, deposited in glow and after glow conditions. Some conventional polymers are shown for comparison (reproduced with permission from [49], Copyright © 1997 Springer).

found both in plants and in insects, cannot be wetted by water since its surface is made of nanobumps covered with a wax. Hence, nano- and micro-texturing is the key to achieve extreme wettability properties, and the understanding of the role of the roughness on CA is extremely important.

The description of the influence of roughness on water contact angle is mainly due to Wenzel [51] and, separately, to Cassie and Baxter [52], within pioneering investigations pushed by the demand for waterproof textiles, using different models. The Wenzel model describes a liquid in contact with the whole solid surface and the enhancement of hydrophobicity (but also hydrophilicity) is ascribed to the increased surface area of the $s-l$ interface according to the following linear relationship:

$$\text{Cos } \theta_w = r \text{ Cos } \theta_{\text{flat}} \tag{9.18}$$

where θ_w is the observed angle on the rough surface, θ_{flat} is the equilibrium contact angle on the corresponding flat surface, and r, with $r > 1$, is the ratio of the actual surface area to the projected area of the surface. This model can account for liquid contact angle higher than 120°, but not for superhydrophobic surfaces on which the drop can freely move: in fact, in such description since the surface contact area is increased the rough asperities will hinder the displacement of the liquid (see Fig. 9.12).

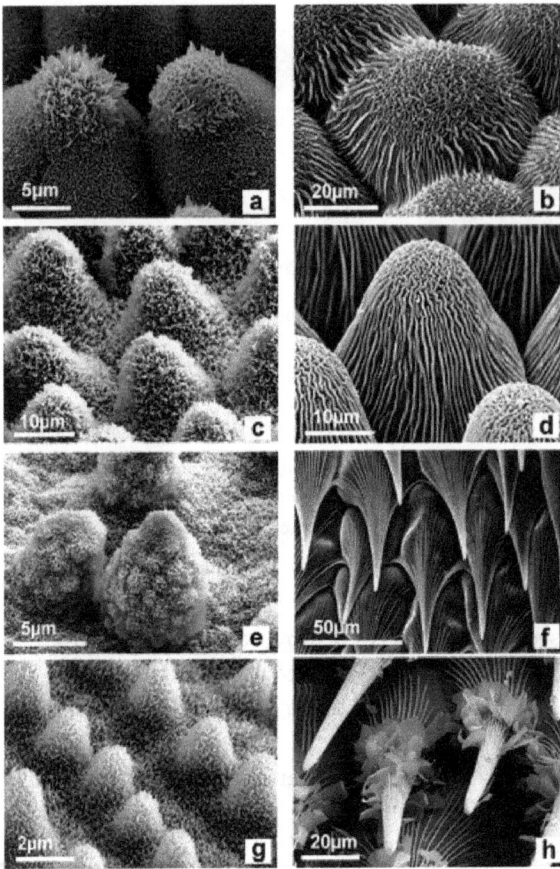

Fig. 9.11: SEM micrographs of nano- and micro-structures on plant surfaces. (a) *Colocasia esculenta*, (b) the seed surface of *Parodia alacriportana*, (c) *Euphorbia myrsinites*, (d) the flower leaf of *Rosa montana*, (e) *Nelumbo nucifera* (lotus leaf), (f) and (h) *Sarracenia leucophylla*, and (g) *Oryza sativa* (Reproduced with permission from [50], Copyright © 2008 Royal Society of Chemistry).

The Cassie–Baxter (CB) model in general describes the liquid contact angle of surfaces composed of two different phases according to

$$Cos\theta_{CB} = fa\, Cos\theta_a + (1-fa)\, Cos\theta_b \qquad (9.19)$$

where f_a is the area fraction of phase a, and θ_a and θ_b are the equilibrium contact angle on the respective pure phases, and it can be extended to the case where the surface is composed of n phases. For a very rough and hydrophobic surface it is possible that the liquid cannot pass through the asperities, but the drops remains pinned onto the tops of the hills, thus probing a composite surface made of solid, s, and air and eq. (9.19) can be written as follows:

Wenzel model   *Cassie-Baxter model*

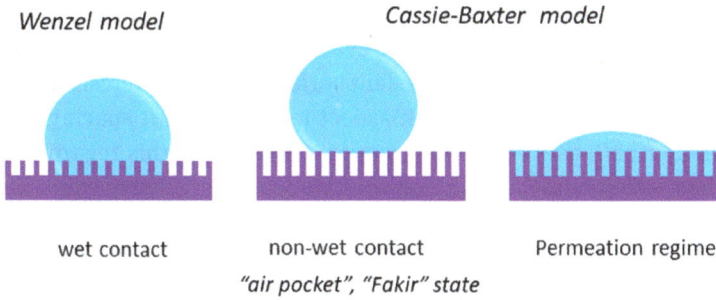

wet contact   non-wet contact   Permeation regime

"air pocket", "Fakir" state

Fig. 9.12: Liquid drop on a rough hydrophobic surface according to Wenzel model, left, and to the Cassie Baxter (CB) one, center. Imbibition/permeation CB state on rough hydrophilic surface on the right.

$$\mathrm{Cos}\theta_{CB} = fs\,\mathrm{Cos}\theta_s + (1-fs)\,\mathrm{Cos}\theta_{air} \qquad (9.20)$$

And since $\mathrm{Cos}\theta_{air}$ is −1, eq. (9.21) is obtained:

$$\mathrm{Cos}\theta_{CB} = fs\,\mathrm{Cos}\theta_s - (1-fs) \qquad (9.21)$$

Equation (9.21), thus, gives the contact angle of a liquid placed on a rough surface in the so-called fakir way, or air pocket one, as depicted in Fig. 9.12 [1].

It is important to highlight that whatever the r value in the Wenzel equation, the contact angle increases only if the value on the corresponding flat surface is higher than 90°. On the other hand, since in the CB equation the factor $(1-fs)$ is present, the measured θ_{CB} angle can be higher than θ_s, even for materials with contact angle lower than 90° (not too much hydrophilic to completely wet the asperities – air should remain trapped between the liquid and the valley, otherwise the composite CB model cannot stand).

As an example, Di Mundo et al. [32] showed that a flat hydrocarbon plasma deposited coating with a water contact angle around 80°, leads to advancing WCA as high as 165° when deposited on a rough PS (polystyrene) sample. Hence water-repellent surfaces can be fabricated playing at the same time with chemistry and texturing to have surfaces with hydrophobic asperities where air is trapped between, as demonstrated by spectroscopic techniques [53]. Further, as a consequence of the non-wet contact such surfaces are characterized by a slippery behavior of water drops, thus by a low WCA hysteresis, besides high water contact angle. In fact, considering a drop on a tilting plate, as reported in Fig. 9.7c, in an air pocket regime the energy barrier of the maxima between the metastable states decreases since most of the drop lies on air: the energy gained when the front of the contact line moves is very close to the energy necessary to leave the wetted region on the rear part. Therefore, to identify a superhydrophobic water-repellent behavior, the measurement of a high static water contact angle is not enough: the roughness and chemistry could be such to get a Wenzel state, a wet contact, preventing the drop motion, and characterized by high hysteresis. A good characterization of a water slippery surface should be based on the evaluation of the advancing and receding WCA: superhydrophobic water-repellent surfaces show hysteresis lower than 15° [54]. In general, features with high aspect ratio and short spacing are needed to avoid water collapsing into the grooves; therefore at this regard the use of the mere rms – roughness as criterion to drive water repellency could be misleading.

Such surface topography, even in the broad variety of the possible specific geometries, generally reveals not only a low angle hysteresis but also a high efficiency in the bouncing drop phenomenon. This kind of surfaces can be considered "robust" since, when subjected to drops impinging on the surface at high speed, they can hinder water penetration in the cavities of the textured layer (below the apexes of surface protrusions), strongly opposing the failure of the CB regime [55]. In a study, more focused on this topic, PTFE surfaces plasma decorated with very dense spherical ending micro-cones have shown a very short (12 ms) time of drop/surface contact upon bouncing, meaning high bouncing efficiency. A longer contact time, or even a partial water penetration in the cavities (drop pinning), has been detected on larger and more spaced micro-cones (and completely no bouncing, under the same conditions, on bare PTFE). Such decorated PTFE surfaces are shown in their SEM images in Fig. 9.13, from A to C with decreasing micro-cones density, and, on the right, the corresponding time lapse of the bouncing event. About 1.43 m/s is the velocity of the impinging drops. In sample D the spacing is too large and the falling drop can leave water residues. [56]. In this example, though plasma-treated surfaces were characterized by a very low WCA hysteresis, the kinetic behavior under harsh conditions (falling drops) was quite different.

Both bottom-up and top-down approaches have been proposed for preparation of nano- and micro-superhydrophobic surfaces. Photoinitiated copolymerization of butyl methacrylate and ethylene dimethacrylate in presence of cyclohexanol and 1-decanol, allow the deposition of a nanoporous coating as shown in Fig. 9.14. P. A. Levkin et al. showed that, during the deposition, the inert solvent acts like a porogen leaving such rough surface and leading to WCA as high as 172° with negligible hysteresis [57]. Many papers deal with patterning via lithographic techniques followed by deposition (PECVD, or self assembling or alike) of a low energy polymeric film [8, 58]. Chang-Hwan Choi and Chang-Jin Kim examined a wide range of nanostructures on silicon, by interferometric lithography and subsequent etching as reported in Fig. 9.15 a–c. The deposition of a thin Teflon-like coating made the samples with the higher aspect ratio structures superhydrophobic [58].

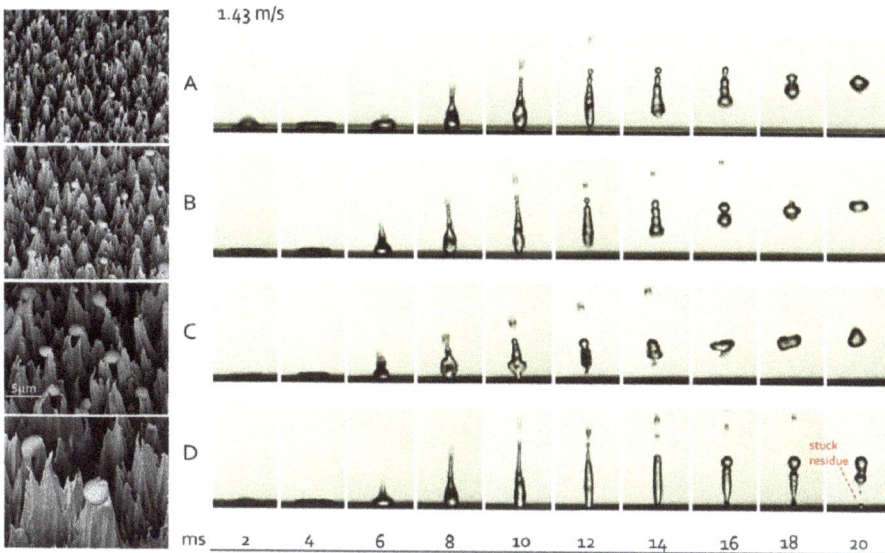

Fig. 9.13: PTFE surfaces textured with a plasma process generating spherical-ending micro-cones. SEM images on the left at the same magnification. From A to C, plasma parameters are tuned to decreasing micro-cones density. On the right, the corresponding timelapse of the bouncing event. 1.43 m/s is the velocity of the impinging drops. In sample D the spacing is too large and the falling drop can leave water residues. (adapted with permission from [56], Copyright © 2015 Elsevier Ltd.).

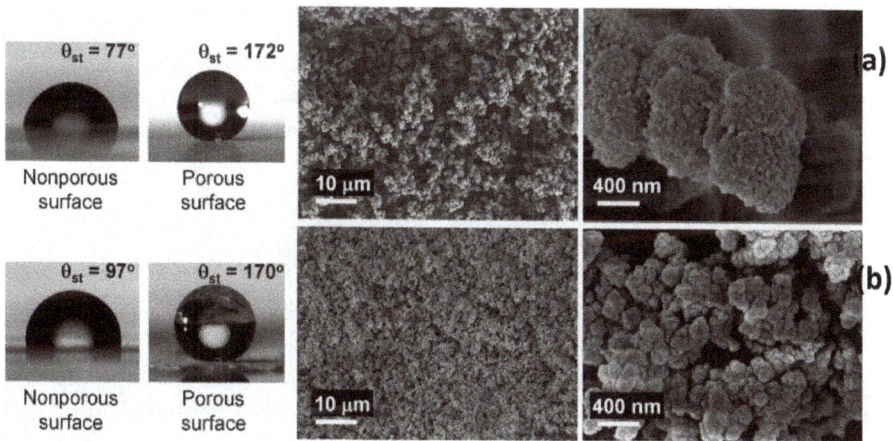

Fig. 9.14: Shape of water droplets formed on porous and nonporous polymer layers and SEM images of the superhydrophobic porous polymers: (a) Poly(butyl methacrylate-co-ethylene dimethacrylate) and (b) Poly(styrene-co-divinylbenzene) (reproduced with permission from [57], Copyright © 2009 WILEY-VCH Verlag GmbH & Co. KGaA, Weinheim).

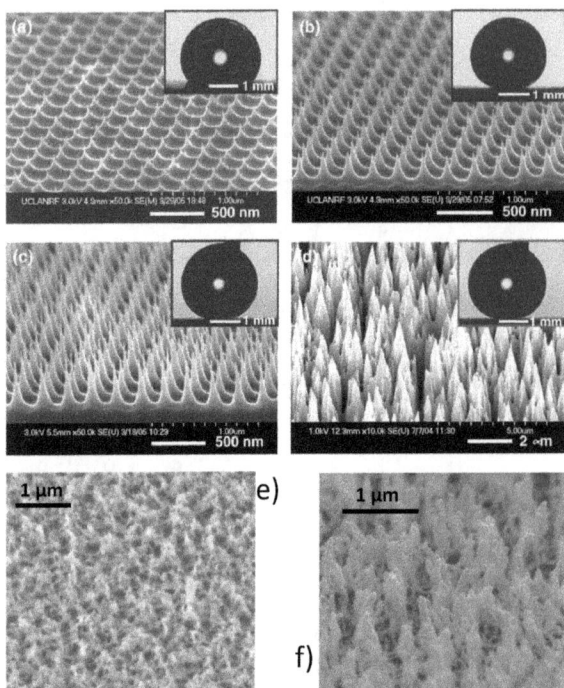

Fig. 9.15: SEM images of nanotextured samples for application to superhydrophobic surfaces. (a)–(c) Regular nanopost structures created on Si by the interference lithography followed by DRIE. (d) Random nanopost structures created by the black silicon method. Nanostructures with height higher than 200 nm, in (b)–(d) show extreme superhydrophobicity, short structures, (a) do not (*reproduced with permission from 58, Copyright 2006 IOP Publishing*). (e) and (f) superhydrophobic plasma nanotextured polystyrene (adapted with permission from [34], Copyright 2008 American Chemical Society), and polycarbonate (adapted with permission from [37], Copyright © 2011 WILEY-VCH Verlag GmbH & Co. KGaA, Weinheim), the latter coated with a plasma-deposited silicone-like thin film.

A widely diffuse single step top-down approach is plasma etching a material under high ion bombardment conditions (reactive ion etching) producing stochastic shaped nanostructures (sometimes in the form of pillars, see Fig. 9.15d–f). CF_4 plasmas have been tested for etching/roughening polybutadiene; [59], CF_4 / O_2 plasmas for polyethylene [60], SF_6/ O_2 plasmas for polydimethylsiloxane [61] and for polystyrene [32, 34]. In general, the grafting of functional groups occurs along the non-homogeneous etching, thus, if the density of the fluorinated ones is optimized, the suitable hydrophobic chemistry can be combined with the nano-texture.

Oleophobic surfaces need a separate description since the surface tension of oil/organic liquids is lower than that of water. Therefore, to get a oleophobic material at a solid-air-oil interface, the surface energy should be very low, and in turn the correct choice of chemistry and texturing is even more difficult. Jung et al. [62] produced hierarchic structures, composed of micrometric patterned pillars covered with nanometric vertical platelets of n-hexatriacontane, and a final coating of n-perfluoroeicosane ($C_{20}F_{42}$) by thermal evaporation. An hexadecane contact angle as high as 115° could be achieved by well designing the structures morphology. Tuteja et al. [63, 64] demonstrated that micrometric re-entrant structures, as those shown in Fig. 9.16, coated with fluorodecyl POSS and alike, can trap air thus leading to octane advancing contact angle of 163° (145° receding), but also to design "omniphobic" surfaces repellent to a wide range of liquids such as acetic acid, toluene and hexadecane. A valuable review on oleophobic surfaces and their applications has been published by Kesong Liu et al. [65].

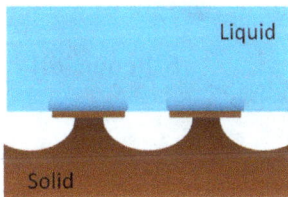

Fig. 9.16: Re-entrant, hoodoo-like microstructures necessary for oleophobic behavior of surfaces, when properly coated with low surface energy film.

However, the nanotexturing approach to water- and oil-repellent surfaces can present some concerns. Under unfavorable ambient conditions (long-term rainfall, continuous liquid flow as in pipelines), the air cushions trapped between the nano-protrusions can be replaced by the liquid. In other cases dirtiness can be mechanically anchored by protrusions making them more difficult to remove than on flat surfaces. Thus, to address these limitations slippery liquid-infused porous surfaces (SLIPS) have been developed. SLIPS consist of a smooth and defect-free layer of lubricating liquid film locked in between micro- and nano-features of a solid substrate [66–68]. In Fig. 9.17 an example of fabrication of SLIP surface is presented.

Robust SLIPS must stand three rules: (i) the lubricating fluid must stably adhere to the substrate filling the porosities of the substrate; (ii) the porous solid must be wetted by the lubricating fluid rather than by the liquid to be repelled; and (iii) the lubricant layer must be immiscible with the working fluid.

Such surfaces have been demonstrated as highly slippery and stable in harsh ambient conditions. In fact, due to the infused liquid mobility, damaged substrates can be repaired through lubricating layer motion. Furthermore, such layer is incompressible (compared to compressible air layers with nanotexture superhydrophobic surfaces), hence these surfaces provide good self-healing properties and pressure stability [69].

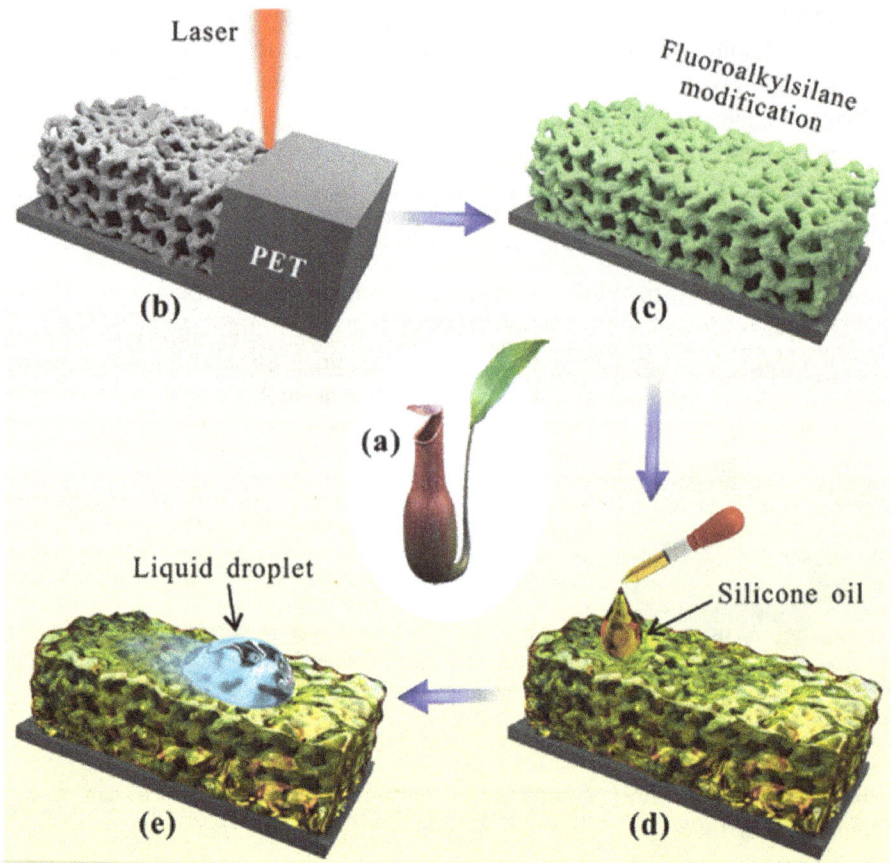

Fig. 9.17: Schematic diagram of the preparation process of SLIPS from polyethyleneterephthalate by femtosecond laser direct writing. (a) Photo of the *Nepenthes* pitcher plant, from which SLIPS are inspired. (b) Femtosecond laser ablation used to generate interconnected porous microstructures. (c) fluoroalkylsilane modification used to lower the surface free energy. Green color denotes the fluorosilane molecular layer. (d) Infusion of the lubricating liquid (silicone oil) into the laser-induced micropores. (e) Foreign liquid droplet sliding down the as-prepared slippery surface. Reproduced with permission from [37], Copyright © 2018 WILEY-VCH Verlag GmbH & Co. KGaA, Weinheim.

9.5.2 Hydrophilic to super-hydrophilic materials

It has been already stressed that polymers are normally characterized by low energy both in their bulk and in their surface, thus generally express high water contact angle. If this property represents an advantage for the application of the polymers in most of the manufacts because it turns into good inertness, it appears as a limit when gluing, painting, dyeing, washing, or functionalizing (for typical device purpose) their surfaces is relevant.

It is well known that surfaces bearing functionalities with asymmetric charge distribution (-OH, -NH$_2$, -CO$_2$H, C = O) can be characterized by a high surface energy. Thus, several surface modification techniques have been addressed to the addition of such groups to polymer surfaces. Plasma treatments, thanks to the low temperature and easily scalable nature of the process, have been thoroughly investigated, and actually used, with O$_2$, NH$_3$, N$_2$/H$_2$, H$_2$O feed gases, for addressing hydrophilicity to PE, PET, PTFE, and others [70, 71], also in the version, very diffused in industry, of corona discharges [72].

Research interest toward superhydrophilic surfaces is particularly recent (last decade), mostly for the emerging applications of microfluidics and antifog reduction. The latter can be obtained through the coalescence of the condensed micro-droplets up to formation of a liquid film when the WCA of the surface is particularly low. Figure 9.18 shows photographs of condensed water vapor on a transparent plastic with a different hydrophilic and superhydrophilic character of the surface [73].

Fig. 9.18: Photographs of condensed water drops on the surface of a transparent plastic (polycarbonate) exposed to water vapor. The material has been treated by a patented plasma process [68]. As the WCA decreases (value reported over each picture) the mean width of the condensed water drops increases, up to coalesce and formation of a liquid film (antifog condition). Under each image the average diameter of drops is reported.

Unlike the superhydrophobic surfaces, a WCA lower than 5° can be achieved also onto flat surfaces if high chemical affinity with water is established (e.g. strong acid–base interactions or strong permanent charge separations). At this regard, it is worth mentioning the natural case (from plant kingdom) of super-hydrophilic leaves which exist both with a relatively smooth surface (those permanently wet) and with micro-bump-rich (aquaplaning) or micro-porous (water absorbing) surface [74]. However, in order to achieve a super-hydrophilic performance the strategy of combining a moderate hydrophilic chemistry with a textured morphology is preferred. This is probably pushed by the necessity of reducing aging effects of the hydrophilic chemistry (see next). In this frame plasma nanotexturing by O_2 etching, which generates nano-pillars with a high O/C surface ratio, has been utilized onto polymers like PMMA [75] or photoresist epoxy novolak SU-8 [76]. This strategy configures the so-called permeation wetting regime, which has been depicted as a hydrophilic CB regime since the liquid permeates the textured solid and the drop faces a composite surface consisting of both solid and liquid [77]. In fact, as shown in Fig. 9.12 (right cartoon), the liquid penetrates underneath and between the asperities and as a matter of fact the drop probe a surface made of water and solid, for which, considering that the cosine of the contact angle of water with itself is equal to 1, an equation similar to eq. (9.22) can be derived:

$$Cos\theta_{CB} = fs\, Cos\theta_s + \ (1-fs) \qquad (9.22)$$

Analyzing eq. (9.22), it can be observed that surfaces that follow this regime are extremely wettable even if the water contact angle on the corresponding flat surface, θ_s, is not particularly low.

9.5.3 Hydrophobic recovery of hydrophilic surfaces

Highly hydrophilic surfaces (WCA lower than 30–40°) are in air intrinsically unstable thus they suffer "hydrophobic recovery," that is the tendency to reach the equilibrium by decreasing the solid/air interfacial energy. This phenomenon occurs through different "aging" mechanisms, which are more or less important depending on the surface characteristics. Among these, the most frequent are:

- reorientation of surface functional groups in order to have the low energy ones facing atmosphere and the high energy ones buried in the bulk of the polymer
- adsorption on the hydrophilic surface of hydrocarbon molecules present in the atmosphere, as a consequence of their low surface energy.

The former mechanism is particularly relevant in polymers since characterized by high chains/bonds mobility and, in fact, demonstrated to be favored by temperature increase [70].

Such instability in hydrophilic surfaces, especially when dealing with polymer modifications, is not just a mere fundamental speculation, it has serious technological concerns. Just thinking to applications fields involving dyes adhesion, metal polymer adhesion or printing onto polymers surfaces.

As an example of such surface modification processes, plasma-treated polymers with hydrophilic modifications show a steep increase (even 30–40°) of the water contact angle in the first 3–5 days after the treatment, then a slighter increase up to a plateau reached in 1–2 weeks. However, though increased, the final value of the contact angle generally is lower than for the untreated polymer. A typical example is shown in Fig. 9.19. Several efforts in optimizing working conditions in

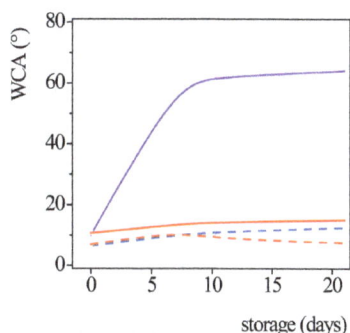

Fig. 9.19: WCA time evolution at room temperature for a polycarbonate sample after oxygen plasma treatment. Blue line corresponds to sample stored in air, while red line to one stored in water. Both advancing, continuous line, and receding angle, dashed line, are reported.

order to contain aging effects have been made by academic and industrial researchers [70, 78].

In case of super-hydrophilic micro/nanostructured modifications, like those above reported [75, 76], different aging kinetics have been observed: the WCA stabilizes in 2–3 months at values between 30° and 60°. That is the reason why, also in the plasma community, new strategies are under study to get stable superhydrophilic polymer surfaces. One of these approaches is the deposition of an inorganic highly hydrophilic coating, such as OH-rich SiO_x film, onto a tailored nanotextured surface [11].

It is worth highlighting that the hydrophobic recovery is not an aging phenomenon intrinsic of the material but related to the interaction between the material and its environment. As a matter of fact, when the surface of a polymer is made more hydrophilic and it is stored in water the WCA shows a negligible variation compared to the as-treated sample, as it can be observed in Fig. 9.19. Another important consideration should be done regarding the hysteresis of the WCA. As already discussed in Section 9.3, the advancing and receding contact angle show a different behavior when the surface is stored (as usually) in air: while the advancing angle show the typical increasing trend, the receding one increases only slightly. Such an effect, rarely reported in literature due to the long unawareness about receding measurements (but already observed [79]), suggests that high energy groups remain within the reorganized surface and are "recalled" by the wetting action.

9.5.4 Wettability of porous surfaces

Porous polymeric materials include mostly membranes or adsorbents; however, concepts related to their wetting can be extended to composite materials where polymers are used as fillers or aggregates such as many ceramic based composites as those used, for instance, in construction applications [80].

When water or a water solution enters in contact with a porous unsaturated material, it can be absorbed as a consequence of a depression/pressure P_{cap} produced by the capillary action (or interaction) between the liquid and the surface of the pores in the material.

The pressure P_{cap} generated in a capillary can be calculated through the Laplace-Washburn equation:

$$P_{cap} = \frac{2\sigma.\cos\vartheta}{r} \tag{9.23}$$

where r is the mean radius of the capillary pore, σ the surface tension of the liquid, ϑ the water contact angle.

Therefore, P_{cap} is higher as the pore size decreases and the water contact angle increases. In a hydrophilic material, with $\theta < 90°$, P_{cap} is positive and it functions as a depression that draws the water in the pore (Fig. 9.20a); in a hydrophobic material, with $\theta > 90°$, P_{cap} is negative and indicates a pressure that tends to expel the water from the pore (Fig. 9.20b).

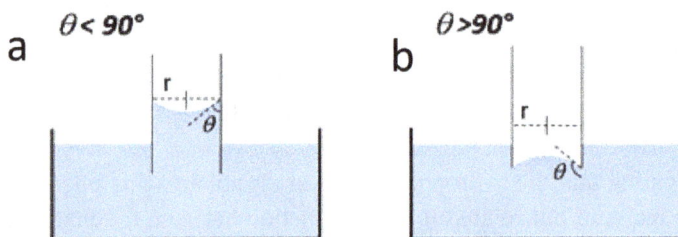

Fig. 9.20: Pressure in a capillary immersed in water for an hydrophilic (a) and an hydrophobic material (b).

Very recently it has been shown that a more complex situation exists when dealing with a small drop in contact with the porous surface [81]. In this case, indeed, the Laplace pressure at the curved drop surface (that facing air), and the gravitational force play also a role. Hence the net force leading the drop to penetrate into the pore will be the result of a balance of those three contributions and, substantially, besides the contact angle ϑ and the pore radius r, will be influenced also by the radius of the drop.

Practically, the direct measurement of the contact angle from the sessile drop method can be very difficult, not only for the aforementioned hysteresis issues but also for absorption and spreading phenomena. Consider what happens on a paper filter once a drop of liquid is placed: the drop contact diameter increases, spreading over the surface, and on the other hand, since paper is porous, water will be absorbed filling the pores and penetrating underneath (see Fig. 9.21). Hence in a period of time, which depends on the nature of the liquid and of the paper, the characteristics of the probing drop (contact angle, volume diameter, and height) will change. If the spreading or absorption rate is too fast compared to the operator observation the measurement of the contact angle is questionable. However, observation of the drop characteristics variation, better with a goniometer equipped with a camera, can be very useful even when fast spreading or absorption kinetics are observed. In fact, recording the drop image evolution, besides wettability evaluation, can give information on the absorbency and spreading behavior of such materials that often are developed for application exploiting such properties: adsorbents, filters, writing papers, and so on. The combination of absorption and spreading is analogous to what is observed on nanotextured hydrophilic materials in the permeation regime described in Section 9.5.2. The main difference is that in this case the observed phenomenon is a bulky one more than limited to the surface.

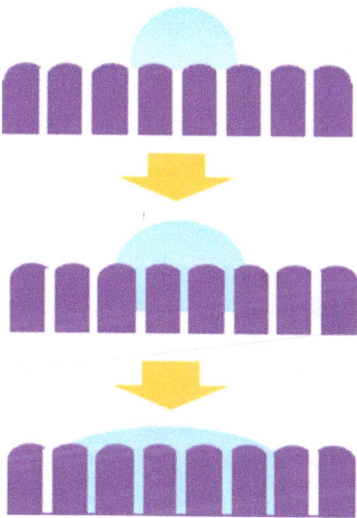

Fig. 9.21: Drop of liquid in contact with a porous surface. During the liquid contact event, absorption (decrease of volume of drop above the surface) and spreading (increase of the liquid/surface contact area) can be observed.

Since nature very rarely helps investigators, it should be considered that also the evaporation of the liquid contributes to drop volume changes. Thus, it is important in such measurements to inhibit the evaporation, using low vapor pressure liquids, or saturating the ambient around the drop. As an alternative, some authors suggest to take into account the effect of evaporation placing a drop of the same liquid in the same conditions onto a not absorbing material (e.g. polytetrafluoroethylene for water). The liquid evaporation rate on such reference sample can be considered for mathematical correction of the volume variation on the surface under investigation. In general both spreading and absorption can develop on a porous surface [82, 83]. However the former does not affect the volume although it widens the contact area, and absorption is commonly slower since the liquid molecules should diffuse beneath the substrate surface.

S. Farris et al. deeply investigated spreading/absorbance behavior over different biopolymers, polysacccarides (chitosan, pullulan, and pectine) and pig skin gelatin-coated PET [82]. They observed water drop evolution on the timescale of 60 s and found that for some materials the increase of liquid contact area (spreading) was more evident than the decrease of volume (absorption). By trigonometric correlation of the drop shape and the contact angle they could separate the spreading contribution from the absorption one and concluded that the considered biopolymers, except chitosan, whose surface presented pores, show negligible absorption characteristics.

Dastjerdi et al. [83] optimized the synthesis of a composite amino-functionalized polysiloxane coating with amphiphilic features. It is interesting to highlight that, with a contribution due to the hierarchical roughness, the water drop evolution of coatings deposited onto TiO_2-coated PET textiles is very peculiar. Though appearing slippery (advancing WCA as high as 150° with an hysteresis lower than 3°) in a matter of few minutes the surface becomes absorbent because of the polymer chains mobility pushing the hydrophilic amino groups toward water.

References

[1]   Garbassi F, Morra M, Occhiello E. Polymer Surfaces: From Physics to Technology, Garbassi F, Morra M, Occhiello E, Eds., Wiley, Vol. 36, 1995.

[2]   Määttänen A, Ihalainen P, Bollström R, Toivakka M, Peltonen J. Wetting and print quality study of an inkjet-printed poly(3-Hexylthiophene) on pigment coated papers. Colloids Surf A Physicochem Eng Asp 2010, 367, 76–84, doi: 10.1016/j.colsurfa.2010.06.019.

[3]   Su B, Li M, Lu Q. Toward understanding whether superhydrophobic surfaces can really decrease fluidic friction drag. Langmuir 2010, 26, 6048–6052, doi: 10.1021/la903771p.

[4]   Becker H, Gärtner C. Polymer microfabrication technologies for microfluidic systems. Anal Bioanal Chem 2008, 390, 89–111.

[5]   Mortazavi M, Nosonovsky M. Adhesion, wetting, and superhydrophobicity of polymeric surfaces. In: Polymer Adhesion, Friction, and Lubrication, Hoboken, NJ, USA, John Wiley & Sons, Inc., 2013, 177–226.

[6]   Barthlott W, Schimmel T, Wiersch S, Koch K, Brede M, Barczewski M, Walheim S, Weis A, Kaltenmaier A, Leder A, et al. The salvinia paradox: superhydrophobic surfaces with hydrophilic pins for air retention under water. Adv Mater 2010, 22, 2325–2328, doi: 10.1002/adma.200904411.

[7]   Plasma Surface Modification of Polymers: Relevance to Adhesion, Mittal KL, Lyond M, Ed., London, CRC Press, 1994.

[8]   Corbella C, Portal S, Rubio-Roy M, Vallvé MA, Ignés-Mullol J, Bertran E, Andújar JL. Surface structuring of diamond-like carbon films by colloidal lithography with silica sub-micron particles. Diam Relat Mater 2010, 19, 1124–1130, doi: 10.1016/j.diamond.2010.03.020.

[9]   Chen Y, Xie Q, Sari A, Brady PV, Saeedi A. Oil/water/rock wettability: influencing factors and implications for low salinity water flooding in carbonate reservoirs. Fuel 2018, 215, 171–177, doi: 10.1016/j.fuel.2017.10.031.

[10]   Thomas Young B. An Essay on the Cohesion of Fluids. Philos Trans R Soc London 1805, 95, 65–87, doi: 10.1098/rstl.1805.0005.

[11] Palumbo F, Di Mundo R, Cappelluti D, d'Agostino R. SuperHydrophobic and superhydrophilic polycarbonate by tailoring chemistry and nano-texture with plasma processing. Plasma Process Polym 2011, 8, 118–126, doi: 10.1002/ppap.201000098.

[12] Di Mundo R, Palumbo F, D'Agostino R. Nanotexturing of polystyrene surface in fluorocarbon plasmas: from sticky to slippery superhydrophobicity. Langmuir 2008, 24, 5044–5051, doi: 10.1021/la800059a.

[13] Wakeham W, Assael M, Marmur A, Coninck J, Blake T, Theron S, Zussman E. Material properties: Measurement and data. In: Springer Handbooks, Springer, 2007, 85–177.

[14] Zisman WA. Relation of the equilibrium contact angle to liquid and solid constitution. In: Fowkes FM, Ed.. Contact Angle, Wettability, and Adhesion, AMERICAN CHEMICAL SOCIETY, 1964.

[15] Girifalco LA, Good RJ. A theory for the estimation of surface and interfacial energies. i. derivation and application to interfacial tension. J Phys Chem 1957, 61, 904–909, doi: 10.1021/j150553a013.

[16] Kwok DY, Neumann AW. Contact angle measurement and contact angle interpretation. Adv Colloid Interface Sci 1999, 81, 167–249, doi: 10.1016/S0001-8686(98)00087-6.

[17] Fowkes FM Dispersion Force Contributions to Surface and Interfacial Tensions, Contact Angles, and Heats of Immersion. In; 1964; pp. 99–111.

[18] Owens DK, Wendt RC. Estimation of the surface free energy of polymers. J Appl Polym Sci 1969, 13, 1741–1747, doi: 10.1002/app.1969.070130815.

[19] Wu S. Calculation of interfacial tension in polymer systems. J Polym Sci Part C Polym Symp 2007, 34, 19–30, doi: 10.1002/polc.5070340105.

[20] Li D, Neumann AW. Contact angles on hydrophobic solid surfaces and their interpretation. J Colloid Interface Sci 1992, 148, 190–200, doi: 10.1016/0021-9797(92)90127-8.

[21] Żenkiewicz M. Methods for the calculation of surface free energy of solids. J Achiev Mater Manuf Eng 2007, 24, 137–145.

[22] Van Oss CJ, Good RJ, Chaudhury MK. The role of van Der Waals forces and hydrogen bonds in "hydrophobic interactions" between biopolymers and low energy surfaces. J Colloid Interface Sci 1986, 111, 378–390, doi: 10.1016/0021-9797(86)90041-X.

[23] Kitazaki Y, Hata T. Surface-chemical criteria for optimum adhesion. J Adhes 1972, 4, 123–132, doi: 10.1080/00218467208072217.

[24] Van Oss CJ, Chaudhury MK, Good RJ. interfacial lifshitz – van der waals and polar interactions in macroscopic systems. Chem Rev 1988, 88, 927–941, doi: 10.1021/cr00088a006.

[25] Good RJ. Contact angle, wetting, and adhesion: A critical review. J Adhes Sci Technol 1992, 6, 1269–1302, doi: 10.1163/156856192X00629.

[26] Marmur A. Solid-surface characterization by wetting. Annu Rev Mater Res 2009, 39, 473–489, doi: 10.1146/annurev.matsci.38.060407.132425.

[27] Furmidge CGL. Studies at phase interfaces. I. The sliding of liquid drops on solid surfaces and a theory for spray retention. J Colloid Sci 1962, 17, 309–324, doi: 10.1016/0095-8522(62)90011-9.

[28] De Gennes PG. Wetting: Statics and dynamics. Rev Mod Phys 1985, 57, 827–863, doi: 10.1103/RevModPhys.57.827.

[29] Morra M, Occhiello E, Garbassi F. Contact angle hysteresis on oxygen plasma treated polypropylene surfaces. J Colloid Interface Sci 1989, 132, 504–508, doi: 10.1016/0021-9797(89)90264-6.

[30] Dwight DW, Riggs WM. Fluoropolymer surface studies. J Colloid Interface Sci 1974, 47, 650–660, doi: 10.1016/0021-9797(74)90242-2.

[31] Gotoh K, Nakata Y, Tagawa M, Tagawa M. Wettability of ultraviolet excimer-exposed PE, PI and PTFE films determined by the contact angle measurements. Colloids Surf A Physicochem Eng Asp 2003, 224, 165–173, doi: 10.1016/S0927-7757(03)00263-2.

[32] Di Mundo R, Palumbo F, D'Agostino R. Influence of chemistry on wetting dynamics of nanotextured hydrophobic surfaces. Langmuir 2010, 26, 5196–5201, doi: 10.1021/la903654n.

[33] Johnson RE, Dettre RH. Contact angle hysteresis. iii. Study of an idealized heterogeneous surface. J Phys Chem 1964, 68, 1744–1750, doi: 10.1021/j100789a012.

[34] Di Mundo R, Palumbo F, D'Agostino R. Nanotexturing of polystyrene surface in fluorocarbon plasmas: from sticky to slippery superhydrophobicity. Langmuir 2008, 24, 5044–5051, doi: 10.1021/la800059a.

[35] Vourdas N, Tserepi A, Gogolides E. Nanotextured super-hydrophobic transparent poly(methyl methacrylate) surfaces using high-density plasma processing. Nanotechnology 2007, 18, doi: 10.1088/0957-4484/18/12/125304.

[36] Palumbo F, Di Mundo R, Cappelluti D. SuperHydrophobic and superhydrophilic polycarbonate by tailoring chemistry and nano-texture with plasma processing. Plasma Process Polym 2011, 8, 118–126, doi: 10.1002/ppap.201000098.

[37] Di Mundo R, De Benedictis V, Palumbo F, d'Agostino R. Fluorocarbon plasmas for nanotexturing of polymers: a route to water-repellent antireflective surfaces. Appl Surf Sci 2009, 255, doi: 10.1016/j.apsusc.2008.09.020.

[38] Coulson SR, Woodward I, Badyal JPS, Brewer SA, Willis C. Super-repellent composite fluoropolymer surfaces. J Phys Chem B 2000, 104, 8836–8840, doi: 10.1021/jp0000174.

[39] Gao L, McCarthy TJ, Hydrophilic TI. Comments on definitions of hydrophobic, shear versus tensile hydrophobicity, and wettability characterization. Langmuir 2008, 24, 9183–9188, doi: 10.1021/la8014578.

[40] Vourdas N, Tserepi A, Gogolides E. Nanotextured super-hydrophobic transparent poly(methyl methacrylate) surfaces using high-density plasma processing. Nanotechnology 2007, 18, 125304–125311, doi: 10.1088/0957-4484/18/12/125304.

[41] Di Mundo R, Labianca C, Carbone G, Notarnicola M. Recent Advances in Hydrophobic and Icephobic Surface Treatments of Concrete. Coatings 2020, 10, 449, doi: 10.3390/coatings10050449.

[42] Meuler AJ, Smith JD, Varanasi KK, Mabry JM, McKinley GH, Cohen RE. Relationships between water wettability and ice adhesion. ACS Appl Mater Interfaces 2010, 2, 3100–3110, doi: 10.1021/am1006035.

[43] Extrand CW, Kumagai Y. Contact angles and hysteresis on soft surfaces. J Colloid Interface Sci 1996, 184, 191–200, doi: 10.1006/jcis.1996.0611.

[44] Pierce E, Carmona FJ, Amirfazli A. Understanding of sliding and contact angle results in tilted plate experiments. Colloids Surf A Physicochem Eng Asp 2008, 323, 73–82, doi: 10.1016/j.colsurfa.2007.09.032.

[45] Yuan Y, Lee TR. Contact angle and wetting properties. Springer Ser Surf Sci 2013, 51, 3–34, doi: 10.1007/978-3-642-34243-1_1.

[46] Chen H, Muros-Cobos JL, Amirfazli A. Contact angle measurement with a smartphone. Rev Sci Instrum 2018, 89, 035117, doi: 10.1063/1.5022370.

[47] Taylor M, Urquhart AJ, Zelzer M, Davies MC, Alexander MR. Picoliter Water Contact Angle Measurement on Polymers. Langmuir 2007, 23, 6875–6878, doi: 10.1021/la070100j.

[48] Sedighi Moghaddam M, Waïslinder MEP, Claesson PM, Swerin A. Multicycle Wilhelmy Plate Method for Wetting Properties, Swelling and Liquid Sorption of Wood. Langmuir 2013, 29, 12145–12153, doi: 10.1021/la402605q.

[49] D'Agostino R, Favia P. Plasma Processing of Polymers, D'Agostino R, Favia P, Eds., NATO ASI., Kluwer Academic/Plenum Publishers, Boston, 1996.

[50] Koch K, Bhushan B, Barthlott W. Diversity of Structure, Morphology and Wetting of Plant Surfaces. Soft Mat 2008, 4, 1943–1963, doi: 10.1039/b804854a.

[51] Wenzel RN. Resistance of Solid Surfaces to Wetting by Water. Ind Eng Chem 1936, 28, 988–994, doi: 10.1021/ie50320a024.

[52] Cassie ABD, Baxter S. Wettability of Porous Surfaces. Trans Faraday Soc 1944, 40, 546–551, doi: 10.1039/tf9444000546.

[53] Checco A, Hofmann T, Dimasi E, Black CT, Ocko BM. Morphology of Air Nanobubbles Trapped at Hydrophobic Nanopatterned Surfaces. Nano Lett 2010, 10, 1354–1358, doi: 10.1021/nl9042246.

[54] McHale G, Shirtcliffe NJ, Newton MI. Contact-angle hysteresis on super-hydrophobic surfaces. Langmuir 2004, 20, 10146–10149, doi: 10.1021/la0486584.

[55] Di Mundo R, Bottiglione F, Carbone G. Cassie State Robustness of Plasma Generated Randomly Nano-Rough Surfaces. Appl Surf Sci 2014, 316, 324–332, doi: 10.1016/j.apsusc.2014.07.184.

[56] Di Mundo R, Bottiglione F, Palumbo F, Favia P, Carbone G. Sphere-on-Cone Microstructures on Teflon Surface: Repulsive Behavior against Impacting Water Droplets. Mater Des 2016, 92, 1052–1061, doi: 10.1016/j.matdes.2015.11.094.

[57] Levkin PA, Svec F, Fréchet JMJ. Porous polymer coatings: a versatile approach to superhydrophobic surfaces. Adv Funct Mater 2009, 19, 1993–1998, doi: 10.1002/adfm.200801916.

[58] Choi CH, Kim CJ. Fabrication of a dense array of tall nanostructures over a large sample area with sidewall profile and tip sharpness control. Nanotechnology 2006, 17, 5326–5333, doi: 10.1088/0957-4484/17/21/007.

[59] Woodward I, Schofield WCE, Roucoules V, Badyal JPS. Super-Hydrophobic Surfaces Produced by Plasma Fluorination of Polybutadiene Films. Langmuir 2003, 19, 3432–3438, doi: 10.1021/la020427e.

[60] Fresnais J, Chapel JP, Poncin-Epaillard F. Synthesis of Transparent Superhydrophobic Polyethylene Surfaces. Surf Coatings Technol 2006, 200, 5296–5305, doi: 10.1016/j.surfcoat.2005.06.022.

[61] Tsougeni K, Tserepi A, Boulousis G, Constantoudis V, Gogolides E. Control of nanotexture and wetting properties of polydimethylsiloxane from very hydrophobic to super-hydrophobic by plasma processing. Plasma Process Polym 2007, 4, 398–405, doi: 10.1002/ppap.200600185.

[62] Jung YC, Bhushan B. Wetting behavior of water and oil droplets in three-phase interfaces for hydrophobicity/philicity and oleophobicity/philicity. Langmuir 2009, 25, 14165–14173, doi: 10.1021/la901906h.

[63] Tuteja A, Choi W, Ma M, Mabry JM, Mazzella SA, Rutledge GC, McKinley GH, Cohen RE. Designing Superoleophobic Surfaces. Science (80-) 2007, 318, 1618–1622, doi: 10.1126/science.1148326.

[64] Pan S, Kota AK, Mabry JM, Tuteja A. Superomniphobic Surfaces for Effective Chemical Shielding. J Am Chem Soc 2013, 135, 578–581, doi: 10.1021/ja310517s.

[65] Liu K, Tian Y, Jiang L. Bio-inspired superoleophobic and smart materials: design, fabrication, and application. Prog Mater Sci 2013, 58, 503–564.

[66] Zhang J, Yao Z. Slippery properties and the robustness of lubricant-impregnated surfaces. J Bionic Eng 2019, 16, 291–298, doi: 10.1007/s42235-019-0024-5.

[67] Li J, Ueda E, Paulssen D, Levkin PA. Slippery Lubricant-Infused Surfaces: Properties and Emerging Applications. Adv Funct Mater 2019, 29, 1802317, doi: 10.1002/adfm.201802317.

[68] Yong J, Huo J, Yang Q, Chen F, Fang Y, Wu X, Liu L, Lu X, Zhang J, Hou X. Femtosecond Laser Direct Writing of Porous Network Microstructures for Fabricating Super-Slippery Surfaces

with Excellent Liquid Repellence and Anti-Cell Proliferation. Adv Mater Interfaces 2018, 5, 1701479, doi: 10.1002/admi.201701479.

[69] Wang C, Guo Z. A comparison between superhydrophobic surfaces (SHS) and slippery liquid-infused porous surfaces (slips) in application. Nanoscale 2020, 12, 22398–22424, doi: 10.1039/d0nr06009g.

[70] Inagaki N, Narushim K, Ejima S. Journal of adhesion science hydrophobic recovery of plasma-modified film surfaces of ethylene-co- tetrafluoroethylene co- polymer 2012, 37–41.

[71] Favia P, Stendardo MV, D'Agostino R. Selective grafting of amine groups on polyethylene by means of NH3-H2 RF glow discharges. Plasmas Polym 1996, 1, 91–112, doi: 10.1007/BF02532821.

[72] Kim J, Chaudhury MK, Owen MJ. Hydrophobicity loss and recovery of silicone hv insulation. IEEE Trans Dielectr Electr Insul 1999, 6, 695–702, doi: 10.1109/94.798126.

[73] WO2012020295A1 – Optical Elements Having Long-Lasting Hydrophilic and Anti-Fog Properties and Method for Their Preparation.

[74] Koch K, Barthlott W. Superhydrophobic and superhydrophilic plant surfaces: an inspiration for biomimetic materials. Philos Trans R Soc A Math Phys Eng Sci 2009, 367, 1487–1509, doi: 10.1098/rsta.2009.0022.

[75] Tsougeni K, Vourdas N, Tserepi A, Gogolides E, Cardinaud C. Mechanisms of oxygen plasma nanotexturing of organic polymer surfaces: from stable super hydrophilic to super hydrophobic surfaces. Langmuir 2009, 25, 11748–11759, doi: 10.1021/la901072z.

[76] Walther F, Davydovskaya P, Zürcher S, Kaiser M, Herberg H, Gigler AM, Stark RW. Stability of the hydrophilic behavior of oxygen plasma activated SU-8. J Micromech Microeng 2007, 17, 524–531, doi: 10.1088/0960-1317/17/3/015.

[77] Quéré D. Non-sticking drops. Reports Prog Phys 2005, 68, 2495–2532, doi: 10.1088/0034-4885/68/11/R01.

[78] Favia P, Milella A, Iacobelli L, d'Agostino R. Plasma pretreatments and treatments on polytetrafluoroethylene for reducing the hydrophobic recovery. In: Plasma Processes and Polymers, Wiley-VCH Verlag GmbH & Co. KGaA, Weinheim, FRG, 2005, 271–280.

[79] Morra M, Occhiello E, Garbassi F. Contact angle hysteresis in oxygen plasma treated poly (tetrafluoroethylene). Langmuir 1989, 5, 872–876.

[80] Di Mundo R, Petrella A, Notarnicola M. Surface and bulk hydrophobic cement composites by tyre rubber addition. Constr Build Mater 2018, 172, 176–184, doi: 10.1016/j.conbuildmat.2018.03.233.

[81] Marmur A. Penetration and displacement in capillary systems of limited size. Adv Colloid Interface Sci 1992, 39, 13–33, doi: 10.1016/0001-8686(92)80053-Z.

[82] Farris S, Introzzi L, Biagioni P, Holz T, Schiraldi A, Piergiovanni L. Wetting of biopolymer coatings: contact angle kinetics and image analysis investigation. Langmuir 2011, 27, 7563–7574, doi: 10.1021/la2017006.

[83] Dastjerdi R, Montazer M, Stegmaier T, Moghadam MB. A smart dynamic self-induced orientable multiple size nano-roughness with amphiphilic feature as a stain-repellent hydrophilic surface. Colloids Surf B Biointerfaces 2012, 91, 280–290, doi: 10.1016/j.colsurfb.2011.11.015.

Elisabetta Tranquillo, Antonio Gloria, Marco Domingos

10 Nanoindentation: characterizing the mechanical properties of polymeric surfaces

10.1 Introduction and background

The past few decades have brought unprecedented technological developments to many industries that have seized the opportunity to explore new and emerging markets, investing in the development of high-performance products for various sectors such as aerospace, automotive, and medical. Faced with stringent regulations and a growing demand for more affordable and sustainable materials/processes, many companies have quickly shifted their focus to polymeric materials. When compared to metals, polymers are usually cost competitive (on a volumetric basis), require less energy to be processed and, despite their lower density, still display a good strength-to-weight ratios, high corrosion resistance, and low electrical and thermal conductivities, which can be advantageous in many of the above mentioned applications [1]. The medical field has probably been the major adopter of this class of materials that are increasingly used to develop drug delivery systems, wound closure and healing products, and surgical implant devices and bioresorbable scaffolds for tissue engineering. This vast range of applications is not exclusive to the medical field and it is intimately related to the easy tunability of polymers' physical, chemical, and mechanical properties [2]. These are usually affected by the macromolecular structure of the polymers that can vary in terms of number of repeating units, branching and number of chemical links [3]. In fact, each of the above properties is also strongly dependent on the polymer structure, including the functional groups present in the polymer backbone and the side groups on the main chain [3]. Therefore, the rational design of polymeric materials with adequate structural properties for technologically relevant applications require, among other things, an in-depth understanding of their mechanical behavior at multiple scales and especially at the nanometer scale. The ability to produce quantitative, absolute measurements of different mechanical properties (e.g., elastic modulus, E) with nanoscale spatial resolution is now widely recognized by materials scientists as key to understand the performance of many engineering materials, including polymers. This has become possible since the moment surface contacts between materials have been found to be highly dependent on their mechanical properties. Since then, many different indentation and impression systems equipped with well-defined contact geometries have been developed to measure hardness and elastic modulus of materials [4–6]. Despite significant advances in terms of sensory and actuation technology, nanoindentation measurements in polymeric materials remain challenging due to their viscoelastic nature and overall

https://doi.org/10.1515/9783110701098-010

weaker mechanical resistance compared to metals [7]. The aim of this chapter is to introduce the reader to the field of nanoindentation by providing a general overview of the principles and applications of this technique to the mechanical characterization of polymeric surfaces. After briefly introducing some important mechanical properties of polymeric materials, we discuss in more detail the basic concepts and approaches underpinning nanoindentation, including the Oliver and Pharr method, types of indenters and factors affecting experimental measurements. We then report on the available nanoindentation systems for polymeric materials, namely depth-sensing indentation (DSI) systems and scanning probe microscopy (SPM) systems. Finally, we discuss in detail the theory and mathematical calculus for the determination of hardness and elastic modulus in polymeric materials and conclude with a brief outlook in terms of the application of nanoindentation for the design of biological constructs.

10.2 Mechanical properties of polymers

The mechanical properties of materials are largely affected by their chemical composition, crystallinity, morphology, molecular weight, cross-linking degree, and molecular orientation. However, external factors such as temperature or thermal treatment, large differences in pressure, and environmental factors such as humidity, solar radiation, or other types of radiation can also influence polymers' mechanical properties [2]. In addition, the mechanical behavior of polymeric materials and the absolute values of their mechanical properties display a sensitivity to the kind of strain that is imposed by the applied force, namely tension, compression, biaxial, or shear. In order to determine and evaluate the mechanical properties of polymers, different tests, as well as instruments, can be employed. A large number of these tests are already standardized by the American Society for Testing and Materials (ASTM), including tensile, flexural, impact, and hardness, and are routinely used by mechanical scientists [8]. In fact, the ductility and strength (measured by a tensile test), related hardness properties, and fracture toughness (or impact resistance) are nowadays essential to characterize the properties of polymers for different technological applications.

Over the years, many attempts to interrogate the mechanical behavior of the materials at ever more reduced penetration depths, probing the mechanical resistance of materials at the very first surface molecular layers, have been reported and seen as critical to understand the correlation between mechanical properties of a material and the surface contacts [9]. In this regard, several types of indentation and impression tests have been proposed to evaluate mechanical properties such as hardness and elastic modulus for material selection and design improvements in a number of practical applications. Indentation testing is an effective method for evaluating the mechanical properties of materials and has been widely adopted in the last century, especially for hard materials. In fact, this technique has found many applications in probing the mechanical properties of small volumes of materials and thin films [10]. Since the early 2000s, with the evolution of testing instruments, techniques, and data analysis, nano-indentation of polymeric materials and composites really took off and became one of the most important techniques used in materials science.

10.3 Nanoindentation

Nanoindentation belongs to a new generation of techniques capable of analyzing samples at a microstructural level, being widely adopted to test thin films and surface mechanical properties. The evaluation of surface and interfacial properties of polymeric materials in general, and biomaterials in particular, is of great importance, as it allows for a deep understanding of the behavior and interaction between engineered constructs and biological tissues in the human body. Therefore, nanoindentation tests in biology and biomaterials research are expected to show a rapid increase in the coming years [11].

The application of nanoindentation in the characterization of polymeric materials and their constituents has been a particularly growing research area [12]. Compared to other types of materials, polymeric materials are characterized by a complex viscoelastic-plastic response. Therefore, the viscoelastic-plastic response of these materials depends on numerous factors, such as contact geometry and depth of penetration (i.e., deformation), loading rate (or deformation rate), and ambient temperature [9]. In order to characterize the mechanical properties of polymers, a hard object, called indenter, with a defined shape and size is usually inserted into the tested material under specific pressure and kept for a predefined time before unloading. Afterwards, the hardness of the tested material can be extracted from the relationship between the total indentation load and displacement or area [13]. The forces involved are generally of the order of millinewton or micronewton and the depth of penetration remains in the order of nanometers [13]. The force F and penetration depth h_{max} are varied or measured, generating loading–unloading curves from which the hardness and elastic response of the material under load is evaluated [14, 15] (Fig. 10.1).

10.3.1 Basic concepts and approach toward the Oliver and Pharr methods

In the early 1990s, Oliver and Pharr proposed a new method to analyze load–displacement data to obtain hardness and elastic modulus values, which is still used nowadays to characterize the mechanical properties of polymeric materials [5]. Nanoindentation is commonly defined as a depth-sensing or instrumented

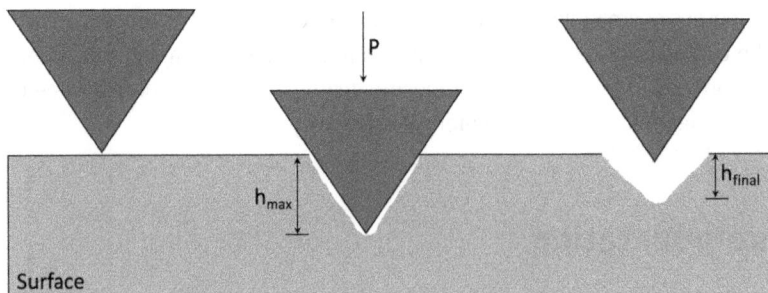

Fig. 10.1: Schematic representation of indentation experiment.

indentation technique by which the application of a controlled load to a material surface results in an induced local surface deformation.

Values of load and displacement (i.e., depth) can be monitored during the loading and unloading phase, thus allowing for the evaluation of properties such as reduced modulus and hardness using the generated load–displacement curves, in conjunction with well-defined equations based on the theory of the elastic contact [16, 17].

Considering the working force (e.g., 1 μN to 500 mN) and displacement ranges (e.g., 1 nm to 20 μm), the nanoindentation technique may clearly bridge the gap between the atomic force microscopy and the macroscale mechanical testing [16–19]. When compared to the traditional indentation techniques, nanoindentation provides better control over the scale of applied forces, displacement and spatial resolutions, which is critical for the analysis of soft biological tissues and advanced polymer-based materials with submicron resolution [16–20]. Taking advantage of the small size of the employed probe, this technique can be used to locally assess material properties in heterogeneous, small and thin samples, which would otherwise be impossible using traditional mechanical testing methods.

In most of the commercially available nanoindenters, capacitance or inductance is usually considered to monitor the displacement, whereas electrostatic force generation, expansion of a piezoelectric element or magnetic coils are employed for the force actuation [16–20].

A comprehensive analytical and experimental approach toward a generalized form of nanoindentation analysis was initially described by Doerner and Nix [21], starting from load–displacement data obtained through the use of nonrigid indenters of different geometries.

A further generalization of the approach consisting on the largely employed compliance method for indentation analysis was later reported by Oliver and Pharr [5]. The modulus may be also determined from load–displacement curves using the method of Field and Swain [22].

The above reported methodological approaches have been widely discussed in the literature, also focusing on limitations and features related to the instrument calibrations [16, 23, 24].

Even though hardness and elastic modulus are usually indicated as the most common output of nanoindentation tests, there is a growing consensus that these parameters are not sufficient to fully characterize the mechanical behavior of polymeric materials. In this vein, several reports have been published, highlighting the importance of determining other mechanical parameters such as creep compliance, dynamic moduli – storage or elastic and loss or viscous moduli, using nanoindentation [16, 17, 25, 26].

As previously mentioned, the Oliver and Pharr model is a widely employed method to assess the nanoindentation parameters starting from an experimental load–displacement (i.e., load–depth) curve [4, 5, 16]. A further schematic representation of the tip–sample interaction [16] during the indentation process is reported in Fig. 10.2, evidencing penetration depth (h_{max}), contact depth (i.e., the height of the contact between tip and sample, h_c), and residual or final depth (h_f).

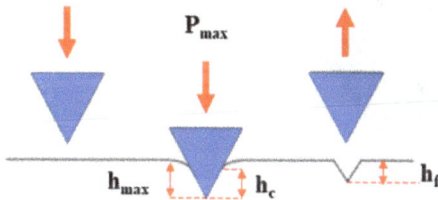

Fig. 10.2: Indentation process: schematic representation of tip–specimen interaction, reporting penetration depth (h_{max}), contact depth (h_c), residual or final depth (h_f) [16].

An example of the approach employed for the evaluation of the hardness and elastic modulus through the Oliver and Pharr method is reported in Section 10.6.

10.3.2 Indenter types

Measuring the mechanical properties of polymeric materials can be performed using indenters with different shapes, including pyramidal, conical, cylindrical, and wedge-shaped (Fig. 10.3). The selection of the indenter shape depends mainly on the type of material and properties being measured. For example, the Berkovich indenter with a triangular pyramidal shape is often preferred for hardness measurements due to its sharp geometry and ease of calibration. The Berkovich indenter (Fig. 10.3a) is particularly useful when the measurement of hardness and elastic modulus of polymeric materials is carried out at a very small scale. On the other hand, Vickers indenters (Fig. 10.3b), characterized by a four-sided pyramid shape, have

proven difficult to use because of the poorly defined, less sharp indentations generated. Conical indenters are attractive in that fully elastic contact with well-defined
stress fields may be obtained.

Comparing the stress–strain field generated by Berkovich indenters with those
of conical shape, it is possible to visualize that the initial contact stress of conical indenter is small, and only elastic deformation occurs, followed by a gradual, smooth
transition from elastic to plastic deformation.

However, the application of conical indenters remains limited, mainly due to difficulty in producing high-quality spherical diamond indenters at a submicron scale [27].

Finally, for testing soft materials, such as polymers, the cylindrical indenter
is often the preferred choice, with the flat contact surface shape allowing for
larger initial contact stiffness and the wedge-shaped surface providing linear
loading.

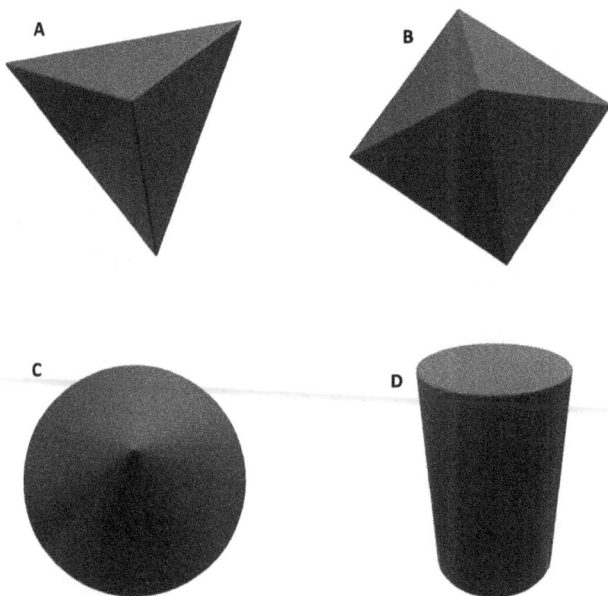

Fig. 10.3: Indenter tip geometries: (a) Berkovich pyramid, (b) Vickers pyramid, (c) conical,
and (d) cylindrical.

10.4 Factors affecting nanoindentation test results

Nanoindentation is a method of measuring the mechanical properties of small volumes of materials. Elastic modulus, hardness, fracture toughness, creep, and dynamic properties such as storage and loss modulus can be measured. However, like

many other techniques, nanoindentation is prone to experimental errors that can arise during the acquisition of mechanical data and directly affect the calculation of elastic modulus and hardness of testing materials. The most important factors influencing these measurements and contributing to nanoindentation errors will be discussed in the next sections [28].

10.4.1 Sample preparation and surface detection

The degree of surface roughness of the sample is very important for contact depth during the nanoindentation test. The roughness can be changed using different polishing methods, which must be performed according to the properties and test acquisitions of the samples. Mechanical and electrochemical polishing can lead to work hardening and an increase in surface roughness, respectively.

Polishing is usually performed by holding the specimen against a rotating polishing wheel covered with a mat previously impregnated with a polishing compound (usually a suspension of fine particles in a lubricant). A progressive decrease in grit size, with thorough rinsing in-between to minimize contamination, is often required. The polishing procedure involves a substantial amount of deformation of the surface of the specimen material, and it is common to encounter an unwanted indentation size effect resulting from the polishing procedure. Importantly, the thickness of the specimen must be at least ten times the depth of the recess or three times the diameter of the recess.

Although material properties can, in theory, be measured anywhere below 10% of a film sample thickness (in order to avoid excessive substrate effects), this condition does not apply when determining elastic modulus.

With regard to thin films, an appropriate test method must be employed to account for the influence of the substrate on the results. In this context, the continuous stiffness measurement (CSM) method can be used, providing additional information on the assessment of the mechanical properties in the most appropriate regions.

A suitable guideline must also be considered to account for the sample surface and its alignment with respect to the vertical axis of the indenter tip during the nanoindentation process.

This guideline plays an important role especially in the case of hard surfaces which may cause the offset angle to be intensified as a consequence of the lateral stiffness of the transducers. The surface detection is essential for nanoindentation tests, in particular for measurements with indentation depth at the nanoscale. No surface is ideally smooth. In fact, even the polished specimens have surface undulations with several nanometers' height. In these cases, at the initial stage of loading, the indenter will be touching the specimen only at a few points. As a consequence, the effective contact area at a certain depth of penetration will be smaller than the area assumed by the theoretical models, causing the apparent stiffness of the material to be lower and the results distorted [24] (Fig 10.4).

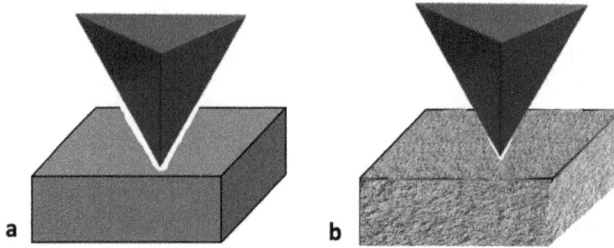

Fig. 10.4: (a) Smooth surface: the indenter touches the surface sample only at a few points; (b) rough surface: the indenter touches the surface sample completely.

As the load increases, the contact surface becomes smoother, due to larger elastic or plastic deformation at the contact points with high stress concentration, and the specimen response begins to correspond to its true properties. In order to overcome this issue, two different methods can be employed: (1) determination of the contact point using the fitting functions as two methods of polynomial extrapolation. The fitting range is between zero and the depth less than 10% of the maximum penetration depth; (2) definition of the contact point as the first increment of the test load or contact stiffness. Both the step scale of the load and indentation depth measurements should be small enough so that the uncertainty of the contact point is less than required limit. The typical step value of small load under nanoscale should be less than 5 μN [13].

Over the past years, nanoindentation testing methods and equipment have been continuously enhanced with the aim of allowing for an efficient evaluation of the mechanical properties (i.e., hardness and modulus). However, although nanoindentation can be employed to assess the mechanical properties of biological tissue and polymers [7, 29–41], a significant uncertainty is often associated with the accuracy of such measurements. Different from ceramics and metals, very low forces are achieved in the case of polymers as a consequence of their high compliance [30].

The theoretical load resolution is generally better than 1 nN for many commercially available nanoindenters, which should be capable to detect the initial contact. However, from a practical point of view, the external influences limit the resolution of these instruments to no better than 100 nN, thus leading to uncertainties in the surface position and clearly affecting the experimental results obtained from nanoindentation tests [30].

In the experimental phase, a basic step consists in the identification of the sample position by setting the measured indentation depth and the contact area to zero. In many cases, a specific amount of preloading (~1 μN) is often applied for the identification of the sample surface regardless of the stiffness of the analyzed sample, which may provide a significant indentation in the case of soft polymers. For this reason, the measured contact depth and area are affected by errors [7, 30, 31, 36, 42].

An incorrect surface detection clearly provides erroneous findings as the determination of the hardness and modulus values is directly related to the contact area.

In the case of soft polymers, the hardness and modulus values may generally be overestimated due to an underestimation of the contact area. This is also corroborated by experimental findings [7, 30, 36] which demonstrated that the values obtained from nanoindentation were higher than those from further testing methods (e.g., dynamic mechanical analysis).

The viscoelastic or time-dependent behavior is a further characteristic of polymers [43–46], leading to the formation of a "nose-shape" of the unloading curve, which introduces a complication in the traditional analysis of the unloading data according to the Oliver and Pharr method, as reported in the Section 10.6. Although some strategies may be adopted to take into consideration the viscoelastic properties of polymers, it is important mentioning that in many cases, errors associated with viscoelasticity, may indeed be lower than those arising from the inaccurate surface detection.

One of the critical issues is to detect the initial point of contact between the specimen and tip. In this scenario, nanoindentation studies on polymeric materials with different elastic moduli were focused on the identification of the initial contact during the testing process [30]. The practical feasibility and accuracy related to surface-detection techniques involving the use of quasistatic loading and further techniques using the dynamic forces associated with the CSM method were investigated [30]. For highly compliant materials, the experimental results were also integrated with the theoretical findings obtained from finite element analysis. Such investigations evidenced that the precise surface detection is crucial to attain accurate results. Values of mechanical properties with higher accuracy were always achieved in the case of surface detection through the use of dynamic testing [30].

The above reported experimental findings clearly stress the potential of adopting appropriate strategies to obtain smaller errors in surface detection and, hence, to avoid a significant underestimation of the contact area, which should affect the results.

10.4.2 The environment controls

The establishment of appropriate environmental controls is of critical importance for nanoindentation testing as it helps mitigating errors and uncertainty associated with the measurement of the specimens' depth. Ambient temperature fluctuations can lead to an expansion and contraction of the sample causing errors in the measurement detection. Even the simple manual manipulation of the specimen or indenter during the pre-test stage can cause an increase in temperature, followed by a slow decrease during the actual mechanical testing. The most critical stage is actually during the unloading stage where a decrease in temperature and consequent volumetric contraction of the specimen can cause a slower withdrawal of the indenter from the material leading to lower apparent specimen compliance and higher modulus as reported by Menčík et al. [24] (Fig. 10.5).

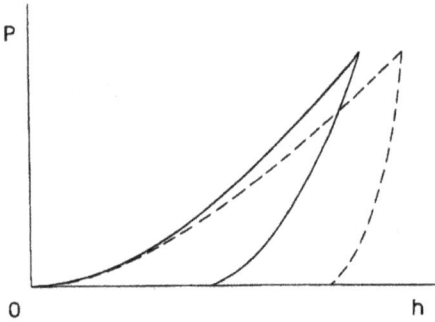

Fig. 10.5: Load-penetration diagram without drift (solid line) and with drift (dashed line) [24].

Therefore, to avoid this, the equipment must be placed in a sealed chamber for a thermal buffer. In this way, an ambient temperature of 23 ± 5 °C and relative humidity of less than 50% will be guaranteed while ensuring that other environmental factors are strictly controlled [13].

10.4.3 The selection of distance

Another very important aspect to take into consideration during nanoindentation tests is the distance between adjacent indentation points which should be at least five times the maximum indentation diameter. In fact, the influence of the interface, free surface, and prior residual indentation on the indentation results depend on the shape of the indenter and the properties of the sample.

The indentation test starts after the indenter contacts the specimen. The indenter penetrates slightly into the specimen due to the initial contact force also known as the minimum force detectable by the instrument (generally 1 μN). However, no matter how small the initial contact force, there will always be a corresponding penetration of the indenter beneath the undisturbed specimen free surface, as shown schematically in Fig. 10.6 [47]. All subsequent displacement measurements taken from this datum will be in error by this small initial penetration depth. Therefore, the initial penetration depth has to be added to all displacement measurements, in order to correct for this initial penetration.

Besides the initial penetration depth, other errors must also be accounted for, including the scatter of measured values of penetration depth due to the instrument "noise" and hysteresis effects in the tested material. Although generally small, this

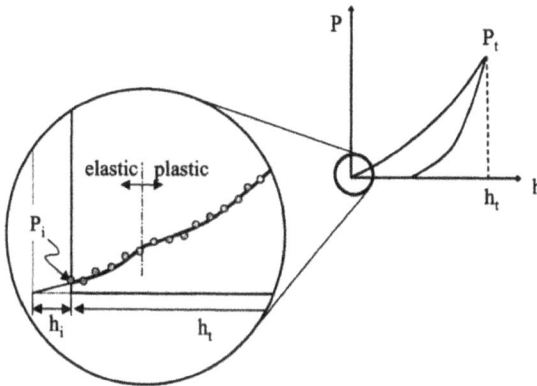

Fig. 10.6: Schematic of the effect of initial penetration depth on load–displacement data for a depth-sensing indentation test.

error can become apparent in the case of very low nominal loads (several millinewtons). In order to prevent this kind of errors, it is recommended that the output data be scrutinized and checked to see whether the initial depths of penetration for all tests carried out with the same initial load are approximately the same, otherwise, it is necessary to assess the genuine initial depth.

10.5 Nanoindentation of polymers

Measurement at the nano- and micro-level provides a number of challenges particularly in the field of polymeric materials. However, nanoindentation has developed to become a relatively simple technique suitable for many other types of materials.

When a nanoindentation tip is pushed into the sample under a controlled load, the force, displacement and time are recorded simultaneously. The applied forces can be as small as a few nanonewtons or as large as several newtons, allowing for the study of mechanical properties at a large range of size scales.

Polymers create significant challenges for measuring elastic modulus by indentation testing due to their surface. In fact, many polymers are soft while others are stiffer, causing the material response not to be measured accurately. Depth-sensing indentation (DSI) represents nowadays one of the principal techniques for the mechanical characterization of materials [48, 49]. DSI allows to measure the amount of penetration of an indenter tip into a thick material using a constant load rate or a constant travel rate [4]. The main goal of DSI methods is to produce absolute quantitative measurements of the elastic modulus. Therefore, DSI methods have limited capabilities for studying polymers or polymer composites and other important polymer systems.

On the contrary, the scanning probe microscopy (SPM) and in particular the atomic force microscope (AFM) can be used to study polymeric materials with nanometer spatial resolution.

The AFM uses a probe consisting of a sharp tip located near the end of a cantilever beam that is scanned across the polymer surface. The AFM can be operated in force mode to perform indentation tests producing a force curve, which is a plot of tip deflection as a function of the vertical motion of the scanner. This curve can be analyzed to provide information on the local mechanical response [28].

10.5.1 Nanoindentation using depth-sensing indentation systems

Nanoindentation using depth sensing determines the mechanical properties of the material from the depth data of the load indentation. There are relatively few indentation studies in the literature that use DSI systems to measure the properties of polymers. This is due to an increase in elastic modulus (E) measurements with decreasing depth of penetration, using DSI system, often referred to as the effect of the indentation size. In fact, the values of E measured for polymers using DSI are significantly higher than values measured using tensile testing or dynamic mechanical analysis (DMA). Lucas et al. [50] reported the comparison between the value E measured by DSI that is 1.2 GPa and those E (tensile test) = 0.4 GPa, and E (DMA) = 0.5 GPa for polytetrafluoroethylene (PTFE).

Therefore, such comparisons can often be misleading, because the values of E for many polymer systems can cover a wide range due to their potential variations in microstructure, semicrystalline morphology, anisotropy, molecular weight, and cross-linking density. However, comparisons of modulus values are more appropriate for polymer samples with identical chemistry, molecular weight, and treatment history [7].

10.5.2 Nanoindentation using scanning probe microscopy systems

SPM nanoindentation systems can be divided into different classes. In the first class, AFM and DSI systems are integrated with the DSI transducer-tip assembly mounted in place of the AFM cantilever probe. The indentation is controlled by the DSI system and contact mode scanning of the sample is permitted using the DSI force signal in the AFM feedback loop [7].

However, the contact force applied during scanning is orders of magnitude larger than typical AFM contact mode forces, and thus scanning with these systems can severely deform and damage polymeric materials [7]. Furthermore, the resolution of the resulting images is limited because the indentation tips have larger tip radii compared to AFM cantilever tips.

The second type of instrumentation is the interfacial force microscope (IFM). The IFM system presents a number of advantages over DSI systems in term of nanoindentation of polymers. The main advantage is the ability to indent using rigid displacement control, which allows tip penetration into the sample to be controlled with subnanometer resolution thus ensuring that the load frame compliance is zero. This attribute results in much lower applied forces compared to DSI such that true nanoscale spatial resolution can be achieved [51]. Nevertheless, the scanning capabilities of the IFM are limited compared to the AFM, but the contact mode scanning forces are significantly less than the integrated DSI–AFM system. Finally, AFM cantilever probes are used for quantitative indentation methods for imaging purposes [52].

The ability to combine nanoindentation test with robust, high-resolution imaging capabilities has led to several advantages for the characterization of polymers and biological materials.

These successes include studies of a wide variety of commercial polymeric materials [28], interphases in fiber-reinforced polymer materials [53], crystalline regions in polyethylene and polyethylene blends [54], and polymer thin films [55].

10.6 Determination of hardness and elastic modulus

The interaction between the tip and the sample during the indentation process provides data which may be used to assess material properties, such as the Young's modulus or elastic modulus (E) and the indentation hardness (H).

As an example, starting from the load–depth curve shown in Fig. 10.7, which is representative of the experimental data obtained from nanoindentation tests carried out on neat polycaprolactone fibers carried out in 1 mN to 5 mN load range using a nanotest platform with a diamond pyramid-shaped Berkovich-type indenter tip, some technical considerations may be made.

The hardness H and the reduced modulus E_r (i.e., the combined modulus of the tip and the sample) can be directly evaluated from analysis of load–displacement (i.e., load–depth) data according to the following equations [4, 5, 16, 56]:

$$H = \frac{P_{\text{max}}}{A_c} \tag{10.1}$$

$$E_r = \frac{\sqrt{\pi}}{2} \frac{1}{\sqrt{A_c}} S \tag{10.2}$$

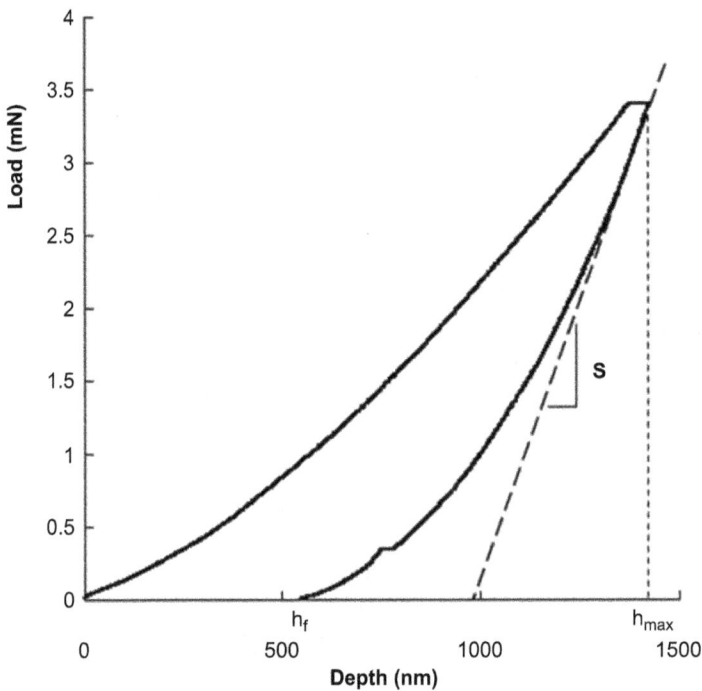

Fig. 10.7: Typical load–displacement (i.e., load–depth) curve obtained from experimental nanoindentation tests on polycaprolactone fibers, evidencing the loading–unloading process related to an applied trapezoidal load function (hold period of 20 s, loading–unloading rate of 300 μN/s). The measurements were performed using a nanotest platform with a diamond pyramid-shaped Berkovich-type indenter tip [16].

with S being the initial unloading stiffness (i.e., the slope of the unloading curve dP/dh, evaluated at the maximum load), P_{max} represents the maximum load, and A_c is the projected contact area between the indenter tip and the sample at maximum load (and, hence, at maximum indentation depth h_{max}).

Based on an ideal Berkovich pyramid geometry, which is perfectly sharp without defects at the tip, the relationship between the projected contact area and contact depth (h_c) is given by [16]

$$A_c = 24.5h_c^2 \qquad (10.3)$$

The contact depth h_c represents the actual value of the displacement, mainly occurring, even if exclusively, in a plastic fashion.

At the beginning of the unloading phase, a material would generally display an elastic response.

The difference ($h_{max} - h_c$) should represent an assessment of the "instantaneous" elastic recovery, as in the case of a flat punch indenter.

In this context, a correction factor (ε) was introduced by Oliver and Pharr [4, 5, 16, 56] in order to avoid the errors related to the flat punch assumption, which is usually employed to determine the contact stiffness.

Benefiting from Sneddon's elasticity theory and experimental results obtained by Oliver and Pharr [4, 5, 16, 56, 57], the contact depth is determined by subtracting the downward elastic displacement of the indented surface from the measured value of the maximum indentation depth as reported below:

$$h_c = h_{max} - \varepsilon \frac{P_{max}}{S} \tag{10.4}$$

where ε represents a constant, which is dependent upon the tip geometry and is assumed to be equal to 0.75 for a Berkovich tip.

The presence of tip imperfections requires appropriate calibrations to determine the area function $A_c(h_c)$ from indentations upon hard and polymer-based materials, also trying to reduce the elastic and viscoelastic effects of the response.

A_c can be a polynomial function of the contact depth h_c with coefficients obtained from a material with known elastic properties (i.e., fused silica) which is indented to different depths.

As an example, for an indenter with a tip imperfection, $A(h_c)$ may be described by [9, 58]

$$A(h_c) = 24.5\ h_c^2 + \sum_{i=0}^{7} a_i h_c^{\frac{1}{2^i}} \tag{10.5}$$

The above reported coefficients (a_i) are usually evaluated by performing indentations at varying depths in a material with an elastic modulus that does not appreciably change with depth and does not possess any further surface "layer," like a surface oxide layer generally present in metals.

Anyway, it is also worth remembering how the tip calibration represents a critical step in interpretation as manufacturer's specifications are usually not accurate and mechanical and chemical imperfections are inevitably present on the tip [9, 58].

The reduced modulus (E_r) is related to the material Young's modulus by the following equation [16, 56, 59]:

$$\frac{1}{E_r} = \frac{\left(1 - v_i^2\right)}{E_i} + \frac{\left(1 - v_s^2\right)}{E_s} \tag{10.6}$$

where the subscripts i and s refer to the tip and substrate material, respectively, whereas v represents Poisson's ratio.

If the indenter material properties (e.g., $E_i = 1,141$ GPa and $v_i = 0.07$ for a diamond Berkovich tip) and the Poisson's ratio of the sample material are known, the material Young's modulus (E_s) may be calculated from the reduced modulus. In general, the plane strain modulus $E' = E/(1 - v^2)$ is reported if the Poisson's ratio of the material is not known [16].

However, the Oliver and Pharr method may be generally summarized as follows.

The data taken from the upper portion of the unloading curve are fitted with the following power law relationship:

$$P = \alpha(h - h_f)^m \tag{10.7}$$

with m and α being the empirical constants, which are determined after fitting the unloading data and h_f is the final (residual) depth.

When the fitting parameters (α and m) are determined, h_f can be evaluated by substituting $h = h_{max}$ and $P = P_{max}$ in the eq. (10.7), as reported below:

$$h_f = h - \left(\frac{P}{\alpha}\right)^{\frac{1}{m}} \tag{10.8}$$

The contact stiffness may be obtained from eq. (10.7) by differentiating P with respect to h:

$$S = \frac{dP}{dh} = m \, \alpha \left(h_{max} - h_f\right)^{m-1} \tag{10.9}$$

The reduced modulus E_r can be calculated as

$$E_r = \frac{\sqrt{\pi}}{2\beta} \frac{S}{\sqrt{A_c}} \tag{10.10}$$

with β representing a correction factor according to the lack of symmetry of the employed indenter tip [60].

Furthermore, an interesting consideration must be made concerning the applied load function during the nanoindentation tests. As frequently reported, polymers exhibit a time-dependent (i.e., viscoelastic) behavior.

The effect of viscoelasticity on indentation is frequently associated with the creep phenomenon as well as a sinking of the tip into the tested specimen under a constant load. When the elastic response of the material is dominated by a creep behavior, then a nose-shaped curve, corresponding to the unloading phase of the nanoindentation test, is usually obtained (Fig. 10.8 – right) [9, 16, 61].

In such case, that means when the loading phase is followed by unloading phase without a hold period at peak load (Fig. 10.8 – left), a slight increase of the displacement occurs in the initial portion of the unloading process, as a consequence of the

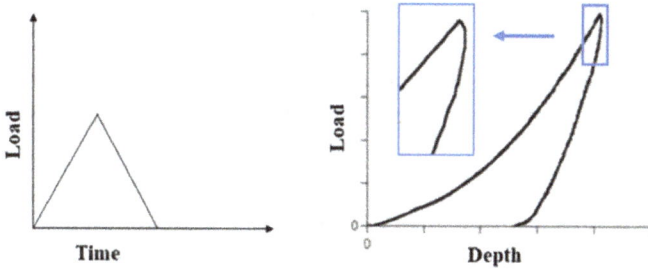

Fig. 10.8: Triangular load function (left) and resulting load–depth curve with "nose effect" for a viscoelastic material (right) [2].

creep rate of the material which is initially greater in comparison to the adopted unloading rate.

Consequently, the initial unloading region displays a negative and changing slope (Fig. 10.8 – right), making the evaluation of the modulus impossible.

For this reason, the applied load functions generally consider a hold period at peak load (Fig. 10.9 – left), thus allowing the material to approach an equilibrium before the unloading phase [9, 16].

Accordingly, instead of triangular load functions, trapezoidal ones must be applied considering appropriate load hold periods and loading–unloading rates.

Figure 10.9 (right) reports the effect of the hold period on the load–depth curve [16].

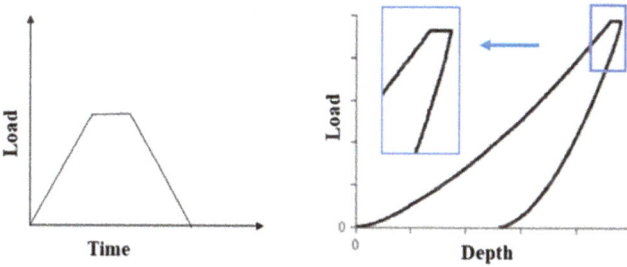

Fig. 10.9: Trapezoidal load function (left) and resulting load–depth curve without "nose effect" for a viscoelastic material (right) [2].

Many studies on polymeric materials have reported hold periods ranging from 3 to 120 s allowing the creep effect to dissipate before the unloading phase [16, 61–63].

In this scenario, further improvements for the analysis of viscoelastic materials were introduced using trapezoidal load functions (Fig. 10.9 – left), avoiding the creep effect on the determination of the modulus.

The proposed numerical improvements were focused on the initial unloading rate and on the creep rate before the unloading phase, which may be assessed from data measured during the indentation process [16, 64–67]. When the creep rate is low enough in comparison to the selected unloading rate, the effect of the corrections is negligible.

Further nanoindentation methods have also been rigorously developed for modelling the time-dependent behavior of viscoelastic materials [16, 17, 25, 26, 64, 68–71].

Some of these approaches involve the use of creep compliance function, viscous-elastic-plastic models three-element standard solid models [16, 25, 26, 68–71] to study the load–displacement data.

Moreover, some commercially available nanoindenters also possess a dynamic testing option, thus providing the possibility to apply sinusoidal excitations for the assessment of the viscoelastic properties as well as the mechanical spectra (i.e., storage or elastic and loss or viscous moduli as a function of frequency) [16, 17, 20].

Such approaches may represent intriguing alternatives to a quasi-static process involving a trapezoidal load function, as they do not measure an equilibrium modulus but directly provide the time-dependent properties.

In polymer-based materials, tip–specimen adhesion may strongly affect the measurement of the modulus [16, 61, 72–74]. The effect of adhesion leads to the presence of a region with negative values of the load in the load–displacement curve in the unloading phase (Fig. 10.10) [16].

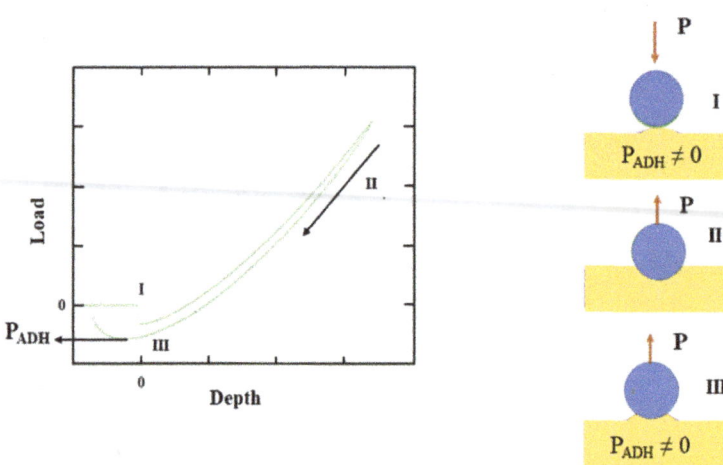

Fig. 10.10: Effect of adhesion: load–depth curve and schematic representation of tip–specimen interaction in different phases [2].

Some investigations on soft polymers have evidenced how a significant tip–specimen adhesion may lead to an overestimation of the material modulus [16, 72, 73].

More accurate values of the modulus have been found using specific adhesion models (e.g., Johnson–Kendall–Roberts, JKR) [16] in the case of indentations carried out through a spherical tip [16, 72, 73, 75].

A combination of viscoelastic analyses and JKR model has been also considered to provide further approaches in the field [16, 75–78].

Finally, nanoindentation technique may be considered a powerful tool to provide a further insight into the structure–property relationship of polymeric materials at different levels.

In brief, hardness is related to polymer chain flexibility and, consequently, to physical and chemical entanglements. The entanglement density among the molecular chains generally plays an important role in providing a contribution to the rigidity of the polymer amorphous region and represents a topological restriction for the molecular motion.

Consequently, molecular chains determine the nanoindentation behavior of polymers.

In addition, as reported in the literature, many functionalization processes may lead to polymer chain scissions and improve the flexibility and mobility of surface molecular chains, providing a reduction of the hardness values in surface-modified polymers.

For this reason, nanoindentation technique was also used to analyze the effect of a surface modification on the fibers of 3D additive manufactured (AM) PCL scaffolds for tissue engineering [56]. Moreover, the effect of poly(ester amide) (PEA) on blends of poly(ε-caprolactone) (PCL)/PEA was also investigated for the fabrication of 3D AM scaffolds with improved properties [18]. In this case nanoindentation allowed to interpret the higher values of hardness obtained for the fibers of PCL/PEA (up to 15% w/w of PEA) scaffolds in comparison to the PCL ones. Such result may be considered as a consequence of an increase in the density of physical entanglements due to the blending process during scaffold manufacturing, which provided a reduction in the mobility and flexibility of the molecular chains. However, beyond a threshold concentration value for PEA (15% w/w) nanoindentation results also demonstrated that the fabrication process did not continue to play an efficient role in increasing the hardness. Subsequent studies from the same group also show the relevance of nanoindentation tests on the development of 3D poly(ester urea) (PEU) scaffolds with adequate mechanical properties for articular cartilage regeneration. By controlling the synthesis of the material, the authors demonstrate the possibility of generating PEU scaffolds with substantially lower hardness compared to PCL, thus improving in vitro chondrogenesis [19].

10.7 Conclusions

Over the past years, nanoindentation has emerged as a powerful tool to provide a further insight into the structure–property relationship of different materials including metals, ceramics and polymers. As a depth sensing technique, nanoindentation has found relevant applications in materials science, especially for probing mechanical properties of nanometer-scaled volume of materials. Although being considered a well-established technique for the characterization of hard materials (i.e., metals and ceramics), its application in polymeric systems has proven quite challenging due to their low hardness, great compliance and time-dependent (or viscoelastic) behavior. Despite these challenges, nanoindentation has become increasingly important and, supported by recent progress in terms of data analysis and technology, the technique is nowadays routinely used to characterize the backbone chemistry, indentation rate, processing technique and surface treatment of polymeric systems. In fact, it's fair saying that nanoindentation has actually made the quantitative evaluation of mechanical properties possible for materials which were previously impossible to measure.

It plays a pivotal role in the characterization of polymer-based materials with a hierarchical structure, measuring properties at small length scales which may be used for a better understanding of the macroscale behavior. A wide range of mechanical properties may be evaluated through a suitable selection of the tip geometries and testing protocols.

Considering its sensitivity, spatial resolution, and flexibility, nanoindentation has been increasingly considered as an important technique for evaluating micro- and nanoscale mechanical properties of polymer-based biomaterials and biological tissues. Such approach has been employed to assess the mechanical properties of microstructural features in teeth and bone, as well as to analyze the variations in mechanical characteristics, according to the changes in tissue composition and structure, in the case of both hard and soft tissues.

In the future, it's expected that significant advances in methodologies and data analysis will lead to a continuous increase in the utility of the nanoindentation technique in the characterization of polymer-based materials for multiple applications, and in particular for the rational design of smart biomaterials for healthcare engineering.

References

[1]   Gowda TY, Sanjay M, Bhat KS, Madhu P, Senthamaraikannan P, Yogesha B. Polymer matrix-natural fiber composites: An overview. Cogent Engineering 2018, 5, 1446667.
[2]   Sadasivuni KK, Saha P, Adhikari J, Deshmukh K, Ahamed MB, Cabibihan JJ. Recent advances in mechanical properties of biopolymer composites: A review. Polym Comp 2020, 41, 32–59.
[3]   Aguilar-Vega M. Structure and mechanical properties of polymers. In: Handbook of Polymer Synthesis, Characterization, and Processing, 2013, 425–434.
[4]   Oliver WC, Pharr GM. Measurement of hardness and elastic modulus by instrumented indentation: Advances in understanding and refinements to methodology. J Mater Res 2004, 19, 3–20.
[5]   Oliver WC, Pharr GM. An improved technique for determining hardness and elastic modulus using load and displacement sensing indentation experiments. J Mater Res 1992, 7, 1564–1583.

[6]   Oliver WC, Pethica JB. Method for Continuous Determination of the Elastic Stiffness of Contact between Two Bodies, Google Patents, 1989.

[7]   VanLandingham MR, Villarrubia JS, Guthrie WF, Meyers GF. Nanoindentation of Polymers: An Overview, Macromolecular Symposia, Wiley Online Library, 2001, 15–44.

[8]   Dizon JRC, Espera Jr AH, Chen Q, Advincula RC. Mechanical characterization of 3D-printed polymers. Addit Manuf 2018, 20, 44–67.

[9]   Briscoe B, Fiori L, Pelillo E. Nano-indentation of polymeric surfaces. J Phys D: Appl Phys 1998, 31, 2395.

[10]   Zhang C, Zhang Y, Zeng K, Shen L. Characterization of mechanical properties of polymers by nanoindentation tests. Philos Mag 2006, 86, 4487–4506.

[11]   Mann A. Nanoindentation, Surfaces and Interfaces for Biomaterials, Elsevier Ltd, 2005, 225–247.

[12]   Gibson RF. A review of recent research on nanoindentation of polymer composites and their constituents. Compos Sci Technol 2014, 105, 51–65.

[13]   Wang H, Zhu L, Xu B. Principle and Methods of Nanoindentation Test, Residual Stresses and Nanoindentation Testing of Films and Coatings, Springer, 2018, 21–36.

[14]   Haque F. Application of nanoindentation development of biomedical to materials. Surf Eng 2003, 19, 255–268.

[15]   Cheng Y-T, Cheng C-M. What is indentation hardness?. Surf Coat Technol 2000, 133, 417–424.

[16]   Ebenstein DM, Pruitt LA. Nanoindentation of biological materials. Nano Today 2006, 1, 26–33.

[17]   Fischer-Cripps AC. Examples of Nanoindentation Testing, Nanoindentation, Springer, 2002, 159–173.

[18]   Gloria A, Frydman B, Lamas ML, Serra AC, Martorelli M, Coelho JF, Fonseca AC, Domingos M. The influence of poly (ester amide) on the structural and functional features of 3D additive manufactured poly (ε-caprolactone) scaffolds. Mater Sci Eng C 2019, 98, 994–1004.

[19]   Moxon SR, Ferreira MJ, Santos PD, Popa B, Gloria A, Katsarava R, Tugushi D, Serra AC, Hooper NM, Kimber SJ. A preliminary evaluation of the pro-chondrogenic potential of 3D-bioprinted poly (ester urea) scaffolds. Polymers 2020, 12, 1478.

[20]   VanLandingham MR. Review of instrumented indentation. J Res Natl Inst Stand Technol 2003, 108, 249.

[21]   Doerner MF, Nix WD. A method for interpreting the data from depth-sensing indentation instruments. J Mater Res 1986, 1, 601–609.

[22]   Field J, Swain M. Determining the mechanical properties of small volumes of material from submicrometer spherical indentations. J Mater Res 1995, 10, 101–112.

[23]   Fischer-Cripps A. A review of analysis methods for sub-micron indentation testing. Vacuum 2000, 58, 569–585.

[24]   Menčík J, Swain MV. Errors associated with depth-sensing microindentation tests. J Mater Res 1995, 10, 1491–1501.

[25]   Lu H, Wang B, Ma J, Huang G, Viswanathan H. Mech. Time-Depend Mater 2003.

[26]   Oyen ML. Spherical indentation creep following ramp loading. J Mater Res 2005, 20, 2094–2100.

[27]   Cripps ACF, Anthony C. Nanoindentation. In: Mechanical Engineering Series, New York, NY, Springer-Verlag, 2002, Cited on pages (48, 51, 52 and 76.) 2004.

[28]   Wang D, Russell TP. Advances in atomic force microscopy for probing polymer structure and properties. Macromolecules 2018, 51, 3–24.

[29] Barbakadze N, Enders S, Gorb S, Arzt E. Local mechanical properties of the head articulation cuticle in the beetle Pachnoda marginata (Coleoptera, Scarabaeidae). J Exp Biol 2006, 209, 722–730.

[30] Deuschle J, Enders S, Arzt E. Surface detection in nanoindentation of soft polymers. J Mater Res 2007, 22, 3107–3119.

[31] Hayes S, Goruppa A, Jones F. Dynamic nanoindentation as a tool for the examination of polymeric materials. J Mater Res 2004, 19, 3298–3306.

[32] Herrmann K, Strobel P, Stibler A. Neues Härtemessverfahren für sehr weiche Elastomere: Untersuchung und Einführung des Verfahrens IRHD SUPERSOFT. Materialprüfung 2002, 44, 83–86.

[33] Li Z, Brokken-Zijp JC, De With G. Determination of the elastic moduli of silicone rubber coatings and films using depth-sensing indentation. Polymer 2004, 45, 5403–5406.

[34] Loubet J, Oliver W, Lucas B. Measurement of the loss tangent of low-density polyethylene with a nanoindentation technique. J Mater Res 2000, 15, 1195–1198.

[35] Loubet J-L. Some measurements of viscoelastic properties with the help of nanoindentation, Extended Abstracts, Proc. Workshop on Indentation Techniques, ICMCTF-95, San Diego, CA, 1995.

[36] Odegard G, Gates T, Herring H. Characterization of viscoelastic properties of polymeric materials through nanoindentation. Exp Mech 2005, 45, 130–136.

[37] White CC, VanLandingham MR, Drzal P, Chang NK, Chang SH. Viscoelastic characterization of polymers using instrumented indentation. II. Dynamic testing. J Polym Sci B Polym Phys 2005, 43, 1812–1824.

[38] Cuy JL, Mann AB, Livi KJ, Teaford MF, Weihs TP. Nanoindentation mapping of the mechanical properties of human molar tooth enamel. Arch Oral Biol 2002, 47, 281–291.

[39] Jämsä T, Rho J-Y, Fan Z, MacKay CA, Marks Jr SC, Tuukkanen J. Mechanical properties in long bones of rat osteopetrotic mutations. J Biomech 2002, 35, 161–165.

[40] Berger E, Tripathy S, Vemaganti K, Kolambkar Y, You H, Courtney K. An atomic force microscopy indentation study of biomaterial properties, World Tribology Congress, 2005, 229–230.

[41] Dickinson ME, Mann AB. Nanomechanics and chemistry of caries-like lesions in dental enamel. MRS Online Proc Lib Arch 2004, 844.

[42] Cao Y, Yang D, Soboyejoy W. Nanoindentation method for determining the initial contact and adhesion characteristics of soft polydimethylsiloxane. J Mater Res 2005, 20, 2004–2011.

[43] Cheng Y-T, Cheng C-M. Relationships between initial unloading slope, contact depth, and mechanical properties for conical indentation in linear viscoelastic solids. J Mater Res 2005, 20, 1046–1053.

[44] Fischer-Cripps AC. A simple phenomenological approach to nanoindentation creep. Mater Sci Eng A 2004, 385, 74–82.

[45] Oyen-Tiesma M, Toivola YA, Cook RF. Load-displacement behavior during sharp indentation of viscous-elastic-plastic materials. MRS Online Proc Lib 2000, 649, 151–156.

[46] Yang S, Zhang Y-W, Zeng K. Analysis of nanoindentation creep for polymeric materials. J Appl Phys 2004, 95, 3655–3666.

[47] Fischer-Cripps AC. Factors affecting nanoindentation test data. Introduc Contact Mech 2000, 61–82.

[48] Bollino F, Armenia E, Tranquillo E. Zirconia/hydroxyapatite composites synthesized via sol-gel: Influence of hydroxyapatite content and heating on their biological properties. Materials 2017, 10.

[49] Díez-Pascual AM, Gómez-Fatou MA, Ania F, Flores A. Nanoindentation in polymer nanocomposites. Prog Mater Sci 2015, 67, 1–94.

[50] Lucas B, Oliver W, Pharr G, Loubet J, Gerberich W, Gao H, Sundgren J, Baker S. Thin films: Stresses and mechanical properties VI. MRS Sympos Proc 1997, 233.
[51] Marrese M, Guarino V, Ambrosio L. Atomic force microscopy: A powerful tool to address scaffold design in tissue engineering. J Funct Biomater 2017, 8, 7.
[52] Garcia R. Nanomechanical mapping of soft materials with the atomic force microscope: Methods, theory and applications. Chem Soc Rev 2020, 49, 5850–5884.
[53] Nikaeen P, Depan D, Khattab A. Surface mechanical characterization of carbon nanofiber reinforced low-density polyethylene by nanoindentation and comparison with bulk properties. Nanomaterials 2019, 9, 1357.
[54] Arévalo-Alquichire S, Morales-Gonzalez M, Navas-Gómez K, Diaz LE, Gómez-Tejedor JA, Serrano M-A, Valero MF. Influence of polyol/crosslinker blend composition on phase separation and thermo-mechanical properties of polyurethane thin films. Polymers 2020, 12, 666.
[55] Caballero-Quintana I, Maldonado JL, Meneses-Nava MA, Barbosa-García O, Valenzuela-Benavides J, Bousseksou A. Semiconducting polymer thin films used in organic solar cells: A scanning tunneling microscopy study. Adv Elect Mater 2019, 5, 1800499.
[56] Gloria A, Causa F, Russo T, Battista E, Della Moglie R, Zeppetelli S, De Santis R, Netti P, Ambrosio L. Three-dimensional poly (ε-caprolactone) bioactive scaffolds with controlled structural and surface properties. Biomacromolecules 2012, 13, 3510–3521.
[57] Chinh NQ, Gubicza J, Kovács Z, Lendvai J. Depth-sensing indentation tests in studying plastic instabilities. J Mater Res 2004, 19, 31–45.
[58] Wornyo E, Gall K, Yang F, King W. Nanoindentation of shape memory polymer networks. Polymer 2007, 48, 3213–3225.
[59] Hu Y, Shen L, Yang H, Wang M, Liu T, Liang T, Zhang J. Nanoindentation studies on Nylon 11/ clay nanocomposites. Polym Test 2006, 25, 492–497.
[60] Franco Jr AR, Pintaúde G, Sinatora A, Pinedo CE, Tschiptschin AP. The use of a Vickers indenter in depth sensing indentation for measuring elastic modulus and Vickers hardness. Mater Res 2004, 7, 483–491.
[61] Klapperich C, Komvopoulos K, Pruitt L. Nanomechanical properties of polymers determined from nanoindentation experiments. J Trib 2001, 123, 624–631.
[62] Rho J, Zioupos P, Currey J, Pharr G. Variations in the individual thick lamellar properties within osteons by nanoindentation. Bone 1999, 25, 295–300.
[63] Rho J-Y, Pharr GM. Effects of drying on the mechanical properties of bovine femur measured by nanoindentation. J Mater Sci Mater Med 1999, 10, 485–488.
[64] Cheng Y-T, Ni W, Cheng C-M. Determining the instantaneous modulus of viscoelastic solids using instrumented indentation measurements. J Mater Res 2005, 20, 3061–3071.
[65] Feng G, Ngan A. Effects of creep and thermal drift on modulus measurement using depth-sensing indentation. J Mater Res 2002, 17, 660–668.
[66] Ngan A, Wang H, Tang B, Sze K. Correcting power-law viscoelastic effects in elastic modulus measurement using depth-sensing indentation. Int J Solids Struct 2005, 42, 1831–1846.
[67] Tang B, Ngan A. Accurate measurement of tip-sample contact size during nanoindentation of viscoelastic materials. J Mater Res 2003, 18, 1141–1148.
[68] Cheng L, Xia X, Scriven L, Gerberich W. Spherical-tip indentation of viscoelastic material. Mech Mater 2005, 37, 213–226.
[69] Cheng L, Xia X, Yu W, Scriven L, Gerberich WW. Flat-punch indentation of viscoelastic material. J Polym Sci B Polym Phys 2000, 38, 10–22.
[70] Cheng Y-T, Cheng C-M. Scaling, dimensional analysis, and indentation measurements. Mater Sci Eng Rep 2004, 44, 91–149.

[71] Oyen ML, Cook RF. Load–displacement behavior during sharp indentation of viscous–elastic–plastic materials. J Mater Res 2003, 18, 139–150.

[72] Carrillo F, Gupta S, Balooch M, Marshall SJ, Marshall GW, Pruitt L, Puttlitz CM. Nanoindentation of polydimethylsiloxane elastomers: Effect of crosslinking, work of adhesion, and fluid environment on elastic modulus. J Mater Res 2005, 20, 2820–2830.

[73] Carrillo F, Gupta S, Balooch M, Marshall SJ, Marshall GW, Pruitt L, Puttlitz CM. Erratum:"Nanoindentation of polydimethylsiloxane elastomers: Effect of crosslinking, work of adhesion, and fluid environment on elastic modulus"[J. Mater. Res. 20, 2820 (2005)]. J Mater Res 2006, 21, 535–537.

[74] Grunlan JC, Xia X, Rowenhorst D, Gerberich WW. Preparation and evaluation of tungsten tips relative to diamond for nanoindentation of soft materials. Rev Sci Ins 2001, 72, 2804–2810.

[75] Ebenstein DM, Wahl KJ. A comparison of JKR-based methods to analyze quasi-static and dynamic indentation force curves. J Colloid Interface Sci 2006, 298, 652–662.

[76] Greenwood J, Johnson K. Oscillatory loading of a viscoelastic adhesive contact. J Colloid Interface Sci 2006, 296, 284–291.

[77] Wahl KJ, Asif SA, Greenwood JA, Johnson KL. Oscillating adhesive contacts between micron-scale tips and compliant polymers. J Colloid Interface Sci 2006, 296, 178–188.

[78] Giri M, Bousfield DB, Unertl WN. Dynamic contacts on viscoelastic films: Work of adhesion. Langmuir 2001, 17, 2973–2981.

Index

https://doi.org/10.1515/9783110701098-011